2nd International Conference on Advances in Energy Resources and Environment Engineering (ICAESEE 2016)

IOP Conference Series: Earth and Environmental Science
Volume 59

Guangzhou, China
30 – 31 December 2016

ISBN: 978-1-5108-3956-4
ISSN: 1755-1307

Printed from e-media with permission by:

Curran Associates, Inc.
57 Morehouse Lane
Red Hook, NY 12571

Some format issues inherent in the e-media version may also appear in this print version.

Printed by Curran Associates, Inc. (2017)

For permission requests, please contact the Institute of Physics
at the address below.

Institute of Physics
Dirac House, Temple Back
Bristol BS1 6BE UK

Phone: 44 1 17 929 7481
Fax: 44 1 17 920 0979

techtracking@iop.org

Additional copies of this publication are available from:

Curran Associates, Inc.
57 Morehouse Lane
Red Hook, NY 12571 USA
Phone: 845-758-0400
Fax: 845-758-2633
Email: curran@proceedings.com
Web: www.proceedings.com

Table of contents

Volume 59

2nd International Conference on Advances in Energy Resources and Environment Engineering

30–31 December 2016, Guangzhou, China

Accepted papers received: 21 March 2017
Published online: 18 April 2017

Preface

Papers

IOP Conf. Series: Earth and Environmental Science **59** (2017) 011001 doi:10.1088/1755-1315/59/1/011001

Preface

The previous First International Conference on Advances in Energy Resources and Environment Engineering (ICAESEE 2015) was organized by Northeast Petroleum University, China and successfully taken place on March 14-15, 2015 in Daqing, China.

The 2016 2nd International Conference on Advances in Energy Resources and Environment Engineering (ICAESEE 2016) has been held on December 30-31, 2016 in Guangzhou, China. ICAESEE 2016 is to bring together innovative academics and industrial experts in the field of energy resources and environment engineering to a common forum. The primary goal of the conference is to promote research and developmental activities in energy resources and environment engineering and another goal is to promote scientific information interchange between researchers, developers, engineers, students, and practitioners working all around the world. The conference will be held every year to make it an ideal platform for people to share views and experiences in energy resources and environment engineering and related areas.

The conference will be held every year to make it an ideal platform for people to share views and experiences in energy resources and environment engineering and related areas. We invite original papers describing an idea or concept, addressing issues and problems, or focusing empirically on potential or realistic fields.

In order to organize ICAESEE 2016, we have sent our invitation to scholars and researchers from all around the world. Eventually, over 200 submissions were submitted to us. These papers have gone through a strict reviewing process performed by our international reviewers and 74 papers have been accepted. All the submissions were double-blind reviewed, both the reviewers and the authors remaining anonymous. First, all the submissions were divided into several chapters according to the topics, and the information of the authors, including name, affiliation, email and so on, removed. Then the TPC chairs assigned the submissions to reviewers according to their research interests. Each submission was reviewed by two reviewers. The review results should be sent to chairs on time. If two reviewers had conflicting opinions, the paper would be transmitted to the third reviewer assigned by the chairs. Only papers which were approved by all reviewers were accepted for publication.

We sincerely express our thanks to all the members of 2016 2nd International Conference on Advances in Energy Resources and Environment Engineering (ICAESEE 2016). Thanks are also given to IOP publisher for producing this volume.

Conference Chair of ICAESEE 2016
Shiquan Zhou
Huazhong University of Science and Technology, China

ICAESEE 2016 Organizing Committees

Conference Chairs

Assoc. Prof. Dr.Low Tang Jung, Universiti Teknologi PETRONAS, Malaysia
Prof.Shiquan Zhou, Huazhong University of Science and Technology, China

Technical Program Committees

Assistant Prof.Chien Chiang Lin, Shih Hsin University, Taiwan
Prof.Alexander Ginzburg, National Research Moscow State University, Russian Federation
A.Prof. Jingcheng Liu, Chongqing Petroleum and Natural gas Society, China
Prof.Onea Alexandru, Junior Development Labs, Romania(Managing Director)
A.Prof.Hemin Du, Xi`an Technological University, China
Prof. Jyhjong Lin, Department of Information Management, Ming Chuan University, Taiwan
Distinguished Professor Li-Der Chou, National Central University, Taiwan
Assoc. Prof. Eng. Krzysztof WITKOWSKI, University of Zielona Gora, Poland
Prof.Mohd Khairol Anuar Mohd Ariffin, Universiti Putra Malaysia, Malaysia
Prof.Mingming Mao, Shandong University of Technology, China
A.Prof.Yu-Ming Fei, Chilhlee University of Technology, Taiwan
Pro.Yongming Wang, Anhui University of Technology, China
Dr.Shuanghong Li, Shanghai Jiao Tong University, China
Prof.Shenggang Li, Shaanxi Normal Universtiy, China
Prof.Genqing Bian, Xi'an University Of Architecture &Technology, China
Prof.Shang Liu, Tianjin University of Finance & Economics, China
Dr.Youqiang Xing, Southeast University, China
Dr.Libiao Bai, Chang' an University, China
Dr.WenbinYang, Xi'an University of Posts & Telecommunications, China
Dr.Dongping Zhao, Northwestern Polytechnical University, China

Peer review statement

All papers published in this volume of *IOP Conference Series: Earth and Environmental Science* have been peer reviewed through processes administered by the proceedings Editors. Reviews were conducted by expert referees to the professional and scientific standards expected of a proceedings journal published by IOP Publishing.

IOP Conf. Series: Earth and Environmental Science **59** (2017) 012001 doi:10.1088/1755-1315/59/1/012001

Emission of VOC Compounds from Ta-Shan Chemical Industrial Park in Middle China

Jian Li[1], Chunjuan Xie[2]*, Qun Cao[1], Liushui Yan[1], Jiaochan Zhong[1]

[1] Key Laboratory of Jiangxi Province for Persistant Pollutants Control and Resources Recycle, School of Environmental and Chemical Engineering, Nanchang Hangkong University, Nanchang, China, 330063

[2] School of Basic Science Departments, East China Jiaotong University, Nanchang, China, 330013

*Emsil: xiechunjuan@126.com

Abstract. VOCs emission from a Ta-Shan Chemical Industrial Park in January 1-8, 2014 and summer (July 16-20, 2014) have been studied in this paper. VOCs were collected by the stainlessteel canisters and following pre-concentrate and analyzing by GC/MSD. The results showed that a total of 33 VOC species in winter samplers and 41 VOC species in summer samplers were detected. Among the 41 VOC species detected in the winter and summer samplers, 1-Butene and Ethane were the two most abundant compounds. Their highest average mean concentrations were 6013±1758 and 3188±1025 pptv in summer (contributed more than 12.5% and 4.8 % of the total VOC concentrations), and about 5895±1246 and 2763±789 pptv in winter (contributed more than12.9% and 7.8% of the total VOC concentrations), respectively. In winter, stack emission was found to be the main souce of VOCs emission, while in summer, fugitive emission become the primary source.

1. Introduction

China has made great economic achievements these years. However, as the sharp increase costs of land utilization and labor, thousands of factories moving from coastal city to the middle and west China such as Jiangxi province, Hunan province, Anhui province, Hubei province and Sichuan province, some of them even moved out of China to Southeast Asia such as Vietnam, Thailand and Malaysia. The incoming factories not only bring the immeasurable economic benefits to the local government, but also bring a successive environment problem such as air quality deterioration. In 2013, the number of national average haze days is 4.7, 2.3 more than the long term average (2.4), and is the highest level since 1961[1]. In northern Jiangsu and central Henan, the haze days were more 10 days even 15 days more than the long term average1. What's worse, in 2014, a wide range of long time haze appeared for the first time in the Middle China and even worse in 2015 and 2016. The composite-regional air pollution has become the prominent environmental problem in China now.

Volatile organic compounds (VOCs) are defined as organic compounds that with high volatility and can easily participate in photochemical reactions [2]. It also play an prominent role in the formation of ozone and secondary organic aerosol (SOA), which two were considered to have strong correlation with the composite-regional air pollution [3,4].Recent studies have shown that chemical industries are a major source of volatile organic compounds in the urban city, like Guangzhou[5], Houston[6], and Hongkong[7]. Zheng et al. found that solvent use-related industrial sources account for more than 80% of the total VOC emission in the Pearl River Delta region in 2011[5]. Chen et al.

suggested that a chemical park was an odorous source that affected the population near the industrial district [8]. Therefore, investigating VOC emission characteristics in the chemistry industrial park is critical for understanding the formation of composite-regional air pollution and devising effective control policies.

2. VOCs Sampling and Analyzing

Ta-Shan chemical industrial park is located at east of Leping City(Longitude 116°53′36″ - 117°32′40″ E, latitude 28°42′14″ -29°23′24″ N), the southern suburbs, with the total planning area of 10 km^2 in the middle-east of Jiangxi province. And the park is one of the largest industrial park in middle-south China.The leading industry were fine chemistry and pharmaceutical chemistry. Leping has a humid subtropical climate with four distinct seasons. Winters are short and fairly mild (average low in January is 2℃ or 35.6F), but with occasional frosts and snow is not unheard of. Summer is long and humid, with amongst the highest temperatures in China (average 34℃ or 93F in July).

Stack emission and fugitive emission samples were collected from 3 selected plants in the Ta-Shan chemical industrial park in winter (January 1-8, 2014) and summer (July 16-20, 2014). The ambient air samplers were all carried out at the core of the each selected plant. Manufacturing workshop, warehouses of raw materials and goods, stack emission also be selected for the sampling sites. VOCs samplers were collected by using a leak-free, two-liter, stainlessteel canister (UCI,USA) equipped with a restricted sampler operating at 38 mL/min (Entech Instrument Inc., California,USA). VOCs were first preconcentrated in the preconcentrator then analyzed by GC-MSD or GC-FID/ECD. The Gas chromatogram of a standard mixture of 55 Non-methane hydrocarbons (NMHCs) showed in Figure 1. And all the VOC compounds showed the good correlation conefficient (R2>0.99).

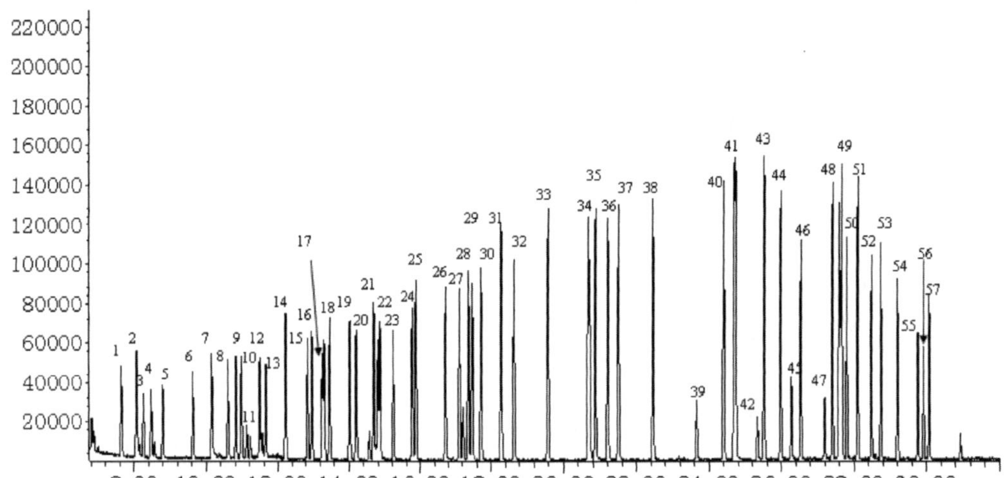

Figure 1. The Gas chromatogram of a standard mixture of 55 Non-methane hydrocarbons (NMHCs)

1.Isobutane,2.1-butene,3.n-Butane,4.trans-2-butene,5.cis-2-Butene,6.3-methyl-1-butene,7.Isopentane,8.1-pentene,9.2-methyl-1-butene,10.n-pentane,11.Isoprene,12.cis-2-pentene,13.2-methyl-2-butene,14. 2,2-dimethyl-butane, 15.cyclopentene, 16.4-methyl-1-pentene, 17.cyclopentane, 18.2-methylpentane, 19.3-methylpentane, 20.1-hexene, 21.n-hexane, 22.2-hexene, 23.3-hexene, 24.methyl-cyclopentane,25.2,4-dimethylpentane, 26.Benzene, 27.Cyclohexane, 8.2-methyl-hexane, 29.2,3-dimethyl-pentane, 30.3-methyl-hexane,31.2,2,4-trimethyl-pentane,32.n-heptane, 33.Methyl-cyclohexane,34. fluorobenzene (IS) 35.Toluene,36.2-methyl-heptane,37.3-methyl-heptane,38.n-octane, 39.chlorobenzene-d5(IS), 40.ethyl-benzene, 41.m/p-Xylene, 42.Styrene, 43.o-Xylene, 44.n-nonane, 45.4-bromofluorobenzene, 46.Isopropylbenzene, 47.α-Pinene, 48.n-propylbenzene,49.p-ethyltoluene,50.1,3,5-trimethyl-benzene,51.o-ethyltoluene, 52.1,2,4-trimethylbenzene, 3.n-Decane, 54.1,2,3-trimethylbenzene, 55.m-diethylbenzene, 56.p-diethylbenzene, 57.o-diethylbenzene

3. Results and Discussion

VOCs concentrations were measured by analyzing the averaging results of samples from different industrial sectors or processes, such as manufacturing workshop, warehouse of goods and raw materials, stack emission and ambient levels. These sectors were divided into "fugitive emission" , "stack emission" and "ambient levels". A total of 33 VOC species in winter samplers and 41 VOC species in summer samplers were detected in this study. Arithmetic mean concentrations of top 10 highest VOC compounds were listed in Table 1 along with the corresponding standard deviation (SD).

Among the 41 VOC species detected in the winter and summer samplers, 1-Butene and Ethane were the two most abundant compounds. Their highest average mean concentrations were 6013±1758 and 3188±1025 pptv in summer (contributed more than 12.5% and 4.8 % of the total VOCs concentrations), and about 5895±1246 and 2763±789 pptv in winter (contributed more than12.9% and 7.8% of the total VOC concentrations), respectively. It was easy to see from Table 1 that the total VOCs concentrations varied widely between the different industrial sectors or processes. Among the thress types of sectors, stack emission showed the highest total VOC concentrations (45765 pptv) in winter. The descending order of total mean concentration was stack emission> fugitive emission> ambient levels and the ratio of the three was 43%: 34%: 23%. While in summer, fugitive emission increased sharply, which contributed more than 48% of the total VOC mean concentration compared to 35% of stack emission. This might be explained by that high temperture in summer might accelerate the speed of VOCs leaking from the Valves, Tank lids or Pumps.The results implied that fugitive emission might be one of the most important VOCs source in the chemical industrial park in summer and inside manufacturing workshop and warehouse were important microenvironment for exposure to VOCs, expecially in summer. In consideration of gaining more economic benefit while avoiding healthy and environmental problems on the same time, some strategies must be adopted for controlling the VOCs Leaking.

4. Conclusions

The characteristic of VOCs emission from Ta-Shan ChemicalIndustrial Park have been discussed above. A total of 33 VOC species in winter samplers and 41 VOC species in summer samplers were detected. The highest total VOCs of 65757 pptv showed fugitive emission was the main source of VOCs in summer, while stack emission was main source in winter (45765 pptv). The manufacturing workshop and warehouse were important microenvironment for exposure to VOCs, expecially in summer. It is now widely recognized that industry must decrease its environmental influence to allow the sustainalbe development and a major contributing factor will be through controlling emissions. This will be driven by a combination of public pressure, environmental legislation and enterprises' social responsibility.

Acknowledgements

This work was financially supported by the National Natural Science Foundation of China (Grants 21467018, 41003057, 41203076), the Natural Science Foundation of Jiangxi Province of China 20142BBG70005, 20151BAB203027), Foundation of Jiangxi Educational Committee (GJJ14511, GJJ14406) and Fund of Environmental Protection Department of Jiangxi Province (2014).

References

[1] Government report 2013 China Meteorological Administration http://www.cma.gov.cn.
[2] Kota S H and Park C.H 2014 Atmospheric Environment 82(1) 24-35.
[3] Chen W H and Persily A K2014 Building and Environment 71(71)204-211.
[4] Wei WJ and Xiong JY 2014 Atmospheric Environmen 82(1) 327-334.
[5] Zheng JY 2013 Science of TheTotal Environment 456(7) 127-136.
[6] Zhao W and Hopke PK 2004 Environment Science and Technolgy 38(5) 1338-1347.
[7] Zheng JY and Zhong L J 2010 Atmospheric Envrionment 44(6) 814-823.
[8] Chen LY and Jeng FT 2000 Environment Science and Technolgy 34(7) 1166-1173.

Table 1 Mean±SD concentration (pptv) of top 10 highest VOC compounds in Ta-Shan chemical industrial park.

VOC species	Winter						Summer					
	F[a] (n=10)		S[b] (n=5)		A[c] (n=5)		F[a] (n=12)		S[b] (n=6)		A[c] (n=6)	
	Mean	Rang	Mean	Rang	Mean	Rang	Mean	Rang	Mean	Rang	Mean	Rang
1-Butene	3548±1236	2360-4586	5895±1246	3315-7951	2903±657	1871-3255	4348±1518	3850-5540	6013±1758	5215-6687	3303±659	2860-3582
Ethane	2763±789	1611-3453	2652±790	1514-2885	1967±562	1298-2469	3188±1025	1491-3712	2568±588	2131-2619	2067±464	1712-2269
Acetylene	1556±552	1310-1798	1782±658	1278-1985	1315±325	978-1463	1765±657	1256-2015	2236±673	1847-2486	1553±427	1457-1679
Isobutane	1170±338	890-1285	1895±552	1668-1980	1284±368	987-1561	1207±485	1045-1453	2506±568	1985-2760	1384±368	1240-1586
n-Hexane	980±257	887-1125	2078±775	1587-2575	1172±178	872-1220	1157±368	961-1256	2116±589	1846-2368	1011±162	875-1155
Acetone	2380±960	1283-2450	1887±661	1771-2103	1003+253	854-1278	2952±882	2668-3360	2612±610	2272-2987	921±158	860-1059
Propane	1879±545	1780-2211	1556±768	1265-1930	741±312	411-821	1986±428	1769-2153	2416±568	2014-2689	894±123	770-968
Ethene	912±146 [d]	856-1126	1257±518	1032-1556	704±64	515-958	1156±321 [e]	1027-1388	1352±542	1158-1470	871±138	856-985
Isopentane	1269±589	779-1332	1563±654	1336-1882	743±361	257-1054	1387±582	975-1421	1891±611	1535-2164	814±341	684-1089
Methylene chloride	1668±687	1258-1826	3305±1442	2750-3995	777±218	478-895	1790±550	1680-1985	3056±785	3116-3998	750±198	615-878
Others	17277 [f]	-	21895 [f]	-	11785 [f]	-	32212 [g]	-	21245 [g]	-	11312 [g]	-
Total VOCs	35402	-	45765	-	24394	-	65757	-	48021	-	24880	-

[a] fugitive emission;[b] stack emission [c] ambient levels;[d] four samples average. [e] six samples average.

[f] Total of 23 other VOC sepcies detected in winter samples (including Propene, Dimethyl ether, n-Butane, n-Pentane, Butanal, 2-methyl propanal, Tetrahydro furan, 1,2-dichloroethane, Benzene, Cyclohexane, trimethyl-2-pentene, Toluene, Chlorobenzene, Ethylbenzene, m/p-Xylene, 5-methyl-3-hexanone, o-Xylene, Benzaldehyde, a-pinene, Octanal, tert-butyl-benzene, 4-methyl-benzaldehyde, Naphthalene).

[g] Total of 31 other VOC sepcies detected in winter samples(Propene, Dimethyl ether, n-Butane, n-Pentane, Butanal, 2-methyl propanal, Tetrahydro furan, 1,2-dichloroethane, Benzene, Cyclohexane, trimethyl-2-pentene, Toluene, Chlorobenzene, Ethylbenzene, m/p-Xylene, 5-methyl-3-hexanone, o-Xylene, Benzaldehyde, a-pinene, Octanal, tert-butyl-benzene, 4-methyl-benzaldehyde, Naphthalene, n-Pentane,n-Octane, Ethyl-benzene,2-Butanone, Styrene, Ethyl acetate, 3-methyl-hexane, 2,3-Dimethyl-pentane).

Purification effect of two typical water source vegetation buffer zones on land-sourced pollutants

Gang Li

Institute of Soil and Water Conservation, Zhejiang Institute of Hydraulics & Estuary

happylglove@126.com

Abstract. Two vegetation buffer zones (tree-shrub-grass pattern and tree-grass pattern) were selected as test objects around Siming reservoir in Yuyao City of China. The effect of the storm runoff intensity (low and high intensity) and the buffer zone width (1 m, 3 m, 5 m, 7 m, 9 m, 12 m, 16 m) on pollutants (suspended solids, ammonium nitrogen and total phosphorus) was studied by the artificial simulation runoff. The results showed that with the increase of the width of buffer zone, the pollutant concentration was decreased. The purification effect of the two buffer zones on suspended solids and total phosphorus was basically stable at 52-55% and 34-37%, respectively. But the purification effect on ammonium nitrogen was the tree-shrub-grass pattern (69.7%) significantly better than that of tree-grass pattern (52.1%). The purification rate at the low runoff intensity was 1.8-2.0 times that at the high runoff intensity. The relationship between the purification rate and buffer zone width can be expressed by the natural logarithm equation, and the model adjustment coefficient was greater than 0.92.

1. Introduction

The buffer zone of reservoir bank plays an important role in soil and water conservation, water purification, flood control [1,2]. A certain width vegetation buffer zone in reservoir bank can decrease pollutants of runoff, and reduce the degree of pollution through filtration, permeation, absorption, retention and deposition [3,4]. A large number of studies show that the type and structure of the buffer zone is an important factor to affect its ecological environment. For example, the Lee's study [5] found that in the natural rainfall condition, 7.1m wide grass strip can intercept more than 92% of the sediment in surface runoff. the same wide switchgrass hedgerow intercepted 95% of sediment, 80% of total nitrogen, 62% of nitro-nitrogen, 78% of total phosphorus and 58% of PO4-3-P. In 2005, Mariet [6] studied the purification effect of the forest and grass riparian buffer zone on the N element, respectively, in 6 countries of Europe. His study results showed that the forest riparian buffer zone can well absorb the N element. And the grass riparian buffer zone absorbed a small amount of the N elements but it played an important role in the whole riparian ecosystem cycle. In the same year, Syversen [7] carried out experiments on 4 different regions of the southern Norway. The results indicated that the purification rate of 10m wide riparian buffer zone was better than that of 5m wide riparian buffer zone. The purification rates of phosphorus and nitrogen were 60%-89% and 37%-81%, respectively.

2. Materials and Methods

2.1. study site

The Siming Lake was located in Yuyao city of China. The upstream catchment area is 103.1 km^2, the total storage capacity is 122.72 million m^3 and the normal capacity is 79.46 million m^3. And it was designated as the one-level drinking water source protection area. The average annual rainfall of

IOP Conf. Series: Earth and Environmental Science **59** (2017) 012002 doi:10.1088/1755-1315/59/1/012002

liangnong hydrologic station, near the reservoir, is 1561.3 mm, and the mean annual runoff is 93.36 million m^3. The land and water ecotone zone is relatively flat, and the main vegetation is weeds. The upper of buffer zone were main cultivated land and residential land.

2.2. experimental design

The experiment designed three levels of pollutant concentrations and two levels of storm runoff intensity. According to the scheme selection, a total of 7 experiments were designed (in table1).

Table 1. runoff intensity and pollutant concentration

Group	SRI $m^3 \cdot h^{-1}$	SSC $mg \cdot L^{-1}$	ANC $mg \cdot L^{-1}$	TPC $mg \cdot L^{-1}$
A	2.0	200	2.0	0.4
	2.0	400	10.0	1.0
	2.0	700	20.0	1.5
	3.9	700	20.0	1.0
B	2.0	200	2.0	0.4
	2.0	400	10.0	1.0
	2.0	700	20.0	1.5

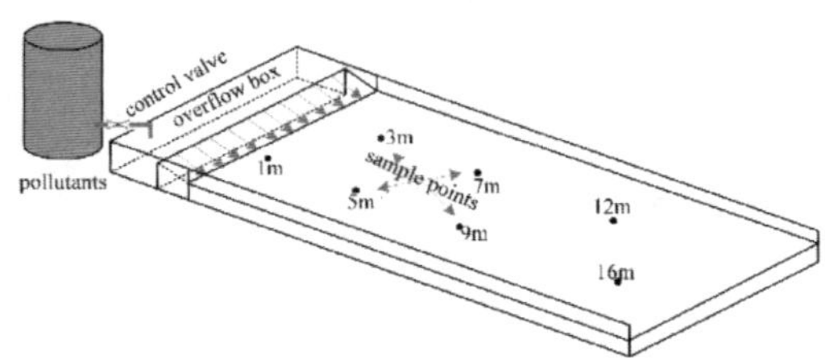

Figure 1. Experimental schematic diagram

Note: The SRI indicates storm runoff intensity.
The SSC indicates suspended solids concentration. The ANC indicates ammonium nitrogen concentration.
The TPC indicates total phosphorus concentration.

The sewage was mixed with chemical fertilizer and dried clay. Then adding the fertilizer and the clay in the bucket (0.5 m^3 plastic bucket), and keeping stirring to make it even. The bottom of plastic bucket was installed control valve. Regulating the control valve switch to control the water discharge, so as to realize the simulation of storm runoff. In order to realize the uniform distribution of the sewage in the test area, the sewage in the plastic bucket was first introduced to the overflow box. In this way, the distribution uniformity of the sewage can be controlled. The sewage sampling points was set at the distance of 1 m, 3 m, 5 m, 7 m, 9 m, 12 m and 16 m. A small pit was dug in each sampling point and the sewage samples were collected from the small pit. This method is called pit sampling. experimental simulation and sampling methods are shown in Figure 1.

3. Results and Analysis

3.1. effect of buffer zone on SSC

The change law that the different buffer zones reducing suspended solids concentration (SSC) was consistent, but the purification capabilities were different. With the width increased, the SSC decreased gradually (in figure 2). Under high SSC (700 mg/L), the width increasing from 0 m to 9 m can significantly reduce the SSC. And the purification rate (SPR) increased from 16.7% to 46.5%. When the width was larger than 10 m, the increasing rang of the SPR was not obvious. Under the condition of medium concentration (400 mg/L), the change law was similar with that of high concentration. From 0 m to 9 m, the buffer zone can significantly reduce the SSC, and the SPR increased from 5.25% to 34.25%. When the width was greater than 10 m, the SPR was stable between 48%-50%. Under the condition of low concentration (200 mg/L), the width had a shorter turning point in 7 m. When the width was larger than 7 m, the SPR was stable between 60%-63%.

3.2. effect of buffer zone on ANC

Figure 2 also showed the effect of different buffer zones on the ammonium nitrogen concentration (ANC). there were differences in the effects of width on the ANC. the tree-shrub-grass pattern (TSG pattern) better than the tree-grass pattern (TG pattern). The former APR was significantly higher than the latter. When the former width was 16 m, the APR can reach 75%, but the latter was only about 45%. Under medium concentration (10.0 mg/L) condition, the two patterns had good effect on the purification of ANC. With the increase of the width, the ANC decreased. The width less than 9 m, the TSG pattern was better than the TG pattern. When more than 9 m, the purification effect was basically no difference.

The width equal to 16 m, the former APR was 60.2% and the latter was 51.5%. At high concentration (20.0 mg/L), the two patterns showed significant differences in the APR. Especially in 5-7 m, the purification capacity of the TSG pattern was higher (5.07 mg/L) than that of the TG pattern. The APR of TSG pattern reached 55.6% at width equal to 5 m, which was equivalent to APR of TG pattern at width 12 m.

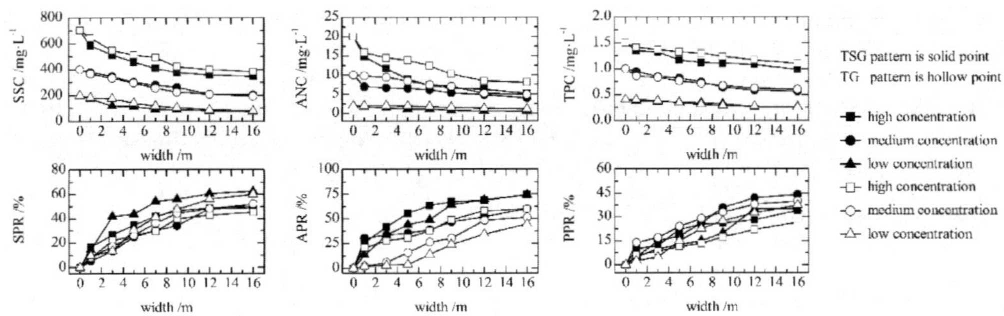

Figure 2. Variation of pollutant concentration and purification rate with width of buffer zone

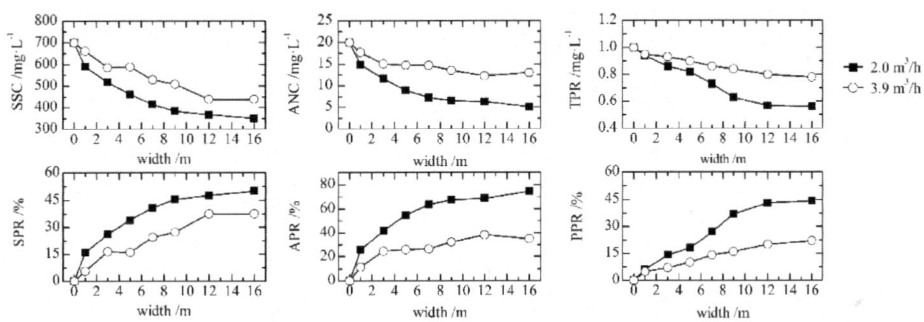

Figure 3. Purification effect under different runoff intensities

Table2. Statistical model parameters and width

group	SSC/SRI	parameters			$adj.R^2$	Width /m		
		a	*b*	*c*		50%	60%	70%
A	200/2.0	12.37	-0.70	29.49	0.978	5.3	11.8	27.2
	400/2.0	36.47	4.76	-59.32	0.967	15.3	21.6	30.0
	700/2.0	16.05	0.72	7.50	0.973	13.4	25.7	48.4
	700/3.9	27.82	5.54	-46.11	0.927	26.3	39.9	59.5
B	200/2.0	42.46	4.40	64.84	0.951	10.6	14.6	19.6
	400/2.0	30.59	3.05	-35.58	0.927	13.4	19.7	28.5
	700/2.0	18.24	1.10	-5.34	0.963	19.7	34.9	61.1

3.3. effect of buffer zone on TPC

According to the effect of the two patterns on total phosphorus concentration (TPC), the purification effects of buffer zone on the low and medium concentration (1.0 mg/L and 0.4 mg/L) were consistent, but it was different on high concentration (1.5 mg/L). Under the conditions of low and medium concentration, the width increased from 0 m to 16 m, the TPC can be reduced by 0.43 mg/L (PPR was 42%) and 0.15 mg/L (PPR was 36%), respectively. In high concentration, especially in 5-7 m, the purification effect of TSG pattern was obvious higher (0.17 mg/L) than that of TG pattern. The PPR was not as good as SPR and APR. When the width was 16 m, the PPR of the two patterns were only 34%

and 27%, respectively. The width less than 8 m, the difference of PPR between the two patterns was not obvious.

3.4. effect of storm runoff intensity on pollutants

Figure 3 reflected the purification effect of TSG pattern on pollutants under different storm runoff conditions. Under high intensity (3.9 m^3/h) and low intensity (2.0 m^3/h) of storm runoff conditions, TSG pattern can purify pollutants at a certain extent. And with the increase of the width, the purification rate was also increasing. Width increased from 0 m to 16 m, under low runoff intensity, the SSC, ANC and TPC were decreased by 350 mg/L, 14.88 mg/L and 0.44 mg/L, respectively. Under high runoff intensity, the SSC, ANC and TPC were decreased by 262 mg/L, 7.01 mg/L and 0.22 mg/L, respectively. In the 16 m of width, the SPR were 50% (2.0 m^3/h) and 37.43% (3.9 m^3/h), respectively. And the APR were 74.4% and 35.05%, respectively. And the PPR were 44% and 22%, respectively.

3.5. relationship between purification effect and width

Table 2 showed the statistical relationship between the purification rate (SPR) and the width (L). the SPR and the L were in accordance with the SPR = $a\ln(L + b) + c$ statistical model. The adjustment coefficient (adj. R^2) of all the model was more than 0.92.

4. Conclusions

Buffer zone on the different pollutant concentrations in runoff had certain purification benefits, and with the increase of the buffer zone width, the purification efficiency increased. But when the buffer zone width was greater than 10m, the increase range of the purification rate was smaller. The purification rate on suspended solids was basically stable at 52-55%, the purification rate on total phosphorus was stabilized at 34-37%. The purification effect of the two patterns on ammonium nitrogen was different. the tree-shrub-grass pattern was better than the tree-grass pattern. Under the condition of low storm runoff intensity, the concentration of pollutant had no obvious influence on the purification rate of the buffer zone. However, under the condition of high concentration of pollutants, the storm runoff intensity had a significant impact on the purification efficiency of buffer zone. The purification rate of buffer zone at low storm runoff intensity was 1.8-2.0 times that of buffer zone at the high storm intensity. With the width of buffer zone change, the purification effect of buffer zone on the suspended solids had obvious law, which can be expressed by the natural logarithm equation, and the adjustment coefficient of the equation is greater than 0.92. Through the calculation of the model, if the purification effect of buffer zone on suspended solid reached more than 50%, the width of buffer zone was about 15 m.

References

[1] Lowrance R., Mcintyre S., Lanee C. Erosion and deposition in a filed/forest system estimated using cesium-137 activity. Journal of soil and water conservation, 1988, 43:195-199.
[2] Wu J.Q. Quantitative study of the damping effect of buffer strips with different slopes on runoff and pollutant removal efficiency. Advances in water science. 2011, 22(1):112-117.
[3] Yan L.F., Shi X.F., Yu L.Z., et al. Elimination effects of riparian vegetation buffer zones on surface water nitrogen and phosphorus in Shenyang suburbs. Chinese Journal of Eco-Agriculture, 2011, 19(2): 403−408
[4] Tang H., Huang S.F., Wang M., et al. Removal of Runoff Pollution by Different Grass-buffers. Environmental Science & Technology, 2009, 32(2):109-113.
[5] Lee K.H., Isenhart Thomas M., Schultz Richard C. Sediment and nutrient removal in an established multi-species riparian buffer. Journal of Soil and Water Conservation, 2003, 58(1): 1-8.
[6] Hefting M.M., Clement J.C., Bienkowski P., et al. The role of vegetation and litter in the nitrogen dynamics of riparian buffer zones in Europe. Ecological Engineering, 2005, 24(5):465-482.
[7] Syversen N. Effect and design of buffer zones in the Nordic climate: The influence of width, amount of surface runoff, seasonal variation and vegetation type on retention efficiency for nutrient and particle runoff. Ecological Engineering, 2005, 24(5):483-490.

The impact of dike construction on the flow and sediment siltation in the project of Meishan Container terminal phase II

Jinhua Wang, Weisheng Zhang

State Key Laboratory of Hydrology-Water Resources and Hydraulic Engineering, Nanjing Hydraulic Research Institute, China

E-mail: jhwang@nhri.cn

Abstract. In order to avoid the adverse impact of Qinglongshan deep groove on the flow pattern of the Meishan Container terminal phase II ensuring the safety of berthing, a submarine dike is proposed along the land front direction toward the Qinglongshan Island. The influence of different dike layout schemes on the water flow in the surrounding sea area was studied through the mathematical model. In addition, the effects of different dike heights and gap width on the surrounding water flow and siltation were analysed. The results show that the crossflow in front of the terminal under the two submerged dike heights and gap widths are basically the same, which is less than that of without the construction of the dike. Appropriate reduction of dike height is conducive to improve the flow at the channel between the Qinglongshan and Meishan Island. After the dike construction, the groove will maintain a certain depth, and will not result a blocking at the top side of dike and the rear channel.

1. Intruduction

Ningbo-Meishan port area has good water and land conditions to construct large-sized container berths. In the southeast of Meishan Island, a container terminal phase I has been built and already put into use [1]. In the north of container terminal phase II, there exist a deep grove between Qinglongshan and Meishan Island (figure 1). The tidal current of the deep groove obliquely pierces the water area in front of the wharf making the flow at the southeast side of the wharf disorder, and the large local cross-flow can influence the use of the dock [2]. In order to ensure the safety of ship berthing and load-unloading operation, the embankment construction along the land front line along the direction of Qinglongshan Island was proposed. The project will smooth the tidal current around the terminal area, and resulting a small crossflow in front of the wharf. However, the impact on the topography for the area behind the Qinglongshan Island is not known. Will the construction of dike causes a damage to the island landscape? This paper tries to analyze this problem from the perspective of sediment movement and landform evolution.

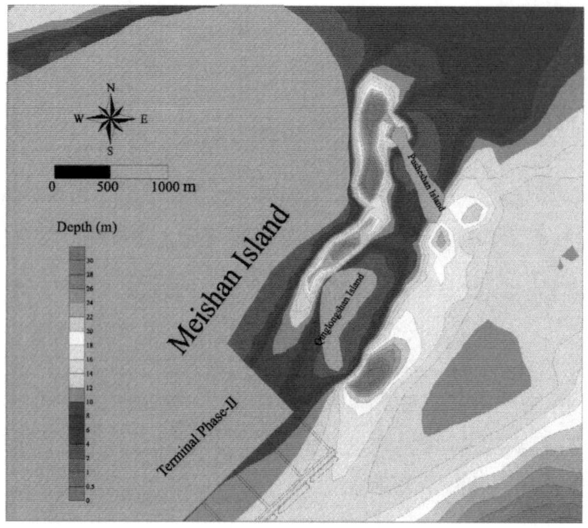

Figure 1. Water depth distribution around Qinglongshan Island (datum-level 85)

The local hydrodynamic conditions will be adjusted after the construction of the dike between the Qinglongshan Island and the terminal. Which can significantly reduce the velocity at the water area behind the Qinglongshan Island. The slow velocity water area will result a rapid deposition of sediment. In this study, the sediment siltation after the project will be modelled using a mathematical model, and the influence of different dike heights and gap widths on the surrounding water flow and siltation will be discussed.

2. Mathematical model establishment and verification

2.1. Model verification

The numerical model is DHI MIKE21 [3]. In order to simulate the tidal dynamics of the sea near the wharf, a model nesting method was adopted to provide the open boundary tidal condition [4]. The water depth distribution of the model is shown in figure 2, and the mesh of the mathematical model calculation is shown in figure 3. The triangular meshes are used to fit the irregular coastlines. The grid resolution is about 20m near the coast, the number of grid cells is 34488, and the number of nodes is 18556.

The mathematic model was verified by comparing the water level and flow velocity to the measured data. The tidal, tidal current locations of the measuring point are shown in figure 2, arranged 10 tidal current and 4 tidal elevation measurement points. Figure 4 shows the results of tide verification at four tidal level sites and it can be seen that the simulation error is less than 0.1m, and the tidal wave phase coincides well with the measured results

Figure 5 shows the verification result of the flow velocity and direction of spring tide. Due to the large number of islands and the complicated topography, the flow pattern in this area is complicated. From the simulation and the measured results, the flow at the points of $1^{\#}$, $2^{\#}$, $5^{\#}$, $6^{\#}$and$7^{\#}$ show the swirling pattern. While the flow of other points show the reciprocating pattern. Points $1^{\#}$, $2^{\#}$, and $3^{\#}$ are different in the flow direction, at the point $3^{\#}$ the flow is reciprocating flow due to the boundary constraint, and the flood velocity is 1 times larger than that of ebb velocity at $3^{\#}$. In the vicinity of the project area ($9^{\#}$), the tidal current is strong and the flood-ebb velocity is about 1.50 m/s during the spring tide, with its flow direction parallel to Meishan terminal Phase I shoreline.

From the verification results we can see the model simulate the tidal current very well compared to the observed data, which indicates that the boundary conditions provided by the mathematical model and the values of the model calculation parameters are reasonable and correct.

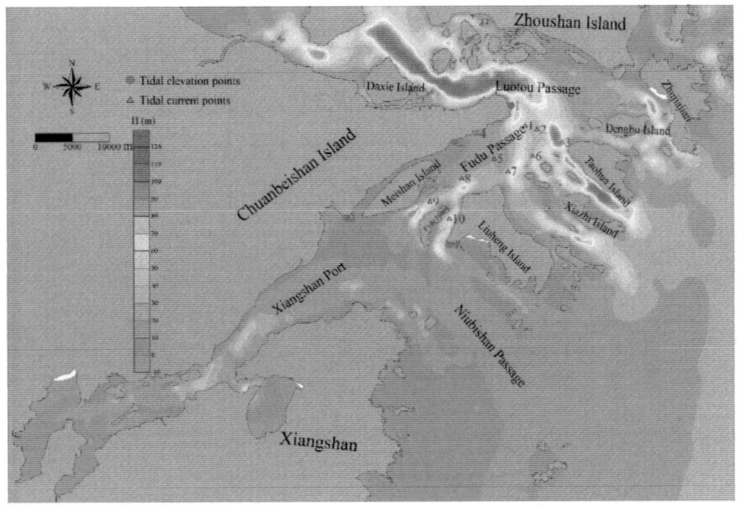

Figure 2. Calculation area and locations of the verification points

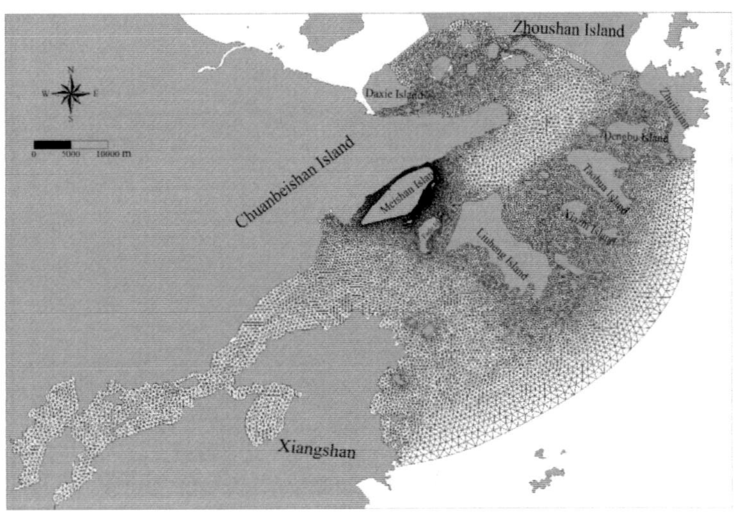

Figure 3. Computational mesh

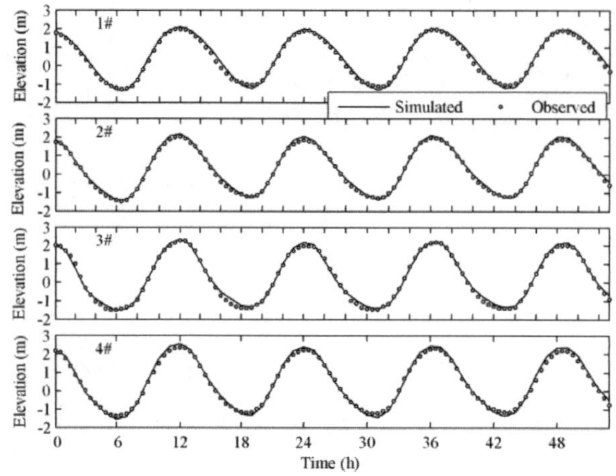

Figure 4. Tidal verification

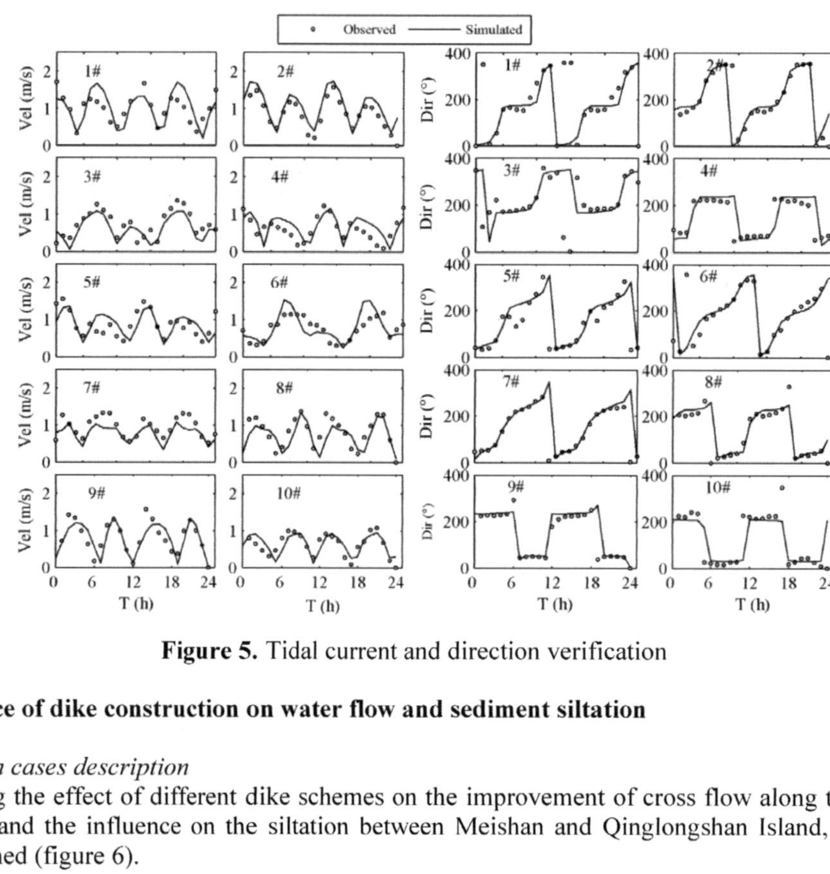

Figure 5. Tidal current and direction verification

3. Influence of dike construction on water flow and sediment siltation

3.1. Design cases description

Considering the effect of different dike schemes on the improvement of cross flow along the front of the wharf, and the influence on the siltation between Meishan and Qinglongshan Island, four cases were designed (figure 6).

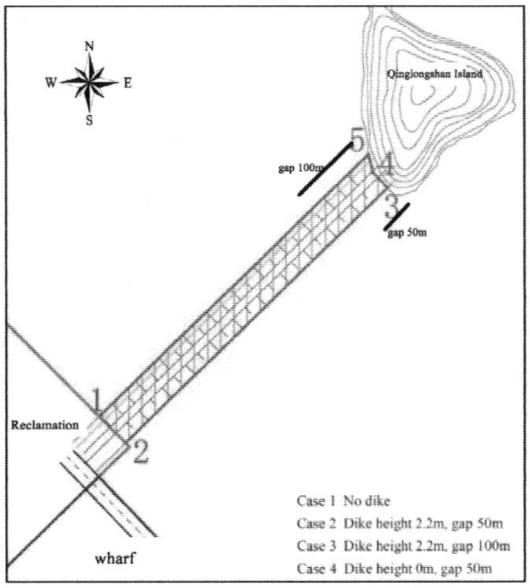

Figure 6. The 4 design cases

3.2. Analysis of water flow characteristics between Qinglongshan and Meishan Island under different cases

Under the land constraints of the second phase terminal, the forefront flow is parallel to the forefront of the pier. In the tidal grove between Qinglongshan and Meishan Island, the flow velocity is relatively large. After the second phase construction, the flow in the channel between Qinglongshan and Meishan Island has weakened and influence scope can reach the north of Meishan Island [5].

The local flow velocity near Meishan Island in Qinglongshan is slightly different from that before the project. The analysis shows that although the embankment dike has been constructed from case 2 to case 4, the flood and ebb flow pathway mainly at the outside Qinglongshan. The flow velocity inside the channel is relatively small.

In order to further understand the influence of the embankment project on the surrounding water flow, the flow pattern of the maximum northeast and the southwest flow velocity in the channel are added. Taking the comparison of the case 1 and 2 as an example (figure 7). From the comparison between the moments of the maximum northeast flow velocity occurs in the channel we can see that as the embankment is not built, the flow velocity in the channel is large and the flow direction is parallel to the channel axis. After the completion of the embankment dike with the top height equals 2.2m, the water flow is smooth in front of the terminal. As the maximum southwest flow occurs in the channel (figure 7b), the construction of the embankment project enhanced the smoothness of the water flow at the frontier of the land area, and the flow velocity in the channel was greatly reduced. The embankment dike construction has little effect on the flood current (figure 7c). In case 1, the water flow is strong at the northeast corner of the newly reclaimed land area and forms a recirculation area at the foreland of the land area (figure 7d). After the closure of the embankment dike, the flow becomes smooth and the flow velocity has increased, which is conducive to reducing the back of the terminal siltation.

In contrast to the instantaneous flow patterns of the case 2 to case 3, it is found that there is little influence on the flow pattern in the other areas except for the local water flow near the gap.

In the case 4, when the height of the embankment dike is lowered (85 datum in elevation of 0 m), the flow velocity in the channel is a little smaller than that without the construction of the dike, but the dike can also smooth the frontier water flow at the terminal resulting a small cross section velocity.

The maximum velocity of northeastern and southwest flow in the channel occurred during the high tide period, and it was inconsistent with the flood and ebb tide time of the outer channel of Meishan Island. The construction of the embankment dike could increase the smoothness of water flow in the foreland of the terminal land area. The decrease of dike height is helpful to increase the water flow in the channel between Qinglongshan and Meishan Island and to reduce the siltation.

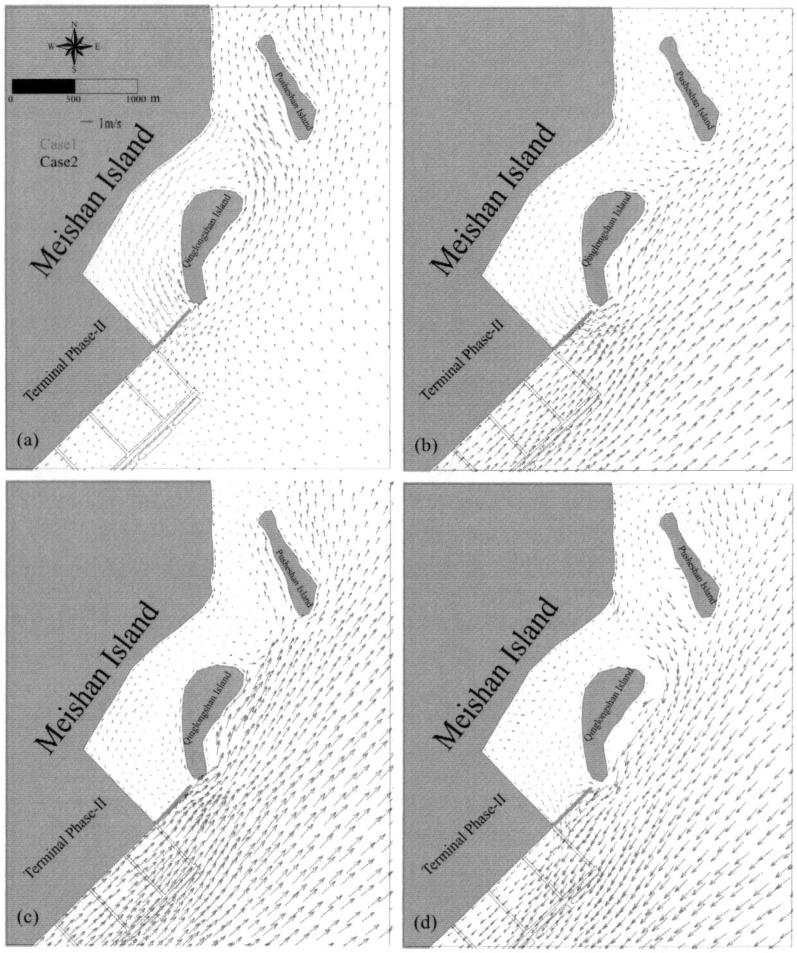

(a) ~ (d) are the maximum instantaneous velocity of northeast and southwest of Qinglongshan and Meishan Island, and the flood, ebb outside the Meishan Island

Figure 7. Comparison of Instantaneous flow between case 1 and 2

3.3. Sedimentation for the 4 cases

The annual silt distribution is calculated by mathematical model [6], and figure 8 shows the annual siltation for the case1~case4 respectively. After the implementation of case 1, the local maximum sedimentation is about 3m in the tidal channel between Qinglongshan and Meishan Island due to the relatively high sediment concentration before the project. After the implementation of case 2, the construction of dike makes the flow velocity within the channel between Qinglongshan and Meishan Island decrease obviously, and the siltation within the channel is strengthened. The distribution of siltation for the case 3 is basically the same as that of case 2, and the degree of siltation is reduced only near the gap inside the embankment. After the implementation of case 4, the dike height decreased from 2.2 m to 0 m, and the flow velocity in the tidal channel between Qinglongshan and Meishan Island increased significantly compared with those in case 2 and 3, which resulted in the siltation within the channel significantly weakened.

In general, with the increase of siltation, the local cross-section area is reduced, resulting the flow velocity increased, and the new cross-section depth will be maintained. The siltation at the two sides of dike will gradually reach the balance. Due to the maintenance of water and wave dynamics, and with a certain amount of tide volume and overhead flow above the dike, the channel between Meishan and Qinglongshan Island will not silt to death.

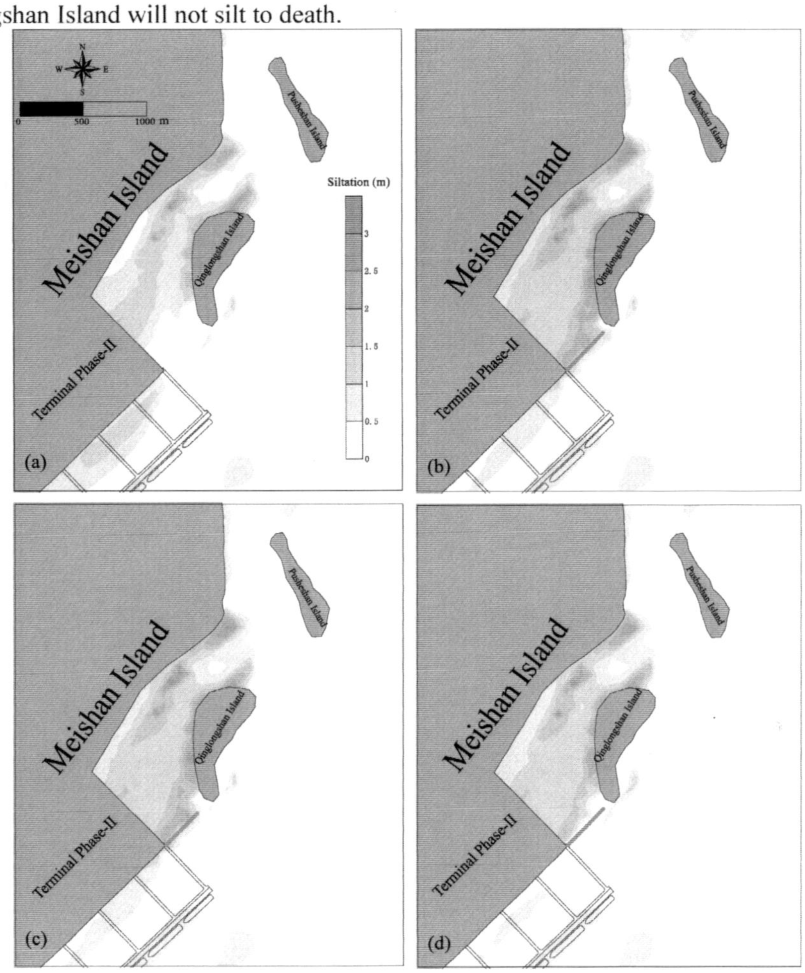

Figure 8. Distribution of siltation for the cases 1 to 4 respectively

4. Conclusions

The influence of the Qinglongshan embankment on the water flow in the surrounding sea area and the siltation between Meishan Island and the Qinglongshan channel on the backside of the embankment are analyzed by mathematical model. The conclusions are as follows:

1) The construction of the embankment dike project slow down the flow velocity in the channel between Qinglongshan and Meishan Island. With the decrease of the height of the embankment dike, the flow velocity in the channel increased. The maximum velocity in the channel is about 0.50 m/s when the elevation is 0m, which increases obviously compared to the elevation of the embankment equals 2.2m.

2) The crossflow after the construction of the dike embankment decrease to 0.23 m/s. The amount of siltation can be decreased after the decline of the dike height. The influence of different gap width on the distribution of back siltation is only limited to the area near the gap.

3) With the increase of siltation, the local cross-section flow velocity will be increased, and thus maintain a new section depth. With the combination effect of the flow and wave dynamics, the grove will not be disappeared after the dike construction.

Acknowledgements
This research is supported by the National Natural Science Foundation of China (51309159) and the China Postdoctoral Science Foundation (2012M511244).

References:

[1] Hu JH. The ship berthing of Meishan container terminal. *Navigation Technology*, 2015(01):8-12.
[2] Wang JH, Zhang WS, Zhang JS. Tidal and sediment Mathematical model study for Phase Ⅱ *Container Terminal of Meishan Port* [R]. NHRI, 2012.
[3] DHI. MIKE21 flow model FM user guide [R].DHI,2014
[4] Zhang WS, Zhang JS. Tidal response of sea level rise in marginal seas near China. *Advances in Water Science*, 2013(02):243-250.
[5] Wang JH, Zhang WS, Zhang JS. The impact of dike construction on the flow and sediment siltation for Phase Ⅱ Container Terminal of Meishan Port [R]. NHRI, 2013.
[6] Wang YG, Lin X, Wu Z. A method for calculating depositional rate after warping the bank in the silty estuary. *Ocean Engineering*, 2000(03):67-70.

IOP Conf. Series: Earth and Environmental Science **59** (2017) 012004 doi:10.1088/1755-1315/59/1/012004

Study on the surface tensions of MDEA-methanol aqueous solutions

S Q Wang[1], L M Wang, F Wang and D Fu

[1] Corresponding author. Address: School of Environmental Science and Engineering, North China Electric Power University, Baoding, 071003, People's Republic of China. Tel.: +86-312-15930719397;

E-mail addresses: 673040151@qq.com

Abstract. The surface tensions (γ) of N-methyldiethanolamine (MDEA)-methanol (MeOH) aqueous solutions were measured by using an automatic surface tension-meter (BZY-1). The temperature ranged from 303.2K to 323.2K. The mass fractions of MeOH and MDEA respectively ranged from 0.05 to 0.15 and 0.2 to 0.4. On the basis of the experimental measurement, the effects of temperature and mass fraction of MDEA and MeOH on surface tensions were analyzed.

1. Introduction

Climate change and environmental problems caused by the emissions of carbon dioxide (CO_2) have attracted increasing attentions worldwide and the reduction of CO_2 has become a global issue[1, 2]. Post combustion CO_2 capture using amine as absorbent is one of the mature technologies for reducing CO_2 emissions from coal-fired power plants[3-8]. Aqueous solutions of MDEA have been widely used for the removal of CO_2 form a variety of gas streams[9, 10].

Besides the amine blends, adding a physical solvent to the aqueous solutions of alkanolamines is considered to be an effective method to lower the regeneration energy cost[11, 12]. For example, *Archane et al.* [11] studied the vapor-liquid equilibria, the concentration of the molecular form of absorbed CO_2, the partial pressure of CO_2 and the liquid phase speciation for CO_2-poly(ethylene oxide)400 (PEG400)-diethanolamine (DEA) aqueous solutions, and demonstrated the influence of PEG400 on the absorption of CO_2. Their results showed that lower energy cost can be achieved in regeneration because at a given CO_2 loading and constant concentration of DEA, the CO_2 molecular concentration increases with the increase of PEG400 concentration, whereas the ion repartition is not significantly influenced by the solvent composition. *Henni and Mather*[13] studied the solubility of CO_2 in a mixed nonaqueous solvent of MDEA and MeOH. The results showed that at low acid gas partial pressures the solubility of CO_2 is higher in the mixed solvent than in pure MeOH.

A knowledge of surface tension is required when designing or simulating an absorption column for CO_2 capture. In recent years, there are many experimental and theoretical works concerning the surface tensions of aqueous solutions containing MDEA and physical solvent. *Fu et al.* [14] measured the surface tensions of PEG400, MEA-PEG400 and DEA-PEG400 aqueous solutions and satisfactorily modeled by using a thermodynamic equation. On the basis of experiments and calculation, they demonstrated the effects of the temperature, the mass fractions of amines and PEG400 on the surface tension. *Fu et al.* [15] studied the surface tensions of MDEA–piperazine (PZ) aqueous solutions. They proposed a theoretical model to correlate the surface tensions and the calculated results agreed well with the experiments. However, experimental work concerning the surface tensions of MDEA-MeOH aqueous solutions is relatively rare.

The main purpose of this work is to investigate the surface tension of MDEA-MeOH aqueous solutions, and then demonstrate the effects of temperature and mass fractions of MDEA and MeOH on the surface tension. To this end, the surface tension was measured at the temperatures from 303.2K to 323.2K. The mass fraction of MeOH and MDEA respectively ranged from 0.05 to 0.15 and 0.2 to 0.4.

2. Experimental Section

2.1. Materials

The samples used in this work are detailed in Table 1. Purities are as stated by the supplier, and no further purification was carried out. Aqueous solutions of MDEA-MeOH were prepared by adding doubly distilled water.

2.2. Apparatus and procedure

The surface tension was measured by using the BZY-1 surface tension meter produced by Shanghai Hengping Instrument Factory. The BZY-1 meter employs the Wilhemy plate principle, i.e., the maximum tensile force competing with the surface tension is measured when the bottom edge is parallel to the interface and just touches the liquid. The measurement ranges for temperature and surface tension are respectively (268.15 to 383.15) K and (0.1–400.0) $mN \cdot m^{-1}$. The uncertainty is ±0.1 $mN \cdot m^{-1}$. The size-volume of the different samples used in the BZY-1 meter is 20 ml. During the experiments, the copper pan in the host of the BZY-1 meter is connected with the thermostatic bath (CH-1006, uncertainty is ±0.1 K). Via the circulation of the water, the temperature of the water in the copper pan is kept the same as that in the thermostatic bath. The aqueous solution is put into the solution container immersed in the copper pan and its temperature can be measured by a thermocouple. The scale reading of the thermocouple has been well calibrated by a mercury thermometer.

3. Results and discussion

The experimental results for the surface tensions of MDEA-MeOH aqueous solutions are respectively shown in Table 1.

Table 1. Surface tensions (γ) of MDEA-MeOH aqueous solutions under different mass fractions of MDEA (w_{MDEA}) and MeOH (w_{MeOH}). Pressure (p) = 101 kPa.

w_{MDEA}	w_{MeOH}	γ / ($mN \cdot m^{-1}$)		
		303.2K	313.2K	323.2K
0.20	0.00	54.2	53.8	51.7
	0.05	52.6	51.3	49.6
	0.10	48.9	47.6	45.9
	0.15	45.8	44.8	43.6
0.30	0.00	51.2	50.2	48.6
	0.05	49.7	48.6	46.5
	0.10	45.6	44.5	43.2
	0.15	42.8	42.6	42.2

0.40	0.00	49.8	48.8	47.1
	0.05	45.4	44.7	43.4
	0.10	44.2	43.3	42.6
	0.15	42.2	40.7	40.1

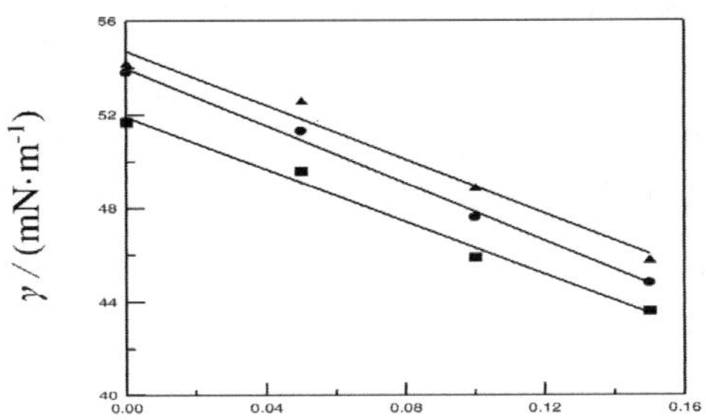

Figure 1. w_{MeOH} dependence of the surface tensions of MDEA-MeOH aqueous solutions at w_{MDEA}=0.20. Symbols: ▲T=303.2K; ●T=313.2K; ■T=323.2K. Lines: trend lines.

Figure 1. shows the influence of w_{MeOH} on the surface tension of MDEA-MeOH aqueous solutions. From this figure, one may find that at given temperature and given w_{MDEA}, the surface tension decreases with the increase of w_{MeOH}.

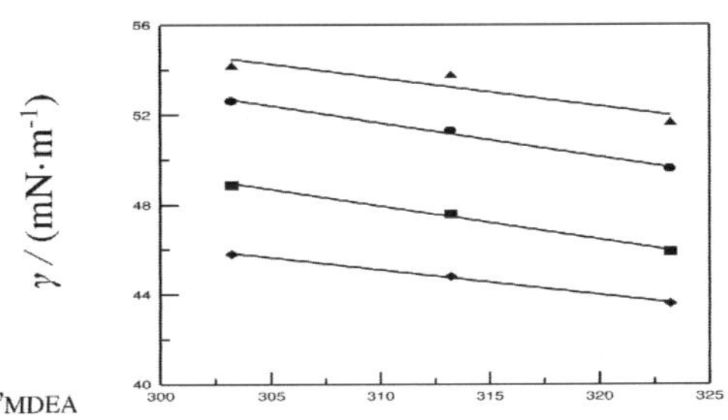

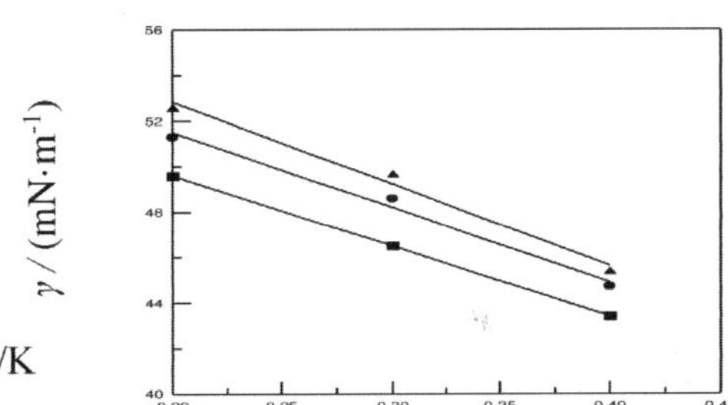

Figure 2. Temperature dependence of the surface tensions of MDEA-MeOH aqueous solutions at w_{MDEA}=0.20. Symbols: ▲w_{MeOH}=0; ●w_{MeOH}=0.05; ■w_{MeOH}=0.10; ◆w_{MeOH}=0.15. Lines: trend lines.

Figure 3. w_{MDEA} dependence of the surface tensions of MDEA-MeOH aqueous solutions at w_{MeOH}=0.05. Symbols: ▲T=303.2K; ●T=313.2K; ■T=323.2K. Lines: trend lines.

Figure 2. shows the temperature dependence of the surface tension of MDEA-MeOH aqueous solutions. One finds that at given w_{MeOH} and w_{MDEA}, the surface tension decreases with the increase of temperature. Figure 3. shows the influence of w_{MDEA} on the surface tension of MDEA-MeOH aqueous solutions. From this figure, one may find that at given temperature and given w_{MeOH}, the surface tension decreases with the increase of w_{MDEA}.

4. Conclusions

In this work, the surface tensions of MDEA-MeOH aqueous solutions were measured by using the BZY-1 surface tension meter. The effects of temperature and mass fractions of MDEA and MeOH on the surface tension were demonstrated. Our results show that:

(1) The increase of temperature tends to decrease the surface tension of MDEA-MeOH aqueous solution.

(2) The increase of the mass fraction of MDEA and MeOH tend to decrease the surface tension of MDEA-MeOH aqueous solution.

References

[1] Dickens GR 2004 Nature 429 513-515
[2] Herzog HJ 2001 Environ. Sci. Technol. 35 148A-153A
[3] Zhang MK and Guo YC 2013 Appl. Energy. 111 142-152
[4] Maham Y, Liew CN and Mather AE 2009 J. Solu. Chem. 31 743-756
[5] Faramarzi L, Kontogeorgis GM, Thomsen K and Stenby EH 2009 Fluid. Phase. Equilibr. 282 121-132
[6] Kim I, Hoff KA, Hessen ET, Haug-Warberg T and Svendsen HF 2009 Chem. Eng. Sci. 64 2027-2038
[7] Maham Y, Mather AE and Mathonat C 2000 K. J. Chem. Thermodyn. 32 229-236
[8] Guo HX, Zhou ZM and Jing GH 2013 Int. J. Greenh. Gas. Con. 16 197-205
[9] Pant KK and Srivastava VK 2007 Chem. Eng. J. 133 229-237
[10] Navaza JM, Gómez-Díaz D and La Rubia MD 2009 Chem. Eng. J. 146 184-188
[11] Archane A, Fürst W and Provost E 2011 J. Chem. Eng. Data 56 1852-1856
[12] Archane A, Gicquel L, Provost E and Fürst W 2008 Chem. Eng. Res. Des. 86 592-599
[13] Henni A and Mather AE 1995 J. Chem. Eng. Data 40 493-495
[14] Fu D, Du LX and Wang HM 2014 J. Chem. Thermodyn. 69 132-136
[15] Fu D, Xu YF and Hua XY 2012 Fluid. Phase. Equilibr. 314 121-127

Research on trading patterns of large users' direct power purchase considering consumption of clean energy

He Guojun[1], Guo Lin[2], Yu Zhicheng[1], Zhu Xiaojun[1], Wang Lei[3,4], Zhao Zhiqiang[2]

[1] Electric Power Research Institute, State Grid Chongqing Electric Power Corporation, Chongqing 401123, China
[2] Power Exchange Center of Chongqing Electric Power Corp., Yuzhong District, Chongqing 400014, China
[3] State Grid Electric Power Research Institute,Nanjing 210003,China
[4] Beijing Kedong Power Control System Co., Ltd., Beijing 100085, China

Abstract. In order to reduce the stochastic volatility of supply and demand, and maintain the electric power system's stability after large scale stochastic renewable energy sources connected to grid, the development and consumption should be promoted by marketing means. Bilateral contract transaction model of large users' direct power purchase conforms to the actual situation of our country. Trading pattern of large users' direct power purchase is analyzed in this paper, characteristics of each power generation are summed up, and centralized matching mode is mainly introduced. Through the establishment of power generation enterprises' priority evaluation index system and the analysis of power generation enterprises' priority based on fuzzy clustering, the sorting method of power generation enterprises' priority in trading patterns of large users' direct power purchase is put forward. Suggestions for trading mechanism of large users' direct power purchase are offered by this method, which is good for expand the promotion of large users' direct power purchase further.

1. Introduction

With the continuous progress of reform process of China's electricity market, and after the separation of plant and network and open competition of generation side, there are growing calls of introducing competitive mechanism in sale side and improving electricity price system. As an important part of opening market competition, the development and implementation of direct purchase of electricity has an important significance.

The current direct purchase transaction mode of large users is negotiated bilateral transactions, centralized matching transactions and centralized bidding transactions. Bilateral contract transaction model of large users' direct power purchase conforms to the actual situation of our country, but large users direct transaction have a negative impact on clean energy consumption. So, comprehensive benefit evaluation technology of clean energy consumption is very important.

2. Trading patterns of large users' direct power purchase

Literature [1], [2] combined with China's current planning forms and the status quo of market transactions, the important principles of the scientific design of large users direct purchase of electricity is put forward. From the perspective of economics and market, the misunderstanding of large users direct purchase is analyzed in literature [3], the problems which existing in the direct purchase of large users in our country at present are explained and the solution is put forward. Two modes of large users direct purchase of electricity trading patterns is proposed in literature [4], and trading rules and algorithms are designed. In literature [5], the influence of time sharing price and power quality is considered, algorithm for the fixed cost allocation problem of large user transfer cost is provided by postage stamp method， which can reduce cross subsidy.

3. Power generation characteristics

3.1 Traditional thermal power generation characteristics

Thermal power generation refers to the heat generated by oil, coal and natural gas to heat water, high temperature and high pressure steam can be formed, and the water vapor can drive generator to generate electricity. Thermal power plant's site is flexible, which can reduce loss of power transmission and distribution. The operation is stable and reliable, which is the backbone of the national grid power supply. Units are barely affected by the environment, climate and other adverse factors. But, utilization rate of traditional thermal power resources is low, operating cost is high, and wasting problem is serious. Besides, the time of start and stop is long, equipment operation and maintenance is difficult.

3.2 Sichuan hydropower generation characteristics

Sichuan gives priority to the development of hydropower, so that hydropower occupy the dominant position of power supply in Sichuan. By the end of 2002, hydropower installed capacity is 11 million 731 thousand and 500 kilowatts, accounting for 64.78% of the province's installed capacity, ranking first in the country. Hydropower became the main power supply power and peak power. In recent years, the national network of Sichuan electric power company has been taken various measures to utilize redundant power. The next step will continue to take various measures to consume surplus hydropower.

3.3 Characteristics of wind power generation

Wind energy resources are rich, clean and renewable. Compared with conventional thermal power units, the construction period of the wind farm is short, and the automation degree is high. It covers a small area, and abundant wind resource places tend to desolate sands or hilly area , which reduces the wind farm land costs in construction period. However, wind power also has certain limitations: the motive forcer of wind power is natural wind, so that wind speed and wind quantity are not controllable, and the wind is difficult to store.

4. Centralized matchmaking transaction mode based on large user priority classification

4.1. Centralized matching transaction mode

Centralized matching transaction mode refers to release the supply and demand information by electricity trading center through a unified trading platform. The two sides declare purchase and sale electricity and electricity price respectively. According to certain rules and security constraints to match . In the centralized matching mode, both sides reported electricity quantity and electricity prices before the deadline. Trading center will match in a fixed time. The core idea is "high-low match", seller in accordance with the offer from low to high priority scheduling. Buyers in accordance with the order from high to low priority, the highest priority between the buyer and the seller will be traded first, the transaction price is average price for both parties, and so on.

4.2. Establishment of priority evaluation index system for power generation enterprises

Big users are more willing to deal with power generation companies with good credit ratings, good social benefits, and good power generation. It is very important to determine the access conditions of the power generation enterprises for the relevant parties.

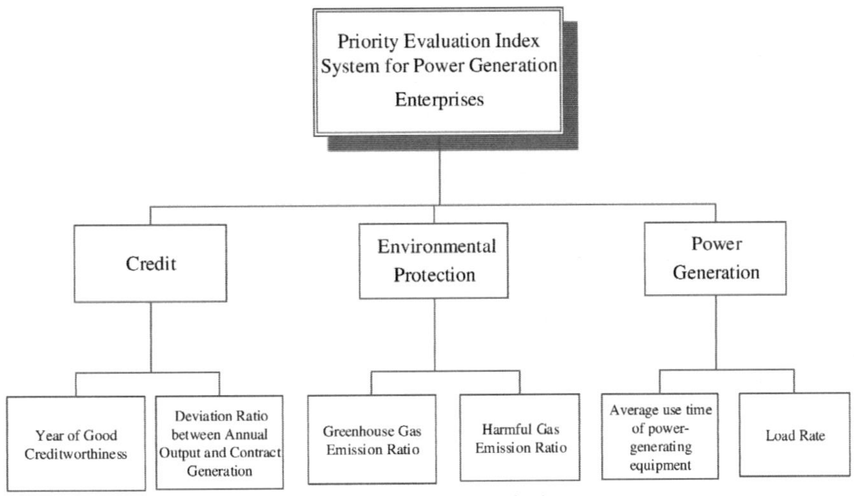

Figure 1..Priority evaluation index system.

4.2.1 Credit:

Credit is the integrity of power-generation firms in their production and management. Transaction risk can be reduced while trading with credible enterprise for both grid companies and customers.
Year of contract-credit: "contract-credit" is the most authoritative comprehensive assessment which declared by the government. Therefore, year of contract-credit is one of index which evaluate user's credit condition.
The deviation ratio between annual output and contract generation: the deviation is the default volume of annual transaction below the 97% of contact generation (power adjustment is included).

4.2.2 Environmental protection

Big users in direct power purchase are all required to conform to environmental policy by relevant files. Greenhouse gas emission ratio:

$$\eta_{ggr} = \frac{m_{gg}}{E_{out}} \tag{1}$$

E_{out} is the total switching energy, m_{gg} is the mass of greenhouse gas emissions.
Harmful gas emission ratio:

$$\eta_{hgr} = \frac{m_{hg}}{E_{out}} \tag{2}$$

E_{out} is the total switching energy, m_{hg} is the mass of harmful gas emissions.

4.2.3 Power generation

Average utilization hours of power generation equipment is the ratio between generating capacity and average capacity. It reflects the utilization level of time. The formula is：

$$T = \frac{Q}{Q_1} \tag{3}$$

Q is the generated energy, Q_1 is the average capacity.

Load rate is the ratio between the average load and the highest load. It reflects the different level of load. Bigger number means production is balance, so the priority is higher.

4.3.The priority classification of power enterprises based on fuzzy clustering

Fuzzy clustering analysis is a method to study the classification of a given object by using the fuzzy mathematics method. Taking into account the market may have a competitive and non- competitive units, bidding unit may also have non-competitive power, so the priority of the power plant needs to be considered in accordance with the two conditions which are bidding energy of the bidding units and non- competitive power in bidding units and non-competitive units' generation.

4.4 Centralized matching transactions that can promote clean energy consumption

Centralized matching transactions is the buyer in accordance with the offer from high to low priority, the seller in accordance with the offer from low to high priority. The first match is the highest priority buyers and sellers , and then match the inferior priority buyers and sellers.

Market clearing use the way of matching. The highest bid in consumptive power manufacturers and the lowest bid of inter provincial clean energy can match, according to the minimum amount of electricity as its turnover. After transaction is confirmed, and quoted price is eliminated, the match can carry on again according to the same way until one side does not offer, or both sides have offer but inter provincial clean energy price is higher than other side.

5. Conclusion

Trading patterns of large users' direct power purchase which considering clean energy consumption are analyzed in this paper, various power generation characteristics are summed up, trading patterns of large users' direct power purchase are introduced. Through the establishment of priority evaluation index system of power generation enterprises and analysis of priority problems of power generation enterprises based on fuzzy clustering, recommendations from the trading mechanism and development stage are put forward, which is conducive to the further expansion of large users' direct power purchase.

Reference:

[1] Xia Qing, Bai Yang, Zhong Haiwang. Institutional Design and Suggestions for Promotion of Direct Electricity Purchase by Large Consumers in China[J], Automation of Electric Power Systems, 2013, 37(20):1-7.

[2] Zhang Xian, Liu Fubin, Peng Tao. A Discussion on Key Issues for Direct Trading Between Power Users and Plants [J]. Automation of Electric Power Systems,2014, 38(13):33-37.

[3] Hu Jianhyi, Chen Xiyin. Discussion on direct power purchase of large consumers[J]. Power System Technology, 2007,31(24):40-45.

[4] Chen Haoyong, Zhang Senlin, Zhang Yao. Model and algorithm of direct power purchase for large customers in electricity market[J]. Power System Technology, 2008,32(21):85-90.

[5] Zhou Ming, Zheng Yanan, Li Gengyin. Integrated allocation method for fixed cost of large customer transfer cost considering time sharing price and power quality[J]. Proceedings of The Chinese Society for Electrical Engineering, 2008, 28(19): 125-130.

[6] Sun Bo, Shi Quansheng. Analysis of the price design model of direct purchase of large consumers under the profit balance[J]. East China Electric Power. 2013, 41(1): 47-49.

[7] Wang Beibei, LIU Xiaocong, LI Yang. A study on the daily scheduling and operation simulation of large capacity wind power access considering user side interaction[J]. Proceedings of The Chinese Society for Electrical Engineering, 2013, 33(22): 35-44.

Production of bio ethanol from waste potatoes

Mohamad Jaber Noufal[*1, 2], **Baizhan Li**[1], **Zena Ali Maalla**[1, 2]

[1]Faculty of Urban Construction and Environmental Engineering, Chongqing University, 174 Shazhengjie, Shapingba District, 400045, Chongqing City, People's Republic of China
[2]Faculty of Civil Engineering and Environmental Engineering, Al-Baath University, Homs City, Syrian Arab Republic

[*]corresponding author's e-mail address: mohmadnofal@hotmail.com

Abstract. In this research, production of ethanol from waste potatoes fermentation was studied using Saccharomyces cerevisiae. Potato Flour prepared from potato tubers after cooking and drying at 85°C. A homogenous slurry of potato flour prepared in water at solid-liquid ratio 1:10. Liquefaction of potato starch slurry was done with α-amylase at 80°C for 40 min followed by saccharification process which was done with glucoamylase at 65°C for two hr. Fermentation of hydrolysate with Saccharomyces cerevisiae at 35°C for two days resulted in the production of 33 g/l ethanol. The following parameters have been analysed: temperature, time of fermentation and pH. It found that Saccharification process is affected by enzyme Amylase 300 concentration and concentration of 1000μl/100ml gives the efficient effect of the process. The best temperature for fermentation process was found to be about 35oC. Also, it noticed that ethanol production increased as a time of fermentation increased but after 48 hr further growth in fermentation time did not have an appreciable effect. Finally, the optimal value of pH for fermentation process was about 5 to 6.

1. Introduction

Bioethanol as an alternative source of energy has received particular attention world over due to depletion of fossil fuels. From the 18th century to the beginning of this century, significant discoveries about the biology and chemistry of fermentation and distillation made it possible to produce cheaper ethanol from a variety of organic materials [1, 2].

To generate ethanol from biomass two critical processes followed. First, the starch or hemicelluloses and cellulose portions of the biomass are broken down into simple sugars through a process called saccharification. Second, the sugars are fermented to produce ethanol [3, 4].

The sugary substrates available are comparatively expensive than molasses but can easily use for ethanol production with some modification in the process. On the other hand, cellulosic materials are cheaper and available in plenty, but their conversion to ethanol involves many steps and is expensive. The starchy substrates are promising due to their economic viability and availability. Starchy crops like corn, barley, wheat, rice and tuber crops viz. potato, sweet potatoes are being exploited for the production of bioethanol worldwide since it is rich in starch, which makes it the second substrate for ethanol production. The problems associated with its processing will also be less than in other grains. It is also semi-perishable food which can store for a considerable period without spoilage. Good quality alcohol can produce from potato which can use for fuel. The purpose of this research was to explore the parameters and operating conditions that give the best conversion of potato starch into ethanol [4, 5, and 6].

2. Experimental work

2.1. Material and Methods

2.1.1. Potato preparation. Waste of potato brought from chips factory and restaurants. 1 kg waste potatoes were cooked with two liters water in a pressure cooker at 85 oC for one hour then it dried overnight in an oven at 70oC and mashed to a fine powder (0.3mm).

2.1.2. Preparation of potato slurry. Slurries prepared from potato powder mixed with water at a ratio (1:20 w/w). The slurry was treated with amylase enzyme (1000µl/100ml) at 80 oC for 40 min under shaking conditions.

2.1.3. Analysis of waste potatoes. Waste of potatoes contained 77% moisture, 18.2% starch, 2.4 % proteins and 0.6% total sugars while the respective hydrolysate of potato was 13.2, 70, 10.8, 1.5 and 4.5% (Table 1).

Table 1. Composition of waste potato and potato hydrolysate [6]

Component	Composition (%)	
	Waste potato	Potato hydrolysate
Moisture	77	13.2
Starch	18.2	70
Proteins	2.4	10.8
Total sugar	0.6	1.5
Others	1.8	4.5

2.1.4. Saccharification of slurry. Saccharification of the slurry was carried out at 65oC for two hr using Amy 300 enzyme of concentration (1000 µl/100ml). The reaction monitored by the yield of total reducing sugars estimated by dinitro salicylic acid method [7].

2.1.5. Effect of enzyme concentration. The liquefied potato flour saccharified with different levels of Enzyme Amy 300 (600, 800, 1000 µl/100ml) at 65°C. The reaction monitored by the yield of total reducing sugars which estimated by dinitro salicylic acid method [8].

2.1.6. Yeast culture. A fast fermentation strain of Saccharomyces cerevisiae was maintained on yeast extract peptone dextrose agar medium containing in grams per 100ml (potato starch 10; peptone 0.1; malt extract 0.1 ; yeast extract 0.2; magnesium chloride 6H2O 0.1; calcium carbonate 0.2 ; ammonium phosphate 0.2 ; and ferrous sulphate .7H2O 0.001). Yeast cells were grown in inoculums medium at 35oC for 20 hr under shaking condition (100 rpm) and centrifuged at 8000 rpm for 20 min. For testing the effect of pH on fermentation few drops of 1N HCl or 1N NaOH were added to this medium to obtain the desired initial pH [8].

2.1.7. Fermentation of potato's hydrolysate. The hydrolysate of potato inoculated with Saccharomyces cerevisiae of 0.8% concentration (w/v) at 35oC. Ammonium sulfate of 0.2% concentration (w/v) added as a source of nitrogen. Flasks incubated at 35°C under the stable condition for 96 hr and ethanol content measured at an interval of 24 h by gas-liquid chromatography (GC 8200) using capillary column and flame ionization detector.

3. Results and discussion

3.1. Effect of Concentration

Fig.(1) shows that as enzyme concentration increased the total reducing sugar also increased and level of 1000 µl/100ml gave practical effect in saccharification. This behavior caused by the higher growth rate of microorganisms at high values of inoculums concentrations which lead to higher rate of organic matters degradation in the process. This finding is in agreement with that found by Nagoda, T [9].

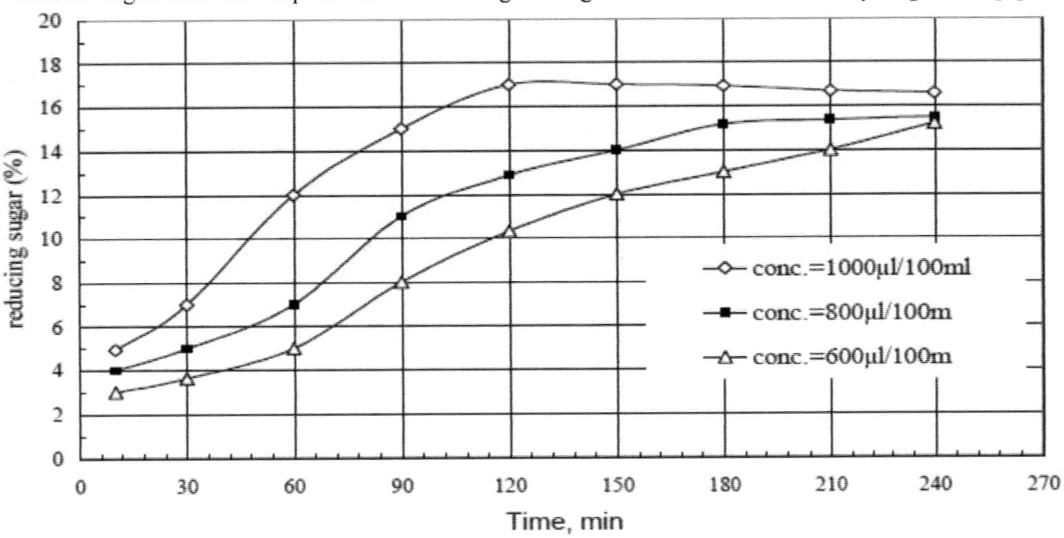

Figure 1. Effect of Amy concentration on Saccharification on hydrolysate at 65 °C

Fermentation of 30% slurry of potato hydrolysate was carried out at different temperatures (25, 30, 35, 40 °C) under stable conditions up to 48 hr. Fig.(2) shows that fermentation at 35oC gives the maximum content of ethanol of 32g/l. Due to the effect of temperature on the activity of the microorganisms.

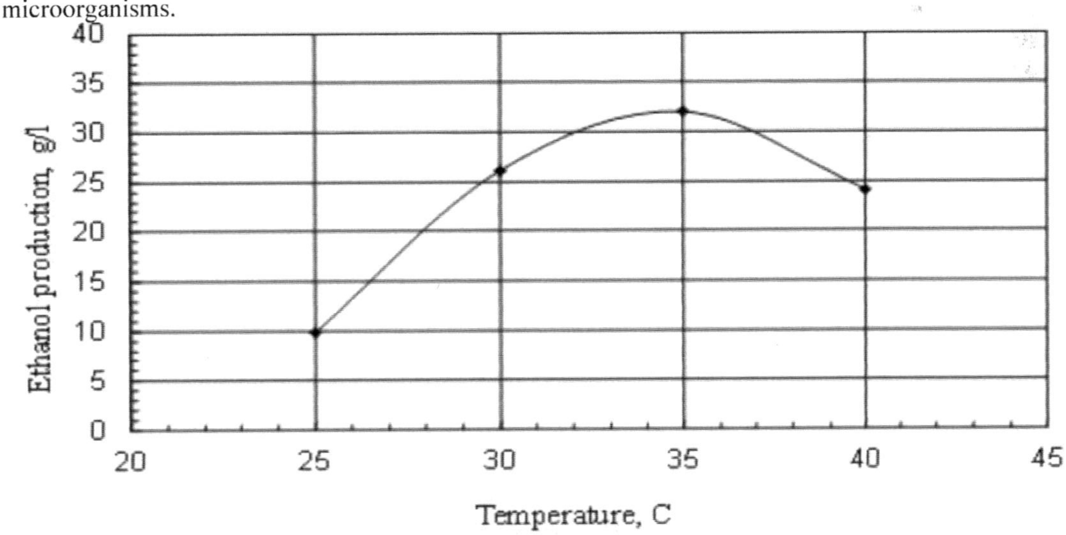

Figure 2. Effect of temperature on ethanol production (Enzyme Conc. =1000µl/100ml,Time of fermentation = 48 hr, pH=5.5)

It knew that temperature above 40oC affects the membrane composition of microorganisms, e.g. the phospholipids fatty acid composition changes with temperature and hence affects the enzymatic system of the organisms [10].

IOP Conf. Series: Earth and Environmental Science **59** (2017) 012006 doi:10.1088/1755-1315/59/1/012006

3.2. Effect of Time Fermentation

Fermentation of potato hydrolysate was carried out at 35 °C for different time intervals using enzyme concentration (1000μl/100ml) of Amy 300 containing amyloglucosidase. Fig.(3) shows that ethanol production increased as a time of fermentation rose from 24 to 48 hr, so it reached to 32g/liter at 48hr; however, further growth in fermentation time did not have an appreciable effect. This result is in agreement with that mentioned by Wang F [11] and Oner [12].

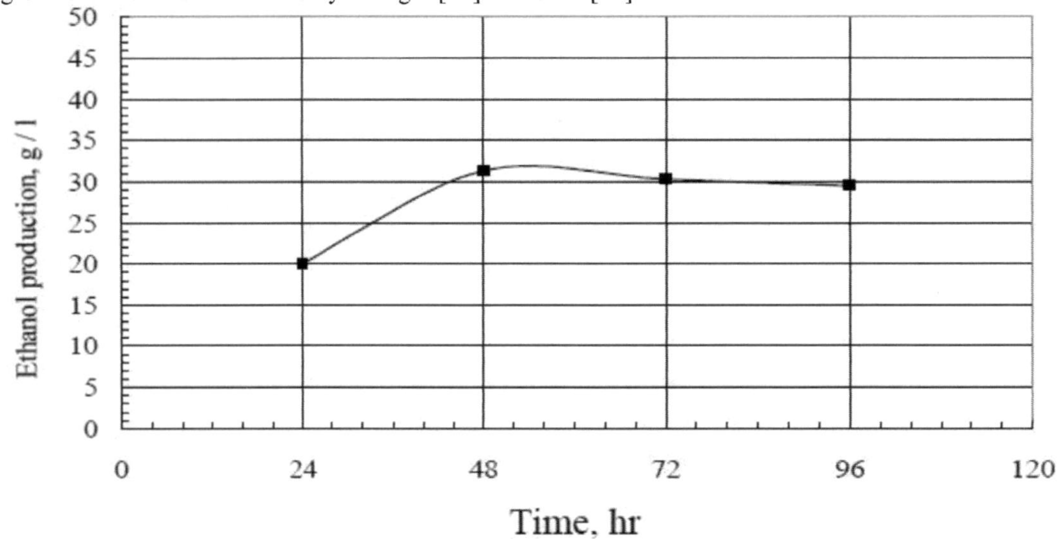

Figure 3. Effect of fermentation time on ethanol production (Enzyme Conc. =1000μl/100ml, Temp =35oC, pH =5.5)

3.3. Effect of pH

Figure (4) shows that the production of ethanol has increased with increasing pH from 3 to 6, the optimal production of ethanol was 32g/l which occurs at pH 6. After that, the production began to decrease with increasing pH and reached to 22 g/l at pH 8. This behavior can interpret by the effect of pH on the activity of α Amy enzyme, since the business of this protein is severely affected by the value of pH especially after the value of pH=6 [8].

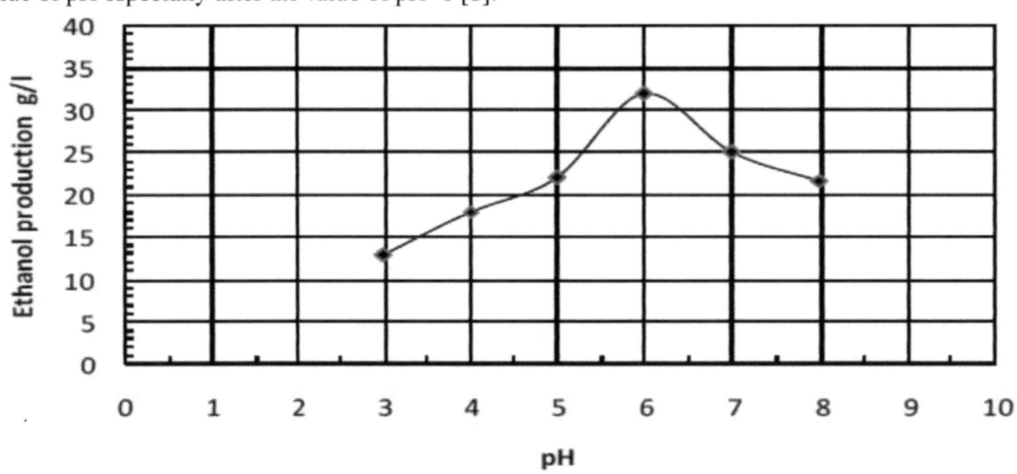

Figure 4. Effect of pH on ethanol production (Enzyme Conc. =1000μl/100ml, Temp. =35oC, time of fermentation=48hr)

4. Conclusions

1) Saccharification process is affected by enzyme Amy 300 concentration and concentration of 1000μl/100ml gives practical effect on the process.
2) The best temperature for fermentation of potato starch to produce ethanol using Saccharomyces cerevisiae is about 35°C.
3) Ethanol production increased as a time of fermentation rose from 24 to 48 hr; however further growth in fermentation time did not have an appreciable effect.
4) The optimal value of pH for the fermentation process to produce ethanol production was about 6.

References

[1] Hill, J.; Nelson, E. ; Tilman, D.; Polasky, S. and Tiffancy, D. "*Environmental, economic and energetic costs and benefits of biodiesel and ethanol biofuels*". Proc.of the Nat. Academy of Sci. USA, 103; 11206-11210 (2006).

[2] Grassi, G. "*Modern Bioenergy in the European Union, Renewable Energy*", Appl. Environ. Biotechnology.16: 985- 990 (1999).

[3] Lemmel, S. A.; Heimsch, R. C. and Korus, R. A. "*Kinetics of growth and amylase production of Saccharomycopsis fibuligera on potato processing wastewater*". Appl. Environ. Microbiol. 39, 387-393 (1980).

[4] Cazetta, M. ; Celligoi, M. ; Buzato, J. and Searmino, I. "*Fermentation of molasses by Zymomonas mobilis*" : Effects of temperature and sugar concentration on ethanol production. Biores. Technol. 98: 2824-2828 (2007).

[5] Jarl, K. "*Production of microbial food from low cost starch materials and purification of industry's waste starch effluents through the Symba yeast process*". Food Technol. 23: 1009-1012 (1996).

[6] Kobayashi, F.; Sawada, T. and Ushiyama, T." *Saccharification and alcohol fermentation in starch of steam exploded potato*". Appl. Biochem. Biotechnol. 69: 177-189 (1998).

[7] Miller, G. L. "*Use of dinitrosalicyclic acid reagent for determination of reducing sugar*". Anal. Chem. 31, 426-428(1959).

[8] Shigechi, H. ; Koh, J.; Fujita, Y. Matsumoto, T. "*Direct production of ethanol from raw corn starch via fermentation by use of a novel surface- engineered yeast strain co-displaying glucoamylase and alphaamylase*". Appl. Environ. Microbiol. 70: 5037- 5040 (2004).

[9] Nagoda, T. W. and Steinkraus, K. H. "*Influence of the rate of ethanol production and accumulation on the viability of Saccharomyces cerevisiae in rapid fermentation*". Appl. Environ. Microbiol. 31: 158-162 (1976).

[10] Chen, H. Xu, J. and ZH, L. "*Temperature cycling to improve the ethanol production with solid state simultaneous saccharification and fermentation*". Prikl Biokhim Mikrobiol. 43:65-68 (2007).

[11] Wang, F. Gao, J. ;Yang, Y. and Xu, P. "*Optimization of an ethanol production medium in very high gravity fermentation*". Biotechnol. Lett. 29: 233-236 (2007).

[12] Oner, E.T., Oliver, S.G. and Kirdar, B., "*Production of ethanol from starch by respiration deficient recombinant Saccharomyces cerevisiae*". Appl. and Environ. Microbiol., 71:6443-6445 (2005).

IOP Conf. Series: Earth and Environmental Science **59** (2017) 012007 doi:10.1088/1755-1315/59/1/012007

Study on the optimization of comprehensive transportation system in China under the perspective of total factor efficiency

Mei SHANG [1], Lu ZHANG [2]

[1,2]School of Management, Xi'an University of Science and Technology, Xi'an Shaanxi, China

[1]940615683@qq.com, [2]709311192@qq.com

Abstract. Using BCC model of DEA technique, this paper estimate the energy effectiveness of China's local transportation industry. Based on data from China's statistical yearbook from 1998 to 2014, operating kilometers of regional railway, highway and water-way industry and employees of civil aviation industry and consumption of transportation energy are chosen as input variables and GDP of regional transportation industry every year as output variable to estimate the operational energy efficiency. It is found that energy efficiency of provinces such as Beijing, Tianjin, Hebei, Shanghai, Jiangsu, Zhejiang, Shandong, Hainan, Qinghai, Ningxia and Xinjiang are relatively effective, compare to national average, provinces such as Shanxi, Anhui, Liaoning, Jiangxi, Jilin, Hunan, Shaanxi, Heilongjiang, Guangxi, Hubei, Guizhou, Inner Mongolia, Gansu, Sichuan, Yunnan are on the low side, which mainly concentrated in the northeast and central and western regions, then, the paper grouped these provinces which energy efficiency on the low side according to the redundancy, the major and minor reasons lead to their lower efficiency are analyzed, and optimization suggestions are put forward.

1 Introduction

With China reforming and opening up to the outside world, transportation industry, as lifebood of domestic economic development, growed quickly.The railway covered all provinces and cities of China, ranking second in the world.By the end of 2014, total mileage of the highway is 104900 kilometers, ranking second in the world. Civil aviation has developed rapidly after the 21st century, Beijing, Shanghai and Guangzhou have built hub airports, connecting airport network of major countries inland. In the aspect of waterway transportation, domestic waterways have been completed by Heilongjiang and Songhua River(serving for the three northeast provinces of China),Yangtze River and Pearl River(serving for the southwest),Huaihe River and the Beijing-Hangzhou Canal(serving for mid-east). Foreign scholars Matthew G.Kariaffis[1] uses data envelopment analysis to study the coordination of transport system in transport efficiency and effectiveness. Ding Xiaodong and Xu Ling[2] use DEA method to evaluate the performance of Chinese transportation industry, finding investment in transportation routs and turnover are key factors affecting transportation efficiency. Chang Yaqing and Song Lai[3] use DEA model to calculate total factor productivity(TFP) of various modes of transport, thinking that transport system are in good condition. Chen Shuai and Sun Youwang[4] analyze energy consumption of transportation system, and propose to built a conservation-oriented integrated transport system.

It is obvious, there are lots of reasearches analyzing comprehensive transport system, but few study taken energy consumption into consideration, which give oppotunity to this paper. In the paper, based on statistic data of regional transport mileage, energy consumption and output of various transportation modes to measure their TFP, analyze causes of total factor inefficiencies, and put forward proposal to optimize the existing comprehensive transport system.

2 Theoretical Review

Data Envelopment Analysis (DEA) first proposed by the United States' famous scientist Charnes and other researchers in 1970's[5,6].The basic idea is to put each appraisal object as a DMU (Decision Making Units),and each DMU has two types (inputs and outputs) of evaluation indexes. By analyzing the input-output ratio, the validity of he DMU is determined.

DEA evaluation model is divided into two categories: one is CCR model (suitable for constant returns to scale), another is BCC model (suitable for variable returns to scale)[7].

3 China provincial TFP of comprehensive transport system measuring with DEA

3.1 Evaluation model, input and output index selection

In the paper, China provincial TFP of comprehensive transport system will be measured using DEA method. In DEA model, input and output generally refers to cost and benefit. Taking into account the data availability and purpose of this study, operating mileage and energy consumption are chosen as input variables, and GDP as output variables for various provincial transportation modes. Taking increased growth of transportation industry into consideration, BCC model is suitable for this study.

3.2 Data analysis and processing

Following data are collected from year 1997 to 2014.

(1) Operating mileage and energy consumption: Using operating mileage of the railway, highway, river transport as one of the input indexs, which reflects the constitution of the regional transportation system. Regional civil aviation employees is chosen to substitute for civil aviation operating mileage for lacking of civil aviation mileage in statistic book . As mentioned above, energy consumption is chosen as another input index.

(2) GDP of various transportation modes :This index could be used to reflect output efficiency of provincial transport systems.

3.3 Empirical Analysis

3.3.1 Empirical results. According to collected statistical data and the BCC model of DEA,TFP of provincials transportation system from year 1997 to 2014 could be measrued as in Figure 1. (if TFP is 1, the provinces will not be listed here). It is obvious that TFP of 11 provincial integrated transport systems (including Beijing, Tianjin, Hebei, Shanghai, Jiangsu, Zhejiang, Shandong, Hainan, Qinghai, Ningxia, Xinjiang) have been relatively effective.Compared with the national average level, Fujian, Henan and Guangdong are one the higher side comparing to average; and Shanxi, Anhui, Liaoning, Jiangxi, Jilin, Hunan, Shaanxi, Helongjiang, Guangxi, Hubei, Guizhou, Neimenggu, Gansu, Sichuan and Yunnan are one the lower side comparing to average.

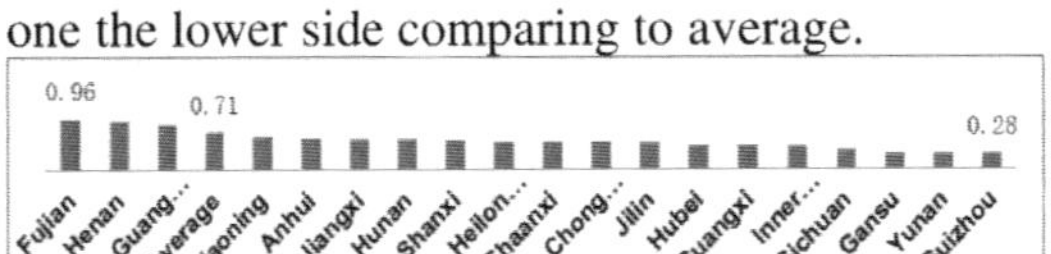

Figure1. Provincial transportation system efficiency

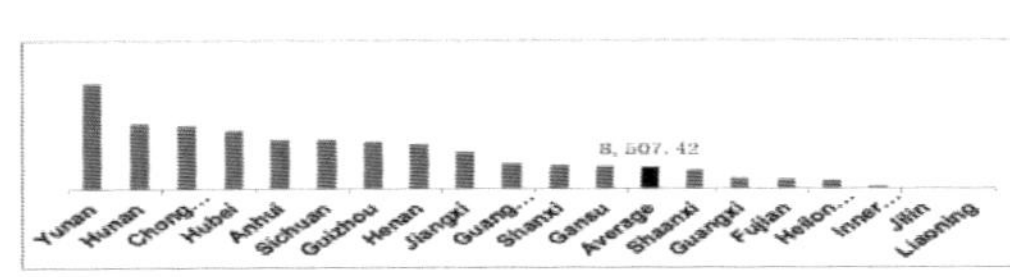

Figure2. The provincial highway redundancy (unit: km)

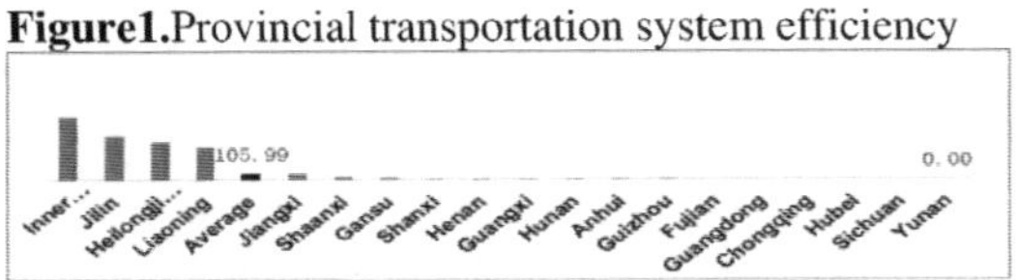

Figure3. The provincial railway redundancy (unit: km)

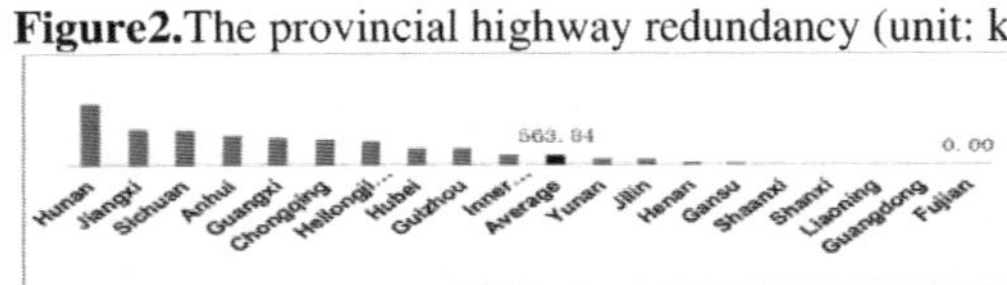

Figure4. The provincial waterway redundancy (unit: km)

IOP Conf. Series: Earth and Environmental Science **59** (2017) 012007 doi:10.1088/1755-1315/59/1/012007

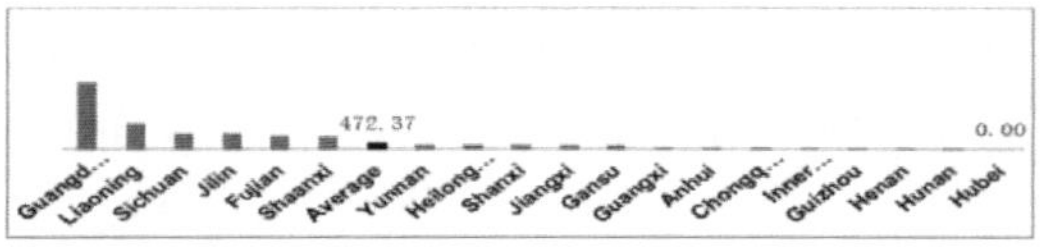

Figure5. The provincial aviation redundancy (unit: person)

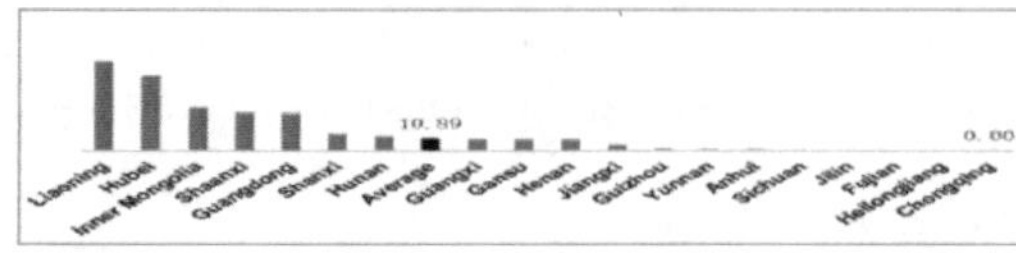

Figure6. The provincial energy redundancy(unit: million tons of standard coal)

3.3.2 Results Analysis. In order to further analyze the reasons for the low TFP provinces of the overall transportation system, various inputs redundancy of provincial transportation system is estimated (Figure 2-6) by the result of DEA. Based on the national average, factors affecting provincial transportation system is grouped into primary cause and secondary causes as listed in table 1.

Table1.domain factors affecting TFP of provincial integrated transport system

Provincial classification	primary cause	secondary cause
Yunnan	highway	—
Shanxi,Henan		energy
Hubei,Chongqing,Guizhou		inland waterway
Inner Mongolia, Heilongjiang	railway	inland waterway
Liaoning,Jilin		civil aviation
Anhui,Jiangxi, Hunan	water transportation	highway
Guangxi		—
Sichuan		civil aviation+highway
Guangdong,Shaanxi	civil aviation	highway
Fujian		—
Gansu	energy	highway

From table 1, it is obvious that provincial TFP is different. Limited by length of paper, primary cause and secondary cause of lower provincial TFP is analyzed for typical provinc in views redundancy of highway, railway, water transportation, vivil aviation and energy.

(1)Approach to solve the problem of highway redundancy: Yunnan highway redundancy is higher than the national average more than 5 times. Chongqing, Hunan, Hubei are about 3 times higher than the average respectively. It is recommended that investment in new highway project shall be decided carefully, lower-use highway shall be redesign or extended in order to make full use of highway resources.

(2)Approach to solve the problem of railway redundancy: the railway redundancy level of four provinces(Inner Mongolia, Liaoning, Jilin and Heilongjiang) is significantly higher than the national average, it is recommended that railway technology innovation shall be used to make the original line meet the current and future transportation needs.

(3)Approach to solve the problem of water transportation redundancy: Water transportation redundancy of Hunan is six times higher than that of the national average. Sichuan, Jiangxi and other provinces are also about three times higher than that of national average. It is recommended that high efficiency waterways shall be make further full use of , and some low efficiency waterway shall be shut down.

(4)Approach to solve the problem of civil aviation redundancy:Civil aviation employees redundencyh of Guangdong Province is 11 times higher than the average. Liaoning is 4 times higher than average.Sichuan, Jilin, Fujian and shaanxi provinces also about 2 times higher than the average.It is recommended that repeat and unused lines shall be cut down, and effort shall be made to establish collaborative relationships among highway,railway and air transportation.

(5)Approach to solve the problem of energy redundancy: There are 14 provinces have some problems in energy redundancy, in which, 5 provinces including Hubei, Liaoning, Shanxi, Shaanxi and Inner Mongolia are in the higher side. Excepting Hubei, other 4 provinces are all resources mining-oriented provinces. It is recommended that passenger and cargo transportation shall be arranged properly, pipe-transportation or multi-purpose transport equipment shall be developed quickly to improve energy efficiency.

4 Conclusions

China's economic development led to the development of highway, railways,waterway and civil aviation industry, "mutual competition and ahead construction" concept is deeply impact the development all modes of transportation, which resulted in incoordination waste of resources. In this paper, referring to DEA method, BCC model is used to evaluate provincial transportation TFP. And following conclusions are drown as follows:

(1) The efficiency of Beijing, Tianjin, Hebei, Shanghai, Jiangsu, Zhejiang, Shandong, Hainan, Qinghai, Ningxia, Xinjiang and other provincial integrated transport system has reached a relatively effective; taking the national average as a standard, Shanxi, Anhui, Liaoning, Jiangxi, Jilin, Hunan, Shaanxi, Heilongjiang, Guangxi, Hubei, Guizhou, Inner Mongolia, Gansu, Sichuan, Yunnan are lower than the average efficiency.Those regions are mainly concentrated in the eastern and western regions.

(2) Classifying by the size of input redundancy for lower comprehensive transportation system TFP, the primary and secondary causes lead to its inefficient TFP is anlyzed , and counter-measures are put forwarded accordingly.

Acknowledgments

I would like to express my gratitude to all those who helped me during the writing of this paper.My deepest gratitude goes first and foremost to Professor Shang Mei, my supervisor, for her constant encouragement and guidance. In the preparation of this paper, she spent much time helping me to finish this paper from preliminary data collection &data analysis to the conclusion. Secondly, I feel grateful to all the scholars who had contributed a lot academic achievements in this field. I finally owe my sincere gratitude to my friends and fellow classmates who gave me their help and time in listening to me and helping me work out my problems during the difficult course of this paper.

References

[1] Matthew.G.K. 2004 DEA Approach for evaluating the effectiveness of urban transit systems. *European Journal of Operational Research*, **152**354-364.

[2] Ding Xiao-dong, XU Ling, YAO Zhi-gang. 2011 Performance Evaluation of Chinese Transportation Industry Based on DEA Method.*Journal of Wuhan University of Technology*, **03**77-81.

[3] Chang Ya-qing,Song Lai 2008 A Study on Relative Efficiency, Technical Progress &TotalFactorProductivity of the TransPortation Industry within Chinese Reforming Period.*Traffic & Transportation*.(Hangzhou:Zhejiang University Press vol 1) chapter 1 pp 34-45.

[4] Chen Shuai, Sun You-wang. 2007 Study on the structure optimization problem of comprehensive transportation system in resource scarce society.*Technology & Economy in Areas of Communications*, **04**97-99.

[5] Odeek J. 2000 Assessing the relative and productivity growth of vehicle inspection services: an application of DEA and Malmquist indices. *European Journal of Operation Research*, **126**01-14.

[6] Pan Ying. 2007 Discussion on Development of Comprehensive Transportation System. *Hebei Jiaotong Science and Technology*, **03**10-14.

[7] Xu Yang. 2010 Study on the Integration of Regional Economic Development and Transportation System of Shaanxi Province.*Chang'an University*.

Ore-controlling factors and prospecting direction of Feimo Cu-Mo Polymetalic deposit in Yuanyang, Yunnan province, China

Tianguo Wang[1, 2], **Jishun Liu**[1, 2], **Zicheng Wu**[1, 2]

[1]Key Laboratory of Metallogenic Prediction of Nonferrous Metals and Geological Environment Monitoring, Ministry of Education, Central South University, Changsha 410083, China
[2]School of Geosciences and Info-Physics, Central South University, Changsha 410083, China

wangtianguo@csu.edu.cn

Abstract. Major ore prospecting breakthroughs have currently been made in the Feimo Cu-Mo polymetallic deposit from the middle section of the Ailaoshan Orogeny Belt. Controlling factors of the ore body have the following distinctive characteristics: NNE and NW trending faults are the major mineralization passageways and ore-containing structures; the orebody exists in the Feimo section of the Along Formation within the Ailaoshan Group; and the concession around the granitic massif has multi-period dyke intrusions and is the preferred prospecting target. Through studies of orebody controlling factors, ore prospect orientations are determined in the Feimo deposit. The Feimo Cu-Mo polymetallic deposit extends along strike and dip directions and has considerable resource potential. In addition, there is considerable potential from the deep-seated Cu-Mo orebody in the area and the possible occurrence of a rich Cu-Mo ore body in surrounding areas such as Caishanping and Niujiaozai.

1. Introduction

The Ailaoshan Orogeny Belt has a distinctive geological background and is rich in mineral resources making it an important source of copper and gold production base in China [1, 2]. Many successful previous studies undertaken in this area focused on determining Cu and Au ore-forming conditions, metallogenic regularity and ore prospecting [3, 4]. The recent discovery, through deeper geological exploration, of the Longbohe Cu deposit, the Daping Au deposit, the Tongchang Cu-Mo deposit, and the Feimo Cu-Mo polymetallic deposit in the Ailaoshan Cu-Au polymetallic belt has caused widespread concern in the mining industry [5, 6]. Furthermore, minimal access to the public for researching the Cu-Mo deposit is restricting exploration work in this area [7]. Thus, this study looks at the factors controlling the formation of the Feimo Cu-Mo polymetallic deposit and provides a directive for prospecting and important guidance for the exploration of similar ore deposits in this area.

2. Geological Background

The Feimo Cu-Mo polymetallic deposit is located at the central part of the Ailaoshan Orogenic Belt (Fig. 1). The strata that lie in the ore area comprised metamorphics that belong to the Ailaoshan Group of Proterozoic rocks. Given that this area is influenced by and located on the western side of Honghe

Fault Zone, folds and faults are relatively developed in the study area. Folds comprise NE-strike monoclinic structures, while the faults comprise NNE, EW, and NW trending groups. Magmatic rocks are observed within the area, particularly the Dapijia massif (binary monzogranite) that outcrops at the western part of the study area. There are also several metamorphosed basic dykes, granitic pegmatite, and rare granite dyke outcrops in the study.

Orebody is influenced and controlled by lithology and structure. The occurrences are stratiform, lentiform, and lense-shaped (Ptla$_2$), with dimensional expansion, contraction, branched composite, and pinch out phenomena. Major metallic minerals include chalcopyrite, bornite, chalcocite, molybdenite, and pyrite. Dominant ore structures are disseminated structure, veinlet-stockwork structure, and massive structure. Observed ore textures include euhedral-subehdral, corona, dissolution, inclusion, and metasomatic relict texture. Wall rock is highly altered by diopsidic alteration, pyritization, silicification, carbonatisation, chloritization, epidotization, actinolitization, tremolitization, and sericitization, with the former three mentioned being closely related to mineralization. The orebody underwent three mineralization phases: the metamorphosed sedimentary period, the hydrothermal period (ore forming period), and the hypergene period. Hence, this is a sedimentary-metamorphism-magmatic hydrothermal overlapping modified Cu-Mo polymetallic deposit.

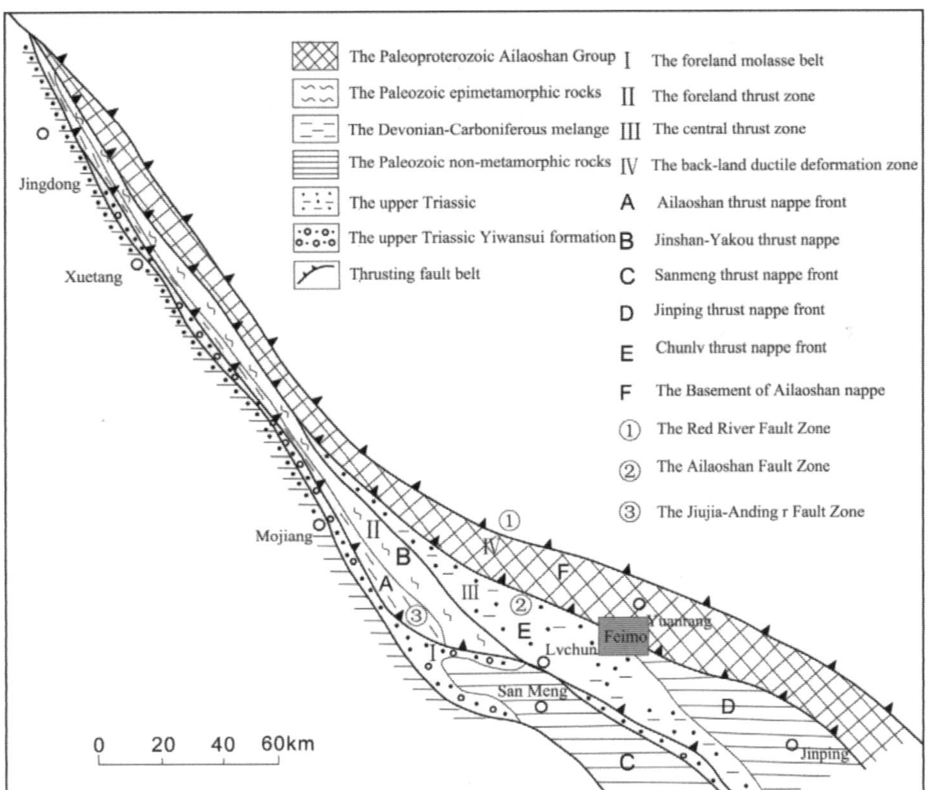

Figure 1. Regional geological sketch map of the Feimo Cu-Mo polymetallic deposit.

3. Ore-controlling factors

The Feimo Cu-Mo polymetallic deposit is located within the secondary fault on the western side of the HongHe fault. The Honghe fault is a large-scale strike-slip movement feature that formed a complicated structural deformation metamorphosed zone. It provided a large pathway and space for ore-fluid large-scale transportation and storage. Granitic magma and ore-bearing hydrothermal fluid are controlled strictly by secondary NW–NWW faults induced, forming one of the largest ore-guiding faults and ore-hosting structures within this zone. In the study area, ore-bearing hydrothermal fluid settled and enriched along the intersection of the NW and the NE trending faults, or at the intersection between a fault bend with a micro fault and inter-bed peelings. These intersections are the major ore-hosting structures for Cu-Mo polymetallic deposits. Micro faults represent one of these important ore occurring structures.

The stratigraphy of the study area consists of the Feimo section of the Along Formation within Lower Proterozoic-aged Ailaoshan Group. The lithologies are characterized by light-gray colored diopsidic marble with intercalated light gray colored bio-plagioclase schist, gneiss, and plagioclase-amphibole granulite. The three latter ones are of volcanic origin with high Cu background value, and provide part of the ore forming materials. The diopsidic marble is chemically active, has good permeability, and is brittle and easily fractured, being easily replaced by ore-bearing hydrothermal fluid or settling of the fluid. It also promotes migration and settling of ore fluid, and together with dark colored metamorphic rocks, they are beneficial to mineralization. Rich orebodies occur within this lithology combination.

Magmatic activity promotes the creation of the ore-bearing layer by modification, superimposition and enrichment processes. Although no economically viable orebody is found within the massif, the granite has a high content of Cu (Cu 0.46%, Au 0.39g/t, Ag 6.6 g/t) benefitting copper mineralization. This clearly demonstrates that granitic magma is also a source of one of the ore-forming materials. Ore-bearing hydrothermal fluid, created by expulsion following a magmatic activity reaction with surrounding wall rock, then settled in the form of Cu-Mo polymetallic orebody around the outer contact zone of the Dapijia granite massif. The intrusion date of the granite massif is estimated to be 57.72 Ma and is related to the ore-forming date of 47.81±0.71 Ma, indicating that the granite massif is involved in controlling the mineralization process. As such, the area surrounding a granite massif, especially with multiple dyke intrusions, is a preferred prospecting target.

In particular, a large number of metamorphosed basic dyke outcrops in the area are closely related to the orebody at both sides of the orebody. The Cu, Pb, Zn, and Ni contents within the metamorphosed basic dyke are higher. Pyritization and Chalcopyrite can also be observed in the dyke, resulting in smaller sized and independent mineable copper orebodies. Hence, metamorphosed basic dykes are closely related to orebody formation.

4. Prospecting Direction

According to the comprehensive analyses of metallogenetic rule and geological information of the Feimo Cu-Mo polymetallic deposits, it is highly suggested that the polymetallic ore deposits of the Feimo Cu-Mo formation and its adjacent outer areas provide good prospecting potential.

(1) The Feimo Cu-Mo polymetallic deposit extends along the strike and dip directions, and the resource potential remains great. Favorable ore bearing zones are found at the Ailaoshan Group-Along Formation that is extensively exposed in the area. Furthermoer, the intersections of the zones with existing structures also make it an excellent prospecting target. In the deeper zones and the surrounding mining area, it is essential to identify the features controlling the deposit formation and to seek the presence of zones with the aforementioned combination of favorable characteristics.

(2) On the southeast side of the mining area, magmatic activity remains active, with large quantities of granitic veins being formed along the tectonic intrusion. In the vicinity of the dyke, mineralized veins are exposed allowing sampling access. Samples taken from outcrops showed metal contents of Cu 1.25%, Au 0.091g/t, and Ag 9.01g/t. The analyses of samples showed that the samples have reached recoverable grade, thereby indicating that prospecting the concealed orebodies in the deeper zones is possible through the integration of observations of geological, geochemical, and other characteristics.

(3) Granitic dykes are found in several areas within the mining area. Additionally, the mineralized outcrop of Dapijia Granite Massif is also found in the adjacent area, implying that there may be concealed ore bearing granitic rocks. It is, therefore, inferred that in the deeper zone of the mining area, there is high likelihood of the existence of large-scale porphyry and contact metasomatism Cu-Mo

polymetallic deposits. Therefore, exploration of deeply concealed porphyry and contact metasomatism of Cu-Mo polymetallic deposits in the region will be the focus of further research and work. Based on the research results of the metallogenic model, the comprehensive study of the current project has revealed information regarding mineralization processes. Electromagnetic sounding and other means adopted to determine the target area as well as the implementation of deep drilling can be used to explore the concealed ore-bearing granite and possibly to find new types of orebodies.
(4) Recoverable grade copper orebodies have been found in the western Niujiaozhai area situated outside the mining area. It is also noted that the region between the copper orebodies and the Feimo Cu-Mo polymetallic deposit is located at the side of the Dapijia Granite Massif, on the western side of the Honghe fault zone, resulting in favorable metallogenic conditions. Furthermore, mineralized outcrops are also found within the area and the adjacent Malishu area. However, due to engineering oversight in the region, there is limited accessibility to geological information.
(5) Recoverable copper orebodies, mostly layered, are also found in the Caishanping area located outside the main mining area. Since the ore layers and lithology are similar to the Cu-Mo polymetallic deposit, the Caishanping area also possesses the prospective potential for mining the superimposed modified orebody of the Feimo deposit.
The central section of the Ailanshan copper-gold metallogenic belt has great prospecting potential, however, due to the lack of similar studies, the understanding of its resource potential is still relatively shallow. Further work o research is needed in order to find the orebody within the contact zones of the rock and other favorable ore-bearing zones.

5. Conclusion

The Feimo Cu-Mo polymetallic deposit is situated within the Ailaoshan Copper-Gold metallogenic belt and provides excellent metallogenic conditions.
(1) The factors controlling ore formation in the Feimo Cu-Mo polymetallic deposit display the following characteristics: NNE, NW trending faults are the major ore-forming channels and ore-bearing structures; orebodies are mainly located within the Ailaoshan Group-Along Formation-Feimo Section; granitic rock areas surrounded by multiple dyke intrusions provide the main prospecting target.
(2) The Feimo Cu-Mo polymetallic deposit and its outer adjacent areas have great prospecting potential particularly as the orebodies extend along the strike and dip direction.; It is likely that deeply buried ore-hosted Cu-Mo polymetallic deposits will occur in the larger porphyry and contact metasomatites. The Caishanping and Niujiaozhai areas that are located outside the main mining area also possess great potential of recovery similar to rich orebodies found in the Feimo mining area.

References

[1] Yumul G P, Zhou M F, Wang CY, Zhao T P and Dimalanta C B 2008 *Journal of Asian Earth Sciences* **32** 385–95
[2] Deng J, Wang Q, Li G,and Santosh M 2014 *Earth-Science Reviews* **138** 268–99
[3] Hou Z, Zaw K, Pan G, Mo X, Xu Q Hu Y and Li X 2007 *Ore Geology Reviews* **31** 48–87
[4] Sun X, Zhang Y, Xiong D, Sun W, Shi G, Zhai W and Wang S 2009 *Ore Geology Reviews* **36** 235–49
[5] Yang L Q, Liu J T, Zhang C, Wang Q F, Ge L S Wang Z L, Zhang J and Gong Q J 2010 *Acta Petrologica Sinica* **26** 1723–39
[6] Fu Y, Sun X, Zhou H and Yang T 2016 *Ore Geology Reviews* **77** 43–56
[7] Tran M D, Liu J L, Nguyen Q L, Chen Y, Ji M and Tang Y 2014 *Journal of Asian Earth Sciences* **79** 858–72

Genesis and prospecting perspective of Xianglushan iron deposit, Sichuan Province, China

Tianguo Wang[1, 2], Jishun Liu[1, 2] and Chaowen Huang[1, 2]

[1]Key Laboratory of Metallogenic Prediction of Nonferrous Metals and Geological Environment Monitoring, Ministry of Education, Central South University, Changsha 410083, China
[2]School of Geosciences and Info-Physics, Central South University, Changsha 410083, China

wangtianguo@csu.edu.cn

Abstract. More than 30 ore bodies have currently been discovered in the Xianglushan iron deposit in the central part of the Huili–Dongchuan Aulacogen. The main orebody is stratiform or stratoid within the volcanic sedimentary metamorphic rock of the Paleo–Mesoproterozoic Yinmin formation. Minor ores intrude as veins in the country rock and the mineral components of the iron ore are mainly magnetite and hematite. The ore evolved through the Paleo–Mesoproterozoic submarine volcanic eruption deposition period and the Neoproterozoic iron slurry superimposed transformation, and is identified as a sedimentation–transformation-type deposit. A study is conducted on the regional metallogenic background and geological characteristics of the ore deposit; results show that the ore-forming setting of the Xianglushan iron ore is excellent and that the potential of finding large-scale rich iron deposit is considerable.

1. Introduction

The Huili–Dongchuan Aulacogen, which is located at the western margin of the Yangzi Plate, is the most important Fe-Cu metallogenic structural belt in Sichuan and Yunnan, with occurrences of large- to giant-scale copper and iron deposits, of which the most famous is the Dongchuan copper deposit[1,2,3,4]. The Xianglushan iron ore is located in the middle of the Fe-Cu metallogenic belt. The Yunnan–Dongchuan Cu-Fe-Au mining area is separated by the Jinsha River and the ore body occurs within the Yinmin Formation of the Kunyang Group. In the course of recent mining exploration, it was discovered that there are still large areas for prospecting, as well as good prospecting potential in the deeper and peripheral zones. This paper has chosen the Xianglushan mining area as the focus of research by using modern regional metallogeny to study the formation mechanism of the deposit in terms of ore-bearing intrusive magma that was superimposed and modified by the Presinian volcanic deposition–metamorphism process, and finally proposes the steps required for future prospecting. It is of great significance for deepening the understanding of Fe-Cu mineralization in the Huili–Dongchuan Aulacogen.

2. Geological Background

The Xianglushan iron deposit is located at the western margin of the Yangzi Plate; specifically, in the middle of the Proterozoic Huili–Dongchuan Aulacogen (Fig. 1-a). It is considered to be the most

IOP Conf. Series: Earth and Environmental Science **59** (2017) 012009 doi:10.1088/1755-1315/59/1/012009

important metallogenic structure in the passive continental margin rift system found in Sichuan–Yunnan[5,6]. The aulacogen has a length and width of approximately 250 km and 80 km, respectively. There are several sets of N–S-trending fault zones, with the major Yimen–Luchang and Puduhe faults separating the entire aulacogen into three parts: Huili–Lixi, Tongan–Lunanshan, and Huidongxiaojie–Dongchuan, respectively (Fig. 1-b).

The aulacogen experienced complicated and varied tectonic, magmatic, deposition, and metamorphism processes, which formed a very thick series of Presinian Na-rich volcanic and sedimentary rocks (Kunyang Group or Huili Group), as well as being intruded by basic and ultrabasic magma, which resulted in an E–W-trending basement structure and thus created good geological background conditions for the formation of layered and layer-like iron and copper deposits[7]. After the closure of the Neoproterozoic rift, the formation of a S–N-trending structural belt and corresponding basic and acidic magmatic activity strongly modified and superimposed the layered/layer-like Fe-Cu deposits found in the Kunyang Group[8,9,10]. The platform stage of the Sinian period formed a set of conglomerates of marine–coastal facies, carbonates of platform–shallow marine facies and sediments of continental–lacustrine facies.

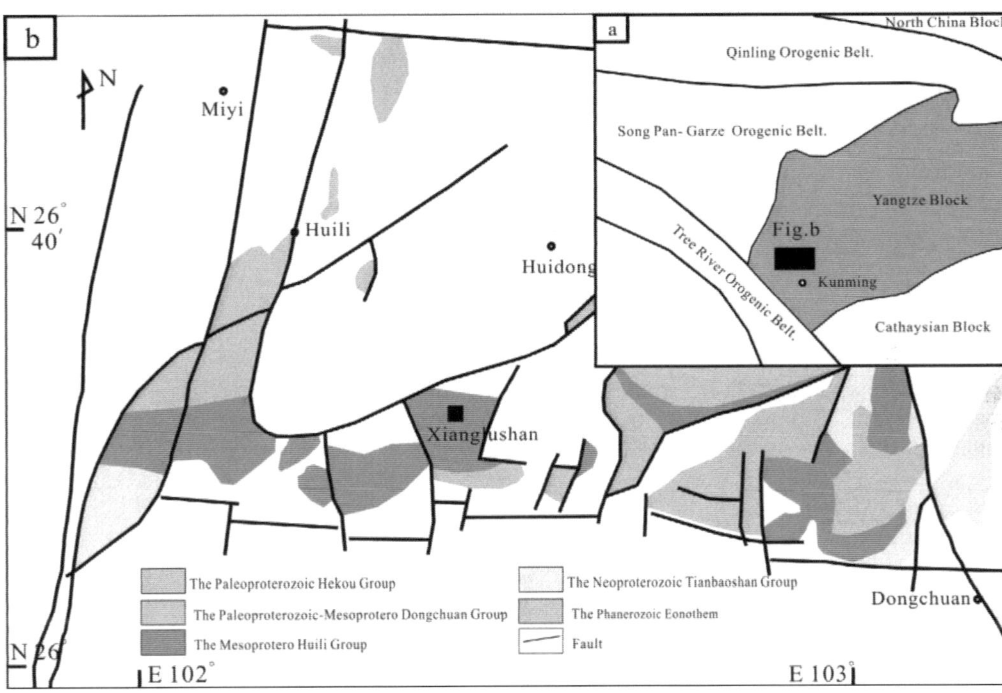

Figure 1. Sketch map of geology in the Huili-Dongchuan Aulacogeosyncline.

3. Geological Setting of Mining Area

The Xianglushan iron deposit is situated in Sichuan Province between the counties of Huili and Huidong. Huili county is governed from Bajioxiang, whereas Huidong county is governed from Longshuxiang. The exposed strata found in the mining area have been determined to be from the Presinian Kunyang Group, which, from bottom to top, is divided into the Yinmin, Luoxue, Heishan, Qinglongshan and Shuangshuijing formations. The volcanic rocks found in the Yinmin Formation have the highest Cu-Fe-S content and also contain exhalative sedimentary iron (copper) deposits. They are known to be the most important Fe-Cu source beds and the main ore-bearing strata found in the area. The structures of the mining area are complicated, with the presence of two distinct regional tectonic lines, of which one is E–W-trending and is represented by the Xianglushan and Tongchangding anticlines; the other is approximately S–N-trending and is mainly represented by the F1 fault. There are other secondary structures with NW and NE trends. The volcanic rocks in the area are mainly from the Yinmin Formation, whereas the tuffaceous sands and slates are from the Luoxue, Heishan, and Shuangshuijing formations. Metamorphism and intrusive dykes are observed.

The Xianglushan magnetite–hematite deposit is found in the Xianglushan, Yaopengzi, and Hejiawan ore sections; these were formed in the metamorphosed series of volcanic and sedimentary rocks of the Yinmin Formation, which contain magnetite and hematite. From the northern end of the ore belt, the Xianglushan ore section is approximately E–W-trending. Then, the ore belt develops a NW or NNW trend on moving southeast to the Yaopengzi ore section, and becomes approximately S–N-trending on moving south to the Hejiawan ore section. The intermittent length of the ore belt reaches 3600 m. At present, more than 30 ore bodies have been discovered, with the larger ore bodies identified as Nos. 1, 7, 8, 9, 22, and 24.

The ore bodies are stratified and lenticular (or lentil-shaped), with their occurrence being similar to that of the surrounding rocks. However, their strike–dip continuity is bad and their tendency starts to be poor, resulting in common cases of reappearance after pinch-out and complex branching phenomena. The average length of an ore body is 100 m, with the longest reaching 170 m. The average thickness was found to be 2–8 m, with the largest ore body reaching 24.57 m. The total iron grade is 25–45%, with the presence of localized rich ore bodies with a total iron grade reaching 60%. The metallic minerals found in the area mainly consist of magnetite, limonite, hematite, and small amounts of chalcopyrite and pyrite. The ore textures are dominated by euhedral, subhedral, and anhedral granular textures; the ore structure is dominated by massive structures. Alterations such as chloritization, epidotization, silicification, calcification, sericitization, and discoloration can be observed in the wall rocks.

Altered diabase dykes that were intruded into the volcanic and sedimentary rock series of the Yinmin Formation were first discovered in the middle of the formation at an elevation of 2512 m (No. 6 cave) and in its lower corresponding part at an elevation of 2434 m (No. 11 cave) in the M64 magnetic anomaly area. In the inside diabase dykes and at the margin of the contact zone are seen irregularly distributed lenticular, irregular, and droplet-like blocks of magnetite-rich ore strips and aggregates. These massive magnetite-rich ore strips and aggregates are identified as having a spongy meteorite-like structure. This could be due to the crystallization of iron slurry. Therefore, this area has the potential to contain formations of ore slurry injection–metasomatism-type magnetite-rich ore bodies.

4. Genesis of deposits

On the basis of the mineralization and geological background of the Xianglushan iron ore and its geological characteristics, it is suggested that the deposits are mainly stratigraphically controlled iron deposits of Proterozoic submarine exhalative sedimentary origin, with secondary iron slurry injection–metasomatism-type magnetite-rich deposits. The ore-forming process is described as follows: activity of the Kunyang Rift during the Paleo–Mesoproterozoic era formed the Huili–Dongchuan E–W-trending aulacogen basin. E–W-trending and S–N-trending faults intersected in the M64–Hejiawan area. Large-scale magmatic activity occurred during the Yinmin period, which resulted in the early Jinningian alkaline ferric Ti-rich volcanic intrusion (ancient volcano structure) in the M64–Hejiawan area, forming the submarine exhalative sedimentary-type magnetite–hematite layer in the Yinmin Formation, which is known as the Xianglushan magnetite–hematite deposit. During the period of inactivity of the submarine volcanoes (Neoproterozoic), gabbro and diabase (dykes) intruded, which was subsequently followed by the closure of the rift. This formed the ore slurry injection–metasomatism-type magnetite-rich ore body and the vein-like rich ore body and at the same time strongly modified, superimposed, and destroyed the existing stratified iron deposits and caused the bad continuity. Therefore, the Xianglushan iron deposit is determined to be a sedimentary–superimposition transformed deposit.

5. Prospecting Direction

The Xianglushan magnetite–hematite deposit basically extends parallel to the bedding layer of basic volcanic rocks within the Yinmin Formation. Owing to the strong structural deformation of the area, the ore-bearing layer and the ore body found within the Yinmin Formation formed a boudin, in which the layers are highly overturned and fragile with poor continuity. The dolomites at the top of the Yinmin, Luoxue, and Qinglongshan formations are mixed, and contrast between horizons is difficult to determine. The current results of surface and tunnel drilling are not good and drilling and shaft explorations have had a low prospecting hit rate, only suggesting millions of tons of recoverable ore reserves.

This situation is not compatible with the excellent metallogenic conditions of the study area, namely, the Xianglushan–M64–Yaopengzi–Hejiawan arcuate structure, which can be compared with those from the Keshu–Sikeshu–Yinmin–Luoxue–Shijiangjun–Lanniping structure in Dongchuan area, with the scale of the former being larger. Toward the southeast, the large-scale Dongchuan Cu-Fe (Au) ore field lies just across the Yangzi River. Toward the west, there are the large-scale Lala Cu-Fe ore deposits, which indicate that this area should have the geological potential to contain large iron-rich deposits. Therefore, the next suggestions for exploration are as follows:

1) Stratigraphic determination: Via tunneling and detailed drill core logging, determine the separation indicators of the Yinmin, Luoxue, and Qinglongshan formations.

2) Structural restoration: Restore the patterns of the structure prior to deformation by using three-dimensional geological mapping, statistics of the occurrence of faults, joints, and fractures, and the relationship between bedding and cleavage.

3) Predict the possible occurrence of the metamorphosed volcanic and sedimentary rocks of the lower Yinmin Formation, design drilling programs and mineshafts, and initiate experimental drilling.

Conclusion

The Xianglushan iron deposits are found within the Huili–Dongchuan metallogenic belt, which has excellent metallogenic conditions. The deposits mainly consist of stratigraphically controlled iron deposits of Proterozoic submarine exhalative sedimentary origin and secondarily consist of iron slurry injection–metasomatism-type magnetite-rich ore deposits. The tectonic deformation of the study area is strong, which leads to the ore body having a shallow section with poor continuity; however, the deeper section of the ore body possesses the potential of containing iron-rich ores on a large scale.

References

[1] Liu J S Wu, Y Z and Duan J R 1996 *Journal of Central South University of Technology* **27** 8–12
[2] He Y T 1996 *Yunnan Geology* **15** 319–29
[3] Dai H G 1997 *Yunnan Geology* **01** 1–39
[4] Hua R M, Wan H C and Ni P 1993 *Contributions to Geology and Mineral Resources Research* **04** 1–8
[5] Wang D B, Sun Z M, Yin F G, Wang L Q, Wang B D and Zhang W P 2012 *Journal of Stratigraphy* **36** 630–35
[6] Zhao X F, Zhou M F, Li J W, Sun M, Gao J F, Sun W H and Yang J H 2010 *Precambrian Research* **182** 57–69
[7] Zhu H P, Fan W Y, Zhou B G, Wang S W, Luo M J, Liao Z W and Guo Y 2011 *Geological Journal of China Universities* **17** 452–61
[8] Yin F G, Sun Z M, Ren G M and Wang D B 2012 *Acta Geologica Sinica* **12** 1917–32
[9] Yin F G, Sun Z M and Zhang Z 2011 *Geological Review* **06** 770–778
[10] Geng Y S, Liu Y Q, Gao L Z, Peng N and Jiang X J 2012 *Acta Geologica Sinica* **86** 1479–90

IOP Conf. Series: Earth and Environmental Science **59** (2017) 012010 doi:10.1088/1755-1315/59/1/012010

Nonlinear dynamics of a cantilevered piezoelectric rectangular plate with voltage excitation

J H Yang[1,2,a], W Zhang [1], M H Yao [1] and G Zhu[2]

[1]College of Mechanical Engineering, Beijing University of Technology, Beijing, China
[2]Beijing Vocational College of Agriculture, Beijing, China

[a]klrs80@sohu.com

Abstract. The nonlinear dynamics of a cantilevered piezoelectric rectangular plate are studied, which are simultaneously forced by the base excitation and the excitation loaded by piezoelectric layers. The nonlinear partial differential governing equations of motion for the system are established by using the Hamilton's principle. The Galerkin approach is used to discretize the partial differential equations to the ordinary differential equations with four-degree-of-freedom. Numerical simulations are presented to investigate the effects of the voltage excitation on the steady-state responses of the cantilevered piezoelectric plate. The bifurcation diagram of the system for w_1, w_2, w_3, w_4 via the voltage excitation amplitude is obtained. The motions of the system can be shown as follows: the chaotic motion → the multiple periodic motion → the periodic motion → the chaotic motion. Based on the above bifurcation diagrams, the voltage excitation amplitude are changed to obtain the waveforms, two-dimensional phase portraits, three-dimensional phase portraits and Poincare maps. The results show that the amplitude of the system can reduce effectively and keep the stability by adjusting the voltage excitation.

1. Introduction

Piezoelectric materials, which include piezoelectric lead-zirconate-titanate (PZT) and piezoelectric polyvinylidene fluoride (PVDF) are new type functional materials and can be utilized as the actuators and sensors in engineering structures [1, 2]. Zhang et al. [3,4] investigated the bifurcations and chaotic dynamics of a simply support symmetric cross-ply composite laminated piezoelectric rectangular plate, which are simultaneously forced by the transverse and in-plane excitations. Correia et al. [5] developed a semi-analytical axisymmetric shell finite element model with embedded or surface bonded piezoelectric actuators or sensors to study active damping vibration control of structures.

2. Equations of motion

As shown in Figure 1, a cantilevered piezoelectric rectangular plate clamped at edge *oa*. The upper and lower layers are made of the PVDF piezoelectric materials as actuators, while substrate layer is made of fiber-reinforced composite material. It is assumed that different layers are perfectly bonded to each other. A Cartesian coordinate *Oxyz* is located in the middle surface of cantilevered piezoelectric plate. The base excitation is represented by $F\cos(\Omega_1 t)$.The dynamic electrical loading is expressed as $V\cos(\Omega_2 t)$.

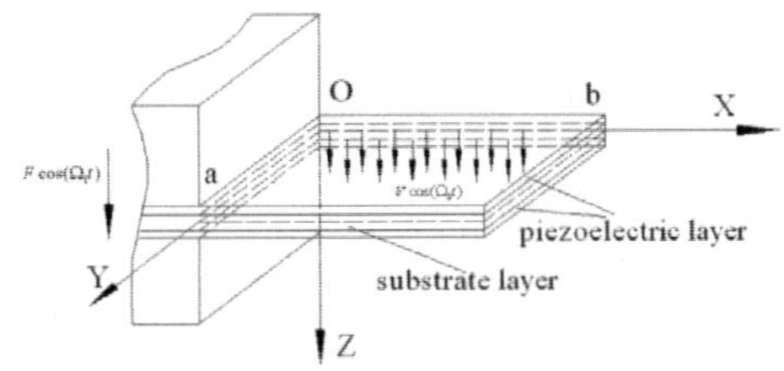

Figure 1 The mathematical model of a cantilevered piezoelectric plate structure

According to the von Karman type equations for the geometric nonlinearity and the Reddy's classic deformation plate theory, we can write the displacement field of the cantilevered piezoelectric plate as followings [6]

$$u_1=u(x,y,z,t)=u_0(x,y,t)-z\frac{\partial w_0}{\partial x},u_2=v(x,y,z,t)=v_0(x,y,t)-z\frac{\partial w_0}{\partial y},u_3=w(x,y,z,t)=w_0(x,y,t) \quad (1)$$

where (u_1,u_2,u_3) are the displacement components along the (x,y,z) directions, (u_0,v_0,w_0) is the deflection of a point on the middle plane ($z=0$).

The nonlinear strain-displacement relations are given as follows

$$\varepsilon_{xx}=\frac{\partial u}{\partial x}+\frac{1}{2}(\frac{\partial w}{\partial x})^2,\varepsilon_{yy}=\frac{\partial v}{\partial y}+\frac{1}{2}(\frac{\partial w}{\partial y})^2,\varepsilon_{zz}=\frac{\partial w}{\partial z},\varepsilon_{xy}=\frac{1}{2}(\frac{\partial u}{\partial y}+\frac{\partial v}{\partial x}+\frac{\partial w}{\partial x}\frac{\partial w}{\partial y}),$$
$$\varepsilon_{yz}=\frac{1}{2}(\frac{\partial w}{\partial y}+\frac{\partial v}{\partial z}),\varepsilon_{xz}=\frac{1}{2}(\frac{\partial u}{\partial z}+\frac{\partial w}{\partial x}) \quad (2)$$

The stress–strain relationship of the cantilevered piezoelectric plate is given

$$\begin{Bmatrix}\sigma_{xx}\\ \sigma_{yy}\\ \sigma_{yz}\\ \sigma_{zx}\\ \sigma_{xy}\end{Bmatrix}=\begin{bmatrix}Q_{11}&Q_{12}&0&0&0\\ Q_{21}&Q_{22}&0&0&0\\ 0&0&Q_{44}&0&0\\ 0&0&0&Q_{55}&0\\ 0&0&0&0&Q_{66}\end{bmatrix}\begin{Bmatrix}\varepsilon_{xx}\\ \varepsilon_{yy}\\ \gamma_{yz}\\ \gamma_{zx}\\ \gamma_{xy}\end{Bmatrix}-\begin{Bmatrix}e_{31}\\ e_{32}\\ 0\\ 0\\ 0\end{Bmatrix}E \quad (3)$$

where Q_{ij} is the elastic stiffness coefficient, e_{ij} is the piezoelectric constant, E is the electric field strength, shown as follows

$$Q_{11}=\frac{E_1}{1-\nu_{12}\nu_{21}},Q_{12}=\frac{\nu_{12}E_2}{1-\nu_{12}\nu_{21}},Q_{21}=Q_{12},Q_{22}=\frac{E_2}{1-\nu_{12}\nu_{21}},Q_{44}=G_{23},Q_{55}=G_{13},Q_{66}=G_{12} \quad (4)$$

where E_i is the elastic modulus, G_{12} is the shear modulus, ν_{12} and ν_{21} are Poisson's ratio.

Using the Hamilton's principle and the second kinds of piezoelectric constitutive equations, the nonlinear governing equations of motion for the cantilevered piezoelectric plate are obtained as

$$\frac{\partial N_{xx}}{\partial x}+\frac{\partial N_{xy}}{\partial y}=I_0\ddot{u}_0-I_1\frac{\partial \ddot{w}_0}{\partial x},\frac{\partial N_{yy}}{\partial y}+\frac{\partial N_{xy}}{\partial x}=I_0\ddot{v}_0-I_1\frac{\partial \ddot{w}_0}{\partial y},$$
$$\frac{\partial N_{xx}}{\partial x}\frac{\partial w_0}{\partial x}+\frac{\partial N_{xy}}{\partial x}\frac{\partial w_0}{\partial y}+\frac{\partial N_{xy}}{\partial y}\frac{\partial w_0}{\partial x}+\frac{\partial N_{yy}}{\partial y}\frac{\partial w_0}{\partial y}+F\cos\left(\Omega_1 t\right)+V\cos(\Omega_2 t)-c_3\dot{w}_0 \quad (5)$$
$$=I_0\ddot{w}_0+I_1\left(\frac{\partial \ddot{u}_0}{\partial x}+\frac{\partial \ddot{v}_0}{\partial y}\right)-I_2(\frac{\partial^2 \ddot{w}_0}{\partial x^2}+\frac{\partial^2 \ddot{w}_0}{\partial y^2})$$

where the dot represents partial differentiation with respect to time t; c_3 is the damping coefficient.

$$N_{\alpha\beta}=\int_{-h/2}^{h/2}\sigma_{\alpha\beta}dz(\alpha=x,y;\beta=x,y),I_i=\int_{-h/2}^{h/2}\rho\left(z\right)^{i}dz,(i=0,1,2) \quad (6)$$

Considering the variation of electric field in thickness, piezoelectric force are given as follows

$$N_{xx}{}^{p} = \int_v e_{31} E dv, N_{yy}{}^{p} = \int_v e_{32} E dv, N_{xy}{}^{p} = N_{xz}{}^{p} = N_{yz}{}^{p} = 0 \tag{7}$$

The stress-strain relations are given as follows

$$\begin{Bmatrix} N_{xx} \\ N_{yy} \\ N_{xy} \end{Bmatrix} = \begin{bmatrix} A_{11} & A_{12} & 0 \\ A_{21} & A_{22} & 0 \\ 0 & 0 & A_{66} \end{bmatrix} \begin{Bmatrix} \varepsilon_{xx} \\ \varepsilon_{yy} \\ \gamma_{xy} \end{Bmatrix} - \begin{Bmatrix} N_{xx}{}^{p} \\ N_{yy}{}^{p} \\ N_{xy}{}^{p} \end{Bmatrix} \tag{8}$$

Substituting equation (6)(7)(8) into equation (5), we obtain the governing equations of motion in terms of generalized displacements as

$$A_{11}\frac{\partial^2 u_0}{\partial x^2} + A_{66}\frac{\partial^2 u_0}{\partial y^2} + (A_{12}+A_{66})\frac{\partial^2 v_0}{\partial x \partial y} + A_{11}\frac{\partial w_0}{\partial x}\frac{\partial^2 w_0}{\partial x^2} + A_{66}\frac{\partial w_0}{\partial x}\frac{\partial^2 w_0}{\partial y^2} + (A_{12}+A_{66})\frac{\partial w_0}{\partial y}\frac{\partial^2 w_0}{\partial x \partial y} = I_0\frac{\partial^2 u_0}{\partial t^2} - I_1\frac{\partial^3 w_0}{\partial x \partial t^2} \tag{9a}$$

$$A_{22}\frac{\partial^2 v_0}{\partial y^2} + A_{66}\frac{\partial^2 v_0}{\partial x^2} + (A_{21}+A_{66})\frac{\partial^2 u_0}{\partial x \partial y} + A_{22}\frac{\partial w_0}{\partial y}\frac{\partial^2 w_0}{\partial y^2} + A_{66}\frac{\partial w_0}{\partial y}\frac{\partial^2 w_0}{\partial x^2} + (A_{21}+A_{66})\frac{\partial w_0}{\partial x}\frac{\partial^2 w_0}{\partial x \partial y} = I_0\frac{\partial^2 v_0}{\partial t^2} - I_1\frac{\partial^3 w_0}{\partial y \partial t^2} \tag{9b}$$

$$\begin{aligned}
&-D_{11}\frac{\partial^4 w_0}{\partial x^4} + (-D_{21} - 4D_{66} - D_{12})\frac{\partial^4 w_0}{\partial x^2 \partial y^2} - D_{22}\frac{\partial^4 w_0}{\partial y^4} + \frac{3A_{11}}{2}\left(\frac{\partial w_0}{\partial x}\right)^2\frac{\partial^2 w_0}{\partial x^2} + \frac{3}{2}A_{22}\left(\frac{\partial w_0}{\partial y}\right)^2\frac{\partial^2 w_0}{\partial y^2} + \left(\frac{1}{2}A_{21} + A_{66}\right)\left(\frac{\partial w_0}{\partial x}\right)^2\frac{\partial^2 w_0}{\partial y^2} \\
&+ A_{11}\frac{\partial u_0}{\partial x}\frac{\partial^2 w_0}{\partial x^2} + \left(\frac{1}{2}A_{12} + A_{66}\right)\left(\frac{\partial w_0}{\partial y}\right)^2\frac{\partial^2 w_0}{\partial x^2} + 2\left(\frac{1}{2}A_{21} + 2A_{66} + \frac{1}{2}A_{12}\right)\frac{\partial w_0}{\partial x}\frac{\partial w_0}{\partial y}\frac{\partial^2 w_0}{\partial x \partial y} + A_{22}\frac{\partial v_0}{\partial y}\frac{\partial^2 w_0}{\partial y^2} + A_{21}\frac{\partial u_0}{\partial x}\frac{\partial^2 w_0}{\partial y^2} \\
&+ A_{12}\frac{\partial v_0}{\partial y}\frac{\partial^2 w_0}{\partial x^2} + 2A_{66}\frac{\partial u_0}{\partial y}\frac{\partial^2 w_0}{\partial x \partial y} + 2A_{66}\frac{\partial v_0}{\partial x}\frac{\partial^2 w_0}{\partial x \partial y} + (A_{21}+A_{66})\frac{\partial w_0}{\partial y}\frac{\partial^2 u_0}{\partial y \partial x} + (A_{12}+A_{66})\frac{\partial w_0}{\partial x}\frac{\partial^2 v_0}{\partial y \partial x} + A_{11}\frac{\partial w_0}{\partial x}\frac{\partial^2 u_0}{\partial x^2} \\
&+ A_{66}\frac{\partial w_0}{\partial x}\frac{\partial^2 u_0}{\partial y^2} + A_{66}\frac{\partial w_0}{\partial y}\frac{\partial^2 v_0}{\partial x^2} + A_{22}\frac{\partial w_0}{\partial y}\frac{\partial^2 v_0}{\partial y^2} - N_{xx}{}^{p}\cos(\Omega_p t)\frac{\partial^2 w_0}{\partial x^2} - N_{yy}{}^{p}\cos(\Omega_p t)\frac{\partial^2 w_0}{\partial y^2}\left(-\frac{\partial N_{xx}{}^{p}}{\partial x}\frac{\partial w_0}{\partial x} - \frac{\partial N_{yy}{}^{p}}{\partial y}\frac{\partial w_0}{\partial y}\right) \\
&+ F\cos(\Omega_1 t) + V\cos(\Omega_2 t) = c_3\dot{w}_0 + I_0\ddot{w}_0 - I_2\frac{\partial^2 \ddot{w}_0}{\partial x^2} - I_2\frac{\partial^2 \ddot{w}_0}{\partial y^2} + \left(I_1\frac{\partial \ddot{u}_0}{\partial x} + I_1\frac{\partial \ddot{v}_0}{\partial y}\right)
\end{aligned} \tag{9c}$$

where (A_{ij}, D_{ij}) respectively are the stiffness elements of the cantilevered plate, which are denoted as

$$(A_{ij}, D_{ij}) = \sum_{k=1}^{N}\int_{z_k}^{z_{k+1}} Q_{ij}{}^{k}(1, z, z^2)dz, (i, j = 1, 2, 6) \tag{10}$$

We consider nonlinear dynamics of the cantilevered piezoelectric plate in the first mode of u_0, v_0 and the first four modes of w_0. We write u_0, v_0, w_0, F in the following forms [7]:

$$\begin{aligned}
&u_0(x, y, t) = u_1(t)\sin\left(\frac{\pi x}{2a}\right)\cos\left(\frac{\pi y}{b}\right), v_0(x, y, t) = v_1(t)\sin\left(\frac{\pi x}{2a}\right)\cos\left(\frac{\pi y}{b}\right), \\
&w_0(x, y, t) = w_1(t)\alpha_1\beta_1 + w_2(t)\alpha_1\beta_2 + w_3(t)\alpha_2\beta_1 + w_4(t)\alpha_2\beta_2, \\
&\alpha_1 = \cosh\frac{k_1 x}{a} - \cos\frac{k_1 x}{a} - \varphi_1\left(\sinh\frac{k_1 x}{a} - \sin\frac{k_1 x}{a}\right), \alpha_2 = \cosh\frac{k_2 x}{a} - \cos\frac{k_2 x}{a} - \varphi_2\left(\sinh\frac{k_2 x}{a} - \sin\frac{k_2 x}{a}\right), \\
&\cos k_i \cosh k_i + 1 = 0, k_i{}^4 = \omega^2\frac{\rho A}{EJ}(i = 1, 2), \varphi_i = \frac{\sinh k_i - \sin k_i}{\cosh k_i + \cos k_i}(i = 1, 2), \beta_1 = 1, \beta_2 = \sqrt{3}\left(1 - \frac{2y}{b}\right)
\end{aligned} \tag{11}$$

For the cantilevered thin plate, the transverse displacement is more obvious than other directions, so all inertia terms in equation (9a) (9b) and equation(9c) about u_0, v_0 are neglected. By means of the Galerkin method, substituting equation (11) into equation (9), the expressions of w_1, w_2, w_3, w_4 are obtained as follows

$$
\begin{aligned}
&\ddot{w}_1 + \mu_1 \dot{w}_1 + \omega_1^2 w_1 + a_1 w_1^3 + a_2 w_2^3 + a_3 w_3^3 + a_4 w_4^3 + a_5 w_1 w_2^2 + a_6 w_2 w_1^2 + a_7 w_3 w_2^2 + a_8 w_3 w_1^2 + a_9 w_1 w_3^2 \\
&\quad + a_{10} w_1 w_4^2 + a_{11} w_2 w_3^2 + a_{12} w_2 w_4^2 + a_{13} w_4 w_2^2 + a_{14} w_3 w_4^2 + a_{15} w_4 w_1^2 + a_{16} w_4 w_3^2 + a_{17} w_1 w_2 w_3 + a_{18} w_1 w_2 w_4 \\
&\quad + a_{19} w_1 w_3 w_4 + a_{20} w_2 w_3 w_4 = a_{21} F \cos\Omega_1 t + a_{22} V \cos(\Omega_2 t) \\
&\ddot{w}_2 + \mu_1 \dot{w}_2 + \omega_2^2 w_2 + b_1 w_1^3 + b_2 w_2^3 + b_3 w_3^3 + b_4 w_4^3 + b_5 w_1 w_2^2 + b_6 w_2 w_1^2 + b_7 w_3 w_2^2 + b_8 w_3 w_1^2 + b_9 w_1 w_3^2 \\
&\quad + b_{10} w_1 w_4^2 + b_{11} w_2 w_3^2 + b_{12} w_2 w_4^2 + b_{13} w_4 w_2^2 + b_{14} w_3 w_4^2 + b_{15} w_4 w_1^2 + b_{16} w_4 w_3^2 + b_{17} w_1 w_2 w_3 + b_{18} w_1 w_2 w_4 \\
&\quad + b_{19} w_1 w_3 w_4 + b_{20} w_2 w_3 w_4 = b_{21} F \cos\Omega_1 t + b_{22} V \cos(\Omega_2 t) \qquad (12) \\
&\ddot{w}_3 + \mu_1 \dot{w}_3 + \omega_3^2 w_3 + c_1 w_1^3 + c_2 w_2^3 + c_3 w_3^3 + c_4 w_4^3 + c_5 w_1 w_2^2 + c_6 w_2 w_1^2 + c_7 w_3 w_2^2 + c_8 w_3 w_1^2 + c_9 w_1 w_3^2 \\
&\quad + c_{10} w_1 w_4^2 + c_{11} w_2 w_3^2 + c_{12} w_2 w_4^2 + c_{13} w_4 w_2^2 + c_{14} w_3 w_4^2 + c_{15} w_4 w_1^2 + c_{16} w_4 w_3^2 + c_{17} w_1 w_2 w_3 + c_{18} w_1 w_2 w_4 \\
&\quad + c_{19} w_1 w_3 w_4 + c_{20} w_2 w_3 w_4 = c_{21} F \cos\Omega_1 t + c_{22} V \cos(\Omega_2 t) \\
&\ddot{w}_4 + \mu_1 \dot{w}_4 + \omega_4^2 w_4 + d_1 w_1^3 + d_2 w_2^3 + d_3 w_3^3 + d_4 w_4^3 + d_5 w_1 w_2^2 + d_6 w_2 w_1^2 + d_7 w_3 w_2^2 + d_8 w_3 w_1^2 + d_9 w_1 w_3^2 \\
&\quad + d_{10} w_1 w_4^2 + d_{11} w_2 w_3^2 + d_{12} w_2 w_4^2 + d_{13} w_4 w_2^2 + d_{14} w_3 w_4^2 + d_{15} w_4 w_1^2 + d_{16} w_4 w_3^2 + d_{17} w_1 w_2 w_3 + d_{18} w_1 w_2 w_4 \\
&\quad + d_{19} w_1 w_3 w_4 + d_{20} w_2 w_3 w_4 = d_{21} F \cos\Omega_1 t + d_{22} V \cos(\Omega_2 t)
\end{aligned}
$$

3. Numerical simulation

We employ the fourth-order Runge-Kutta algorithm to analyze numerically the nonlinear dynamic responses of the system. We chose the physical parameters respectively as Table 1, where a denotes the length of plate, b denotes the width of plate, h denotes the thickness of the plate, the subscript s denotes the substrate layer, the subscript p denotes the piezoelectric layer.

Table 1 The physical parameters of the cantilevered piezoelectric plate

Physical quantity	value	Physical quantity	value	Physical quantity	value
a	1.5m	G_{12s}	4.0Gpa	G_{12p}	1.0Gpa
b	0.8m	ν_{12s}	0.33	ν_{12p}	0.35
h_s	0.005m	ρ_s	1570kg/m^3	ρ_p	2500
h_p	0.001m	c_3	2.83×10^{-5}	e_{31}	$17c^o/m^2$
E_{1s}	125Gpa	E_{1p}	2.5Gpa	e_{32}	$6c^o/m^2$
E_{2s}	7.2Gpa	E_{2p}	2.5Gpa		

When the voltage excitation amplitude V is located in the interval $0V$–$2000V$ and the base excitation amplitude $F=1000N$, substituting the physical parameters to the equations（12）, the bifurcation diagrams for w_1, w_2, w_3, w_4 via the voltage excitation V are shown in Figure 5. Figure 5 shows that the cantilevered piezoelectric plate change from the chaotic motion to the periodic motion, and then to the chaotic motion with the increase of the voltage excitation amplitude V.

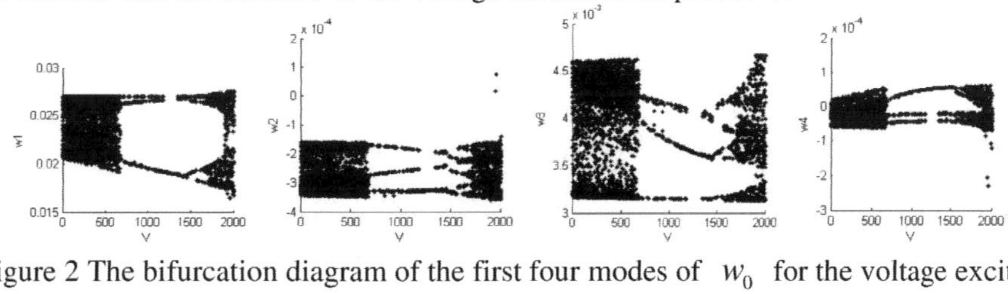

Figure 2 The bifurcation diagram of the first four modes of w_0 for the voltage excitation V

We change the voltage excitation amplitude V to obtain the waveforms (a) (b), the two-dimensional phase portraits (c) (d), Poincare maps (e) (f) and three-dimensional phase portraits (g). Figure 3 illustrates that the chaotic motion of the cantilevered piezoelectric plate exists when the voltage excitation changes to V=10V.

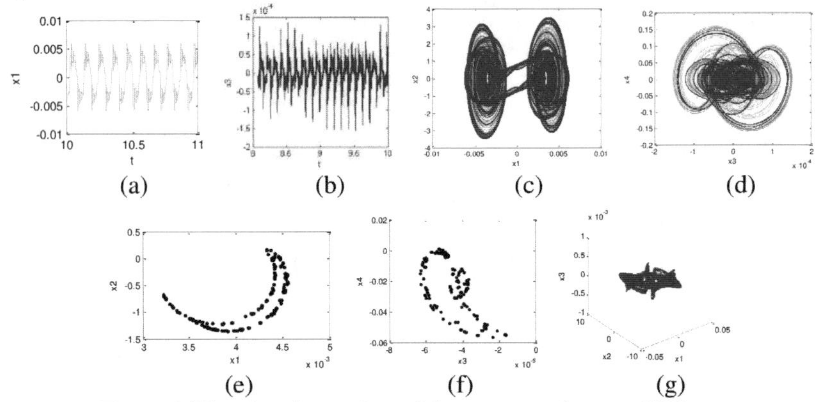

Figure 3 The chaotic motion of the system when V=10V

When the voltage excitation changes to V=800V, Figure 4 illustrates that the periodic motion of the composite laminated cantilevered piezoelectric plate exists.

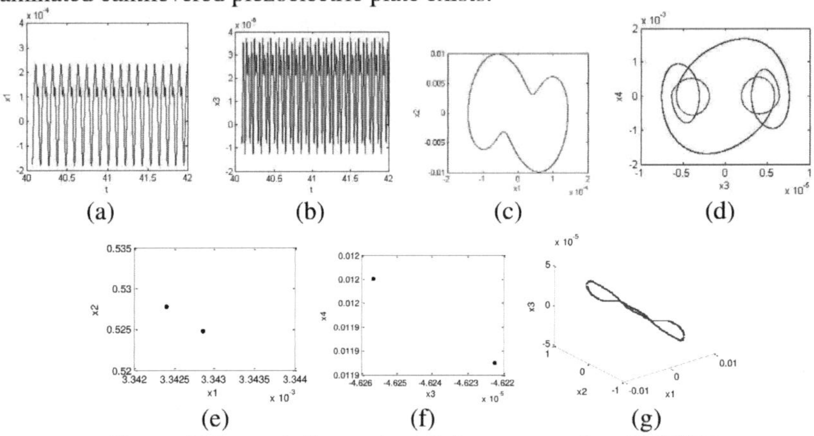

Figure 4 The periodic motion of the system when V=800V

4. Conclusions

We investigate the nonlinear dynamics responses of a cantilevered piezoelectric plate subjected to the base and voltage excitations. The governing equations of motion for the cantilevered piezoelectric plate are established. Analytical study is given by using the fourth-order Runge-Kutta algorithm.

The bifurcation diagram of the cantilevered piezoelectric plate for w_1, w_2, w_3, w_4 via the voltage excitation V is obtained. The bifurcation diagram shows that voltage excitation can significantly reduce the vibration amplitude of the cantilevered piezoelectric plate. By changing the voltage excitation amplitude V, we can control the nonlinear dynamic responses of the cantilevered piezoelectric plate from the chaotic motion to the periodic motion.

Acknowledgments

This work was partially supported by National Natural Science Foundation of China (No.11072008) and Research Project from Beijing Vocational College of Agriculture (No. XY-YF-14-42).

References

[1] Hurlebaus S Gaul L 2006 *Smart structure dynamics* (Mech Syst Sign Proc) p255
[2] Sheta E F Moses R W Huttsell L J 2006 *J Sound Vib* **292** 854
[3] Zhang W Yao ZG Yao MH 2009 *Sci China Ser E Technol Sci* **52** 731
[4] W. Zhang M. H. Zhao and X. Y. Guo 2013*Composite Structures* **100** 554-565
[5] Correia I F Mota Soares C M Mota Soares C A 2002 *Compute Structure* **80** 2265
[6] Pai P F Nayfeh A H 1991 *Nonlinear Dynamics* **2** 445
[7] W Zhang M H Zhao and X Y Guo 2013 *Composite Structures* **100** 554

Absorption performance for CO_2 capture process using MDEA-AMP aqueous solution

Gang Liu[1,a], Liqing Kou[2] and [3]Chao Li

[1] Zhongkan Metallurgical Investigation Design Research Institute Co. Ltd, 1285 Dongfeng Zhong Road, Baoding, 071003, People's Republic of China.
[2] Hydrology and Water Resources Survey Bureau of Baoding Hebei, 239 Xingye Road, Baoding, 071003, People's Republic of China.
[3] School of Environmental Science and Engineering, North China Electric Power University, Baoding, 071003, People's Republic of China.

[a]hpslg@126.com

Abstract. The absorption capacity and the absorption rate of CO_2 in 2-amino-2-methyl-1-propanol (AMP)-N-methyldiethanolamine (MDEA) aqueous solution were measured. The temperatures ranged from 303.2K to 323.2K. The mass fractions of AMP and MDEA respectively ranged from 0 to 0.03 and 0.2 to 0.3. The influence of temperature and w_{AMP} on the absorption capacity and absorption rate of CO_2 was illustrated.

1. Introduction

The concentration of atmospheric carbon dioxide which is considered to be primary greenhouse gas has been increasing at an alarming rate [1, 2]. Large emissions and stability of carbon dioxide lead to the global warming and climatic change [3]. Chemical absorption using amine solvent is the current technology for carbon dioxide capture [4, 5]. Alkanolamines (MEA, DEA, MDEA, PZ, etc.) have been extensively applied to remove CO_2 [6, 7]. However, there are still some drawbacks in their use such as low absorption rate, high energy consumption for CO_2 regeneration [8-10].

In recent decades, interest increased rapidly in the development of new absorbent that gives an optimal trade-off between fast kinetics and low energy requirements. Such new absorbent may be achieved by blended amines [11]. AMP is considered to be an attractive solvent for the removal of CO_2 due to its absorption capacity, absorption rate, selectivity and degradation resistance advantages [12-14]. The blended aqueous solution of AMP and MDEA, preserving the good performances of both AMP and MDEA, is considered to be an attractive solvent for the removal of CO_2. However, studies concerning the absorption capacity of CO_2 in MDEA–AMP aqueous solutions are rarely reported by far. Silkenbäumer et al. [15] measured the solubility of CO_2 in an aqueous solution containing AMP (w_{AMP}=0.11) and MDEA (w_{MDEA}=0.15) at 313 K. Shokouhi et al. [16] measured solubility of CO_2 in aqueous solution of 40 wt% MDEA +5 wt% AMP, 22.5 wt% MDEA + 22.5 wt% AMP, and 10 wt% MDEA + 35 wt% AMP at (313.15, 333.15, and 353.15) K and pressure from vapor pressure of solutions up to 2.5 Mpa. Their results showed that AMP affects efficiently on solubility of CO_2 in MDEA solution.

The main purpose of this work is to determine an appropriate addition of AMP to MDEA aqueous solution, so that the high absorption rate and large CO_2 loading can be simultaneously achieved. To this end, we measured the absorption capacity of CO_2 in AMP promoted MDEA aqueous solution at 1atm and series of temperatures, and demonstrated the influence of temperature and the mass fraction of AMP on the absorption capacity and absorption rate.

2. Experimental Section

2.1. Materials

Both MDEA and AMP were purchased from Huaxin chemical Co., with mass purity≥ 99%. They were used without further purification. Aqueous solutions of MDEA-AMP were prepared by adding doubly distilled water. The uncertainty of the electronic balance was ± 0.1mg.

2.2. Apparatus and procedure

The absorption performance of aqueous solutions was measured by the equipment composed of one high-pressure CO_2 tank, two mass flow controllers (MFC), one absorption bottle, one constant temperature water bath, one desiccator and one CO_2 analyzer (Advanced Gasmitter by Germany Sensors Europe GmbH, the accuracy is ±2%). The flask was immersed into the thermostatic bath and the temperature of the solution can be regulated within 0.1K. During the experiment, CO_2 from a highpressure tank was inlet into the first MFC to maintain a constant flow rate and then into the absorption bottle and absorbed by the solution. The residual and unabsorbed gas firstly flowed into the desiccator and then into the CO_2 analyzer. The gas concentration was measured by the CO_2 analyzer, and the flow rate was measured by the second MFC. Both the data of gas concentration and flow rate were recorded by the computer. In all the experiments, the rotational speed of the magnetic stirrer is fixed as 1000 rpm. The schematic diagram for the measurements of absorption performance was shown in Fig.1.

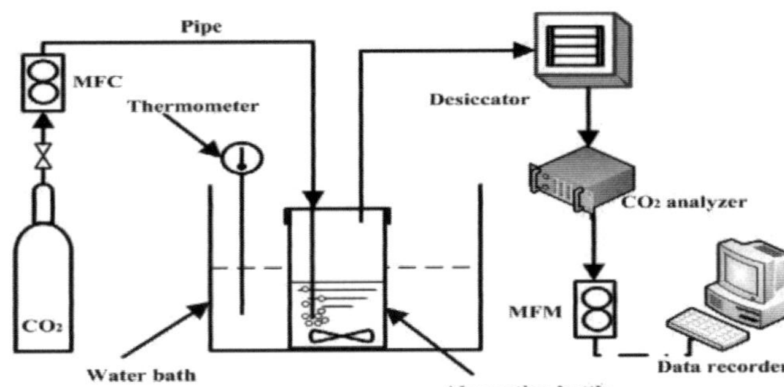

Figure 1 The schematic diagram

3. Results and discussion

The absorption performance of CO_2 in MDEA-AMP aqueous solutions were measured at atmospheric pressure with temperatures ranging from 303.2 K to 323.2 K. The absorption capacity including the mass of the absorbed CO_2 (m) and CO_2 loading (α) are shown in Table 1. One may find from this table that at given mass fraction of MDEA (w_{MDEA}) and AMP (w_{AMP}), the absorption capacity decreases with increasing temperature. At given temperature and given w_{MDEA}, the mass of the absorbed CO_2 and CO_2 loading respectively increase and decrease with the increase of w_{AMP}.

Table 1. Equilibrium absorption amount (m) of CO_2 in MDEA-AMP aqueous solutions and the corresponding CO_2 loadings (α) under different w_{MDEA} and w_{AMP}.

w_{MDEA}	w_{AMP}	α/(molCO_2/{mol MDEA+mol AMP})			m/(gCO_2/100g aqueous solution)		
		303.2K	313.2K	323.2K	303.2K	313.2K	323.2K
	0	5.91	5.78	5.38	0.80	0.78	0.73
0.2	0.03	6.64	6.06	5.57	0.75	0.68	0.64
	0.06	7.19	6.97	6.17	0.69	0.67	0.61

	0.09	7.90	7.70	7.04	0.67	0.65	0.60
0.3	0	8.27	7.57	7.16	0.73	0.68	0.65
	0.03	8.65	8.39	7.80	0.69	0.67	0.62
	0.06	8.92	8.58	8.10	0.64	0.61	0.58
	0.09	9.76	9.39	8.55	0.63	0.60	0.55

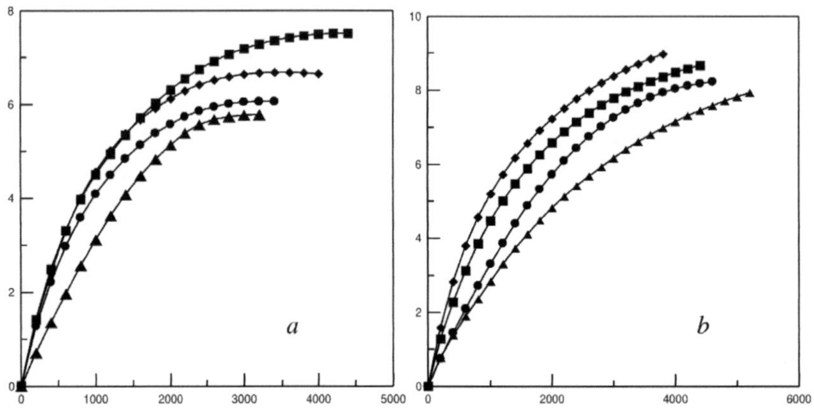

Figure 2 w_{AMP} dependence of the absorption rate of CO_2 in MDEA-AMP aqueous solutions. Symbols: ▲w_{AMP}=0; ●w_{AMP}=0.03; ◆w_{AMP}=0.06; ■w_{AMP}=0.09, (*a*): w_{MDEA}=0.2; T=313.2K, (*b*): w_{MDEA}=0.3; T=303.2K.

Figure 2 shows the influence of w_{AMP} on the absorption rate of CO_2 in MDEA-AMP aqueous solutions, the absorption rate increases with the increase of w_{AMP}. Figure 3 shows the influence of the temperature on the absorption rate of CO_2 in MDEA-AMP aqueous solutions. The absorption rate slightly increases with increasing temperature when the temperatures ranged from 303.2 K to 313.2 K. However, the absorption rate at 323.2 K seems to be very close to that at 313.2 K, indicating that the temperature has little influence on the absorption rate.

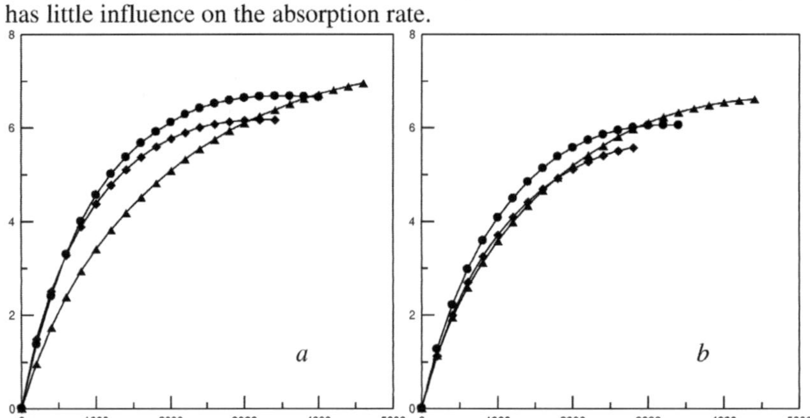

Figure 3 Temperature dependence of the absorption rate of CO_2 in MDEA-AMP aqueous solutions. Symbols: ▲T=303.2K; ●T=313.2K; ◆T=323.2K, (*a*) w_{MDEA}=0.2; w_{AMP}=0.06, (*b*) w_{MDEA}=0.2; w_{AMP}=0.03.

4. Conclusions

In this work, the absorption capacity of CO_2 in AMP promoted MDEA aqueous solution was measured at normal pressure with temperatures ranging from 303.2K to 323.2K. The influence of temperature and w_{AMP} on the absorption capacity and absorption rate of CO_2 was illustrated. Our results showed that: AMP promotes the absorption rate of CO_2 and affects efficiently on absorption capacity of CO_2 in

MDEA aqueous solution. The temperature has little influence on the absorption rate.

Acknowledgements
The authors appreciate the financial support from the National Natural Science Foundation of China (Nos. 21276072), the Natural Science Funds for Distinguished Young Scholar of Hebei Province (No. B2012502076) and the Fundamental Research Funds for the Central Universities (13ZD16 and No. 2016XS108).

References
[1] Solomon S, Plattner G K, Knutti R and Friedlingstein P 2009 *Proc. Natl. Acad. Sci.* **106** 1704–09.
[2] Herzog H 2001 *Environ. Sci. Technol.* **35** 148A–53A.
[3] Cuffey K M and Vimeux F 2001 *Nature* **412** 523–7.
[4] Torralba-Calleja E, Skinner J and Gutiérrez-Tauste D 2013 *J. Chem.* **2013** 1–16.
[5] Mondal M K, Balsora H K and Varshney P 2012 *Energy* **46** 431–41.
[6] Bishnoi S and Rochelle G T 2002 *AIChE J.* **48** 2788–99.
[7] Dubois L and Thomas D 2009 *Chem. Eng. Technol.* **32** 710–718.
[8] Idem R, Wilson M, Tontiwachwuthikul P, Chakma A, Veawab A, Aroonwilas A and Gelowitz D 2006 *Ind. Eng. Chem. Res.* **45** 2414–20.
[9] Fu D, Du L X and Wang H M 2014 *J. Chem. Thermodyn.* **69** 132–6.
[10] Fu D, Zhang P, Du L X and Dai J 2014 *J. Chem. Thermodyn.* **78** 109–13.
[11] Fu D, Wang L M, Mi C L and Zhang P 2016 *J. Chem. Thermodyn.* **101** 123–9.
[12] Al-Masabi F H and Castier M 2011 *Int. J. Greenh. Gas Con.* **5** 1478–88.
[13] Xu S, Wang Y W, Otto F D and Mather A E 1996 *Chem. Eng. Sci.* **51** 841–50.
[14] Yao Z S, Jianwen W, Wei Z and Qing Y 2008 *J. Environ. Sci.* **20** 39–44.
[15] Silkenbäumer D, Rumpf B and Lichtenthaler R N 1998 *Ind. Eng. Chem. Res.* **37** 3133-41.
[16] Shokouhi M, Zoghi A T, Vahidi M and Moshtari B 2015 *J. Chem. Eng. Data* **60** 1250-8.

Prediction of sedimentary facies of x-oilfield in northwest of China by geostatistical inversion

Zhao Lei[1] , Ke Ling[2],He Tingting[1]

[1]Sinopec Exploration & Production Research Institute, Beijing,China;
[2]Research Institute China National Offshore Oil CORP,Beijing,China

Abstract. In the early stage of oilfield development, there are only a few wells and well spacing can reach several kilometers. for the alluvial fans and other heterogeneous reservoirs, information from wells alone is not sufficient to derive detailed reservoir information. In this paper, the method of calculating sand thickness through geostatistics inversion is studied, and quantitative relationships between each sedimentary micro-facies are analyzed by combining with single well sedimentary facies. Further, the sedimentary facies plane distribution based on seismic inversion is obtained by combining with sedimentary model, providing the geological basis for the next exploration and deployment.

1. Introduction

The heterogeneity of alluvial fan reservoirs is strong, and it is necessary to study the sedimentary facies in the development stage of oil and gas fields[1-5]. In recent years, a great deal of research has been done on alluvial fan reservoir, and more understanding has been made on the distribution of the configurational units and depositional characteristics of alluvial fan. Wu Shenghe (2013)discussed the inner layer division system of alluvial fan in the Lower Triassic Kelamayi oil field in Xinjiang, and established the corresponding "general connecting sand body" as the characteristics of the sedimentary configuration model. Chen et al. (2015) established the seismic identification method of alluvial fan reservoirs. Feng et al. (2015) thought that the fan-root reservoirs were mainly flow-channel systems, forming a "tree-branch" combination.

X field is located in the northwest margin of Junggar Basin. The reservoir is about 17.5km away from the Guai-5 area of the Xiaoguai Oilfield in the north and 15.8km away from the south area of the No.45 well. The administrative area is under the jurisdiction of Karamay City. Work area are mostly desert and sandbags.

The X field is a triangular fault block formed by two fractures, and the target zone(P2X3) is divided into three sand groups P2X31, P2X32 and P2X33 from the top to the bottom, and the top structural is a west-to-east dip of the monoclinic.

2. Geostatistical seismic inversion prediction

Seismic data, logging, and cores are the most important sources of information for the analysis of sedimentary facies. Although logging can provide information only near well trajectories, this information often has high resolution and accuracy. Seismic data is not as accurate as logging data and core data. However, seismic data could cover a wider area and have three-dimensional characteristics. Seismic inversion technology can make full use of information provided by logging and seismic data to predict the lithology and properties of reservoir, and which also could provide more macro-information for sedimentary facies analysis.

IOP Conf. Series: Earth and Environmental Science **59** (2017) 012012 doi:10.1088/1755-1315/59/1/012012

The constrained sparse spike inversion—CSSI could obtain the wave impedance data volume at the seismic resolution, which is faithful to the response of the seismic data, but its horizontal resolution is limited by seismic resolution. For the further geological study, when the seismic data can not meet the needs of reservoir study, it is necessary to integrate the well information and the seismic data, then realize the high predictability of fully utilizing the seismic data in the horizontal direction and fully utilize the high resolution of the well in the vertical direction.

The geostatistical seismic inversion workflow fully utilizes logging data and geostatistical data to generate high-resolution impedance models and lithology bodies, and the geostatistical inversion can also solve the problems of low seismic resolutions and have certain help to solve the lithology multiplicity of single seismic attribute superposition [6-8].

As is shown in Fig.1, the trend of geostatistics inversion and post-stack CSSI inversion is consistent, which indicates that both inversion are faithful to seismic information and the two inversion results support each other. The geostatistical inversion has a significantly higher resolution in the vertical direction, and generates a relatively reasonable impedance by means of the statistical PDF parameters, therefore, the sedimentary facies research depends mainly on the analysis results from geostatistical inversion.

The study of sedimentary facies carryed out by the well distribution characteristics of reservoirs and the plane and profile of geostatistical inversion results, the warm represents the low impedance, the cool color represents the high impedance, the warm of the well profile represents the low shale content, and the cool color represents high clay content. By analyzing the characteristics of well reservoirs, the values of well impedance and the distribution of well logs are obtained, then could calculate the sand time thickness distribution (Figure 2), that lay the foundation for the study of sedimentary facies.

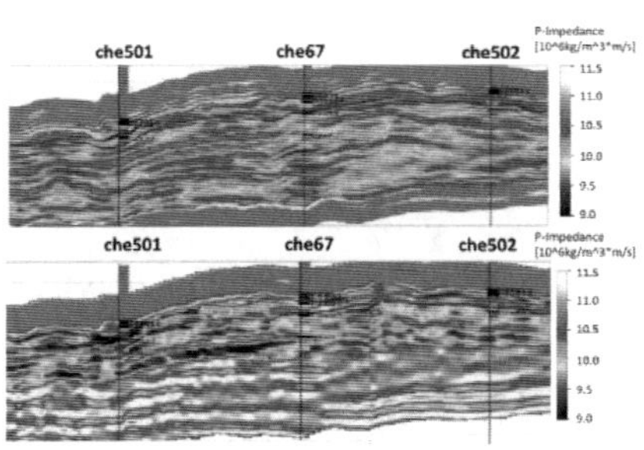

Figure 1. (Upper)Geostatistical inversion profile (Lower)poststack acoustic impedance inversion(Lower)

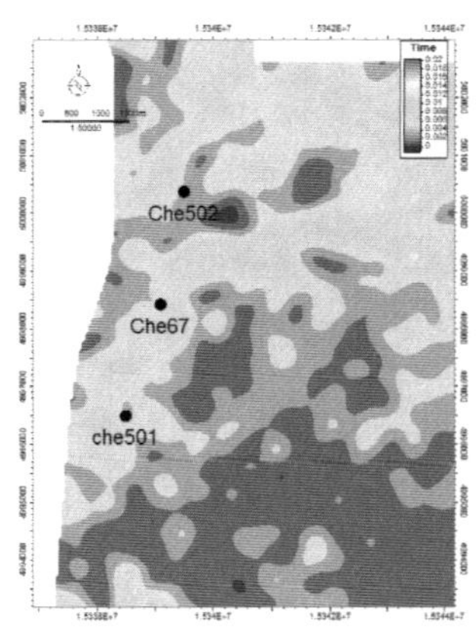

Figure 2. Sand thickness distribution

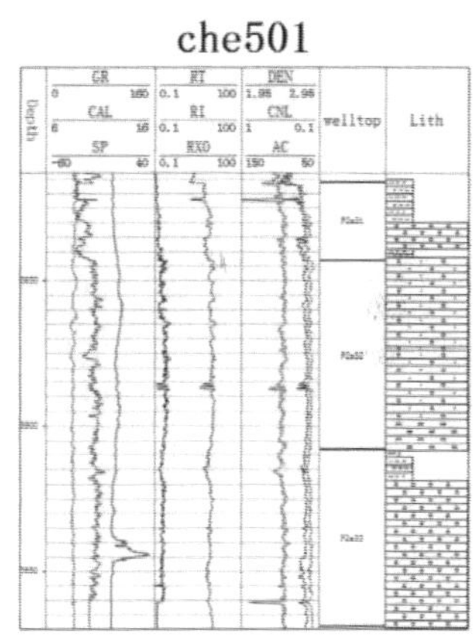

Figure 3. Single well histogram

3. Characteristics of sedimentary facies

3.1. The types of sedimentary facies

The fan-root facies is mainly developed in this area. The overall sediment grain size is coarse, and coarse conglomerates and macadamies are the main sediment, the percent of conglomerate rock is more than 90%. In the northern region close to the provenance, this fan-root deposition system is essentially not developed fan-mid facies, and thus reveal a high resistivity (20-408Ω·m or more) with little fluctuation. The fan-root are subdivided into three sedimentary micro-facies, main channel, swale, and floodplain.

3.2. *Single Well Sedimentary Facies*

In the well Che501, the lithology and sedimentary facies are analyzed (Fig. 3), P2x31(the first sand group), gray-brown silty mudstone, the lower part is gray-brown fine conglomerate, the sedimentary facies is the main channel and floodplain micro-facies. P2x32(the second sand group), brown glutenite and small conglomerates, sedimentary facies is the main channel, swale micro-facies. P2x33(the third sand group), gray-brown fine conglomerate and sandy mudstone, sedimentary micro-facies are the main channel and floodplain.

3.3. *Distribution of sedimentary facies*

Based on the results of sand thickness distribution, the sedimentary facies maps of P2X31, P2X32 and P2X33 are analyzed (Figs. 4). On the whole, the sedimentary micro-facies of the main channel corresponds to the thickest part of the inversion sand thickness map with channel-shape. The swale is distributed around the main channle and begins to diverge in the lower part of the main channel, that is, with the direction of the main channel, the thickness of the swale is gradually thinning, and the scale becomes larger gradually. The floodplain micro-facies are the largest micro-facies in the region, which filling the interspace of the main channel and the swale micro-facies. The sedimentary micro-facies with the best reservoir properties are the main channel micro-facies and the worst ones are the floodplain micro-facies. From the top to the bottom, the main channels are the most developed in the P2X31 sand layer group, and the P2X33 sand group most developed the floodplain micro-facies, therefore, the best property layer is the upper P2X31 sand group.

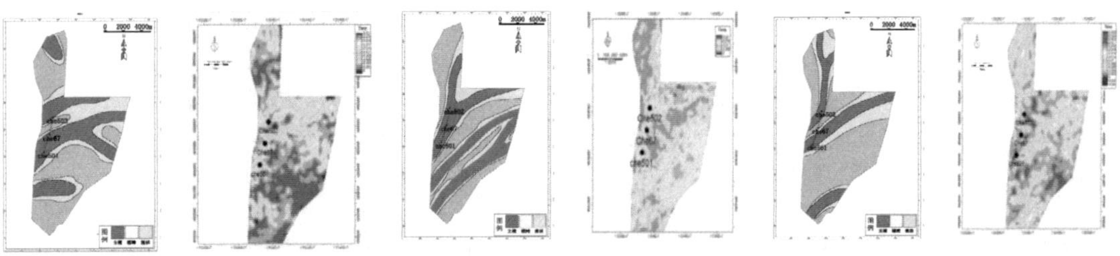

a.Sedimentary Map b.Inversion Map c.Sedimentary Map d.Inversion Map e.Sedimentary Map f.Inversion Map

Figure 4.$P_2x_3^1$ $P_2x_3^2$ $P_2x_3^3$Sedimentary Map and Inversion Map

4. Conclusion

(1) As well data are relatively limited in the exploration area, seismic data can be used to assist the analysis of sedimentary facies. Both constrained sparse spike inversion and geostatistical inversion methods are tried. It has been proved that geostatistical inversion has a higher vertical resolution and will be used.

(2) The Northwest X field belongs to fan-root sub-facies, which is subdivided into three kinds of sedimentary micro-facies, namely main channel, swale and floodplain micro-facies. The reservoir properties are better in main channel and worse in floodplain.

(3) The inversion reservoir thickness map is used to analyze the plane distribution of sedimentary facies. The main channel corresponds to the thicker parts of the thickness map, while the floodplain micro-facies corresponds to the thinner parts. From top to the bottom zone, the main channel is more developed with better reservoir quality. Based on the statistical characteristics, the paper summarizes the sedimentary facies pattern of the X oilfield, and provides guidance for the sedimentary facies analysis of the similar reservoirs.

References

[1] Mu Zhiquan. Application of seismic geologic integrative reservoir prediction technique in Gu

533 block of Daqing Oil field. OGP, 2009, 44(3) :323~ 330

[2] Wang Xiaoguang, He Luming, Lv Jianrong, et al. Configuration of alluvial fan and control mode on remaining oil in Karamay Oilfield[J]. Fault-Block Oil & Gas Field, 2012, 19 (4): 493-496.

[3] Chen Yukun, Wang Yanjie. Architecture characterization of alluvial fan reservoir of Lower Karamay Formation in east district 7(1) of Karamay Oilfield[J]. LITHOLOGIC RESERVOIRS. 2015,27(5):92-97.

[4] Zhaofan. Lacustrine algal limestone reservoir in western Qaidam Basin,China[J]. Carbonates Evaporites. 2013, 25(11): 101-107.

[5] Liu Peng-yuan, LIU Cheng-zhi. Sedimentary Micro-facies Characteristics and Evolution of Quantou Formation in Jijialing Profile in the Southeast Margin of the Songliao Basin: A Typical Profile R eflecting the Evolution of Alluvial Fan to Meandering R iver[J]. GEOSCIENCE. 2015,29(5):1338-1347

[6] Liu Wenling. Geological modeling technique for reservoir constrained by seismic data[J]. ACTA PETROLEI SINICA.2008,29(1):64-68

[7] Dou Song-jiang,ZHAO Yu-wen. Difficulties and Countermeasures of Geological Modeling in Chenghai Oilfield[J]. NATURALGAS GEOSCIENE.2010,21(4): 652-656

[8] Shang Jian-lin, LI Tai-hui. Application of Geostatistical Inversion to Prediction of Thin Sand Body of Karamay Formation in Bai-34 Well Area in Baikouquan Oil Field[J]. XINJIANG PETROLEUM GEOLOGY. 2009,30(06): 758-760.

A brief comparison of existing regional green building design standards in China

J Wang[1], Y Liu[1], J Ren[2] and S Cho[3]

[1]Northwestern Polytechnical University, 127 Youyixi Road, Xi'an 710072,China
[2]Changan University, Middle-section of Nan'er Huan Road, Xi'an 710064, China
[3]Yokohama National University, 79-1 Tokiwadai, Hodogaya-ku, Yokohama 240-8501, Japan

liuyu@nwpu.edu.cn

Abstract. A large country with a variety of regional natural, cultural and economic conditions, China has established a number of green building design (GBD) standards both at national and regional (provincial and municipal) levels. Some researches have been conducted to review and compare such standards. The main aim was to provide valuable references for the establishment of new regionally specific GBD standards in different regions of the country. This paper introduces the preliminary results of the researches. The distribution, frameworks and content of the existing regional GBD standards are introduced and compared in relating to the regionally specific climate, resource, economic and cultural conditions. Conclusions are provided and further researches are recommended.

1. Introduction

To regulate and guide green building development, China established a national green building design (GBD) standard in 2010[1]. Since then, ten regional GBD standards have been developed in five provinces/autonomous regions (Guangxi[2], Zhejiang[3], Jiangsu[4], Shaanxi[5,6] and Qinghai[7]) and five cities (Beijing[8], Tianjin[9], Shanghai[10,11], Wuhan[12] and Chongqing[13]) of the country, which distribute in four of the five climate zones for building thermal design in China (Table 1). Comparisons of such standards were conducted in some funded projects, the results of which are briefly introduced in this paper.

Table 1. Climate zones of the existing regional GBD standards in China

	Names of the Regions	Provinces/ Autonomous Regions	Cities	Climate Zones				
				Severely Cold	Cold	Cold in Summer & Hot in Winter	Hot in Summer & Warm in Winter	Temperate
1	Guangxi	√					√	
2	Zhejiang	√				√		
3	Jiangsu	√				√		
4	Shaanxi	√		√	√	√		
5	Qinghai	√		√	√			
6	Beijing		√		√			
7	Tianjin		√		√			
8	Shanghai		√			√		

9	Wuhan	√	√
10	Chongqing	√	√

2. Comparison of the general frameworks

2.1. Extension of professional areas
The general frameworks of the existing regional GBD standards all consists of 5-7 chapters, each involves one of the building professional areas (e.g. planning, architecture, structure, water, heating & cooling, electric). Comparing to the national standard, all the regional ones have extended their frameworks into more professional areas (e.g. landscape design, interior design, etc)

2.2. Emphasis of process management
Comparing to the national standard, the existing regional ones put more emphasis on the process management. For example, most of the regional standards add a special chapter named "green building scheme", which requires for an overall planning of GBD project to be prepared at the beginning stage of a building design project.

2.3. Relating to green building assessment standards
Some regional GBD standards (e.g. Beijing and Chongqing standards) are directly related to green building assessment (GBA) standards. Some lists all the main indicators of a GBA standard in their appendix, others establish a set of items that are closely related to the key GBA criteria.

2.4. Attention to special technologies
Some regional standards (e.g. Beiijng, Wuhan, Zhejiang and Shaanxi standards) set specific requirements for some special technologies, e.g. technologies used in the curtain wall system, rainwater reclaim system, solar heating and photovoltaic system, heat pump system, etc.

2.5. Provision of assistant information
Most regional standards provide assistant information for GB design. For example, Beijing, Tianjin, Jiangsu, Chongqing, Shaanxi, Qinghai and Zhejiang standards all provide lists of commonly used local materials and or plants, meteorological parameters for design and boundary conditions for software simulations, etc.

2.6. Establishment of inner zoning
Large provinces with obviously different environmental conditions mostly established inner zoning in their regional standards. For example, Guangxi, Jiangsu, Shaanxi and Qinghai standards have established 2,3,3 or 5 zones respectively.

3. Comparison of the content

3.1. Frequency count of four symbolic words
"Climate", "resource", "economy" and "culture" were selected as four symbolic words related to the regionally specific conditions. Frequency count of these words in the existing regional GBD standards show the following: a) "resource" accounts for the highest frequency of appearance, followed by "economic", "climate" and "culture"; b) "climate" and "resource" appear more in the less developed inland regions, while "economy" and "culture" appear more in the more developed coastal regions.

3.2. Description of regional conditions

3.2.1. Climate conditions. While all the existing regional standards use professional terms in the meteorology area, some standards (e.g. Chongqing standard) also uses some popular sayings, such as "mountainous city", "fog town" and "stove place", for the description of regional climate conditions.

IOP Conf. Series: Earth and Environmental Science **59** (2017) 012013 doi:10.1088/1755-1315/59/1/012013

3.2.2. Resource conditions. A large variety of resource are mentioned in the existing regional GBD standards. They not only involve some widely acknowledged ones, such as renewable energy, water, land, materials, wildlife, wastes; but also some not normally recognized ones, such as existing buildings, infrastructure, space and information, etc. The descriptions of resource conditions in different regional standards vary largely and are hardly comparable. For example: "XX is a water shortage city"[2]; "the average annual natural lighting resource in XX is less than other places in north China"[12]; "thick and soft soil is distributed in its east coastal areas"[3]; "XX belongs to a solar energy sub-abundant area)"[8]; "XX possesses special mountainous environmental resources"[13]; "its solar energy resource is rich in the north and poor in the south"[4]; "there are collapsed loess areas in XX" [5-7]; "the precipitation in XX is unbalanced in space/time"[7], etc.

3.2.3. Economic and cultural conditions. There is a lack of description of regional economic and cultural conditions in all of the existing regional GBD standards.

3.3. Design requirements relating to regional conditions

3.3.1. Design requirements relating to regional climate conditions. In the existing regional GBD standards, design requirements relating to regional climate conditions normally involve the following aspects: site plan, outdoor environment, sunlight accessibility, solar thermal utilization, natural ventilation, local vegetation, permeability of site surface, water system, rainwater utilization, water for landscape, water for irrigation, wetland, etc. Some standards also provide more specific design requirements/suggestions that are closely related to the special regional climate conditions. For example, standard in the mountainous areas calls for attention to the relationship between hillside buildings and their surrounding micro-climate[13]; standards in the hot and rainy regions emphasizes waterproof and shading design of the exterior walls, etc[2].

3.3.2. Design requirements relating to regional resource conditions. Some regional GBD standards provide requirements that are closely related to their specific regional resource conditions. For example, Guangxi standard "encourages the use of reclaimed water in its inland areas" and suggests "application of seawater for toilet flush in the water shortage island/coastal areas". For regions where underground water is a precious resource (e.g.Zhejiang province), their standards require "not to apply heat pump systems that uses underground water as a resource". In the mountainous regions (e.g.Chongqing city), their standards suggest "utilization of the mountainous environment and consideration of the terrace space, etc". Some standards provide specific information relating to solar energy resource. For example, Tianjin standard advises that "10°southwest is the best orientation for building solar energy utilization in the Tianjin area."

3.3.3. Design requirements relating to regional economic and cultural conditions. There are few requirement relating to regional economic and cultural conditions in the existing GBD standards. The economic related requirements only involve "technical and economic analysis of buildings"; the culture related requirements mainly refer to "cultural buildings" as a necessary component of public facilities.

4. Conclusions and further research

Overall speaking, while all the existing regional GBD standard share similar structure as that in the national one, their frameworks and content have been enriched and customized mainly in the following aspects:

- *The frameworks are enriched.* New chapters are added and more professional areas (e.g. landscape design, interior design) are involved; articles regarding process management, lists of assessment indicators and assistant information regarding new/special GB technologies are also supplemented.

- *Inner zoning are established.* For regions that cover a large area and possess very different natural, economic and cultural conditions, inner zoning has been established and more specific requirements are provided for each of the inner zones.
- *Description of regionally specific climate conditions and suggestions for the climate adaptive design are provided.* Both meteorology terms and popular sayings are used for the description of regionally specific climate conditions. Specific suggestions for the climate adaptive design are also provided in some of the regional standards.
- *Description of regionally specific resource conditions and suggestions for the resource adaptive design are provided.* A large variety of regional resources are mentioned and some resource specific design suggestions are also provided.

Further researches are recommended to explore the detailed effects and influences of the above GBD standards in the real GBD projects in China.

Acknowledgment
The research of this paper was supported by the National 12th Five-Year Science & Technology Support Project (2015BAL03B04-2), the National 13th Five-Year Key R & D Project of China (2016YFC0700208), and the 2016 International Science & Technology Collaboration and Communication Project of Shaanxi Province (No.2016KW-031) of China.

References

[1] *JGJ T229-2010 Code for Green Design of Civil Building*, Ministry of Housing and Urban Rural Development, China
[2] *DBJ/T45-001-2011 Green Building Design Standard.* Department of Housing and Urban Rural construction of Guangxi Zhuang Autonomous Region
[3] *DB331092-2013 Green Design Standard of Civil Buildings.* Department of Housing and Urban Rural Construction of Zhejiang Province.
[4] *DGJ32J 173-2014 Jiangsu Design Standard for Green Building*, Jiangsu Ministry of Housing and Urban Rural Development, China.
[5] *DBJ 61/T-80-2014 Standard for Green Design of Public Buildings in Shaanxi*, Shaanxi Ministry of Housing and Urban Rural Development, China.
[6] *DBJ 61/T 81-2014 Residential Green Design standards*, Shaanxi Ministry of Housing and Urban Rural Development, China.
[7] *DB63/T 1340-2015 Green Building Design Standard for Qinghai Province*, Qinghai Ministry of Housing and Urban Rural Development, China.
[8] *DB11/938-2012 Design Standard of Green Building*, Beijing Municipal Commission of Urban Planning, China
[9] *DB29-205-2015 Tianjin Green Building Design Standard*, Tianjin Urban and Rural Construction Commission.
[10] *DG/TJ 08-2143-2014 Green Design Standard for Public Building.* Shanghai Municipal Commission of Urban Planning, China.
[11] DGJ 08-2139-2014 Green Design Standard for Residential Building. Shanghai Municipal Commission of Urban Planning, China.
[12] *WJG 117-2011 Design Specification for Green Building of Wuhan*, Wuhan Urban and Rural Construction Commission.
[13] *DBJ50/T-214-2015 Design Standard of Green Buildings*, Chongqing Urban and Rural Construction Commission.

Stress analysis and mitigation measures for floating pipeline

Guo Wenpeng[1] **Liu Yuqing**[2] **Li Chao**[2]

1:Shanxi Natural Gas Co.,Ltd., Taiyuan, Shanxi, China 030032
2: China Petroleum Pipeline Engineering Co.,Ltd., Langfang, Hebei, China 065000

liu_yuqing@cnpc.com.cn

Abstract. Pipeline-floating is a kind of accident with contingency and uncertainty associated to natural gas pipeline occurring during rainy season, which is significantly harmful to the safety of pipeline. Treatment measures against pipeline floating accident are summarized in this paper on the basis of practical project cases. Stress states of pipeline upon floating are analyzed by means of Finite Element Calculation method. The effectiveness of prevention ways and subsequent mitigation measures upon pipeline-floating are verified for giving guidance to the mitigation of such accidents.

1. Introduction

Natural gas pipelines in mountains, hills, and river valley area may likely be washed out of the trenches to make the pipes bare due to rainstorm or running water in the rainy season. In addition, collection of many streams is easy to incur pipeline-floating accident at locations where the pipelines are exposed. Upon pipeline-floating, the changed position of pipeline imposes additional stresses to pipeline, which make the pipeline in bad stress state.

For example, after a heavy storm, a 150 meters long pipeline section was washed out from the trench and pipeline-floating event shown in figure 1 occurred.

Fig.1 Site photo of floating pipeline

2.Basic parameters of the pipeline

Basic parameters of floated pipeline are shown in table 1.

Table 1 Basic parameters of pipeline

Name		Unit	Value
Pipe diameter		mm	711
Wall thickness	General section	mm	10.3
	Cold bend	mm	10.3

Cold bending Angle	/	9
Pipe material	/	API-5L X60
Design pressure	MPa	4.0
Operation pressure upon floating	MPa	0.3

3 Stress analysis on floated pipeline

3.1 Pipeline displacement data

For buried pipeline, spiral ground anchors may be used to fix the pipeline on the bottom of trench with belted anchors to thereby prevent the pipeline from floating[1][2].

Five coordinates are measured on site. By comparing the position coordinates before and after floating, horizontal and vertical displacement of each measured point can be obtained and summarized in table 2. In table 2, displacement component, ΔX, refers to horizontal displacement component along axial direction of pipeline (the reference direction is supposed as from point 5 to point 1); ΔY refers to the horizontal displacement component in direction perpendicular to the pipe axis (the reference direction is supposed as to the arc side); ΔH refers to the pipeline displacements in vertical direction (the reference direction is supposed as upward).

Table 2 Pipeline displacement, in m

SN.	Position description	Displacement component			horizontal component	Total displacement
		ΔX	ΔY	ΔH		
1	Unearthed point	0.072	-0.364	0.524	0.371	0.642
2	/	0.165	-0.091	0.589	0.188	0.619
3	Cold bend	-0.553	-0.105	0.532	0.563	0.775
4	/	0.071	-0.097	0.512	0.120	0.526
5	/	0.269	-0.151	0.461	0.309	0.555
6	Unearthed point	0	0	0	0	0

3.2 Finite element calculation model

ANSYS finite element software is used for analyzing and calculating the stress state of pipeline. The calculation procedure adopts following mechanical models: (1)the pipeline is simulated with elastic-plastic nonlinear material, soil body is simulated with nonlinear soil spring and the pipeline can subject to large deformation; (2)pipe-soil interaction is simulated with spring, the selection of parameters of soil spring is as per ASCE[3][4] and GB50470-2008[5];(3)200m long of pipeline model is established along the outside of exposed section of floated pipeline to simulate the constraint action of buried pipeline to floated pipeline section; (4)rigid constraints are imposed on both ends of model; applying displacement is applied to the model to simulate displacement of the deformed pipeline; (5)two operation conditions,4MPa of design pressure 0.3MPa of operation pressure after floating, is employed for checking.

3.3 Checking Specification

The exposed floating pipeline becomes unconstraint. It can be checked according to the definition and methodology for unconstraint pipeline in Gas Transmission and Distribution Piping Systems (ASME B31.8-2010)[6] (hereinafter referred to as ASME B31.8).

3.4 Stress checking results

Table 3 shows the stress states of pipeline after checking using the finite element software ANSYS according to ASME B31.8.

Table 3 Checking results for 4Mpa of design pressure

Positions	The axial stress (MPa)	Allowable stress (MPa)	Check results	S_E (MPa)	S_A (MPa)	Checking results
Unearthed point 1	394.385	311.25	Don't meet the requirements	5.910	34.615	Meet the requirements
Cold band	143.883	311.25	Meet the requirements	5.509	285.117	Meet the requirements
Unearthed point 2	100.657	311.25	Meet the requirements	4.441	329.341	Meet the requirements
The most adverse point	394.385	311.25	Don't meet the requirements	5.910	34.615	Meet the requirements

According to the checking result, the stresses of pipeline under 4 Mpa of pressure(design pressure) exceed the stress requirement for constraint pipeline in ASME B31.8 such that measures should be taken to reduce the stress. The results indicate that decompression operation after floating is helpful for reducing piping stress and ensuring the safety of pipeline.

3.5 Mitigation measures

The additional stress is caused by displacement. In order to reduce the stress and displacement of pipeline should be resumed as soon as possible[7]. Following measures can be taken:

(1) Excavation

Carry out manual excavation under the exposed pipeline section. During excavation, measures should be taken to prevent the coating and the pipeline from injuring by excavation tools. At the same time, close attention should be paid on the change of the piping to lest damage of personnel during pipeline relocation.

(2) Pipeline relocation.

The personnel simultaneously excavate the residual soil in the trench at left side of the flow direction of media in the pipeline to make pipeline naturally lower to original position with gravity.

(3) Coordinates relocation

After relocation of pipeline, the coordinates of pipeline should be measured again. Each run of joint should subject to RT and UT test again after confirming the coordinate data. Coating should be patched after qualification and the pipeline should be backfilled in layers.

3.6 Pipeline processing effect

By comparing the coordinates of pipeline after relocation with original coordinates, it is shown that the displacement of pipeline is obviously reduced, as shown in table 4.

Table 4 pipeline displacement after relocation, in m

SN. of coordinate points	Instructions	Displacement component			Horizontal displacement	Total displacement
		ΔX	ΔY	ΔH		
1	Unearthed point	-0.017	-0.052	0.067	0.055	0.086
2	/	0.030	-0.069	0.042	0.075	0.086
3	Elbow	/	/	/	/	/
4	/	0.169	0.017	0.098	0.170	0.196
5	/	0.238	0.208	0.052	0.316	0.320
6	Unearthed point	0	0	0	0	0

It is indicated from the checking calculation according to ASME B31.8 that the stresses of pipeline under 4MPa of design pressure can meet the stress requirement for constraint pipeline after relocation. Table 5 summarizes the checking results.

Table 5 Checking results under 4Mpa of design pressure after relocation

Position	Aaxial stress (MPa)	Allowable stress (MPa)	Checking results	S_E (MPa)	(MPa)	Checking results
Unearthed point 1	151	311	Meet the requirements	6.590	278	Meet the requirements
Cold bend	190	311	Meet the requirements	4.739	239	Meet the requirements
Unearthed point 2	133	311	Meet the requirements	0.597	323	Meet the requirements
The most adverse point	267	311	Meet the requirements	6.590	278	Meet the requirements

4. Conclusion

Mitigation measures for pipeline-floating accident are summarized in this paper on the basis of project cases. The effectiveness of these measure are verified through finite element analysis and calculation, and following conclusions are drawn:(1) pipeline relocation is helpful for reducing the stress of pipeline;(2)detailed construction measures should be made during pipeline relocation and close attention should be paid on the tendency of pipeline deformation prevent the uncontrolled pipeline deformation.

Reference

[1] GB50251-2015.Code for Design of Gas Transmission Pipeline Engineering. Beijing: China plan press, 2015.

[2] GB50253-2014.Code for Design of Oil Transmission Pipeline Engineering. Beijing: China plan press, 2015.

[3]ASCE. Guidelines for the Design of Buried Steel Pipe, 2001.

[4] ASCE Technical Council on Lifeline Earthquake Engineering. Guideline for the Seismic Design of Oil and Gas Pipeline System. New York. ASCE, 1984.

[5] Oil and gas field and pipeline construction design professional committee for standardization, GB 50470-2008. Beijing: China plan press, 2009.

[6] American Society of Mechanical Engineers (ASME).2010. ASME B31.8, Gas Transmission and Distribution Piping Systems. The American Society of Mechanical Engineers, New York, NY.

[7] LIU Yuqing, ZHANG Zhenyong, ZHAO Zifeng. Design and application of the heavy-duty pressure test program forpipelines with high design factor[J]. Oil & Gas Storage and Transportation. 2015,34(5):557-561.

Analysis of the water flooded degree in X oilfield and its impact on remaining oil distribution

Xin Yuxia [1,a] , **Tian Zuoji** [1,a] , **Zhang Fanqin** [1,a], **Shan Fachao** [1,a],
Wang Zhizhang [2,a]

[1]PetroChina Research Institute of Petroleum Exploration & Development, Beijing 100083,China
[2]China University of Petroleum (Beijing)

[a]xinyuxia@petrochina.com.cn

Abstract. To the study of remaining oil distribution in old oilfields, the key is the understanding of the water flooded degree of the production wells. In addition, the injection-production plan must be rearranged or adjusted according to the water flooded degree. This paper has determined the important part of potential tapping for high water cut oil reservoirs through the analysis of the water flooded degree of target layers in X oilfield, so as to provide guidance for the further development of the X oilfield.

1. Analysis of the water flooded degree in X formation of X oilfield

To the study of remaining oil distribution in old oilfields, the key is the understanding of the water flooded degree of the production wells. In addition, the injection-production plan must be rearranged or adjusted according to the water flooded degree [1]-[3].

The logging response of the water flooded zones can be seen more clearly through the study of pairs of wells (wells A and B, for example). Well A was put into production in Jan. 1964, and well B was put into production in Aug. 2005 (Fig.1). After water flooded, the resistivity value changed obviously and decreased by 10Ω•m~20Ω•m, and negative resistivity differences can be occasionally found (RXO>RT>=RI). In the process of the old oilfield development, the water washing and water flooded situation of old oil wells can be determined very accurately according to the adjacent old wells and new wells or through comparative interpretation of "inspection logging" data and logging data of old wells. They are the most reliable basic data for determining the water flooded degree of oil reservoirs.

The water flooded degree in X oilfield is high. The water flooded degree has reached 78% and the washing degree has reached more than 90%, which indicates that the remaining oil reservoirs have reached the middle to late stage on the whole. Because of the high water flooded degree and many remaining oil influence factors, the fine analysis of water disturbance characteristics of single sand bodies controlled by microfacies is very necessary. Only according to this, the distribution characteristics of the remaining oil can be really understood, which can create conditions for the effective development of remaining oil [4]-[7].

IOP Conf. Series: Earth and Environmental Science **59** (2017) 012015 doi:10.1088/1755-1315/59/1/012015

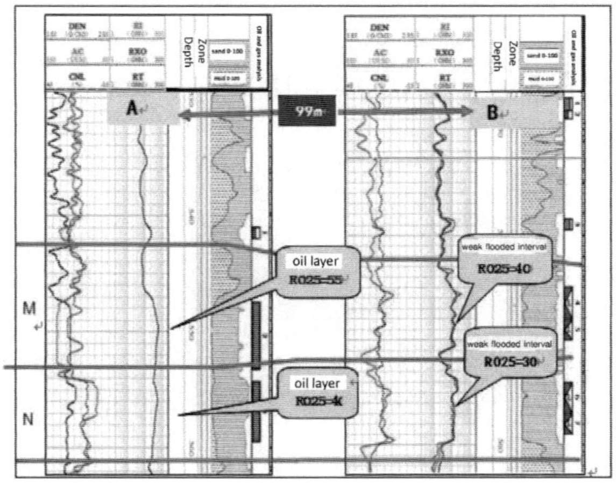

Fig. 1 The Water Flooded Degree Comparison Between Well A and Well B

2. Analysis of remaining oil distribution according to water flooded degree

The main pay zones are N^1 and N^2 in the area. In addition, the higher the development degree of sand bodies is, the higher the water flooded degree is. Below is the description of the water flooded degree, taking main pay zones N^1 and N^2 as an example.

The Fig.2 is the single sand bodies distribution in N^1 time unit. The time unit of N^1 belongs to the inner front of fan delta. It develops microfacies such as underwater branch channel, channel side edge, thin layer sand, inter-channel bay mud, etc. The Fig.3 is the water ratio map of N^1 time unit. Through the analysis and comparison of the two maps, we can know that the majority of the reservoirs with the water ratio >70% are distributed in sand bodies controlled by underwater branch channels. At present, the planar distribution area with the lowest water ratio basically belongs to the underwater inter-channel bay mud facies area, and the area with the water ratio of 60%~70% is distributed mainly in the single channel inter-connection part of underwater branch channels and the channel side edge sand bodies with weak hydrodynamic force in the intermission period of branch channels. This part shall be the most important part of potential tapping for these high water cut reservoirs.

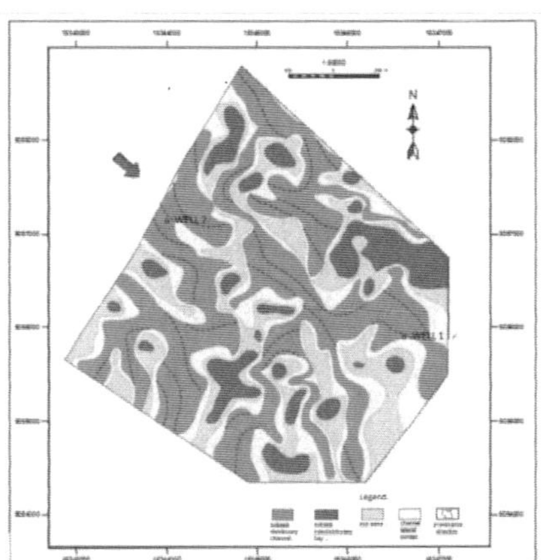

Fig . 2 The Single Sand Bodies Distribution Map of N^1 Time Unit

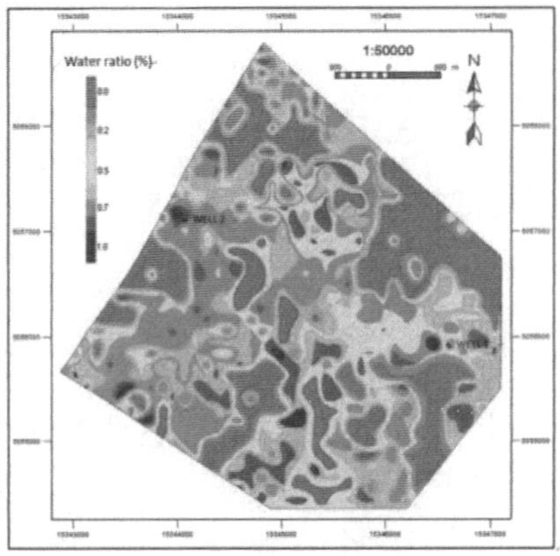

Fig .3 The Water Ratio Map of N^1 Time Unit

The Fig.4 is the single sand body distribution map of N^2 time unit. The time unit of N^2 belongs to inner front of fan delta too. It also develops microfacies such as underwater branch channel, channel side edge, thin layer sand, inter-channel bay mud, etc. The Fig.5 is the water ratio map of N^2 time unit.

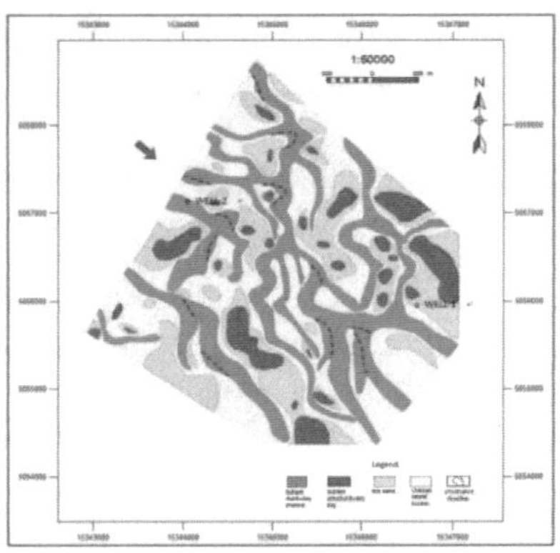

Fig.4 The Single Sand Body Distribution Map of N^2 Time Unit

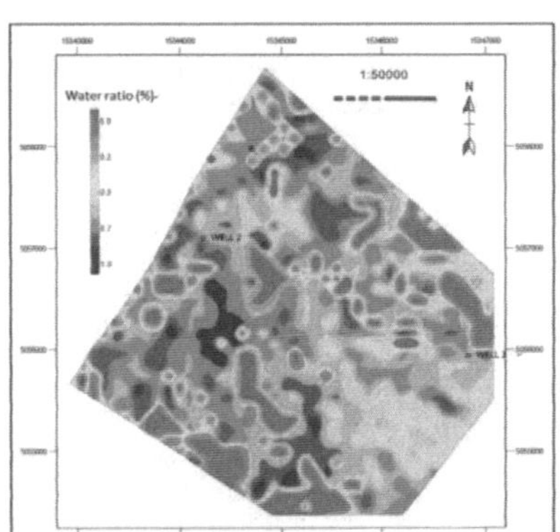

Fig.5 The water ratio plan of N^2 time unit

Through the analysis and comparison of the two maps, we know that the water flooded condition of the time unit of N^2 is more complex. The majority of the NNE-SSW braided underwater branch channels sand bodies distributed in intermediate strips belongs to the 60%~70% water ratio area, and

local water ratio is>80%. This indicates that such underwater branch channels sands have very strong planar heterogeneity. High water ratio areas also appear in the end or side edge of some branch channels or the part of thin layer sands, as shown in Fig.5.
According to the comprehensive comparative analysis of the single sand body distribution map and water ratio contour map of each time unit, we know that the area with the water ratio of 60%~70% is mainly distributed in the single channel inter-connection part of underwater branch channels and the channel side edge sand bodies and thin layer sand bodies with weak hydrodynamic force in the intermission period of distributary channels. This part shall be the most important part of potential tapping for high water ratio reservoirs.

Conclusion

According to the analysis of the water flooded degree in X formation of X oilfield as well as the single sand body distribution map, the conclusion has been obtained as follows: the area with the water ratio of 60%~70% is distributed mainly in the single channel inter-connection part of underwater branch channels and the channel side edge sand bodies and thin layer sand bodies with weak hydrodynamic force in the intermission period of distributary channels. This part shall be the important part of potential tapping for high water ratio reservoirs.

References

[1] Yong Shihe et al. *Processing and Comprehensive Interpretation of Logging Data,* China University of Petroleum Press, 1996.
[2] Han Zhiming, Wang Xiancheng and Li Shengjun. *Research on Interpretation Methods for Water flooded Zones in High Water Cut Period.* Logging Technology, 2000, 24(5): 333～336.
[3] Zhang Hengshu, Luan Mengjie, Cui Guofan, Qi Xiuli, and Liu Jiang. *Water flooded Zone Judgment and Analysis Methods and Their Application.* Journal of Huaihua University, 2016, 35(5): 30~33.
[4] Li Zhen et al. S*ummary of Logging Interpretation Method for Water Flooded Layer.* Chinese Journal of Engineering Geophysics, 2016, 3(4)： 288～294.
[5] Ci Jianfa et al. *Development Status and Future of Water Flooded Layer Logging* [J]. Natural Gas Industry, 2005 , 25(7):44～46.
[6] Han Zhiming et al. *Research on Water Flooded Layer Interpretation Method at High Water Cut Stage* [J], Well Logging Technology, 2000 , 24(5):333～336.
[7] Fan Xiyan, Zhou Yuanqi, et al. *Well logging Interpretation and application of flooded zones in Hu Zhuangji oilfield* [J]. Inner Mongolia Petrochemical Industry, 2001, 27 (5):165～169.

A research of the community's opening to the outside world

Lan Xu[1], Xiangzhuo Liu[2]

[1] College of Surveying and Mapping institute of science and technology, Xi'an University of Science and Technology, Xi'an, Shaanxi 710054, China
[2] School of Resources and Environment, University of Electronic Science and Technology of China, Chengdu, Sichuan 611731, China

E-mail: 1183436081@qq.com

Abstract. Closed residential areas, called community, the traffic network and result in various degrees of traffic congestion such as amputating, dead ends and T-shaped roads. In order to reveal the mechanism of the congestion, establish an effective evaluation index system and finally provide theoretical basis for the study of traffic congestion, we have done researches on factors for traffic congestion and have established a scientific evaluation index system combining experiences home and abroad, based on domestic congestion status. Firstly, we analyse the traffic network as the entry point, and then establish the evaluation model of road capacity with the method of AHP index system. Secondly, we divide the condition of urban congestion into 5 levels from congestion to smoothness. Besides, with VISSIM software, simulations about traffic capacity before and after community opening are carried out. Finally, we provide forward reasonable suggestions upon the combination of models and reality.

1. Introduction

The traffic congestion has been a common problem of numerous mega-cities around the world. It affects the healthy development of cities in many ways [1-3]. At present, the main characteristics of closed communities in China are: closure, large area, big population, single function, out of touch with each other, etc. These characteristics have reduced the density and reachability of traffic network. Urban roads are also affected by traffic flow within the community. Meanwhile, this increases the traffic pressure around traffic network [4-6].

In China, with the policy which allows no construction of closed community, open communities have become mainstream. However, in terms of construction and management about open communities, there are no researches on community security, protection of property rights, environmental quality as well as design rationality. Hence, it is urgent to improve the management, legal principle, technology, and ideas of design [7-8]. With the existing road resources, how to better alleviate the traffic congestion has become one of the most important subjects on solving urban transportation problems [7]. As a result, it is urgent to study and establish a scientific evaluation index system of traffic congestion [1]. Only after further investigation on the mechanism of formation and propagation features, can we set up a scientific scheme to meet travellers' needs. From these, we can see that it's significant to study the problems of traffic congestion for the development of cities [8-10].

2. Evaluation methods

Cities, which has its essence of diversity is its, is a complex of multiple functions. Open community is developed on the basis of this diversity. From the perspective of planning and construction, open community has open space and private living rooms. Open space is taken for daily activities and communications. While the latter are living places, which are secure and quiet. Seen from the road system, the organic combination of open residential roads and surroundings can help ease traffic problems in cities [13]. By using AHP model, we have merged the information of each indicator. Finally, we figure out the extent of impact of district opening on the road capacity (Fig. 1).

According to the difference of road traffic, the operational condition can be divided into 5 grades: high degree of delay, moderate delay, general delay, low delay and no delay [2]. The evaluation factor sets which conclude to $U = \{u_1 u_2 u_3\}$ are corresponding to the number of nodes, the road area and the degree of street. At the same time, we've determined the sets of judgments $V = \{v_1, v_2, v_3, v_4, v_5\}$ which correspond to non-congestion, low-level congestion, general congestion, moderate congestion, high degree of congestion for five evaluation levels. Then, we integrate the three indexes with AHP and figure out that $W = (0.55, 0.18, 0.27)$ And the fuzzy evaluation matrix is established: $R = (r_{ij})_{3\times 5}$.

Based on the evaluation matrix, we can combine the resulting matrix and Weight vector to the final computation.

$$B = W \cdot R = (0.55, 0.18, 0.27) \cdot R = (b_1, b_2, b_3, b_4, b_5) \quad (1)$$

$$b_j = \sum_{i=1}^{n} w_i r_{ij} \quad (2)$$

The degree of congestion can be judged from vector b. From the number of nodes, we can see that the road area rates weigh of the large. Therefore, this paper mainly analyzes the current traffic situation of the important intersections (road nodes) around the district, so that we can obtain the relationship between the average traffic velocity and delay time. In fact, the number of road nodes and the road area rate have a significant effect on the average speed of traffic. That is to say, the more nodes and road area there are, the larger V average will be. Therefore, we will study the impact of V average on road capacity instead of the number of nodes and road area rates.

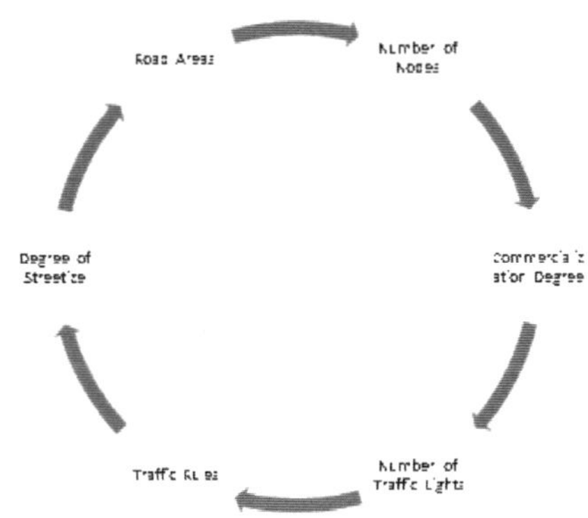

Figure 1. Road traffic indicators evaluation system

3. Comparison of the road capacity based on VISSIM simulation

The opening of the closed cell id a good way resolve traffic congestion. In terms of impacts of two types of residential areas on the traffic before and after opening of community, as well as the operational condition, we need to compare current condition with previous condition. Here, the evaluation function of VISSIM is used to simulate the traffic network. Based on the results, we can judge whether or not the traffic network should be opened.

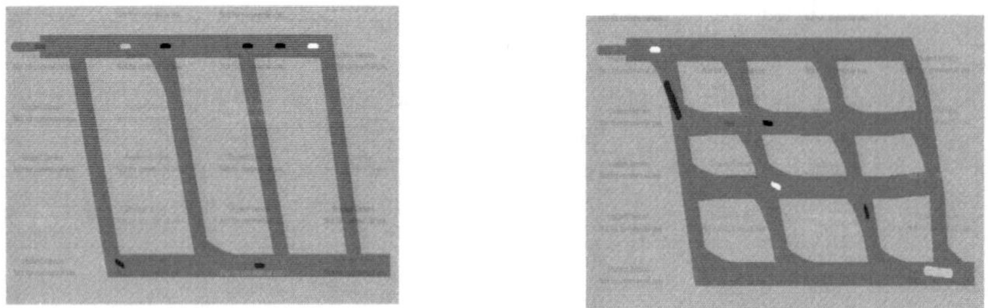

Before Opening After Opening

Figure 2 Comparison of traffic volume before and after community opening

Then we get the following data:

Table 1. Operational Condition before Opening

Vehicle class	Number of vehicle	Time(h)	Distance (km)	Delay (h)	Average speed (km h^{-1})	Average delay(s)
Small car	156	21.94	175	58.67	7.98	46.2
Large Bus	11	2.38	104.8	3.57	44.03	20.16

Table 2. Operational Condition after Opening

Vehicle class	Number of vehicle	Time(h)	Distance (km)	Delay (h)	Average speed (km h^{-1})	Average delay(s)
Small car	156	20.02	175	58.36	8.66	45.89
Large Bus	11	2.18	104.8	3.56	48.07	20.16

From the above model, we calculate the basic road capacity before and after opening. Assuming that, the same type vehicle speed before and after opening the communities is equal to each other. Compared to road capacity before opening, the road capacity after opening has increased by

$$\frac{C_{After}}{C_{Before}}=1-\frac{1000v_{After}(I_0)^{-1}}{1000v_{Before}(I_0)^{-1}}=\left(\frac{v_{After}-v_{Before}}{v_{Before}}\right)\times 100\% \qquad (3)$$

Based on the data in the table with this formula, we can find that the road capacity for small cars has increased by 8.5%, while for large passengers that has increased by 9.2%.

4. Conclusion

To conclude, we provide five suggestions, there are as follows: Firstly, make substantial increases of the road density, set multiple entrances and exits so as not to be directly connected with the trunk roads. Besides, the circles around the roads can resolve the impact of left turn traffic on local road networks. At the same time, improving the capacity of the joint between entrances and channels, mainly depends on the spatial position, area of joint, the capacity of adjoining road and the ways of dealing with traffic flow at the junction [14-15].

Secondly, the location of passageway should ensure the safety, effectiveness of vehicles and avoid non-project traffic. The size of passageways must be able to match the size of vehicles; The road area should be appropriately enlarged when it comes to the compression of streets and roads; Review and demonstrate the width motor vehicle lane standards and then reduce the lower limit of vehicle width; Closely restrain the conditions and range of intersection channelizing broaden.; Properly use the distance of boundary line area; Provide access for prevailing conditions and requirements about enhancing city vitality.

Thirdly, auxiliary roads and adjust ways of traffic flow. Such as the implementation of one-way traffic and restricted access [10]. It is necessary to shrink the width of motorway and emphasize rights of walking and cycling.

Fourthly, we could separate car streams and pedestrian flow in both time and space through signal lights, signs, overpass and tunnels. And then, get through amputating roads and T-shaped roads. Improve the connectivity of traffic networks. Besides, roads in public space such as parks and greenbelts should be available to pedestrian and bicycles. Generally, we can reduce new traffic volumes through restraint of construction of closed communities as well as transferring the land utilization. That means to transfer the residential lands into traffic lands. In this way, public space in our streets can be enlarged. Finally, For the communities, which have no developed internal traffic networks, its opening will not help to ease traffic congestion but bring about negative impacts. So, this kind of community should not be opened.

References

[1] Zhu F L 2006 Research on Urban Traffic Congestion Evaluation Index System (Nanjing: Southeast University) 24

[2] Jiang J L, Song R and Li J 2011 Evaluation of Urban Road Congestion Based on DEA *Traffic Information and Security* **29** 10-14

[3] Li M 2016 *Study on Traffic Flow Assignment Model of Road Network with Congested Roads and Its Application* (Beijing: Beijing Jiao Tong University) p35

[4] Li X P 2014 *Study on Urban Traffic Congestion Countermeasures – Closed Cell Traffic Opening* (Changsha: Changsha University of Science and Technology) 36

[5] Yan B, Cai R D, Hu Y C, Cao M X and Zhan B 2016 Strategic Studies on Community Opening based on Fragility of Traffic Networks *Logistics Technology* **35** 98-101

[6] Gao F, Yang S L, Bian M H, Yuan S J and Yan Y 2016 The Impact of Community Opening on Traffic Passage *Engineering Technology* **10** 136

[7] Liu Z Q 2016 *Transformation Response to Residential Design after the Areas' Opening* China Conference

[8] Yan F, Pei Y L and Cao C H 2007 Application of VISSIM in the public transportation priority signal timing *Journal of Harbin Institute of Technology* **1** 138

[9] Lu Z N, Mao P, Ji Z Y, Qin R X and Chen L M 2016 Study on the Impact of Community Opening on Traffic Passage. *Economic Management* **7** 136

[10] Zu B 2013 *Study on Simulation of City Traffic Control based on VISSM* (Nanjing: Southeast University) 28

[11] Huang Y G 2015 *The Mechanism of Traffic Congestion and Control Method* (Guangzhou: South China University of Technology) 23

[12] Wang B B 2015 *The Mechanism of Urban Traffic Congestion and Dissipation Control Strategy* (Lanzhou: Lanzhou Jiao tong University) 34

[13] Peng Y 2008 *Development Research on Open Structure of the Residential area* (Chongqing: Chongqing University) 28

[14] Tang Z Q 2006 *Study on Urban Large-scale Construction Project Analysis* (Changsha: Changsha University of Science and Technology) 36

[15] Shang Z H 2006 *Analysis of Impact of Residential Area Development on traffic* (Xi'an: Chang'an University) 29

IOP Conf. Series: Earth and Environmental Science **59** (2017) 012017 doi:10.1088/1755-1315/59/1/012017

Metallogenic regularity of the Masiluo–Zhangjiapingzi area, Western Sichuan, China

Chaowen Huang[1, 2], Shugen Zhang[1, 2], Yalong Kang[3]

[1]Key Laboratory of Metallogenic Prediction of Nonferrous Metals and Geological Environment Monitoring, Ministry of Education, Central South University, Changsha 410083, China

[2]School of Geosciences and Info-Physics, Central South University, Changsha 410083, China

[3]Sichuan Institute of Nuclear Geology, Chengdu 610061, China

huangchaowen@csu.edu.cn

Abstract. The Masiluo–Zhangjiapingzi area is one of the most important target areas for ore exploration in western Sichuan, which has abundant gold and copper ore deposits. Substantial ore deposits in this area include the Masiluo gold deposit, Zhangjiapingzi gold deposit, Dongjia wuji specular hematite deposit, Gonglikong Cu–Au deposit, and Bazhe Cu–Au–Fe deposit. This study characterized the metallogenic regularity of the Masiluo–Zhangjiapingzi area using observations of the regional geological background and the characteristics of a typical deposit in this area. The ore deposits were found to be controlled by stratigraphy, geological structures, and magma with distinctive zoning characteristics—gold mineralization on the west part, Cu–Au mineralization on the northeastern part, and Fe–Cu–Au mineralization on the southeastern part.

1. Introduction

The Masiluo–Zhangjiapingzi area is located in the southwestern China transition zone encompassing the Sanjiang and Kangdian metallogenic belts that are located at the southwestern margin of Yangtze Plate [1, 2]. In recent years, many gold, copper, and iron ore deposits (spots) has been found and make this region a major target area in western Sichuan [3, 4,]. However, geological research and ore exploration in the area is still nascent and full of uncertainties regarding metallogenic regularity and directions. Hence, this study aims to determine the metallogenic regularity of the Masiluo–Zhangjiapingzi area to solve this key challenge and achieve a breakthrough in ore exploration.

2. Geological Background

The Masiluo–Zhangjiapingzi area is located at the Jinpingshan thrust nappe belt in the center of the Longmenshan–Jinpingshan orogenic belt belonging to the southwestern margin of Yangtze Plate. It is located at the juncture between the Southwestern China Sanjiang structure belt, Songpan–Ganzi orogenic belt, and Yangtze Plate [7, 8]. The regional geological setting of this area reflects complicated structural alteration and frequent magma activity. The substrate formation phase occurred in the Precambrian. The passive continental margin phase occurred in the Sinian–middle Triassic. During the early to middle Variscan, the earth's crust in this area was subsiding and dipping toward the sea, resulting in the sedimentation of a series of Devonian–Permian terrestrial clastic rocks and limestone, whose lithology and thickness were controlled by the Caledonian northwestern-trend tectonic uplift and rifting. During the late Variscan to early Indosinian, the crustal rifting movements caused massive-scale central fissure eruptions of basic magma and formed Permian–middle Triassic-aged basalt, overlain by middle Triassic marine limestone facies (Fig. 1). The orogeny phase occurred in the late Triassic–Cretaceous. During the late Triassic, the study area was thrust napped and uplifted toward the east owing to the eastward extrusion of the Tibetan Plateau amid the late Traissic Indosinian movement. This resulted in the formation of the parallel beds of the Devonian–mid-Triassic strata involved in the folding that underwent a series of deformation and metamorphism phases. This formed an NNE-trending, mostly dry, compression–shear fault that comprises the basic outline of the regional structures. The later Yanshan movement further reinforced this fault. During the Cenozoic era, this region entered an intracontinental orogenic phase. The Himalayan movement intensified the nappe structure in the study area and entered the collisional orogeny phase at around 78–52 Ma. At approximately 44–28.7 Ma, the detach-thrust phase of the suture zone occurred following the collision phase, after which the structural outlines were further fixed and the revival of compression–shear faults was clear; after 18.9 Ma, the region entered a rapid uplift phase and formed the current deeply cut, high-mountain topography [5, 6].

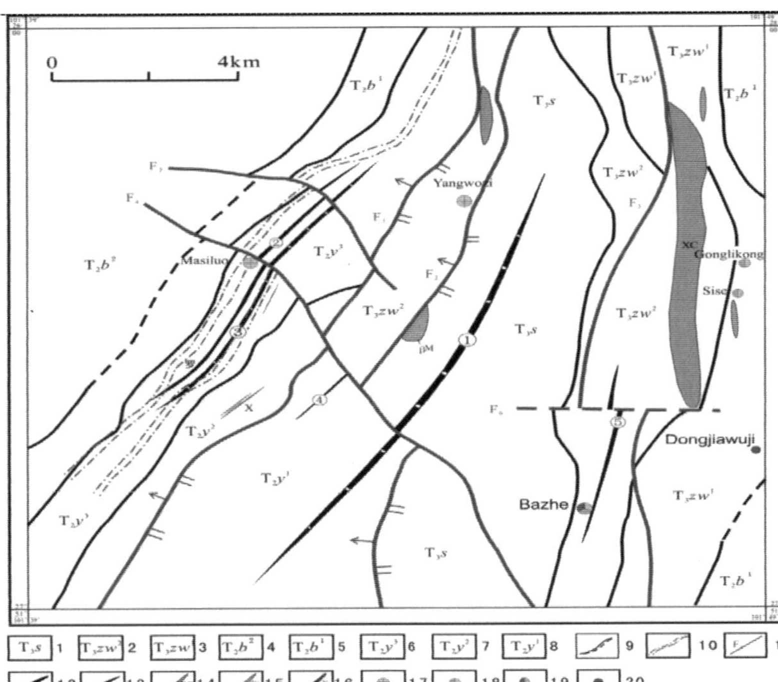

Figure 1. Regional geologic map of Masiluo- Zhangjiapingzi area.
1- Songgui formation; 2- The first menber of Zhongwozu formation; 3- The second member of Zhongwozu formation; 4- The second menber of Baishan formation; 5- The second menber of Baishan formation; 6- The third menber of Yantang formation; 7- The second menber of Yantang formation; 8- The first menber of Yantang formation; 9- Geological boundary; 10- Mylonite; 11- Fault; 12- Anticline; 13- Syncline; 14- Carbonatite dykes; 15- Diabase dikes; 16- Lamprophyre dikes; 17- Au occurrence; 18- Cu-Au occurrence; 19- Fe-Cu-Au occurrence; 20- Fe occurrence

3. Metallogeny of typical ore deposits

The Masiluo–Zhangjiapingzi area has a long history of mining and good metallogenic geological conditions. Based on incomplete statistics, there are 4 gold deposits, 33 gold or mineralized spots,

IOP Conf. Series: Earth and Environmental Science **59** (2017) 012017 doi:10.1088/1755-1315/59/1/012017

more than 20 Au geochemistry and mineral sand anomalies in this area, and there are also some massive alterations and mineralization which related to the process of gold mineralization [7, 8, 9]. Furthermore, uranium(U), rare-earth elements(REE), fluorite(CaF_2), and polymetallic deposits are distributed throughout the study area. Some substantial ore deposits in this area include the Masiluo gold deposit, Zhangjiapingzi gold deposit, Dongjia Wuji specular hematite deposit, Gonglikong Cu–Au deposit, and Bazhe Cu–Au–Fe deposit [10, 11].

(1) Masiluo gold deposit: A hydrothermal gold deposit, where the Au-mineralized stratum is the limestone layer of the Yantang formation, which has the highest Au grade in this region. To the north of the mine, the deposits comprise selective metasomatism bedding of altered rock ore or mineralized rocks, while the south comprises structurally altered rock ore. The wall rock alteration, comprising metallic mineral assemblages and ore-controlling factors of the Yangwozi Au-mineralized spot, was discovered to the east of the Masiluo gold deposit and is similar to the Masiluo deposit, also a hydrothermal gold deposit.

(2) Dongjia Wuji Iron deposit: The occurrence of orebody is strictly controlled by faulting. It is formed by deep magmatism-related hydrothermal fluid injection into extension faults.

(3) Gonglikong Cu–Au deposit: The Gonglikong Cu–Au deposit is a hydrothermal Cu–Au deposit. Cu–Au mineralization occurred in the shear fracture zone and is related to intermediate felsic magma fluid. The recent discoveries of Cu–Au mineralization in the Sishe and Gangou areas within this region have similar wall rock alteration and metallic mineral assemblages as the Gonglikong Cu–Au deposit, all of which are also potentially hydrothermal Cu–Au deposits.

(4) Bazhe Cu–Au–Fe deposit: The orebodies occur within the interlayer faults and intense cleavage zone, which are clearly controlled by faulting. It is a magmatic fluid– hydrothermal fluid superimposed reformation-type deposit. Hydrothermal-type siderite deposits were formed earlier by magmatism-related fluids, and later superimposed by hydrothermal Cu–Au mineralization. During the superficial oxidation period, part of the siderite and sulfides oxidized into limonite.

4. Metallogenic regularity

Mineralization of the Masiluo–Zhangjiapingzi area has distinctive zoning characteristics. The carbonatization–quartz vein-type gold ore series (I- Au series) is dominant in the west, typical of the Masiluo gold deposit and the Yangwozi Au-mineralized spot. The metallogeny underwent two stages of gold mineralization, namely carbonatization-related and pyrite–quartz vein-related. The hydrothermal Cu–Au ore series (II- Cu, Au series) dominates the northeastern part and is represented by the Gonglikong deposit and Sishe Cu–Au mineralized spot. The metallogeny underwent Cu–Au-bearing hydrothermal fluid activity along the fractures, a process in which the fluid fills up the joints beside the faults and forms Cu–Au ore veins. The southeastern part is dominated by a late-stage magmatic fluid–hydrothermal fluid with superimposed Fe, Cu, and Au ore series (III- Fe, Cu, and Au series), typical of the Dongjia Wuji specular hematite and the Bazhe Fe, Cu, and Au polymetallic deposits. The metallogeny was mostly characterized by deep-seated specular hematite–siderite hydrothermal fluid injection into extension faults, and later superimposed by hydrothermal Cu–Au mineralization.

The metallic ore deposits in this region are structurally controlled. Magma and ore-forming hydrothermal activities are characteristic of fault-developed areas, where intense mineralization areas are found on both sides of the strata of regional and secondary faults. Examples of such include the Yangwozi Au-mineralized spot distributed along the NNE-trending fault as well as the Gonglikong and the Sishe Cu–Au mineralized spots distributed along the NS-strike fault. Substantially mineralized areas are typical of secondary faults along the major faults. For example, the highly mineralized areas of the Masiluo Au deposit are along the F_1 and F_2 faults, while the strongest mineralized area is found at the intersection of the F_3 fault with the F_1 and the F_2 faults. At the same time, some extensive faults are ore-bearing ones such as the major orebodies of the Dongjia Wuji specular hematite and Bazhe hematite deposits.

Magmatism controlled the metallic deposit formation in this region. It constitutes ore-bearing magmatic fluid injection into extensive faults forming an injection-type rich orebody such as the

Dongjia Wuji specular hematite and Bazhe Fe–Cu–Au polymetallic deposits. They were formed via deep-seated, ore-bearing magmatic fluid injection into the nearby NS-strike extensive fault. Concomitantly, the hydrothermal fluid formed by magma that is rich in metal elements metasomatized and enriched the strata by leaving the ore-forming metal elements in the strata, such as the high Au-rich carbonatized strata in the gold mine of this region. The metasomatism of sedimentary strata by magmatic fluid from deep-seated carbonate magma intrusions caused the massive high Au background value in the carbonatized strata of this region. The results from the geochemistry survey indicated a close relation between the Au anomaly zone and the carbonatization zone in the Masiluo gold deposit. From a dimensional distribution perspective, the known occurrences of such hydrothermal deposits are closely related to large-scale, deep regional faults, and acidic rocks.

Strata control the formation of metallic ore deposits in this region. The gold deposits in this region primarily occur within the second limestone layer of the Triassic-aged Yantang Formation, indicating that mineralization is controlled by strata.

The brittle limestone has many good characteristics such as high chemical reactivity, high permeability, easy fracturing, which make the rock can be metasomatized and precipitated deposits from ore-bearing hydrothermal fluid. It is also beneficial to precipitation and migration of ore-bearing fluids. Among these properties it helps a major proportion of the economically viable gold orebodies to occurred in the altered marble.

5. Conclusion

The Masiluo–Zhangjiapingzi area is one of the most important and prosperous target areas for ore exploration in western Sichuan. The ore deposits of this region show distinctive zonal characteristics. The western part is a gold metallogeny zone typical of the Masiluo gold deposit and the Yangwozi gold mineralized spot. The northeastern part is a Cu–Au metallogeny zone typical of the Gonglikong Cu–Au deposit and the Sishe Cu–Au mineralized spot, whereas the Dongjia Wuji specular hematite deposit and the Bazhe Fe–Cu–Au polymetallic deposit typical of the Cu–Au–Fe metallogeny zone are present in the southeastern part. The ore deposits in the study area are controlled by strata, geological structures, and magmatism.

References

[1] Wang X C 2000 *Geology and Prospecting* **01** 19–21+24
[2] Wang X C and Zhang Z R 1999 *International Geology Review* **241** 1099–113
[3] Wang D H, Chen Y C, Luo Y N, Ying H L, Li Z W, Fu X F and Yang J M 2002 *Mineral Deposits* **21** 237–40
[4] Wang X C 1994 Zhang Y F 2003 Journal of Mineralogy and Petrology **11** 74–82
[5] Yan S H, Yang J M, Wang D H, Chen Y C and Xu Y 2002 *Acta Geologica Sinica* **76** 384–87
[6] Gao Z B and Wang X C 1992 *Mineral Deposits* **02** 97–105
[7] Teng Y G, Zhang C J, Ni S J and Wu X Y 2001 *Geotectonica et Metallogenia* **01** 195–101
[8] Mao J W, Chen Y C, Zhou J X and Yang B C 1995 *Acta Geoscientica Sinica* **03** 276–290
[9] Mao J W, Chen Y C and Li H Y 1997 *Acta Geoscientica Sinica* **04** 62–4
[10] Zhang Z B, Chen W H, Kang Y L, Yang Z F, Ju Z and Zhao F 2013 Sedimentary Geology and Tethyan Geology **03** 94–100
[11] Zhang Z M, Zhang Z B and Cui W L 2014 *Geotectonica et Metallogenia* **01** 131–9

Research history and prospecting potential of gold deposits in Dongchuan Area, Yunnan, China

Chaowen Huang[1, 2], Shugen Zhang[1, 2]

[1]Key Laboratory of Metallogenic Prediction of Nonferrous Metals and Geological Environment Monitoring, Ministry of Education, Central South University, Changsha 410083, China

[2]School of Geosciences and Info-Physics, Central South University, Changsha 410083, China

huangchaowen@csu.edu.cn

Abstract. The Dongchuan area in China is part of the Sichuan–Yunnan–Guizhou polymetallic metallogenic field and contains extensive ore-deposits. Although gold was first discovered in the area in the 1990s, gold exploration and research are currently limited; only five areas have currently been identified as containing gold deposits with reserves of more than 20 tons. However, the area is believed to have considerable prospecting potential. Gold deposits mainly occurring in the wall of the rock mass have been found in the upper-middle section of the Pingdingshan Group to the lower-bottom section of the Yinming Group in relation to the intrusive action and development of sodium diorite. It has been determined that the core of the anticline complex and its nearby fault zone are the major ore hosting structures and that gold-deposit prospecting should be conducted around the N–S compressive shear zone. In this study, sites are determined that have certain characteristics accompanied by a high degree of rock formation and complete assemblages of lithological composition; these are considered to be locations conducive to the prospecting of gold deposits.

1. Introduction

The Dongchuan area is the third largest ore-deposit production region in China. It is famous for Dongchuan-type ore deposits and its rich resources, attracting both local and international geological experts [1, 2, 3]. Since the Boka gold mine was discovered in the 1990s, geologists have had been working in the Dongchuan area and have gathered a substantial amount of geological information [4, 5, 6]. However, research on gold exploration in the Dongchuan area remains immature compared to research on copper exploration in the same region. The Dongchuan area's gold deposits are the main research focus of this article, which reviews the research history and analyzes future ore prospecting in order to provide further insights into gold exploration in the Dongchuan area.

2. Geological background

The Dongchuan area is in the middle of the southwestern margin of the Yangzi Plate, which has a complicated crustal movement evolution history [7, 8, 9]. Exposed strata in the area are mainly composed of low-grade green schist facies layers of the Mesoproterozoic Dongchuan and Kunyang groups. The Dongchuan area is bounded by the Xiaojiang fault to the east, the Puduhe fault to the west, the near E–W trending Baojiu fault to the north, and the Matang fault to the south, forming a trapezoidal rift basin.

The Dongchuan area is situated in the southern part of the Sichuan–Yunnan–Guizhou polymetallic metallogenic domain, so it has an excellent metallogenic geological setting. Because of its special crustal evolution and dynamic background, tectonic deformation in the area is extremely complex. Fold deformation is great, fault structures are extremely developed, and the area experiences multi-phase, multi-type magmatic activities [10].

3. History of gold exploration

The complicated topography, geology, and mineralization characteristics of the Dongchuan area have limited gold exploration in the area.

(1) During a period from the 1950s to the 1980s, the Dongchuan Mining Bureau's geological exploration team conducted basic geological and prospecting research of the Dongchuan area. They started by exploring the Yinmin, Luoxue, Tangdan, and other large-scale copper deposits. However, there are few reports of the discovery of independent gold deposits, only reports on gold deposits associated with copper deposits. As of 1988, the total reserve of gold-associated copper ores was 21.07 tons.

(2) During a period from the 1990s to 2000s, the Boka and Xiaoliukou gold mines were discovered in the area. In 2003, No. 807 geological brigade submitted its Dongchuan Area Gold Mine Research Report, showing D and E class geological reserves of more than 20 tons and an average gold ore grade of 7.95g/t.

(3) From 2003 to 2007, Yunnan Jinshan Mining Co. Ltd. initiated gold exploration research work in the Boka gold mine. Soil investigation detected 26 geochemical gold and copper anomalies and high-precision magnetic surveys detected 6 anomaly areas and 53 individual anomalies. The main gold and copper anomaly areas detected usually consisted of valuable gold and copper mineralized spots. Geochemical anomalies often become direct indicators for gold prospecting, whereas magnetic anomalies may be used as indirect prospect indicators.

(4) From December 2008 to August 2009, No. 809 geological brigade initiated a large-scale geological survey of gold deposits in the northern section, copper deposits in southern section, and multiple ore and mineralized spots found throughout the Boka mine.

(5) In 2011, Beijing Zhongxing Research Technology Co. Ltd. conducted research in the Yuanbaoshan District. They have completed a 1:2000 geological survey of 1.60 km^2, and through trenching and drilling, acquired an estimated total of 715,200 tons of ore and 5390.84 tons of copper metal.

As of 2005, related geological researches and prospecting analyses of the Dongchuan area have been published in academic journals, therefore strengthening the exploration prospect potential of the area. Previous researchers have had issues regarding the genesis type and occurrence conditions of gold deposits found in the area. The lack of a complete understanding of the genesis, ore-forming horizons, and metallogenic age of these deposits has restricted prospecting of the area.

4. Existing issues

The degree of exploration and research of gold deposits found in the Dongchuan area is relatively low. Gold exploration in the area still faces problems, such as understanding the area's basic geology, ore deposits, and metallogenic theory, as well as finding suitable research methods.

(1) The stratigraphic system and distribution of ore deposits remains unclear. The Dongchuan and Tangdan groups exposed in the Dongchuan area are barren strata with similar lithology types and changes. They are generally characterized by low-grade metamorphism with complex structural changes, thus there are many differences views on the stratigraphic horizons and distributions of reach area. Therefore, stratigraphic determination and generalization of its distributions are important steps for effective mineralization forecasting and ore prospecting.

(2) The formation period, degree of maturity, mechanical properties, and distribution pattern of structurally deformed by-products and the relationship between gold and copper mineralization found in the area have not been effectively determined. The Dongchuan area is situated in a structural unit with strong structural deformation and a complicated distribution. At present, gold mineralization found in the Dongchuan area is structurally controlled. Multi-group and multi-type faults and fractures

are important direct ore-hosting spaces. Therefore, studying the history of tectonic deformation, the period of deformation, and the deformation dynamic background is an important way to understand the ore-controlling structure pattern and for ore prospecting.
(3) There is a lack of effective research on the magmatic activity period, genetic type, diagenesis mechanism, and depositional pattern and their relationships with gold and copper mineralization. Investigation results show that gold mineralization in the area is associated with the distribution of multi-stage and multi-type magmatic activities. The lithofacies type and distribution of in-situ forms are complicated. They are mostly subjected to complex metamorphism and have undergone superficial changes. Most of the characteristics of magmatic rocks became undefined, affecting the correct determination of genesis, period of mineralization, material source, and pattern of in-situ mineralization.
(4) There is comparatively little research regarding the geological and mineralization characteristics of gold deposits. There are no reliable conclusions on mineralization theory and ore prospecting basic information, such as ore-formation age, genetic types, ore-controlling factors, and in-situ mineralization patterns. In the Dongchuan area, only the Xinshan-Majiagou gold mine has reached a detailed investigation and exploration stage. However, the geological mineralization environment, morphology and occurrence of ore deposits, and in-situ distribution patterns are not fully understood. It is difficult to effectively compare different parts of the mining area with ore deposits found in adjacent areas, in terms of their geological conditions, to establish a conceptual model of mineralization and a geological exploration model.

5. Prospecting potential

Good and effective prospecting work requires that we have a comprehensive understanding of the metallogenic elements of the study area. In addition to regional fault zones, development of volcanic rock mass, gold anomaly mineralization, and other geological evidence, this paper also focuses on the following:
(1) Stratigraphic position
In the Dongchuan area, gold and ore deposits are close to the bottom of the Yinmin Formation and its underlying strata. For example, the Xinma-Majiagou mining area has ore deposits in the Tangdan Group of the upper-middle section of the Pingdingsan Formation and in the Dongchuan Group of the lower Yinmin Formation. The Sichuan Huidong Xintian gold deposit, which is adjacent to the Xinshan-Majiagou mine, is also found in the Tangdan Group of the Pingdingshan Formation. Xiaoliukou gold deposits are found within the Dongchuan Luoxue-Yinmin area, with ore-hosting horizons in the underlying strata of the Yinmin Formation. Mineralization analysis of ore-hosting strata in the Boka gold deposit also supports this conclusion.
(2) Lithological structure
The Dongchuan and Tangdan groups are barren strata with similar lithology and changing patterns. They are generally characterized by low-grade metamorphism with complex structural changes, thus there are more differences in terms of stratigraphic horizons and distributions. Ore-hosting strata, such as those of the upper-middle section of the Pingdingshan Formation, have complex lithology, composition, and rock structure. Low-rigidity shale and silty shale, which are chemically inert and have low permeability, act as chemical barriers. The strata also contain lenticular crystalized sandy limestone and dolomitic sandstone that have high porosity and fracture connectivity, a high degree of rigidity, good permeability, and are chemically active. Additionally, the contact between the Yinmin and Pingdingshan formations is a small-angled unconformity. The Pingdingshan Formation is an important source of fluid/rock reaction-extraction for ore-forming material and acts as an ore-forming fluid transport channel, allowing physiochemical property exchange and aggregation precipitation to occur.
The mid-upper part of the Pingdingshan Formation to the lower part of the Yinmin Formation is the main ore-hosting horizon for gold deposits and is important for prospecting indicators of the gold deposits. The higher the degree of development of related rock strata, the more complete the lithologic assemblage types and the more conducive it is for the enrichment of gold mineralization. It is also an

important reference index for identifying the degree of enrichment of gold mineralization. At the same time, the development of the core of the anticline complex and the near N–S compression-shear fracture zone is also a necessary condition for the spatial location of gold deposits. Intersections between fault zones and the interface layers between the Dongchuan and Tangdan groups are the most favorable structural sites for mineralization enrichment of gold deposits. In these areas, sodium diorite magma intruding along the fracture zone is the basic condition of hydrothermal gold mineralization. Small rock-body or larger-scale dike development sites are central for hydrothermal gold mineralization activity and are the sites of gold mineralized enrichment.

6. Conclusion

The Dongchuan area is one of the most important copper deposit areas in China. It is also a part of the Yunguichuan polymetallic metallogenic province. Since the discovery of gold in the 1990s, geologists have been working in this area, gathering a substantial amount of geological information and producing large amounts of literature. However, most of the ore (chemical) sites have not been subjected to systematic evaluation work, thus the presence of deeply-hosted or potentially-hosted ore deposits remain unclear. Gold mineralization in the Dongchuan area is related to the intrusive action and development of sodium diorite and is controlled by complex anticlines and faults. Gold-deposit prospecting should be carried out around the N–S compressive shear zone. Sites with the characteristics mentioned in this study and that are accompanied by high degrees of rock formation and complete assemblages of lithological composition will be locations conducive to prospecting gold deposits.

References

[1] Hua R M, Wan H C and Ni P 1993 *Contributions to Geology and Mineral Resources Research* **04** 1–8
[2] Liu J S Wu, Y Z and Duan J R 1996 *Journal of Central South University of Technology* **27** 8–12
[3] He Y T 1996 *Yunnan Geology* **15** 319–29
[4] Zhang Y F 2003 *Yunnan Geology* **04** 360–70
[5] Li Z W, Qian X G and Tian M 2000 *Geotectonica et Metallogenia* **S1** 37–43
[6] Zhou B G, Lin M, Guo R, Wang Z Z, Shen Z W and Luo M J 2013 *Sedimentary Geology and Tethyan Geology* **01** 93–8
[7] Zhao X F, Zhou M F, Li J W, Sun M, Gao J F, Sun W H and Yang J H 2010 *Precambrian Research* **182** 57–69
[8] Zhu H P, Fan W Y, Zhou B G, Wang S W, Luo M J, Liao Z W and Guo Y 2011 *Geological Journal of China Universities* **17** 452–461
[9] Yin F G, Sun Z M, Ren G M and Wang D B 2012 *Acta Geologica Sinica* **12** 1917–1932
[10] Xue B G 2005 *Yunnan Geology* **03** 243–53

IOP Conf. Series: Earth and Environmental Science **59** (2017) 012019 doi:10.1088/1755-1315/59/1/012019

Real-time crankshaft angular speed tracking and indicated torque estimation via optimized Luenberger sliding mode observer

Y Zhang[1], R Tan[1], T X Zheng[1,3], T L Zhou[2], W M Han[2] and Y J Wang[2]

[1]School of Advanced Manufacturing Engineering, Chongqing University of Posts and Telecommunications, Chongqing 400065, China
[2]School of Automation, Chongqing University of Posts and Telecommunications, Chongqing 400065, China

[3]zhengtx@cqupt.edu.cn

Abstract. The interest in engine indicated torque estimation plays an important role in the automotive industry. In this study, an optimized Luenberger sliding mode observer is proposed based on easily available crankshaft angular speed of a four-cylinder spark ignition (SI) engine. Especially, the new observer is applied to track crankshaft angular speed and estimate engine indicated torque in New European Drive Cycle (NEDC). Convergence is proven through Lyapunov stability theory. The experimental results show that the proposed estimated technique can effectively track speed and has a higher accuracy in steady state.

1. Introduction

The engine indicated torque is one of the most significant performance parameters of internal combustion engines (ICEs). The real-time indicated torque monitoring can grasp the running condition of the vehicle and improve the working performance of engine [1]. The engine indicated torque has been widely applied in diagnostic field including misfire detection, combustion analysis and torque balancing [2-3], etc. Moreover, the importance of engine indicated torque can be discovered by the projected increase in vehicle related emissions worldwide [4]. Actual engine torque is usually hard to be measured due to the deficiencies of high cost and inconvenient of torque sensors. Hence, in order to reconstruct indicated torque based on crankshaft angular speed which is easier and cheaper to measure, an observer can be used instead of expensive sensors [5].

Numerous torque estimation techniques have been addressed previously in the literature [3, 6-7]. Most of the proposed approaches have their foundations on observer design including discrete non-stationary linear approximation observer [3], discrete angular domain observer [6], nonlinear torque observer [7], etc. Further, sliding mode observer has a good robustness and accuracy in real-time torque estimation [1]. However, the problem in dealing with dynamic torque estimation is needed to solve. Other focuses on filtering algorithm by using crankshaft speed [5], such as Kalman filtering [2].

For the design of an observer of engine indicated torque, a stiff crankshaft model is used based on tracking error of crankshaft angular speed. An optimized Luenberger sliding mode observer (OLSMO) is proposed and validated both theoretically (the convergence of observer) and experimentally (on the four-cylinder engine). Compared with sliding mode observer (SMO) and Luenberger sliding mode observer (LSMO), OLSMO has a higher accuracy in tracking speed and it can also effectively improve the dynamic performance of estimation of torque in transient condition. Additionally, the designed OLSMO greatly weakens system chattering.

2. Stiff crankshaft model

In this part, the dynamic system of engine stiff crankshaft model is briefly described. According to the experience equation of SI engine proposed by Crossley and Cook [8], the engine torque, T_{eng}, produced by the engine cylinders can be expressed as

$$\begin{aligned} T_{eng} &= -181.3 + 379.36m_a + 21.91(A/F) - 0.85(A/F)^2 + 0.26A_f + 0.0028A_f^{\,2} \\ &+0.027\dot{\theta} - 0.000107\dot{\theta}^2 + 0.00048\dot{\theta}A_f + 2.55A_f m_a - 0.05A_f^{\,2}m_a \end{aligned} \tag{1}$$

where A/F is air to fuel ratio, A_f is spark advance (degrees before top-dead-center), θ is the crank angle and m_a is the mass of air entering into intake manifold for combustion in cylinder. The engine torque less the net load torque results in acceleration.

$$J_e\ddot{\theta} = T_{eng} - T_{load} \tag{2}$$

where J_e is engine crankshaft effective rotational moment of inertia. In this study, T_{load} is engine load torque, which can be approximately described as three main factors including aerodynamic resistance torque T_{aero}, road grade and rolling resistance torque T_β, and friction brake torque T_{fb} [9]. Further, the rotation dynamic equation of the crankshaft can also be described as

$$J_e\ddot{\theta} = T_{ind} - T_r - T_{frc} - T_{load} \tag{3}$$

where T_{ind} is engine indicated torque. T_r is reciprocating inertia torque of piston and part of the connecting rod. T_{frc} is the average friction torque created by sliding friction of piston and ring and the pumping action of the engine. If we define $x_1 = \theta, x_2 = \dot{\theta}$, then state space can be obtained

$$\dot{x}_1 = x_2 \tag{4}$$

$$J_e\dot{x}_2 = T_{eng} - T_{load} \tag{5}$$

Substitute equation (1) into the state space, and the formula (5) can be represented as

$$\begin{aligned} J_e\dot{x}_2 &= -181.3 + 379.36m_a + 21.91(A/F) - 0.85(A/F)^2 + 0.26A_f + 0.0028A_f^{\,2} \\ &+0.027x_2 - 0.000107x_2^{\,2} + 0.00048x_2A_f + 2.55A_f m_a - 0.05A_f^{\,2}m_a - T_{load} \end{aligned} \tag{6}$$

3. Engine indicated torque estimation

In this study, the sliding surface is chosen as $s = x_2 - \hat{x}_2$, where $\hat{x}_2$ is the estimation of the measured speed x_2. Based on the principle of sliding mode variable structure, sliding surface should meet the accessibility when $s\dot{s} < 0$. In order to satisfy the sliding mode condition, the effective injection signal is necessarily chosen. During the sliding mode motion, discontinuous injection signals can make the well-chosen switching function equal to zero. Therefore, the system yields excellent robustness to disturbance signals and uncertainty. Additionally, the better sharp saturation function is applied to weaken chattering of switching function in tracking crankshaft angular speed. Meanwhile, in order to ensure that the switching function has a good performance of fast convergence, sliding mode gain should be as big as possible. However, overlarge value of sliding mode gain will intensify chattering of switching function and decrease accuracy in tracking speed. Hence, a Luenberger observer is further introduced, which can accelerate estimating speed of estimated states. Then, we can improved injection signal from the derivative of switching function

$$J_e\dot{s} = \dot{x}_2 - \dot{\hat{x}}_2 = E(s) - \left[Ls + \mathrm{K}sat(s/\varepsilon)\right] \qquad (\mathrm{K}, L>0) \tag{7}$$

where $E(s)$ represents error function between designed observer and stiff crankshaft model. K and L are sliding mode gain and Luenberger gain respectively. The second item $Ls + \mathrm{K}sat(s/\varepsilon)$ is a designed injection signal. $sat(\bullet)$ is a saturation function. And ε is a positive constant, the slope of the linear part in $sat(s/\varepsilon)$ is $1/\varepsilon$ that should be selected large enough to approach to sign function. According to the estimated speed $\hat{x}_2$, the crankshaft dynamics in equation (6) can be rewritten as

$$J_e\dot{\hat{x}}_2 = -181.3 + 379.36\hat{m}_a + 21.91(A/F) - 0.85(A/F)^2 + 0.26A_f + 0.0028A_f^{\ 2} + 0.027\hat{x}_2 - 0.000107\hat{x}_2^{\ 2} + 0.00048\hat{x}_2 A_f + 2.55A_f\hat{m}_a - 0.05A_f^{\ 2}\hat{m}_a - T_{load} + Ls + \mathrm{K}sat(s/\varepsilon) \quad (8)$$

Substitute equation (8) into the equation (3), estimated indicated torque can be obtained

$$\hat{T}_{ind} = -181.3 + 379.36\hat{m}_a + 21.91(A/F) - 0.85(A/F)^2 + 0.26A_f + 0.0028A_f^{\ 2} + 0.027\hat{x}_2 - 0.000107\hat{x}_2^{\ 2} + 0.00048\hat{x}_2 A_f + 2.55A_f\hat{m}_a - 0.05A_f^{\ 2}\hat{m}_a + \hat{T}_r + \hat{T}_{frc} + Ls + \mathrm{K}sat(s/\varepsilon) \quad (9)$$

Based on Lyapunov stability theory, if characteristic matrix of Luenberger observer satisfies Hurwize condition, the gradual stability is assured. Additionally, a Lyapunov function exists when error dynamic system of sliding mode observer meets $s\dot{s} < 0$, it is concluded that the error dynamics of designed observer is stable and model states can ultimately converge to the actual system states.

4. Experimental results and discussion

In the experiment, a four-cylinder SI engine (details listed in Table 1) is applied in real-time torque estimation study. In order to verify the proposed observer, a dynamic engine platform is conducted in an unsteady condition and it can be used to measure crankshaft angular speed. The full running condition of vehicle is set as NEDC, which has been widely applied to exam transient emission. Additionally, the crankshaft speed signal is extracted by crankshaft position sensor, and sampling frequency of signal is chosen to 20 kHz. The values of observer gain are chosen as $\mathrm{K} = 60,\ L = 120$.

Table 1. Features and specifications of turbocharged SI engine

Parameter	Value and Unit	Parameter	Value and Unit
Engine Type	4-stroke in-line	Engine Capacity	2.0 L
Compression Ratio	10.5:1	Bore	83 mm
Crank Radius	93/2 mm	Connecting Rod Length	145 mm

According to previous study, the comparison of performance of the designed observer (OLSMO, LSMO, SMO) is presented in next few figures. The tracking results of crankshaft angular speed are given in Figure 1, which displays a good overall performance of OLSMO in speed tracking during 25 seconds under NEDC condition and estimated speed can follow the actual speed signal well without strong chattering. However, dynamic speed tracking result of SMO is bad. Additionally, the speed tracking error of OLSMO is given in Figure 2. The steady state error (during 7 to 10 seconds) is closed to zero, and the speed tracking error is only 3% in transient state (such as 11 to 15 seconds). Therefore, Figure 2 indicates that the proposed observer possesses good stability.

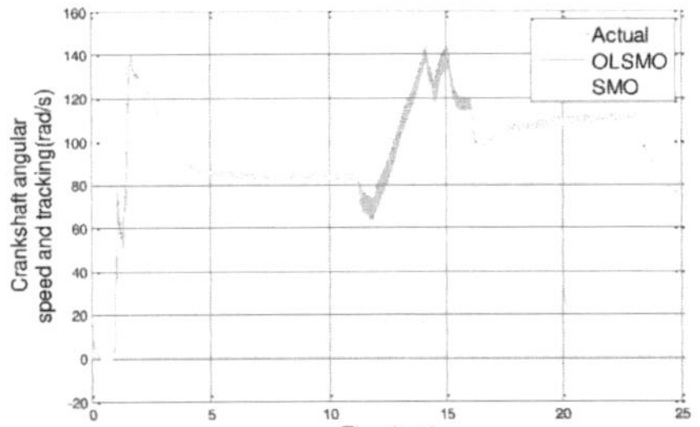

Figure 1. Speed tracking results

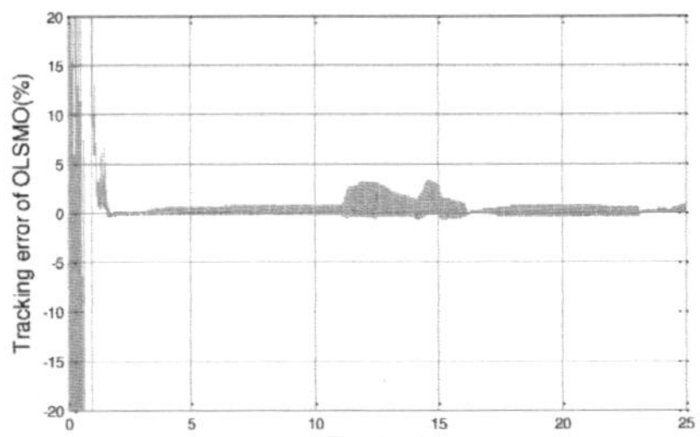

Figure 2. Speed tracking error of OLSMO

It is observed that the SMO cannot reconstruct actual engine indicated torque in transient working condition in Figure 3, but the estimated indicated torque of the SI engine has a better effect using OLSMO than SMO. Compared with LSMO and SMO, the designed OLSMO in Figure 4 can effectively estimate the indicated torque in steady condition. Further, OLSMO greatly weakens the severe chattering existed in the LSMO and SMO. So it is concluded that the proposed OLSMO has better dynamic performance in estimating indicated torque of engine than LSMO and SMO.

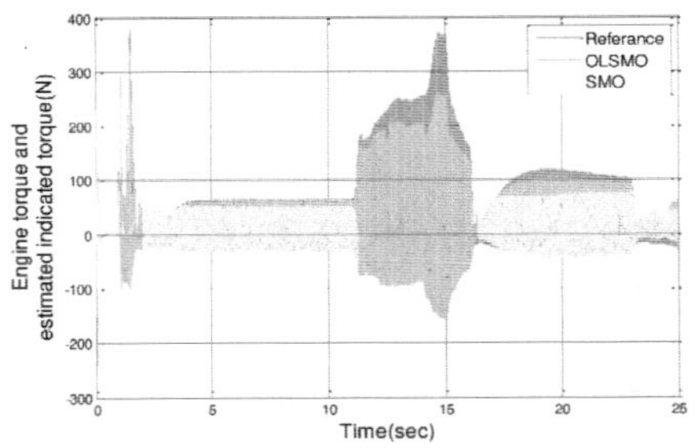

Figure 3. Indicated torque estimation results

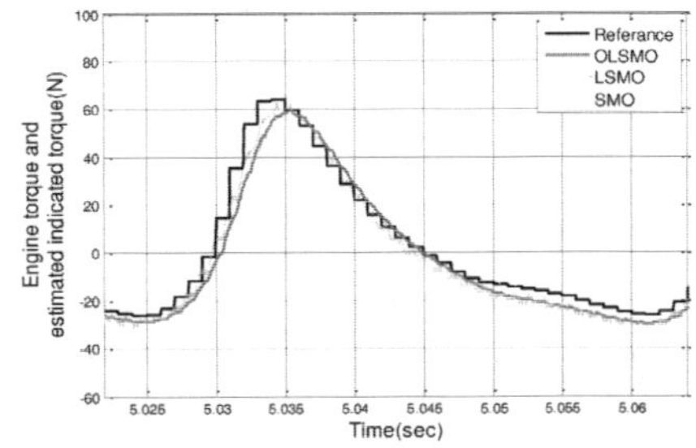

Figure 4. Detail in torque estimation

5. Conclusion

This study presents an optimized Luenberger sliding mode observer that can be applied to estimate indicated torque of SI engine. Compared with LSMO and SMO, OLSMO achieves a higher accuracy in tracking crankshaft angular speed and it can also effectively reconstruct engine indicated torque in non-stationary working condition. Moreover, the designed OLSMO greatly weakens system chattering. Furthermore, we plan to develop the estimated torque which can be used to misfire detection, torque balancing and combustion analysis. Additionally, the estimation accuracy of engine indicated torque is also needed to be considered.

Acknowledgments

This work is jointly supported by Scientific and Technological Innovation Program of Chongqing (Grant No. CYS16168), Chongqing Natural Science Foundation (Grant No. CSTC2014JCYJA60003), Chongqing Natural Science Foundation of Education Commission (Grant No. KJ120511).

References

[1] Wang Y S, Chu F L 2005 Real-Time Misfire Detection via Sliding Mode Observer *Mechanical Systems and Signal Processing* **19**(4) pp 900-912

[2] Helm S, Kozek M, Jakubek S 2012 Combustion Torque Estimation and Misfire Detection for Calibration of Combustion Engines by Parametric Kalman Filtering *IEEE Transactions on Industrial Electronics* **59**(11) pp 4326-4337

[3] Chauvin J, Petit N, Rouchon P, Corde G, Moulin P, Castagne M 2004 Observer Design for Torque Balancing on a DI Engine *SAE Technical Paper*

[4] Fontaras G, Samaras Z 2010 On the Way to 130g CO2/km-Estimating the Future Characteristics of the Average European Passenger Car *Energy Policy* **38**(4) pp 1826-1833

[5] Stotsky A 2008 New Filtering Technique for on-Board Torque Estimation *International Journal of Automotive Technology* **9**(3) pp 289-295

[6] Losero R, Lauber J, Guerra T M 2015 Discrete Angular Torque Observer Applicated to the Engine Torque and Clutch Torque Estimation via a Dual-Mass Flywheel *Proceeding of IEEE 10th Conference on Industrial Electronics and Applications* pp 1020-1025

[7] Kim J S, Choi S B 2015 Nonlinear Torque Observer for a Self-Energizing Clutch Actuator Using a Reaction Torque Estimation Approach *IEEE/ASME Transactions on Mechatronics* **20**(6) pp 3071-3084

[8] MathWork 1998 Simulink/Stateflow Technical Example - Using Simulink and StateflowTMin Automotive Application *Electronic Version of the Report*

[9] Vahidi A, Druzhinina M, Stefanopoulou A, Peng H 2003 Simultaneous Mass and Time-Varying Grade Estimation for Heavy-Duty Vehicles *Proceedings of American Control Conference* **6** pp 4591-4956

IOP Conf. Series: Earth and Environmental Science **59** (2017) 012020 doi:10.1088/1755-1315/59/1/012020

Study on viscosity of MDEA- MeOH aqueous solutions

F Wang[1], L M Wang, S Q Wang and D Fu

[1] Corresponding author. Address: School of Environmental Science and Engineering, North China Electric Power University, Baoding, 071003, People's Republic of China. Tel.: +86-312-1593575212; E-mail addresses:464750393@qq.com

Abstract. The viscosities of the N-methyldiethanolamine (MDEA)-methanol (MeOH) aqueous solutions were measured at temperatures ranging from (303.2 to 323.2) K. The mass fraction of MDEA and MeOH respectively ranged from 0.2 to 0.4 and 0 to 0.15. On the basis of experimental measurement, the effects of temperature, mass fraction of MDEA and MeOH on viscosities were demonstrated.

1. Introduction

In recent decades, atmospheric levels of CO_2 have increased rapidly due to the utilization of grand amount of fossil fuel. The reduction of CO_2 emission became a global issue [1, 2]. Currently, aqueous solutions of alkanol amines are popularly used for the removal of CO_2 from a variety of gas streams [3-5]. MDEA takes the advantages of high resistance to thermal and chemical degradation, low solution vapor pressure, and low enthalpy of absorption [6, 7]. However, the absorption of CO_2 in MDEA aqueous solution is quite slow, in industrial applications this shortcoming can be overcome by adding activators, including MEA, piperazine (PZ), or DEA in MDEA aqueous solution [8-9]. Besides, adding a physical solvent in MDEA aqueous solution is considered to be an effective method to lower the regeneration energy cost [10, 11].

The use of methanol as physical solvent in these kinds of chemical absorption processes could be interesting. *Henni and Mather* [12] studied the solubility of CO_2 in a mixed nonaqueous solvent of MDEA and MeOH. The results showed that at low acid gas partial pressures the solubility of CO_2 is higher in the mixed solvent than in pure MeOH.

The viscosities of alkanol amine aqueous solutions are required when designing or simulating an absorption column for CO_2 absorption. In particular, solution viscosity is important in the mass transfer rate modeling of absorbers and regenerators because these properties affect the liquid film coefficient for mass transfer. By far, there are some experiments concerning the viscosities of aqueous solutions containing MDEA and its blends [13-16]. However, the experiments for the viscosities of MDEA-MeOH aqueous solutions are rare.

The main purposes of this work are to experimentally determine the viscosities of MDEA-MeOH aqueous solution and demonstrate the effects of temperature and mass fractions on the viscosity.

2. Experimental Section

2.1. Materials

The MDEA and MeOH were purchased from HuaXin Chemical Co, with mass purity$\geq$ 99.5%. They were used without further purification. Aqueous solutions of MDEA-MeOH were prepared by adding doubly distilled water.

2.2. Apparatus and procedure

The viscosities of the MDEA-MeOH aqueous solutions were measured at temperatures from 303.2 K to 323.2 K by using the NDJ-5S digital rotational viscometer produced by the Shanghai Changji Geological Instrument factory. The measurement range and measurement error for viscosity are respectively (0.1 to 105) mPa·s and ±1% (for a Newtonian fluid). The instrument was designed and manufactured with 5 cylindrical rotors (0#, 1#, 2#, 3# and 4#) and 4 different velocities (6 rpm, 12 rpm, 30 rpm and 60 rpm). In our experiments, the combination (0-60) rpm correspond to the upper limits of 10 mPa s.The uncertainty of the temperature is ±0.1 K.

3.Results and discussion

The experimental results for the viscosities of MDEA-MeOH aqueous solutions are shown in Table 1.

Table 1. Viscosities (η) of MDEA-MeOH aqueous solutions under different w_{MDEA} and w_{MeOH}. Pressure (p) = 101 kPa.

w_{MDEA}	w_{MeOH}	η / (mPa·s)		
		303.2K	313.2K	323.2K
0.20	0.00	1.72	1.36	1.14
	0.05	1.93	1.56	1.28
	0.10	2.18	1.71	1.38
	0.15	2.10	1.57	1.30
0.30	0.00	2.81	2.02	1.71
	0.05	2.96	2.20	1.80
	0.10	3.24	2.44	2.00
	0.15	3.02	2.32	1.88
0.40	0.00	4.24	3.13	2.31
	0.05	4.77	3.48	2.64
	0.10	5.18	3.69	2.78
	0.15	5.02	3.53	2.67

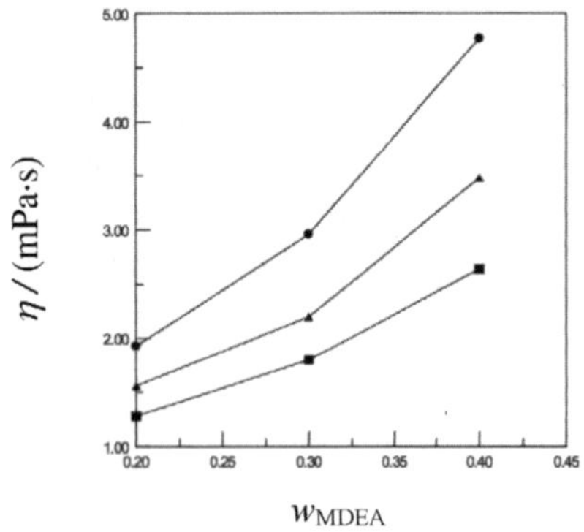

Figure 1. w_{MDEA} dependence of the viscosities of MDEA-MeOH aqueous solutions at w_{MeOH}= 0.05. Symbols: experimental data; ▲T=303.2K; ●T=313.2K; ■T=323.2K; lines: for guiding the eyes.

Figure 1 shows the influence of w_{MDEA} on the viscosities of MDEA-MeOH aqueous solutions. One may find that at given temperature and w_{MeOH}, the viscosity increases with the increasing of w_{MEDA}.

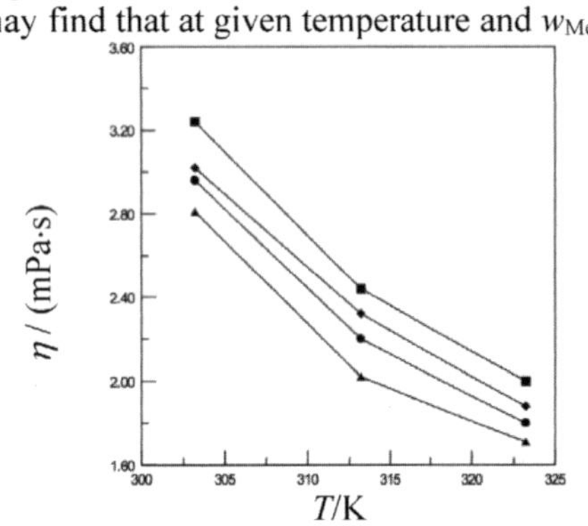

Figure 2. Temperature dependence of the viscosities of MDEA-MeOH aqueous solutions at w_{MDEA}= 0.20. Symbols: experimental data; ▲w_{MeOH}=0; ●w_{MeOH}=0.05; ■w_{MeOH}=0.10; ◆ w_{MeOH}=0.15. Lines: for guiding the eyes.

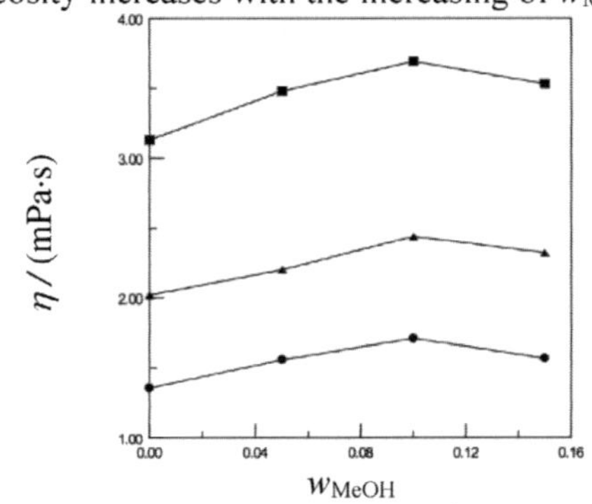

Figure 3. w_{MeOH} dependence of the viscosities of MDEA-MeOH aqueous solutions at T=313.2K. Symbols: experimental data; ●MDEA=0.2; ▲MDEA=0.3; ■MDEA=0.4. Lines: for guiding the eyes.

Figure 2 shows the temperature dependence of the viscosities of MDEA-MeOH aqueous solutions. One may find that at given w_{MeOH} and w_{MDEA}, the viscosity decreases with the increase of temperature. Figure 3 shows the influence of w_{MeOH} on the viscosities of MDEA-MeOH aqueous solutions. One may find that at given temperature and given w_{MDEA}, the viscosty firstly increase and then decrease, one maximum was observed at w_{MeOH}=0.10.

4.Conclusions

In this work, the viscosities of MDEA-MeOH aqueous solutions were measured. The effects of temperature and mass fractions of MDEA and MeOH on the viscosities were demonstrated. Our results show that:

(1) The solution viscosity increases with the increase of w_{MDEA} at given temperature and given w_{MeOH}, and decreases with the increase of temperature at given solution composition;

(2) At given temperature and given w_{MDEA}, the viscosty firstly increase and then decrease, one maximum was observed at w_{MeOH}=0.10.

Acknowledgments
The authors appreciate the financial support from the NationalNatural Science Foundation of China (No. 21276072) and the Fun-damental Research Funds for the Central Universities (No.13ZD16 and No.2016XS108).

References
[1] Nahicenovic N and John A 1991 *Energy* **16** 1347–1377
[2] Steeneveldt R, Berger B and Torp TA 2006 *Chem. Eng. Res. Des.* **84** 739–763
[3] Knudsen JN, Jensen JN and Vilhelmsen PJ 2009 *Energy Proc.* **1** 783–790
[4] Raynal L, Bouillon PA and Gomez A 2011 *Chem. Eng. J.* **171** 742–752
[5] Chakravarty T, Phukan UK and Weilund RH 1985 *Chem. Eng. Prog.* **81** 32–36
[6] Pant KK and Srivastava VK 2007 *Chem. Eng. J.* **133** 229-237
[7] Navaza JM, Gómez-Díaz D and Rubia MDL 2009 *Chem. Eng. J.* **146** 184-188
[8] Kadiwala S, Rayer AV and Henni A 2010 *Fluid Phase Equilib.* **292** 20–28
[9] Derks PWJ, Hogendoorn JA and Versteeg GF 2010 *J. Chem. Thermodyn.* **42** 151–163
[10] Fu D, Chen LH and Qin LG 2012 *Fluid Phase Equilibria* **319** 42-47
[11] Fu D, Xu Y and Hua X 2012 *Fluid Phase Equilibria* **314** 121–127
[12] Henni A and Mather AE 1995 *J. Chem. Eng. Data* **40** 493-495
[13] Muhammad A, Mutalib MIA and Wilfred CD 2008 *Journal of Chem. & Eng. Data* **53** 2226-2229
[14] Fu D, Chen LH and Qin LG 2012 *Fluid Phase Equilibria* **319** 42-47
[15] Fu D, Hao H and Qin L 2013 *Journal of Molecular Liquids* **181** 105-109
[16] Fu D, Qin L and Hao H 2013 *Journal of Molecular Liquids* **186** 81-84

The effect of monoculture peanut and cassava/peanut intercropping on physical and chemical properties in peanut rhizosphere soil under the biochar application and straw mulching

X CHEN, Y TIAN, X F GUO, G K CHEN, H Z HE, H S LI

Key Laboratory of Agro-Environment in the Tropics, Ministry of Agriculture, P.R. China, South China Agricultural University; Guangdong Engineering Research Center for Modern Eco-agriculture and Circular Agriculture, Guangzhou 510642, China

lihuashou@scau.edu.cn

Abstract. Cassava/peanut intercropping is a popular cultivation method in the south China, with the advantage of apparent yield increase. In order to analyze the effect of cassava/peanut intercropping on physical and chemical properties in peanut rhizosphere soil, the physical and chemical properties were investigated under the biochar application and straw mulching. The result showed that the Ph, organic materials content, available phosphorus content, available potassium content in peanut rhizosphere under the biochar application increased by 7.06%, 94.52%, 17.53%, 25.08% (monoculture peanut) and 8.47%, 89.94%, 17.93%, 22.87% (cassava/peanut intercropping) compared with Ck in the same planting patterns. In addition, the available nitrogen content, organic materials content, available phosphorus content, and available potassium content in peanut rhizosphere under the straw mulching increased by 89.80%, 60.92%, 5.95%, 9.98% (monoculture peanut) and 67.09%, 52.34%, 6.96%, 11.94% (cassava/peanut intercropping) compared with Ck in the same planting patterns. In the same treatment conditions, bulk density in peanut rhizosphere soil decreased and porosity and saturated permeability coefficient increased slightly. But there was no significant difference between the two. At the same time, cassava/peanut intercropping could increase soil nutrients. Therefore, it is beneficial to apply biochar and straw mulching, and the suitable intercropping row spacing is more beneficial to increase soil nutrient contents.

1. Introduction

The sustainable utilization of farming land is the base of actualizing sustainable development of agricultural production, and the basis of the agricultural development relies on plowing. Plenty of scientific researches[1-2] indicated that a variety of conservation tillage could improve organic matter content in the soil and soil structure, which was represented by biochar and straw mulching.

Peanut/cassava intercropping is a special planting pattern in Southern China. It can improve multiple crop indexes and the utilization ratio of arable land and it can also increase planting benefit. Currently, some studies indicated that peanut/cassava intercropping could raise the output. However, the influence of soil micro-ecological environment is still unknown. Therefore, the effect of monoculture peanut and cassava/peanut intercropping on physical and chemical properties in peanut rhizosphere soil under the biochar application and straw mulching were studied in this paper to offer the theory support and practice guidance for improvement of soil structure in Guangzhou.

2. Materials and methods

2.1 Test materials

The experiment was carried out at the end of March 2016 to the end of July in Zengcheng research base of South China Agricultural University (113°29′4″~113°59′44″E, 23°4′42″~23°37′20″N) and climate was typical marine monsoon climate. The Yuyou 7 (Peanut) and Huanan 205 (Cassava) was tested in the field experiment. Biochar was from Sanli corporation, and straw was from the experimental base of South China Agricultural University. At the beginning of experiment, the average value of pH in soil was 5.5, the content of organic matter was 12.13mg/kg, the content of total nitrogen was 317.60mg/kg, the content of total phosphorus was 292.68mg/kg, the content of total potassium was 2236.61kg/kg, the content of alkali hydrolyzable nitrogen was 27.13mg/kg, the content of available phosphor was 1.14mg/kg and the content of available potassium was 64.14mg/kg.

2.2 Test design

The trial began on July 8, 2016, and ended on March 8, 2016, four months, in Zengcheng research base of South China Agricultural University. Three treatments were established, including blank control (Ck), straw mulching (4500kg/ hm^2) and biochar application (1000kg/ hm^2). Plot area was $4m^2$ with the random arrangement and three replications. P, K and 70%N were applied as base fertilizers. The application of biochar and straw were directly covered on the soil surface. Four rows peanuts were planted in the peanut monoculture treatment, where the distance from the edge was 10cm, the distance of the edge row was 23cm while that of the internal row was 40cm. A peanut seed per hole was planted at a distance of 20cm. Peanut intercropping included one-row cassava and two-rows peanuts, cassava row was in the middle of the plot, peanut row was on the edge of the plot with a distance of 10cm to edge, and the distance between peanut and cassava was 40cm. A cassava seedling was planted at a distance of 1m, and a peanut seed per hole was planted at a distance of 20cm. At the end of the trial period, collecting the rhizosphere of peanut. The soil was air-dried and sifted.

2.3 Analysis and the method of determination

2.3.1 Determination of soil physical properties. Soil bulk density was determined by cutting-ring method[4]. Porosity was calculated by the bulk density and specific gravity of the soil[4]. Soil saturated permeability coefficient was calculated according to Darcy's theorem[4].

2.3.2 Determination of soil chemical properties. Soil pH was measured by the electric potential method and the water-soil ratio was 2.5:1[3]. Available nitrogen content adopted alkaline hydrolysis diffusion[4]. Organic matter content adopted potassium dichromate-volumetric method[4]. Available phosphorus content adopted extraction-molybdenum antimony colorimetric method[4]. Available potassium content was determined by flame atomic absorption spectrophotometry[4].

2.4 Data analysis

Using Excel 2007 to analyze the data and draw the chart and the data in all graphs was the average of the three repetitions. All the data were treated with SPSS 17.0 statistical package, The descriptive statistic analysis was used according to the type of the data, significance test used t-test and variance analysis.

3. Result analysis

3.1 The effect of monoculture peanut and cassava/peanut intercropping on physical properties in peanut rhizosphere soil under the biochar application and straw mulching

Table 1 showed that in the same planting patterns, biochar application and straw mulching could reduce bulk density in peanut rhizosphere soil and increase porosity in peanut rhizosphere soil. Furthermore, to a certain extent, both could improve the permeability in peanut rhizosphere soil. The results showed that bulk density and porosity and saturated permeability coefficient didn't reach a significant level between monoculture peanut and cassava/peanut intercropping($P>0.05$).

Table1. The effect of monoculture peanut and cassava/peanut intercropping on physical properties in rhizosphere soil under the biochar application and straw mulching

Planting pattern	Treatments	Bulk density (g/cm^3)	Porosity (%)	Saturated permeability coefficient ($10^{-3}cm/s$)
Monoculture peanut	Ck	1.239±0.061a	61.627±0.939b	2.650±0.327c
	Biochar application	0.884±0.054b	67.987±0.704a	8.117±0.622a
	Straw mulching	0.952±0.024b	66.500±0.576a	4.783±0.510b
Cassava/peanut intercropping	Ck	1.203±0.117a	62.533±0.887b	2.673±0.100c
	Biochar application	0.840±0.089b	68.420±0.997a	8.307±0.169a
	Straw mulching	0.857±0.095b	67.660±0.885a	4.953±0.074b

3.2 The effect of monoculture peanut and cassava/peanut intercropping on chemical properties in peanut rhizosphere soil under the biochar application and straw mulching

3.2.1 The effect of monoculture peanut and cassava/peanut intercropping on pH in peanut rhizosphere soil under the biochar application and straw mulching. Figure 1 showed that in the same planting patterns, pH of the plots of biochar application in peanut rhizosphere soil increased by 7.06%(monoculture peanut) and 8.47%(cassava/peanut intercropping) respectively compared with Ck($P<0.05$). PH of the plots of straw mulching in peanut rhizosphere soil reduced by 9.78% (monoculture peanut) and 8.46%(cassava/peanut intercropping) respectively compared with Ck($P<0.05$). Furthermore, under the same treatment conditions, pH in peanut rhizosphere under intercropping was lower than that under monoculture.

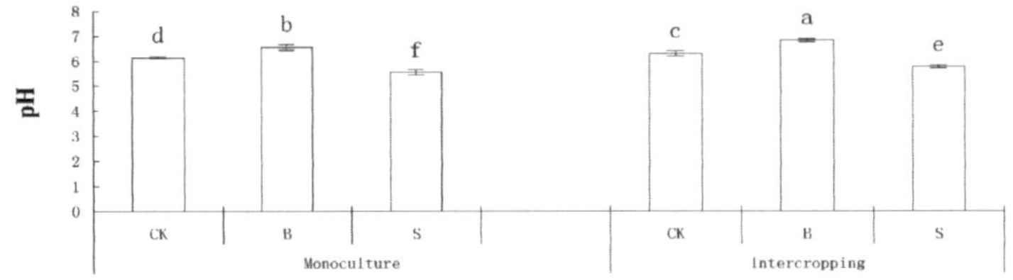

Note:the capital letters indicate different treatment conditions, Ck represents blank treatment, B represents biochar application, S represents straw mulching. The data with different little letters show a significant difference($P<0.05$). The same below.

Fig.1 The effect of monoculture peanut and cassava/peanut intercropping on pH in rhizosphere soil under the biochar application and straw mulching

3.2.2 The effect of monoculture peanut and cassava/peanut intercropping on carbon and nitrogen in peanut rhizosphere soil under the biochar application and straw mulching

3.2.2.1 The effect of monoculture peanut and cassava/peanut intercropping on available nitrogen in

peanut rhizosphere soil under the biochar application and straw mulching. Figure 2 showed that under the same planting patterns, available nitrogen content of the plots of biochar application in peanut rhizosphere soil reduced by 54.69%(monoculture peanut) and 59.49%(cassava/peanut intercropping) respectively compared with Ck($P<0.05$). Available nitrogen content of the plots of straw mulching in peanut rhizosphere soil increased by 89.80%(monoculture peanut) and 67.09%(cassava/peanut intercropping) respectively compared with Ck($P<0.05$). Furthermore, under the same treatment conditions, available nitrogen content in peanut rhizosphere soil under intercropping was higher than that under monoculture($P<0.05$).

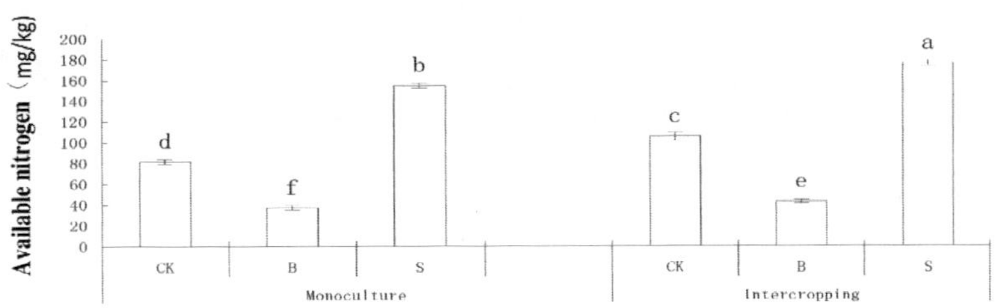

Fig.2 The effect of monoculture peanut and cassava/peanut intercropping on available nitrogen in rhizosphere soil under the biochar application and straw mulching

3.2.2.2 The effect of monoculture peanut and cassava/peanut intercropping on organic matter in peanut rhizosphere soil under the biochar application and straw mulching. Figure 3 showed that in the same planting patterns, organic matter content of the plots of biochar application in peanut rhizosphere soil increased by 94.52%(monoculture peanut) and 89.94%(cassava/peanut intercropping) respectively compared with Ck($P<0.05$). Organic matter content of the plots of straw mulching in peanut rhizosphere soil increased by 60.92%(monoculture peanut) and 52.34%(cassava/peanut intercropping) respectively compared with Ck($P<0.05$). Furthermore, under the same treatment conditions, organic matter content in peanut rhizosphere soil under intercropping was higher than that under monoculture($P<0.05$).

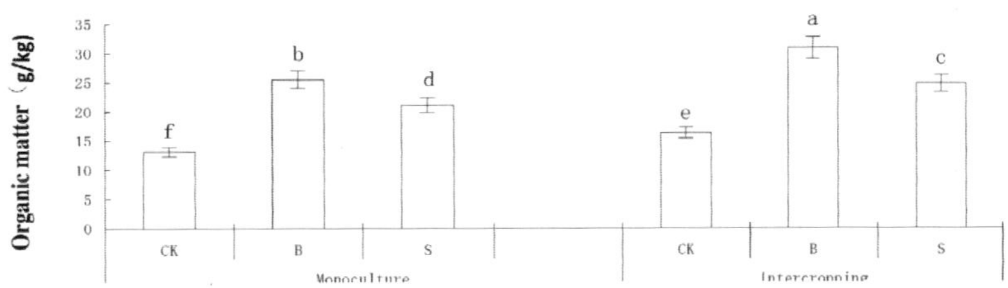

Fig.3 The effect of monoculture peanut and cassava/peanut intercropping on organic matter in rhizosphere soil under the biochar application and straw mulching

3.2.3 The effect of monoculture peanut and cassava/peanut intercropping on available phosphorus in peanut rhizosphere soil under the biochar application and straw mulching. Figure 4 showed that in the same planting patterns, biochar application and straw mulching can prominently improve available phosphorus content in peanut rhizosphere soil. The results showed that available phosphorus content of the plots of biochar application in peanut rhizosphere soil increased by 17.53%(monoculture peanut) and 17.93%(cassava/peanut intercropping) respectively compared with Ck($P<0.05$). Available phosphorus content of the plots of straw mulching in peanut rhizosphere soil increased by 5.95%(monoculture peanut) and 6.96%(cassava/peanut intercropping) respectively compared with

Ck(P<0.05). Furthermore, In the same treatment conditions, Available phosphorus content in peanut rhizosphere soil under intercropping was higher than that under monoculture(P<0.05).

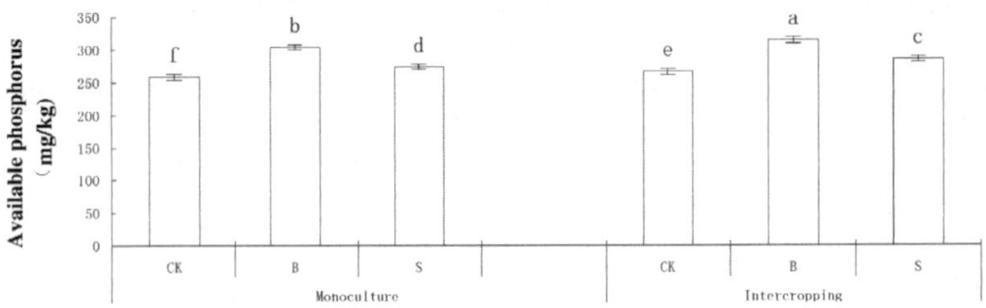

Fig.4 The effect of monoculture peanut and cassava/peanut intercropping on available phosphorus in rhizosphere soil under the biochar application and straw mulching

3.2.4 The effect of monoculture peanut and cassava/peanut intercropping on available potassium in peanut rhizosphere soil under the biochar application and straw mulching. Figure 5 showed that in the same planting patterns, biochar application and straw mulching can prominently improve available potassium content in peanut rhizosphere soil. The results showed that available potassium content of the plots of biochar application in peanut rhizosphere soil increased by 25.08%(monoculture peanut) and 22.87%(cassava/peanut intercropping) respectively compared with Ck(P<0.05). Available potassium content of the plots of straw mulching in peanut rhizosphere soil increased by 9.98%(monoculture peanut) and 11.94%(cassava/peanut intercropping) respectively compared with Ck(P<0.05). Furthermore, in the same treatment conditions, Available potassium content in peanut rhizosphere soil under intercropping was higher than that under monoculture(P<0.05).

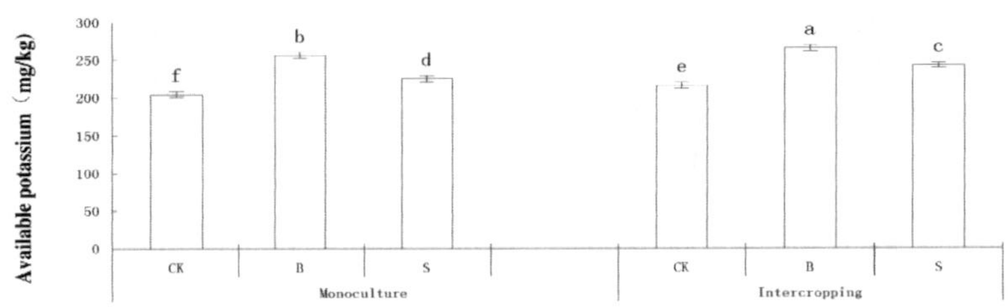

Fig.5 The effect of monoculture peanut and cassava/peanut intercropping on available potassium in rhizosphere soil under the biochar application and straw mulching

4. Discussion

4.1 The effect of monoculture peanut and cassava/peanut intercropping on physical properties in peanut rhizosphere soil under the biochar application and straw mulching

Biochar possesses loose texture, so it can directly improve firmness and reduce bulk density in peanut rhizosphere soil. Eastman's study[5] showed that bulk density decreased by 7.8% and total porosity increased by 6.4% in the soil after using biochar 25t/hm^{-2}. This is consistent with results of this study. Straw mulching can reduce bulk density in peanut rhizosphere soil and increase the total porosity and permeability, because the humus is porous, its viscosity is far less than the viscous soil clay, so it can make soil loose and reduce tillage resistance[6].

4.2 The effect of monoculture peanut and cassava/peanut intercropping on chemical properties in peanut rhizosphere soil under the biochar application and straw mulching

4.2.1 The effect of monoculture peanut and cassava/peanut intercropping on pH in peanut rhizosphere

soil under the biochar application and straw mulching. PH in peanut rhizosphere soil under cassava/peanut intercropping is significantly lower than that under monoculture, because after intercropping of cassava and peanut, the interaction between root systems will induce the secretion of H^+ and OH^- in the root system[7]. Applying biochar to increase peanut rhizosphere soil pH may be due to the different concentrations of alkaline substances in the ash of the biomass, such as K,Ca,Na, magnesium oxides, hydroxides, carbonates, etc. After the biochar application in the soil, it can improve soil base saturation and reduce the level of exchangeable aluminum[8]. Straw mulching can reduce pH in peanut rhizosphere soil. Because acid gas produced by microorganisms can reduce pH in peanut rhizosphere soil in the metabolism of straw[9-10].

4.2.2 The effect of monoculture peanut and cassava/peanut intercropping on carbon and nitrogen in peanut rhizosphere soil under the biochar application and straw mulching

4.2.2.1 The effect of monoculture peanut and cassava/peanut intercropping on available nitrogen in peanut rhizosphere soil under the biochar application and straw mulching. Biochar application leads to the ratio of C/N increased, that will cause biological fixation of soil alkaline hydrolysis nitrogen[11]. Straw mulching can increase available nitrogen content in peanut rhizosphere soil. Because straw mulching can increase soil microbial activity and continuously increase the content of soil microbial biomass nitrogen[12]. Cassava/peanut intercropping can increase available nitrogen content in peanut rhizosphere soil. Because effects of cassava root zone on transfer and utilization of nitrogen in peanut root zone can stimulate and promote nitrogen fixation in the root zone of peanut[13].

4.2.2.2 The effect of monoculture peanut and cassava/peanut intercropping on organic matter in peanut rhizosphere soil under the biochar application and straw mulching. Biochar application can increase organic matter content in peanut rhizosphere soil. Because the formation of soil organo-mineral aggregates improves aggregate stability and reduces the leaching of organic matter[14]. Straw mulching can increase organic matter content in peanut rhizosphere soil. Because there are a large number of microbes around the straw, these microorganisms can accelerate the decomposition and release of organic nutrients in straw[15]. Due to the weakening of microbial activity in organo-mineral aggregates, cassava/peanut intercropping can increase organic matter content in peanut rhizosphere soil[16].

4.2.3 The effect of monoculture peanut and cassava/peanut intercropping on available phosphorus in peanut rhizosphere soil under the biochar application and straw mulching. Biochar and straw contain more available phosphorus, not only that, due to generating positive charge, so biochar and straw can absorb phosphorus nutrients which are not absorbed by the organic matter[17]. Cassava/peanut intercropping can increase available phosphorus content in peanut rhizosphere soil. Because root exudates are more abundant under intercropping and these secretions can activate the soil nutrients[18].

4.2.4 The effect of monoculture peanut and cassava/peanut intercropping on available potassium in peanut rhizosphere soil under the biochar application and straw mulching. Biochar contains more available potassium and reduces the dissolution and migration of potassium nutrients[19]. Straw mulching can increase available potassium content in peanut rhizosphere soil. Because straw mulching can reduce adsorption of potassium in soil and increase its desorption rate[20]. Cassava/peanut intercropping can increase available potassium content in peanut rhizosphere soil. Because intercropping can increase microbial quantity and diversity, so that can promote the transformation of soil organic nutrients into inorganic nutrients[20].

5. Conclusion

(1) Under the conditions of this study, biochar application could increase the total soil porosity, permeability, pH, organic matter content, available phosphorus content and available potassium content in peanut rhizosphere soil and reduce bulk density and available nitrogen content in peanut

rhizosphere soil. Straw mulching could increase the total soil porosity, permeability, available nitrogen content, organic matter content, available phosphorus content and available potassium content in peanut rhizosphere soil and reduce bulk density and pH in peanut rhizosphere soil. In short, biochar application and straw mulching can increase the nutrient content in peanut rhizosphere soil and improve soil fertility.
(2) In the same treatment conditions, there was no difference in physical properties between cassava/peanut intercropping and monoculture peanut. In chemical properties, available nitrogen content, organic matter content, available phosphorus content and available potassium content in peanut rhizosphere soil under intercropping were higher than that under monoculture, but pH in peanut rhizosphere soil under intercropping was slightly lower than that under monoculture. In a word, cassava/peanut intercropping can increase nutrient content to improve the micro-ecological environment in peanut rhizosphere soil, and the suitable intercropping spacing is more conducive to increase nutrient content in peanut rhizosphere soil.

Acknowledgements
This research was supported by "Research and application of key techniques for low carbon, high yield and high quality cultivation of peanut"; "Selection and application of cassava varieties with different resistance and accumulation capacity to heavy metals"; "Science and technology project of Guangdong Province (2013B020303001, 2015B090903077)".

References
[1] HE Shu-lin. The rational use and development of straw counters-field technology in agriculture[J]. *Science Information*, 2006(4): 181-182.
[2] SHA Hong-lin, TONG Shi, ZHANG Wei-you, et al. Analysis on current situation of producing and comprehensive utilization of stalks of agricultural crops[J]. *Journal of Jilin Agricultural Sciences,* 2010(4): 51-55.
[3] LU Ru-kun, XIONG Li-ming, SHI Zheng-yuan. Study of cadmium in soil and crop ecosystem[J]. *Soils*, 1992, 24(3): 129-132, 137-141.
[4] Institute of Soil Science, Chinese Academy of Sciences. Analysis of soil physico-chemical properties. Shanghai: Shanghai Scientific & Technical Publishers, 1978.
[5] Eastman CM. Soil physical characteristics of an Aeric ochraqualf amended with biochar[D]. Columbus:Ohio State University, 2011:41-45.
[6] Effect of nitrogen and mineralization of soil nitrogen on growth and quality of flue-cured tobacco[D]. Henan: Henan Agricultural University, 2004.
[7] ZHANG De-shan, WANG Yu-yun, TANG Li, et al. Effects of wheat and fababean intercropping on available phosphorus of red soils and its relationship with rhizosphere soil Ph[J]. *Plant Nutrition and Fertilizer Science*, 2013, 19(1): 127-133.
[8] HE Xu-sheng, ZHANG Shu-qing, SHE Diao, et al. Effects of biochar on soil and fertilizer and future research[J]. *Chinese Agricultural Science Bulletin,* 2011, 27(15): 16-25.
[9] Setyanto P, Makarim AK, Fagi AM, et al. Crop management affecting methane emissions from irrigated and rainfed rice in Central Java(Indonsia)[J]. *Nutrient Cycling in Agroecosystems*, 2000, 58（58）:85-93.
[10] XIAO Xiao-ping, WU Fen-lin, HUANG Feng-qiu, et al. Greenhouse airemission under different pattern of rice-straw returned to field in double rice area[J]. *Research of Agricultural Modernization*, 2007, 28(5): 629-632.
[11] Lehmann J, Joseph S. Biochar for environmental management science and technology[M]. London: Earthscan, Ltd, London, UK.2009.
[12] LI Cai-hong, ZHANG Yong, YUAN Han-min, et al. Effects of straw returning to field on soil carbon, nitrogen and phosphorus contents[J]. *Ningxia Journal of Agri and Fores Sci & Tech*, 2016, 57(2): 12-15.
[13] LV Yue, WU Pu-te, CHEN Xiao-li, et al. Effect of above and below ground interactions on

maize/soybean intercropping advantage[J]. *Transactions of The Chinese Society of Agricultural Machinery*, 2014, 45(1): 129-136, 142.

[14] Glaser B, Haumaier L, Guggenberger G, et al. The'Terra Preta'phenomenon: a model for sustainable agriculture in the humid tropics[J]. *Naturwissenschaften*, 2001, 88(1):37-41.

[15] WANG Ying, YUAN Jian-guo. Straw on the soil organic matter enhance the exploration and research soil and fertilizer[J]. *J. Shanxi Agric. Univ*, 2007, 27(6): 120-121, 126.

[16] ZHANG Jia-en, GAO Ai-xia, XU Hua-qin, et al. Effects of maize/peanut intercropping on rhizosphere soil and nutrient contents[J]. *Chinese Journal of Applied Ecology*, 2009, 20(7): 1597-1602.

[17] Lehmann J. Bio-energy in the black[J]. *Frontiers in Ecology and the Environment*, 2007, 5(7):381-387.

[18] SHI Wei-min. The root exudates and nutrient availability[J]. *Soils*, 1993, 25(3): 252-256.

[19] FANG Bin, LI Xin-qing, ZHAO Bin, et al. Influence of biochar on soil physical and chemical properties and crop yields in rainfed field[J]. *Ecology and Environmental Sciences*, 2014, 23(8): 1292-1297.

[20] Li XY, Wang QF, Xu FL. Effect of returning application of straw on soil K, P, Zn adsorption-desorption and their availability[J]. *Journal of Huazhong Agricultural University*, 2000, 19(3):227-232.

Study on long-term aerosol distribution over central plains economic region using MODIS data

Zhangluo[1,2], Xushu[3], Bai Lianhong[3] and Caikun[1,2,*]

[1] State Key Laboratory of Remote Sensing Science, Institute of Remote Sensing and Digital Earth, Chinese Academy of Sciences, Beijing, China
[2] Computer and Information Engineering College, Henan University, Kaifeng, China
[3] The School of Mathematics and Computer Science, Xinyang Vocational and Technical College, Xinyang, China

caikun@henu.edu.cn

Abstract. The long-term (2000-2013) Moderate Resolution Imaging (MODIS) level 2 aerosol products were used to study the spatial and temporal distributions of both aerosol optical depths (AOD) over Central Plains Economic Region (CPER), which is critical for a thorough understanding of its formation, transport and accumulation in the atmosphere. The results indicate that the spatial distribution of AOD with higher values locate in Xinxiang city, Kaifeng city, Zhengzhou city and Jiaozuo city. The lower ones locate in the south of Shanxi province and the west of Henan province, which are all mountainous regions. Spring and summer are the two seasons with the lowest AOD in a year. The minimum AOD in CPER occurred in winter. In the Central Plains Urban Agglomeration (CPUA), Kaifeng city, Xinxiang city and Zhengzhou city are the areas with the worst air pollution of particulate matter. The AOD over CPER significantly reduce during the period from 2011 to 2013. In the long term, however, the condition of particulate matter pollution has not been improved.

1. Introduction

In recent years, with the rapid expansion of city scope, the rapid development of industry, and the increase of urban population and vehicles, accompanied urban atmospheric pollution problems, especially the particulate pollution is getting worse, air quality continues to deteriorate become the main problems of the harm to human health, restricting urban development, and harm to harmonious society construction. In 2014, the World Health Organization released a report that each year about 700 million people died of air pollution of the world; Air pollution has become the world's largest single environmental health risks [1]. Therefore, how to accurately obtain the space-time distribution, source and transmission path of aerosol has become one of the world's hot spots; meanwhile, it is also an important guarantee for weighting particulate pollution effect, formulating particulate reduction, prevention and control policies.

Presently, aerosol's remote sensing measurement methods are mainly included satellite remote sensing and ground-based remote sensing. Although ground monitoring instruments have a number of advantages in routine particle monitoring, it can only be carried out on limited ground stations, and the station is rarely in vast rural and underdevelopment areas. Satellite remote sensing has advantages of wide coverage, provide macroscopic change information, and reflect large scale and regional scale transportation of pollutions, which can make up for the shortage of spatial distribution of ground monitoring stations [2].

In recent years, at home and abroad have a lot of new results about air pollution research using satellite remote sensing data. Using the NASA's Level2 aerosol optical thickness of 10 km, Chu et al. [3] demonstrated the very significant application values of MODIS data in the monitoring of global, regional and local air pollution. Engel-Cox et al. studied the application of AOD in research of air pollution by using the MODIS true color image and the aerosol optical thickness data, respectively in qualitatively and quantitatively. Through comparing the satellite remote sensing data with the ground pollutant concentration, it was found that those in the United States, the center region and eastern region have high correlation than the western region [4]. In the aspect of long time series data analysis, Zhai Tianyong used the aerosol optical depth data of Terra MODIS from 2000 to 2013 to analyze the spatial and temporal characteristics of AOD in the Yangtze River Delta and the influence of meteorological factors on its distribution [5]. Wang Zhao et al. used the MODIS C5 aerosol product from 2000 to 2010 to analyze the multi-year variation of aerosol optical thickness and small particle aerosol contribution to total optical thickness in the Central Plains basin [6]. Although there are many studies on the spatial and temporal distribution of aerosols in the major cities such as Beijing-Tianjin-Hebei, the Pearl River Delta and the Yangtze River Delta, few studies have been done on the Central Plains urban agglomeration, which is one of China's seven national-level urban agglomerations.

2. Study area and data

2.1. Overview of the study area

In this paper, the Central Plains Economic Zone (CPER), which takes the Zheng-Bian-Luo urban area as the core, the Central Plains city group as the support, covering all Henan province and surrounding areas of the economic area, located in the center of China, latitude and longitude range of roughly 32 ° N ~ 38 ° N, 110 ° E ~ 118 ° E (as shown in Figure 1). The Central Plains Economic Zone consists of 28 prefecture-level cities (18 prefecture-level cities and 10 provincial-level cities and counties) in Henan, and 12 prefecture-level cities and three prefecture-level cities around Shandong, Hebei, Anhui and Shanxi, those with a total area of about 289,000 square kilometers. This area is relatively flat terrain, plain, and it is a temperate monsoon climate, the weather characters is high temperature and rainy in summer, cold and dry in winter.

The Central Plains urban agglomeration as the core of the Central Plains Economic Zone, it takes Zhengzhou as the center, and Luoyang as deputy center, including Kaifeng, Xinxiang, Jiaozuo, Xuchang, Luohe, Pingdingshan, Jiyuan, those nine regional central cities for the nodes constitute a close ties. In recent years, with the continuous and rapid development of economy in the Central Plains region, especially the Central Plains urban agglomerations, the region's air pollution problem is increasingly prominent. Recently, Henan Province Environmental Quality Bulletin shows that the Central Plains urban agglomerations atmospheric environment pollutants are mainly respirable particulate matters. Therefore, through study the pollution level and longtime spatiotemporal variation characteristics of aerosol particles in the region, it can provides scientific basis for government decision-making, which is of great practical significance to realize the sustainable development of the Central Plains Economic Zone.

2.2. Satellite data sources and data processing

Terra and Aqua satellites are part of NASA's Earth Observing System (EOS) program, which was launched on 18 December 1999 and 4 May 2002, respectively, transit time is for daily locations at 10:30 am and 13:30 pm, which are known as AM and PM satellites, they are all equipped with a Moderate Resolution Imaging Spectroradiometer sensor. In this paper, Using the MODIS / Terra 10 km aerosol optical depth data (MOD04) to analysis the aerosol long time series of the Central Plains Economic Zone, with accuracy of $\pm$ 0.05 $\pm$ 0.15τ and $\pm$ 0.03 $\pm$ 0.05τ on the land and sea respectively [7], those data can be obtained directly from NASA's MODIS website

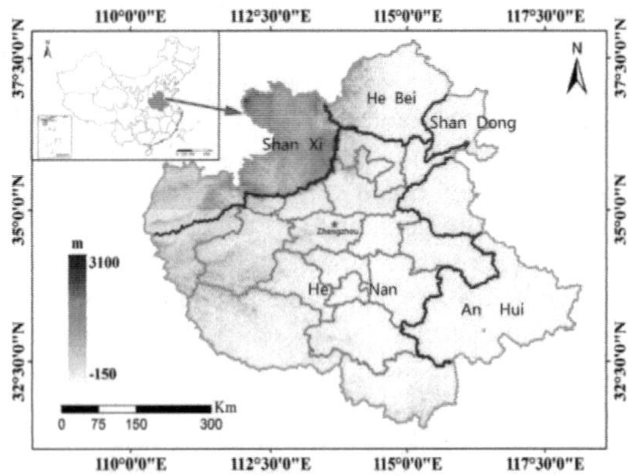

Figure 1. The location and terrain of CPER.

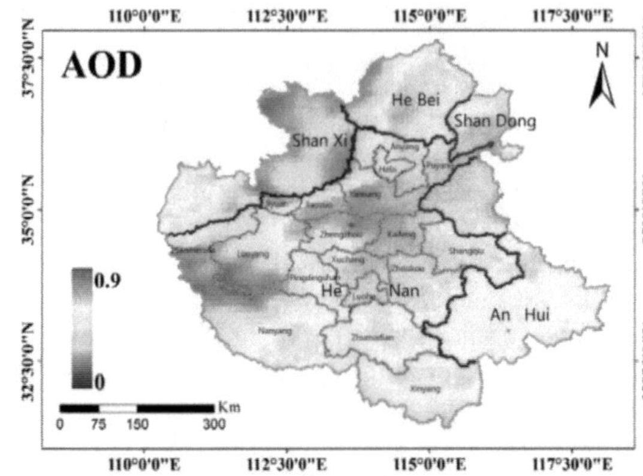

Figure 2. The distribution of AOD over CPER averaged from MOD04 products in the period of 2000–2013.

(http://ladsweb.nascom.nasa.gov/data/search.html). The MOD04 aerosol optical thickness of the product is currently one of the most widely used and the highest reliability of the AOT data. In this paper, we validate the MOD04 data from February 2000 to December 2013 by pixel-by-pixel and then combine the mean values to obtain the monthly and annual mean distributions.

3. Retrieval Algorithm of Aerosol Remote Sensing

The physical meaning of the aerosol optical thickness AOD is the integral of the extinction coefficient along the radiation propagation direction.

$$\tau_\lambda = \int_0^\infty \delta_\lambda N(z)dz \tag{1}$$

Among them, τ_λ means the aerosol optical thickness, λ means the wavelength, δ_λ means the aerosol extinction cross section, $N(z)$ means the aerosol particle number density vertical distribution. The MODIS 10 km resolution aerosol products currently used in the mainstream algorithm for dark pixel inversion algorithm, and the algorithm theory as follows: According to the vegetation - intensive and low albedo of the surface ground (dark pixels), there is a good linear correlation between near - infrared channel reflectance and red - blue channel reflectance. The effect of aerosols is 15 to 30 times smaller in the short-wave infrared band than in the visible band, thus it can be ignored. Therefore, the surface reflectivity of the red-blue band can be obtained from the apparent reflectivity of the short-wave infrared band, and then remove the surface contribution from the apparent reflectivity of the red-blue band, obtain the hemispheric reflectivity of the lower atmosphere, the equivalent reflectance of radiation term in the atmosphere and other atmospheric parameters, then get the aerosol optical thickness.

In order to increase the applicability of the dark pixel inversion algorithm, the new inversion algorithm of MOD04 C005 product uses a new inversion idea: In the dark target region, the relationship between the visible red bands and near infrared band is a function of the scattering angle and the vegetation index, and there is a fixed linear relationship between the visible red band and the blue band surface reflectivity. At the same time, MOD04C005 product inversion algorithm updated the aerosol model based on geographical distribution [8]. Based on the results of Dubovik et al., the MOD04 C005 product identified five aerosol models, namely continental, dust, no absorbable, moderately absorbing and severely absorbing, and these aerosol patterns varied with the season and geographical Change [9].

4. Result Analysis

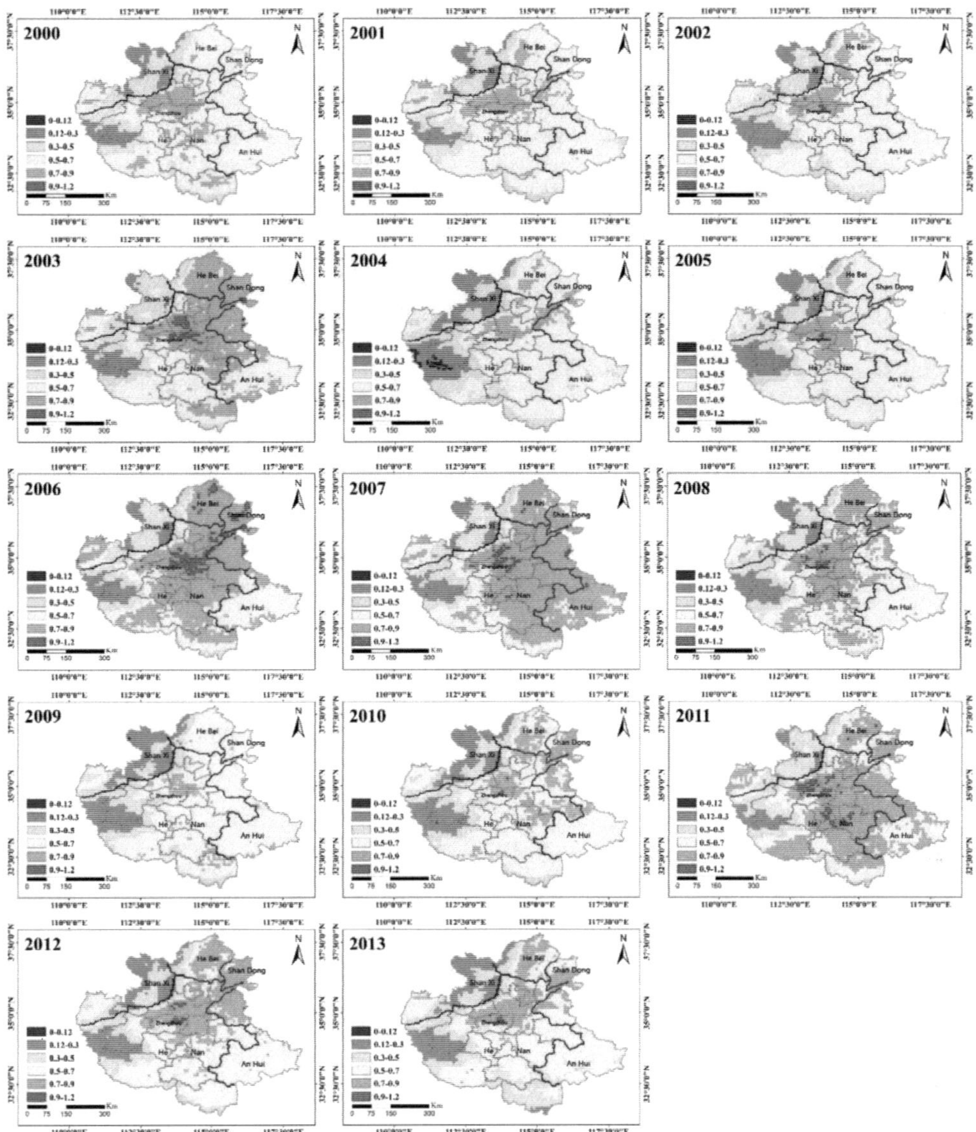

Figure 3. The distribution of annual means AOD over CPER during 2000–2013.

4.1. The temporal and spatial distribution of annual AOD over CPER during 2000-2013

Figure 2 shows the multi-year mean distribution of AOD in the CPER from 2000 to 2013. In Figure 2, the results indicate that the spatial distribution of AOD with higher values locate in Xinxiang, Kaifeng, the north of Zhengzhou and the south of Jiaozuo, which are relatively densely populated and economically developed urban areas. Locating in the southern foot of Taihang Mountains, the dominant wind direction of these regions is northeast and southwest, and sub-dominant wind direction is northwest wind. The high atmospheric stability of these areas leads to high static wind frequency. The topography and meteorological conditions aren't conducive to the dilution and diffusion of air pollutants, causing heavily polluted weather. In addition, the energy structures of these areas mainly dominated by coal. The irrational energy structure and lagging environmental infrastructure increased the pressure of atmospheric environmental quality improvement from the source. On the contrary, the low AOD areas locate in the west of Henan and the south of Shanxi, which are all mountainous regions with high vegetation coverage and without high-density factories and mines.

In terms of interannual variation trend, from 2000 to 2013, the CPER experienced three fluctuations (Figure 3): the first fluctuation from 2000 to 2004, the second from 2004 to 2009, and the last from 2009 to 2013. Among these fluctuations, the peak value of AOD respectively appeared in 2003, 2006 and 2011. In addition, the mean annual AOD and the area of high AOD region (AOD>0.7) in the CPER experienced the process of first increase and then decrease during each fluctuation. It's important to note that the degree of particulate pollution in CPER is obviously reduced relative to the previous fluctuating period (2006~2008), which except for affected by meteorological elements, but also is closely relate to the policies of adjustment of industrial structure and strict supervision of air pollution in Henan province, and even in the entire Central Plains in recent years.

4.2. The interannual trend of AOD means for key cities of Central Plains Urban Agglomeration during 2000-2013

The nine key cities (Zhengzhou, Xinxiang, Kaifeng, Jiaozuo, Xuchang, Luohe, Pingdingshan and Jiyuan) of Central Plains Urban Agglomeration are not only representing the economic core of CPER, but also represent the region of serious particulate pollution. In this paper, we specialize in conducting a statistical analysis of the annual variability of AOD within its administrative area of these nine key cities (Figure 4). As shown in Figure 4, from 2000 to 2013, we find that the three cities with the highest mean of AOD in Central Urban Agglomeration are Kaifeng, Xinxiang and Zhengzhou. Relatively, the three cities with the lowest mean are Luoyang, Jiyuan and Pingdingshan. Among them, the mean value of AOD in Kaifeng, as the city with highest level of particulate pollution reaches 0.74, and the value of Luoyang is only 0.412. In addition, many cities such as Kaifeng, Xinxiang etc. also experienced three obvious fluctuations as referred in 3.1 section. Meanwhile, from 2000 to 2013, except Jiyuan, the AOD's average value of the other eight key cities shows the trend of slow increase or constant during each fluctuation. But in the period of 2011~2013, the mean value of each city appears the trend of significant downward. Therefore, we conclude that the annual average of AOD in Central Plains Urban Agglomeration and CPER always has been in the fluctuations of continuous improvement and rebound. In the short term, the level of particulate pollution in the whole region truly shows the trend of decreasing year by year after 2011. But in the long term, both Central Plains Urban Agglomeration and CPER, the particulate pollution level has not been significantly improved, or even deteriorated after 2000.

4.3. The temporal and spatial distribution of seasonal AOD over CPER during 2000-2013

To analyze the seasonal variation of CPER, we calculated the monthly mean AOD and its standard deviation from 2000 to 2013. As shown in Figure 5, the seasonal trend of CPER is significant, spring and summer are the two seasons with the lowest AOD, while the value of summer and winter are lower relatively. Throughout the year, the highest value of AOD generally appears in the June of summer (AOD years mean = 0.704), the lowest value often exists in the December of winter (AOD years mean = 0.358). In the same year, the difference of AOD in different months can be as high as 0.97. In this paper, we only present the spatial distribution of the seasonal mean of AOD in 2013 because the spatial and temporal distribution of AOD in CPER has good consistency and regularity (Figure 6). The results of the above analysis are different from the case of low visibility events in autumn and winter. Because the AOD represents the extinction effect of the vertical layer aerosol, while ground visibility represents the horizontal extinction effect of the aerosol on the ground surface. In spring, the long-distance transport of coarse dust particles from Inner Mongolia and other places is one of the main reasons for the increase of AOD in the CPER. In summer, the relative humidity in the atmosphere is too high to leads the phenomenon of aerosol particle's hygroscopic growth becomes more obvious. Therefore, AOD also changes bigger relatively.

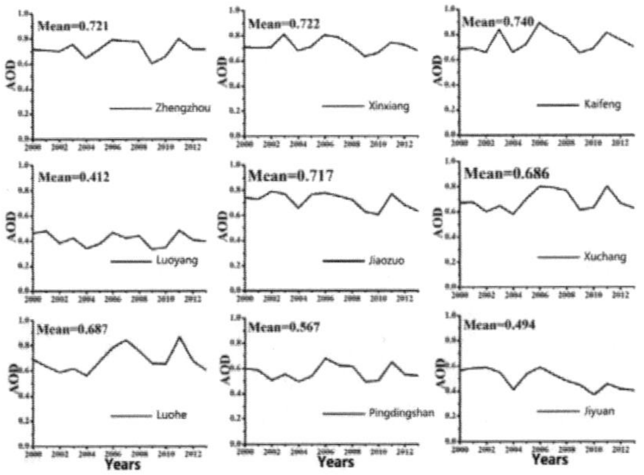

Figure 4. The trend of annual mean AOD for nine cities of Central Plains Urban Agglomeration in the period of 2000–2013.

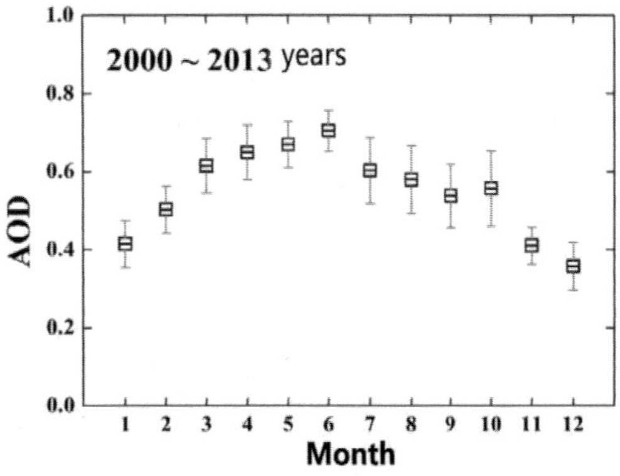

Figure 5. The temporal variations of monthly mean AOD over CPER from 2000 to 2013.The error-bars represent the standard deviations.

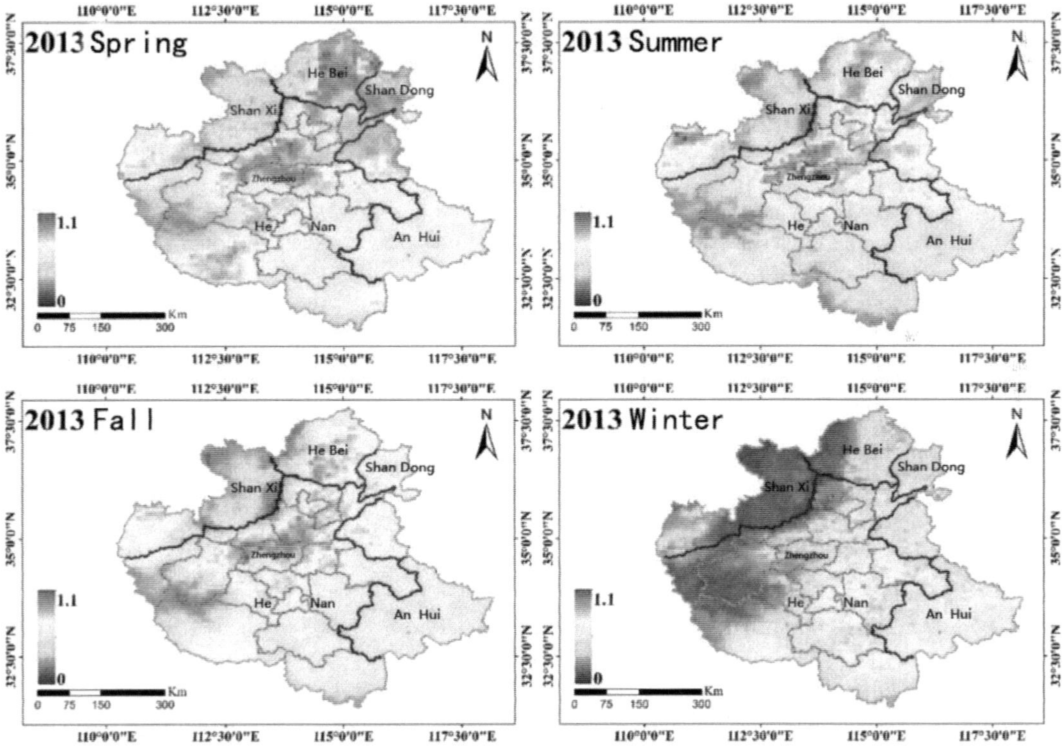

Figure 6. The distributions of seasonal mean AOD over CPER during 2013.

5. Conclusion and discussion

Central Plains Economic Zone as one of the seven state-level urban agglomeration, the air pollution problem has been caused increasing importance attention of the Ministry of Environmental Protection and other departments. Based on MOD04 aerosol product data from 2000 to 2013, this paper analyzes the temporal and spatial variation characteristics of AOD in the Central Plains Economic Zone, and focuses on the nine key cities in the Central Plains urban agglomeration. Research indicates:

1) The AOD high-value areas of Central Plains Economic Zone appear in Xinxiang, Kaifeng, northern Zhengzhou and southern Jiaozuo. The low-value areas are mainly mountainous areas in western Henan and southern Shanxi.

2) From 2000 to 2013, AOD of Central Plains Economic Zone experienced three wave periods, the peak of which was 2003, 2006 and 2011 respectively.

3) The highest value of AOD in the Central Plains Economic Zone usually occurs in spring, summer, and autumn followed, and winter is the smallest in the region.

4) The three cities with the most serious level of particulate pollution are Kaifeng, Xinxiang and Zhengzhou, the lightest three cities are Luoyang, Jiyuan and Pingdingshan.

5) In the short term, after 2011, the particulate pollution level of the Central Plains Economic Zone has obviously decreased year by year, but in the long run, after 2000, neither the Central Plains urban agglomeration nor all the Central Plains Economic Zone, the pollution level of the particulate matter has not been significantly improved; Industrial structure adjustment, air pollution's prevention and control, and other measures still need to strictly enforce.

References

[1] World Health Organization. Review of evidence on health aspects of air pollution – REVIHAAP [R]. 2014.

[2] Zhang Y, Pun B, Wu S, Vijayaraghavan K and Seigneur C. Application and evaluation of two air quality models for particulate matter for a southeastern U.S. episode, *Journal of the Air & Waste Management Association*, 2004, 54: 1478–1493.

[3] Chu D A, Kaufman Y J, Zibordi G, et a1. Global monitoring of air pollution over land from the Earth Observing System-Terra Moderate Resolution Imaging Spectroradiometer(MODIS), *J Geophys Res*, 2003, 108(D21) :Art. No. 4661.

[4] Engel-Cox J A, Christopher H H, Basil w C, et a1. Qualitative and quantitative evaluation of MODIS satellite sensor data for regional and urban scale air quality, *Atmospheric Environment*, 2004, 38:2495-2509.

[5] ZhaiTianyong. Analysis of Spatio-temporal Variability of Aerosol Optical Depth in the Changjiang River Delta, China, Shanghai: East China Normal University, *thesis*, 2014.

[6] Wang Z, Peng Y, Che H, et al. Analyses on Spatial and Temporal Characteristics of AOD in Guanzhong Region of Shaanxi Province Using Long Term MODIS Data, *Plateau Meteorology*, 2013, 32 (1): 234-242.

[7] Remer L A, Kaufman Y J, Tanré D, et al. The MODIS aerosol algorithm, products, and validation, *Journal of Atmospheric Sciences*, 2005, 62: 947-973.

[8] Levy R C, Remer L A, MattooS, et al. Second-generation operational algorithm: Retrieval ofaerosol properties over land from inversion of moderate resolutionimaging spectroradiometer spectral reflectance, *J.Geophys. Res.*, 2007, 112: D13211.

[9] Dubovik O, Holben B N, Eck T F, et al. Variability of absorptionand optical properties of key aerosol types observedin worldwide locations. *Journal of the Atmospheric Sciences*, 2002, 59(3): 590-608.

IOP Conf. Series: Earth and Environmental Science **59** (2017) 012023 doi:10.1088/1755-1315/59/1/012023

Using multiple tracers to quantify groundwater discharge to Yellow river in Weining plain

Y S Wang, X X Chen, M N Zhang and H X Li

Center for Hydrogeology and Environmental Geology Survey, China Geological Survey, Baoding, Hebei 071051, China

E-mail : cug_wys@sina.com

Abstract. It is important to quantify the hydraulic relationship between surface water and groundwater for developing water resources sustainably. In this study, thirty-six samples were taken and the interaction between river water and groundwater was analyzed on the basis of hydrogeochemical characteristics. Environment tracers such as EC and Cl- were used to calculate groundwater inflows along river flow path by the mass balance method, then it provided a detailed longitudinal profile of groundwater discharge rates. The results show that groundwater discharged to Yellow River mainly in dry season and varied spatially along river length with larger rates in two areas. The groundwater inflows calculated by Cl- and EC mass balances differentiated with each other, which indicated potential uncertainties. Thus, it was necessary to use multiple tracers to quantify groundwater discharge in order to improve accuracy. The representation of the groundwater end-member for environment tracers, the completeness of water balance items and sample resolution were shown to be of vital importance for calculating groundwater inflows using environmental trace method.

1. Introduction

In arid and semi-arid areas, groundwater discharge as essential components of river baseflow is important not only in the hydrological cycle [1], but also from an ecological perspectives, especially for the river ecosystems dependent on groundwater [2]. It is basis work to determine groundwater discharge quantitively for the evaluation and sustainable development of water resources in a basin. In recent years, a number of studies have evaluated groundwater discharge rates to rivers by longitudinal river chemistry surveys and lots of environment tracers such as Cl^-, ^{18}O, ^{222}Rn and EC had been used to establish end-member model for theirs water cycle significances. The Weining plain is located in the western of Ningxia province in the Yellow river valley, in where, the industrial and agriculture production mainly depend on surface water. For a long period of irrigation, the interaction between groundwater and surface water is so complex to identify. In this study, geochemistry of groundwater and surface water were analyzed during dry season and Cl^- and EC were used to calculate groundwater discharge inflows using mass balance method with regards of evaporation rates. The potential uncertainties were discussed as well as the factors affecting evaluation.

2. Study area and Methodology

2.1. Study area

Weining basin is a faulted basin developed in Cenozoic, to the north and south of which are Xiang mountain and North Weining mountain. Mean annual rainfall for the study area is 273.5 mm and

1888.4 mm for evapotranspiration. The study focus on a segment of Yellow river in Weining plain, with average width of 600 m, depth of 2 to 4 m and discharge of 1039.8m^3/s. A tributary named Qingshui river is located in the Eastern of basin.

The Quaternary groundwater systems can be divided into 2 groups: upper phreatic aquifer and lower confined aquifer. The vertical extent of upper aquifer consisting of fine sand, coarse and thin silt clay is from 6 to 48 m, in which the buried depth of groundwater is less than 5 m. The vertical extent of lower confined aquifer is from 40 to 150 m, consisting of coarse and silt sand. Groundwater is mainly recharged by leaked water from channels, irrigation return flow, infiltrating meteoric water and laterally penetrating fracture water from rocks along mountain front. It is mainly discharged via vertical evapotranspiration, artificial abstract and inflows to Yellow river. Naturally, groundwater flows from two sides to the middle of the basin.

2.2. Sampling and analytical methods

Thirty-six samples including 24 groundwater samples plus 12 river water samples were taken along 69.7 Km length of Yellow river from Xiaheyan Hydrologic Station to Mingsha Village in 26 March, 2014 (Fig 1). River water were sampled at the depth more than 0.5 m to insure that groundwater and surface water had mixed fully. Water samples were filtered through 0.45 um membranes after pumping more than 10 minutes. Electrical conductivity was measured field when sampling and major elements were determined using ion chromatograph.

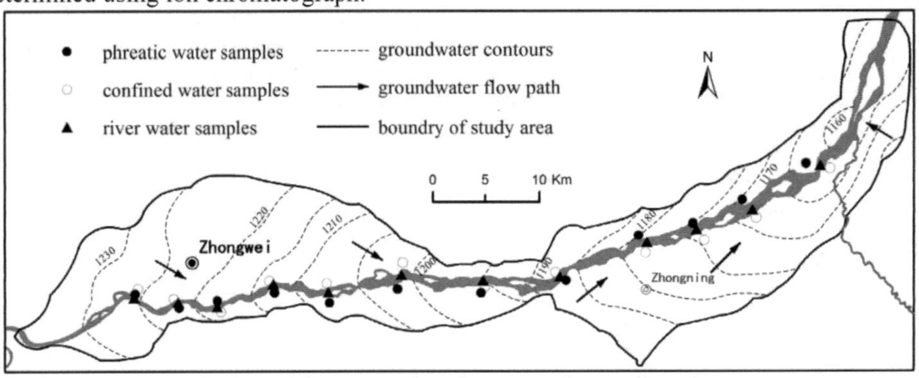

Fig.1 Location map of study area and sampling sites

2.3. Methodology

The end-member model is established on the basis of mass balance that takes account into groundwater discharge to river, evaporation and upstream flow. The change in flow with distance and tracer concentration change along Yellow river that receives groundwater discharge are given as below [3].

$$Q\frac{\partial c}{\partial x} = I(C_i - C) + wEC \quad (1)$$

$$Q = Q_0 + Ix - wEx \quad (2)$$

Where Q is river discharge (m^3/d), C is the concentration of a given tracer in river (mg/L), x is the distance in river flow direction (m), I is groundwater discharge rates (m^2/d), C_i is the concentration of a given tracer in groundwater (mg/L), w is width of a river (m), E is the evaporation rate (m/d), Q_0 is river discharge at the beginning (m^3/d).

3. Results and discussion

3.1. Hydrogeochemistry

The main ions in upper phreatic aquifer were Ca^{2+}, Mg^{2+}, HCO_3^- and SO_4^{2-} with the concentration of 76.4-148.1 mg/L, 36.1-77.28 mg/L, 340.9-568.2 mg/L, 112.7-288.4 mg/L respectively and water types of $HCO_3{\cdot}SO_4$—Ca·Mg mainly, $HCO_3{\cdot}SO_4$—Mg·Ca and $HCO_3{\cdot}SO_4{\cdot}Cl$—Ca·Mg secondly (Fig 2). TDS was 596.5-1162.6 mg/L with a mean of 867.7 mg/L. Water types in lower confined aquifer were similar to that in upper aquifer and TDS was 338.6-928.0 mg/L with a mean of 725.9 mg/L. River water had complex water types of $HCO_3{\cdot}SO_4$—Ca·Mg, HCO_3—Ca·Mg and HCO_3—Ca·Na. TDS was 321.5-375.9 mg/L with a mean of 341.9 mg/L, Which was lower than that in groundwater.

The electrical conductivity (EC) reflects the total contents of all dissolved ions, which can be an effective indictor of flow paths and duration in water cycle. As a conservative ion, chloride is almost impossible to involve with precipitation, ion exchange and biological processes, while its concentration is only affected by mixing and halite weathering [4]. Thus, EC and Cl^- were used to trace terrestrial water cycle in a number of studies. The interaction between groundwater and surface water can be determined effectively by a detailed understanding of spatial distribution and variation of EC and Cl^- in different types of water (Fig 3). The concentration of Cl^- in phreatic groundwater, confined groundwater and surface water were 91.2-178.9 mg/L, 73.5-162.7 mg/L and 39.0-42.2 mg/L respectively, and EC were 1269.2-2469.6 us/cm, 741.7-1974.58 us/cm and 685.6-735.9 us/cm respectively. It was obvious that the concentration of EC and Cl^- were lower in confined water than that in phreatic water arose from more evaporation, whiles lowest in river water among three types of water. Along Yellow river, tracer concentration changed irregularly in groundwater, it was that samples were all located in discharge area and experienced different flow paths and water rock interactions. Tracer concentration had an increasing trend downstream in river water, which identified that surface water had been recharged by groundwater with higher concentration of EC and Cl^- from two sides continuously. Especially, groundwater discharged to the river more strongly in two areas suggested by a rapid increase of Cl^- concentration in river water.

3.2. Groundwater inflows

It can be seen that groundwater discharged to Yellow river and had higher concentration of EC and Cl^- than river water. In general, environmental tracer method will be more accurate when the tracer concentration in groundwater is more distinct from that in river water. Environment tracers such as EC and Cl^- were used to calculate groundwater inflows along river flow path by the mass balance method in this study. A hypothesis was given that tracer concentration and groundwater discharge rates remained constantly during two sample sites and tracer concentration in groundwater averaged that in phreatic water and confined water nearby. The parameters were given as below: Q_0 is 316 m^3/s monitored in Xiaheyan Hydrological Station in 26 March, 2014, river evaporation rate is 5 mm/d, the average width of Yellow river is 600 m and the initial values of EC and Cl^- concentration in river water are 39.0 mg/L and 687.6 uS/cm respectively.

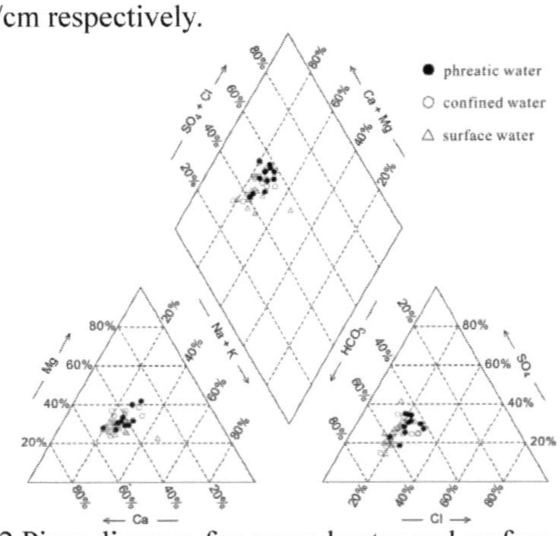

Fig.2 Piper diagram for groundwater and surface water

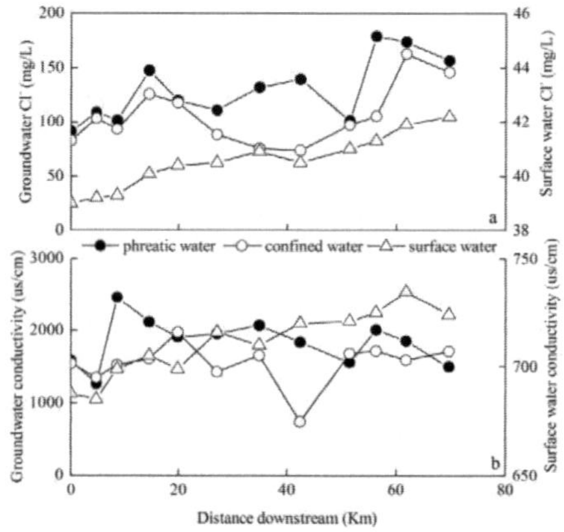

Fig.3 Cl^- concentrations and EC of groundwater and surface water along Yellow River (a. Cl^-, b. EC)

The results inferred from EC and Cl^- mass balance show that groundwater discharge rates varied spatially along river length with greater rates in two areas (Fig 4). In the length of 14.6-19.9 Km, it had a rate more than 10 m^2/d, more than 15 m^2/d in the length of 56.2-61.8 Km, and 0-0.4 m^2/d in the length of 42.4 Km which showed a lower discharge rate here. Despite of continuous evaporation, river discharge mainly increased along flow path except in the areas where it received minor groundwater discharge.

The groundwater inflows calculated by Cl^- and EC mass balances differentiated with each other. For example, in the length of 14.6 Km, groundwater inflows derived from two methods were 37.6 m^2/d and 22.4 m^2/d respectively and in the length of 51.3 Km, 24.9 m^2/d and 1.1 m^2/d respectively. Using Cl^- mass balance, the evaluated river discharge at the end of length was 327.7 m^3/s and 0-37.6 m^2/d for groundwater discharge rates with cumulative flows of 14.1 m^3/s which account for 4.3 percent of river discharge, while the respective values were 322.9 m^3/s, 1.1-29.4 m^2/d、 9.3 m^3/s and 2.9 percent when using EC mass balance. It was obvious that there was an over estimation of groundwater inflow using Cl^- mass balance than EC mass balance.

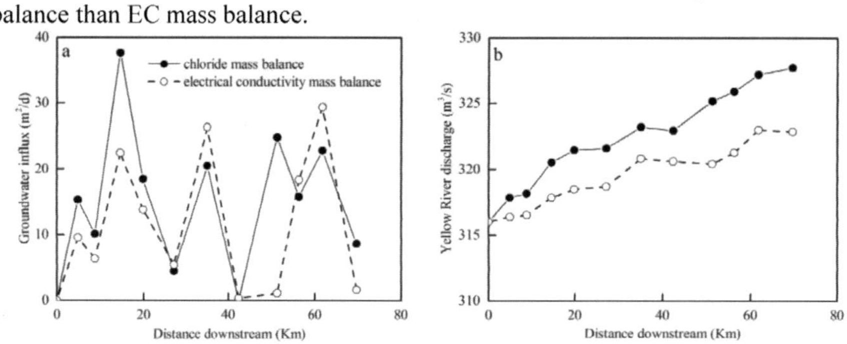

Fig.4 Calculated groundwater inflows and river discharge using Cl^- and EC mass balance (a. groundwater inflows, b. river discharge)

3.3. Uncertainties

It is credible for the cumulative groundwater discharge inferred from this study in contrast with the inflow of 6.7m^3/s using base flow separation method in previous study. The wide spatial variation of groundwater discharge was controlled by the hydrogeological settings in the study area. The quaternary aquifer had a narrow distribution toward west and east. Groundwater flowed from two sides to the middle of plain and discharged to Yellow river. The aquifer thickness and the spatial scale were bigger to benefit for groundwater discharge where there groundwater inflow was higher. The environmental tracer method to quantify groundwater inflow from longitudinal groundwater and surface water samples is attractive in remote areas where hydrological stations are limited. When correctly applied, it is able to provide information on the spatial distribution of groundwater discharge at a scale and accuracy that is impossible with most of other methods.

It may mislead estimates when using a single tracer method indicated by the large differences of groundwater inflows derived from two methods. The method on the basis of mass balance assumes the knowledge of groundwater end-member chemistry [5]. However, the assumption is always invalid that groundwater sampled some distances from the river represents the groundwater discharged to river, which may lead potential uncertainties. For the difficulty of accurate determination of groundwater end-member chemistry, it is necessary to use multiple tracers to quantify groundwater discharge in order to improve accuracy. Secondly, the uncertainties of mass balance items involved with artificial activities can cause large uncertainties in groundwater inflows. For the lack of information of surface water abstraction for industry and domestic needs, these items were without considered in this study. Another factor limiting the method is sampling resolution. Despite of wide spatial variation of groundwater tracer concentration, there was an assumption that groundwater chemistry in one sampling point represented that in entire flow path between two sampling locations in brief. In theory, the more the water samples, the higher the accuracy of groundwater inflows evaluation. In this study, a small resolution of one sample per 6 Km was taken, thus reducing the accuracy of the method.

4. Conclusions

When correctly applied, environmental tracer method is able to provide information on the spatial distribution of groundwater discharge at a scale and accuracy that is impossible with most of other methods. The assumption that tracer concentration in groundwater sampled some distances from the river represented that in the groundwater discharged to river may lead potential uncertainties, thus it is necessary to use multiple tracers to quantify groundwater discharge in order to improve accuracy. The representation of the groundwater end-member, the completeness of water balance items and sample resolution are shown to be important for evaluating groundwater inflows using this method.

References

[1] Cook P G. 2013. Estimating groundwater discharge to rivers from river chemistry surveys. *Hydrol. Processes*. **27** 3694-3707.

[2] Smerdon B D, Gardner W P, Harrington G A, et al. 2012. Identifying the contribution of regional groundwater to the baseflow of a tropical river (Daly River, Australian). *J. Hydrol.* **464-465** 107-115.

[3] Aguilar J B, Harrington G A, Leblanc M, et al. 2015. Chemistry of groundwater discharge inferred from longitudinal river sampling. *Water Resour. Res.* **50** 150-1568.

[4] Mccallum J L, Cook P G, Berhane D, et al. 2012. Quantifying groundwater flows to streams using differential flow gaugings and water chemistry. *J. Hydrol.* **416–417** 118-132.

[5] Gilfedder B S, Frei S, Hofmann H, et al. 2015. Groundwater discharge to wetlands driven by storm and flood events: quantification using continuous Radon-222 and electrical conductivity measurements and dynamic mass-balance modeling. *Geochim. Cosmochim. Acta.* **165** 161-177.

IOP Conf. Series: Earth and Environmental Science **59** (2017) 012024 doi:10.1088/1755-1315/59/1/012024

The international research progress of Ammonia(NH3) emissions and emissions reduction technology in farmland ecosystem

W Z Yang[1,2] and Y Jiao[2]*

1Water-saving Agricultural Engineering Research Center, Inner Mongolia Normal University, Hohhot 010022, China;
2Chemistry and Environment Science College, Inner Mongolia Normal University, Hohhot 010022, China;

Abstract. NH_3 is the important factor leading to the grey haze, and one of the main causes of environmental problems of serious ecological imbalance, such as acid rain and air quality deterioration. The fertilizer excessive application of the current farmland results NH_3 emissions intensity greatly. In order to clear the farmland NH_3 emissions research status and achievements, the literature of farmland NH_3 emission related were retrievaled by the SCI journals and Chinese science citation database. Some factors of NH_3 emission were analyzed such as soil factors, climate factors and farmland management measures. The research progress was inductived on farmland NH_3 emission reduction technology. The results will help to clarify farmland NH_3 emissions research progress. The theoretical guidance was provided on the future of farmland NH_3 emissions research.

1. Introduction

NH_3 emissions is the largest in the gas of containing N[1]. NH_3 is the only source of particulate NH_4^+ and alkaline gaseous substance in the gas. The NH_3 90 percent was converted to NH_4^+, which was reacted with the acid in the atmosphere. There are NH_4^+ 84 percent photochemical reactions. NH_3 exists in aerosol particles with NH_4NO_3, $(NH_4)_2SO_4$ and NH_4Cl binding mode(such as PM2.5)[2,3]. especially the haze formation process, they accounted for 30 percent to 70 percent of PM2.5 mass concentration[4]. Simultaneously, the NH_3 accumulating in the atmosphere will be oxidized into N_2O and NO, etc. The N_2O is not only an important greenhouse gas, but also participate in the stratospheric optical process, destroy the ozone layer, enhance ultraviolet radiation to the earth's surface, causing environmental pollution and huge economic losses[5]. The increase of NH_3 concentration in atmosphere is closely related to the change of agricultural production mode. Agricultural production is an important source of NH_3 emissions[6].

2. NH3 emissions research progress at home and abroad

In recent years, a lot of field data have been accumulated at home and abroad on the NH_3 emissions. Some researchers have done a lot of quantitative research on the NH_3 emission rate and influencing factors under field conditions. There are 3 main factors affecting the NH_3-N loss of nitrogenous fertilizer, such as soil properties, meteorological conditions and agricultural technical measures[7,8].

2.1. Soil factors

Soil pH is one of the important factors affecting NH_3 emissions. The NH_3 emissions increase with the increase of soil pH value[9].Urea applied into the soil, the soil pH increase after a short period of time, and reach maximum on the second day after fertigation, then fell rapidly, and maintain between 7.6-8.0. Some is absorbed by crops in the hydrolysis product NH_4^+, part are transformed into NO_3^--N, part are adsorbed by soil, the rest are released into atmosphere in the form of NH_3 emission loss.

The soil cation exchange capacity (CEC) changes H^+ concentration in soil solution, adsorb NH_4^+ and NH_3 strongly, decrease the concentration of NH_3, inhibit NH_3 emission, reduce NH_3 emission[10].

Xiao(2016) studied soil's clay content was much and organic matter content was high in winter wheat field, and cumulative NH_3 emissions is 3.40-5.06 kg N/hm^2 in different fertilization treatment. The reason is that the texture of clay soil has a strong adsorption capacity of NH_4^+ generally, reduce the NH_4^+ content in the soil liquid phase effectively, thus reduce the loss of NH_3 emission[11,12].

$CaCO_3$ content and total salt content in soil were closely related to the NH_3 emission of nitrogen fertilizer. Zhu(1989) researched that urea is in calcareous soil contained more calcium carbonate, NH_3 emission was more than that in acid paddy soil. Dry land, especially, on calcareous soil, NH_3 emissions is a great loss on application of urea and ammonium bicarbonate[13].

2.2. *Meteorological factors*

Soil temperature directly affects the $NH_4^+ \Leftrightarrow NH_3 + H^+$ reaction equilibrium constant and NH_3 diffusion rate from the soil to the atmosphere. Temperatures elevated accelerate the emission rate of NH_3.On the one hand, Temperature increased the proportion of the liquid NH_4^+-N in the total ammonia significantly and promotes NH_4^+ conversion to NH_3. On the other hand, the soil urease activity increased with the increase of temperature[15].

There is a close relationship between the NH_3 emission and precipitation and other environmental factors in the field of nitrogen fertilizer. Nastr(2000) studied that fertilizer was brought into the deep soil with the rain infiltration, thereby reducing the loss of NH_3 emissions[16]. The rainfall reduced NH_3 emissions a week after fertilization significantly. Rainfall duration affects NH_3 emission[17]. However, a small amount of rain affected NH_3 emissions under irrigation conditions[18,19].

2.3. *Agricultural management measures*

It is difficult that the researchers realize the NH_3 emission reduction in agricultural production practice by changing the physical and chemical properties of soil and climate environment. How to reduce the NH_3 emission losses in agricultural production, it is the key for adjustment of agricultural management measures. Agricultural management practices influence NH_3 emissions significantly.

2.3.1. Fertilizer amount. NH_3 emissions were positively correlated with the amount of nitrogen application[20]. Zhang(2007) studied that NH_3 emissions were 8.2-28.7kg/hm^2 and 21.8-62.1kg/hm^2 on different fertilizer levels of rice season. NH_3 emissions accounted for 3.7%-8.8% and 10.0%-18.9% of urea application rate[21]. Wang(2006) show that increasing N application rate significantly increased soil NH_3 emission rate[22]. The peak of NH_3 emission rate in different nitrogen application treatments appeared at third days (N 140~165 g/hm^{2}h) and fifth days (N 100~400g/hm^{2}h). The effective way to reduce the loss of nitrogen fertilizer was to control nitrogen fertilizer application amount[4,23].

2.3.2. Tillage and fertilization methods. A variety of N fertilizer deep application technology to reduce the effect of NH_3 emission is better by reducing the time of NH_4^+-N fertilizer exposed to the air than surface application. NH_3 emission is the strongest in the case of surface application of N fertilizer, NH_3 emissions on a mixed mode of surface and deep application fertilization are less than that of surface application, and that is weakest on deep application.

2.3.3. Fertilization application period. Fertilizer application period significantly affected NH_3 emission losses. NH_3 emission rate was the highest in tillering fertilizer, and panicle fertilizer was the lowest. Wang Xiubin studied the total amount of NH_3 emission in different fertilization application period. The results showed that the method of reducing the amount of nitrogen application and nitrogen fertilizer application delayed backwards in time can reduce the loss of NH_3 emission and increase the yield[24,25,26]. NO_3^--N and NH_4^+-N seasonal dynamic changes are different on different nitrogen treatment of soil, and mainly depends on the different nitrogen fertilizer N release characteristics, fertilizer use time and Fertilizer use by several times, etc[27].

3. Conclusions

In fact, to take effective control measures, this problem is not confined to China, but worldwide. To strengthen the following several aspects:

In recent years, the domestic scientific research workers studied more NH_3 emission factors, however, most of the existing research focused on paddy and corn field ecosystem. NH_3 emissions vary for the planting patterns and soil environment. To achieve accurate quantitative evaluation model of NH_3 emissions, it is necessary for NH_3 emissions affecting factors and mechanism research on different ecological conditions, different planting patterns and different types of farmland soil.

Previous studies have a lot of potted plants and greenhouse cultivation on NH_3 emissions research work. Their environmental factors control is different on field crops compared in situ observation study,such as moisture and temperature. To strengthen the wild farmland NH_3 emission research, measures of NH_3 emission reduction can help farmland the practical application.

Temperature effect on NH_3 emissions research results are inconsistent. Some think temperature and NH_3 emissions have significant positive correlation, however, some studies suggest temperature is not NH_3 emissions limits. the following work need study temperature to influence the NH_3 emission, to clear temperature influence mechanism of NH_3 emissions on different ecological environment.

To agricultural sustainable development, future research direction must adhere to the general principle of "high-yield, high-quality, high efficiency" and "low consumption, no pollution", actively research the effective and easy technology to reduce NH_3 emission loss.

Acknowledgements

This work was supported by the National Natural Science Foundation of China (41375144, 41565009, 41675140); Youth Innovative Talents Plan in Inner Mongolia for 2016.

References

[1] Webb J H and Menzi 2005 *Environ. Pollut.* **135** 399
[2] Su C G, Yin B, Zhu Z L and Shen Q R 2003 *Chinese J. Appl. Ecol.* **14** 1888 (in chinese)
[3] Tsuruta H, Kanda K and Hirose T 1997 *Nutr. Cycl. Agroecosystems* **49** 58
[4] Ye S C, Lin Z C, Dai Q G, Jia Y S, Gu H Y, Chen J D, Xu L S, Wu F G, Zhang H C, Huo Z Y, *etal.* 2011 *Chinese J. Rice Sci.* **25** 78 (in chinese)
[5] Hutchings N J, Sommer S G, Andersen J M and Asman W A H 2001 *Atmos. Environ.* **35** 1968
[6] M 1998 *Nutr. Cycl. Agroecosystems* **51** 8
[7] Yan X Y, Hajime A and Toshimasa O 2003 *Global Change Biol.* **9** 1080
[8] Tian G M, Cai Z C, Cao J L and Li X P 2001 *Chemosphere* **42** 128
[9] Dancer W S, Peterson L A and Chesters G 1973 *Soil Sci. Soc.Am. P.* **37** 68
[10] Zhang Q L,Zhang M,Ye G P and Lu J F 2002 *Chinese J Soil Sci.* **23** 31 (in chinese)
[11] Xiao J, Fan J L, Ye G P, Liu D Y, Yan J, Luo J F, Houlbrooke D J and Ding W X 2016 *J. Agro-Eviron. Sci.* **35** 2018 (in chinese)
[12] Duan Z H, ZhouY L and Wu S R 1990 *Chinese J. Soil Sci.* **3** 138 (in chinese)
[13] Zhu Z L, Simposon J R,Zhang S L, Cai G X, Chen D L, Freney J R and Jackson A V 1989 *Acta Pedologica Sinaca* **26** 338 (in chinese)
[14] Ju X T, Xing G X, Chen X P, Zhang S L, Zhang L J, Liu X J, Cui Z L, Yin B, Christiea P, Zhu Z L, *etal.* 2009 *P. Natl. Acad. Sci. US A* **106** 3046
[15] Freney J R and Simpson J R 1983 *Gaseous loss of nitrogen from plant-soil systems* (Martin Nijhoff: Dr.W.Junk Publishers)
[16] Nastri A, Toderi G and Bernati E 2000 *Agrochimica* **5-6** 238
[17] Li Q Q, Li Y F, Gao Q, Li S Q, Chen X P, Zhang F S and Liu X J 2015 *J Plant Nutr. Fert.* **21** 578 (in chinese)

[18] Zhai X X, Wang Z L, Dai Z M, Liu X, Wang P, Liu H T, Yin Y C, Cao L, Xu C L and Cui Z Y, *etal.* 2013 *Plant Nutr. Fert. Sci.* **19** 56 (in chinese)
[19] Zhu Z L 2000 *Soil Environ. Sci.* **9** 6 (in chinese)
[20] Wang G L, Chen X P, Cui Z L, Yue S C and Zhang F S 2014 *Agr. Ecosyst. Environ.* **197** 298
[21] Zhang J and Wang D J 2007 *Chinese J. Eco-Agric.* **15** 86 (in chinese)
[22] Wang D, Yu Z W, Yu W M, Shi Y and Zhou Z X 2006 *Chinese J. Appl. Ecol.* **17** 1598. (in chinese)
[23] Xiao X, Yang L L, Deng Y P and Wang J F 2012 *Chinese J. Eco-Agric.* **31** 2066 (in chinese)
[24] Li H, Wang E J and Yang G F 2008 *Mod. Agric. Sci. Technol.* **10** 111 (in chinese)
[25] Sang M M, Fan H, Jiang S S and Jiang J Y 2015 *Environ. Sci.* 36 3358 (in chinese)
[26] Zhang H, Yang Z L and Luo L G 2011 *Plant Nutr. Fert.r Sci.* **17** 1138
[27] Fan H, Jiang S S, Wei Y and Jiang J Y 2016 *Eviron. Sci.* **37** 2906 (in chinese)

Reasonable Width of Narrow Coal Pillar of Gob-side Entry Driving in Large Mining Height

Zhao Ming, Zhang Shuai and Chen Yong

College of Resource and Safety Engineering, China University of Mining and Technology(Beijing), Beijing, No. 11, Xueyuan Road, Haidian district, China

Email:252797905@qq.com

Abstract. To determine the reasonable width of the gob-side entry driving coal pillar in Xieqiao Mine, firstly, the theoretical analysis method was applied to construct the mechanical model for limit equilibrium area of coal seam edge, and the distance obtained via analysis between the bearing pressure extremum and the coal edge was 3.34 m. Secondly, the numerical simulation method was applied to analyze the roadway rock mechanics with different coal pillar widths, and it is concluded that the width of the coal pillar was staying in a chaos area which was not conducive to the stability of the system when the peak stress position of the roof changed suddenly and the reasonable width was determined at 5 m. Finally, the high-strength, high pre-stressed bolt and cable anchor supporting test was carried out in the roadway, with 60 mm of roof subsidence, 207 mm of two sides displacement, 305 mm of floor heave, 135 mm of lateral displacement of coal pillar, 7 d duration of strong roadway rock deformation in pressure, in which the deformation meet the safety production of mine and increased the technical and economic benefit remarkably.

1. Introduction

Reasonable width of the coal pillar is an important topic of coal mining. The small coal pillars in gob-side entry driving can not only make the roadway surrounding rocks in the low stressed state, but also can maximize the resource recovery, therefore, this method are widely adopted. Determining the coal pillar width and stability main focus on the analysis of roadway surrounding rock stress field, displacement field and plastic zone distribution, and such research methods as theoretical analysis, numerical simulation, analog simulation and engineering analogy are widely used [1-5]. Recently, scholars made a lot of researches, and the achievement has guided the engineering practice successfully. This paper is based on the No. 1232 (3) panel of Xieqiao Mine, the method of theoretical analysis and numerical simulation was used to determine the reasonable width of the coal pillar and the application was carried on in the engineering field. The research result will provide the theory and practice basis for the coal pillar width in similar engineering project.

2. Distribution rule of the gob-side coal inclining to the bearing pressure

2.1. Profile of the panel

The No.1232 (3) panel of Xieqiao Mine adopted the large mining height with fully-mechanized and full-seam mining method, and the panel was at an elevation of -463.6 m ~- 555.3 m, and average buried depth was 448 m. The average thickness of coal seam was 4.65 m with 15° of the average angle; the strike length was 3015.2 m and the inclined length was 119.0 m ~ 234.3 m. As shown in figure 1, the No.1222 (3) panel of the upper section had finished mining. The immediate roof of the No.1232 (3) panel was mudstone and No.13-2 coal seam, with 2.05 m of average thickness, and the upper roof was fine sandstone, with 5.26 m of average thickness, and the floor was sandy mudstone, with 4.1 m of average thickness.

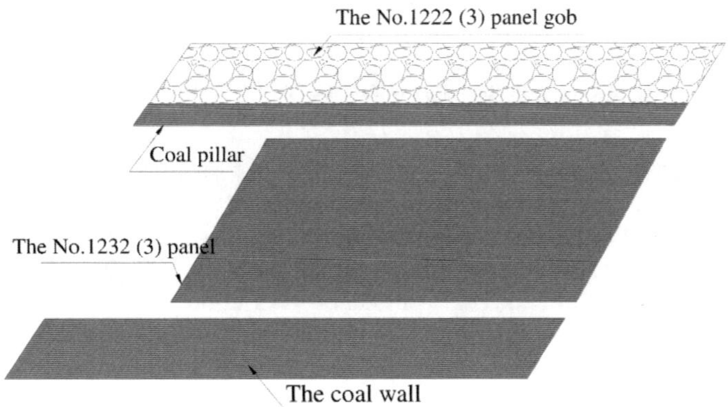

Figure 1. The No.1232 (3) panel layout.

2.2. Limit equilibrium analysis of the gob-side coal mechanical state

Before the gob-side entry driving, the movement of the gob overlying strata tends to be stable while the gob-side coal under the remaining bearing pressure is in limit equilibrium state. The mechanical model is established as the figure 2.

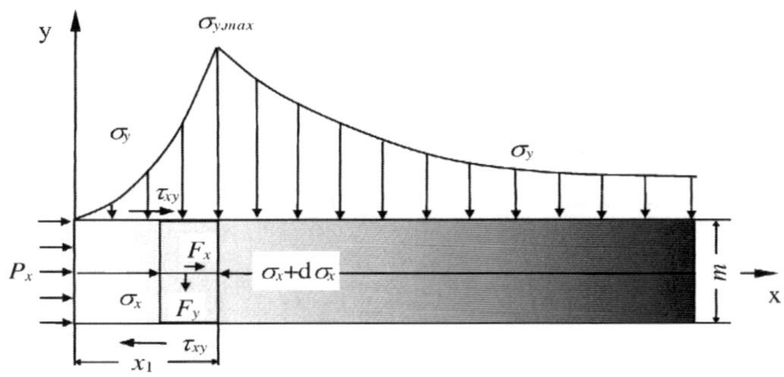

Figure 2. The mechanical model for limit equilibrium area for coal seam edge.

F_x、 F_y are the body forces (MPa) in horizontal and vertical direction respectively; p_x is the coal wall lateral restriction; τ_{xy} is the shear stress (MPa) between the coal seam and the roof-floor interface; m (m) is the coal-seam thickness; x_1 is the distance from the extremum point of bearing pressure to the coal edge. The coal is regarded as homogeneous continuum, and the balance equation can be obtained from the integration of the limit equilibrium theory and the Mohr-Coulomb strength criterion under the plane strain condition as follows [6]:

$$\left.\begin{aligned} &\frac{\partial\sigma_x}{\partial x}+\frac{\partial\tau_{xy}}{\partial y}+F_x=0\\ &\frac{\partial\sigma_y}{\partial y}+\frac{\partial\tau_{xy}}{\partial x}+F_y=0\\ &\tau_{xy}=\sigma_y\cdot\tan\varphi_0+c_0\\ &\sigma_x=\lambda\cdot\sigma_y \end{aligned}\right\} \tag{1}$$

$y=\frac{m}{2}, x=x_1, \sigma_y\big|_{x=x_1}=\sigma_{y1}\cos\alpha$. The solution of equation (1) was as follow:

$$x_1=\frac{m\lambda}{2\tan\varphi_0}\ln\left[\frac{\sigma_{y1}\cos\alpha\tan\varphi_0+2c_0-m\gamma_0\sin\alpha}{2c_0-m\gamma_0\sin\alpha+2\frac{p_x}{\lambda}\tan\varphi_0}\right] \tag{2}$$

The bearing pressure extremum σ_{y1} can be calculated according to the following formula:

$$\sigma_{y1}=2.729(\eta\sigma_c)^{0.729} \tag{3}$$

σ_c is the uniaxial compressive strength of the coal sample and σ_c is 12 MPa; the rheological coefficient η is 0.45; c_0、φ_0 are the cohesive force (2 MPa) and the friction angle (25°) between the coal seam and the roof-floor respectively; λ is the lateral pressure coefficient, λ=1; γ_0 is the average volume weight of the overlying strata, and γ_0 is 25.75 KN/m^3; α is the dip angle of the coal seam and α is 15°; the average thickness m of the coal seam is 4.65 m; coal wall lateral restriction p_x is 0.2 MPa. From integration of (2) and (3), the distance from the position of the bearing pressure extremum to the coal edge is x_1=3.34m.

3. Research on the reasonable width of the entry protection coal-pillar

3.1. *Process of research and analysis*

After excavating, the distribution of the bearing pressure in gob coal changes further with a new unloading loose zone and plastic zone in the roadway edge, while the bearing pressure moves to the deep coal part. Meanwhile, the different width of coal pillar also leads to the corresponding changes in mine pressure distribution of the coal pillar and the roadway roof. Based on the surrounding rock mechanics of gob-side entry driving, the 2D finite-difference calculation program FLAC^{2D} was applied to simulate the mechanics form of the coal pillar and roof after excavating. According to the geological conditions of No.1232 (3) panel, the simulation adopted Mohr coulomb constitutive model with 230 m wide and 192 m high and the vertical load of upper model was in accordance with the mining depth of 448 m. The coal rock mechanics calculation parameters are shown in table 1. Based on the limit equilibrium analysis of the gob coal, the simulating width of the narrow coal pillar is 3 m, 5 m, 7 m respectively, while the comparison stands at 10 m, 15 m, 20 m.

Table 1. Parameters for different rocks and coal tested.

lithology	Volume weight /(kg/m^3)	Tensile strength /MPa	Cohesive force /MPa	Friction angle /(°)	Bulk modulus /MPa	Shear modulus /MPa
fine sandstone	2620	2.14	3.98	41	6964	5223
siltstone	2610	1.82	3.2	39	7039	4846
sandy mudstone	2560	1.56	1.5	37	5355	3526
mudstone	2540	0.92	1.6	29	4792	1716
coal	1350	0.67	1.3	30	2149	935

3.2. *Stress distribution of the coal seam roof with different coal pillar widths*

Figure 3 represents the vertical stress distribution curve of the roadway roof in different coal pillar widths. The figure shows that the vertical stress above the roadway roof is small, and there is a vertical stress peak in the two coal walls (pillar) of the roadway roof. When the coal pillar width stands at 3 m, 5 m and 7 m, the peak stress of coal pillar roof is less than that of coal wall roof; when the coal pillar width stands at 10 m, 15 m and 20 m, it goes to the opposite direction.

When the coal pillar width increases from 3 m to 10 m, the peak stress of the coal pillar roof increases quickly from 6 MPa to 35 MPa; when the coal pillar width increases from 10 m to 20 m, the peak stress of the coal pillar roof changes little and remains stable in high stress level; the peak stress in the coal wall roof decreases slowly with the increase of the coal pillar width, and the peak stress gradually closes to the roadway coal wall with the increase of the coal pillar width. When the coal pillar width changes from 7 m to 10 m, in which the roof peak stress transit to the upper coal pillar from the coal wall.

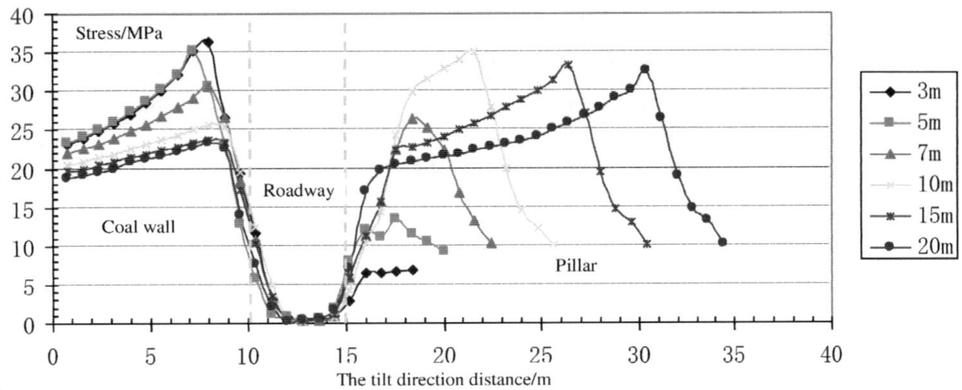

Figure 3. The roof stress distribution curves of different coal pillar widths.

3.3. Characteristics of wall rock deformation with different coal pillar widths

While the width of coal pillar changes, the deformation distribution characteristic of the coal seam and roadway floor also changes. As shown in figure 4, when the coal pillar width stands at 3 m, the roadway is affected by the mining drastically, and the shear failure occurs in the two coal walls and the floor of the roadway, and the failure zone covers throughout coal pillars which are in loose and broken state; when the coal pillar width stands at 5 m, the range of shear failure shrinks. Because the coal pillar is in low stress state, the failure zone covers small, and the appropriate supporting measures should be adopted so that the roadway remains stable; when the coal pillar width stands at 7 m, the coal pillar is in high stress, and the failure zone covers throughout the coal pillar, so the coal pillar stability is poor and it is hard to maintain.

When the coal pillar width stands at 10 m, the shear failure inordinately and a small scale of tensile failure appeared in the roof-floor of the roadway and two sides of the coal pillar, but the failure zone in the coal pillar is not well covered. When the appropriate supporting measures are adopted, the roadway can remain stable. When the coal pillar width reaches 15 m and 20 m, the failure zone around the roadway shrinks apparently and the tunnel can remain stable well in mining.

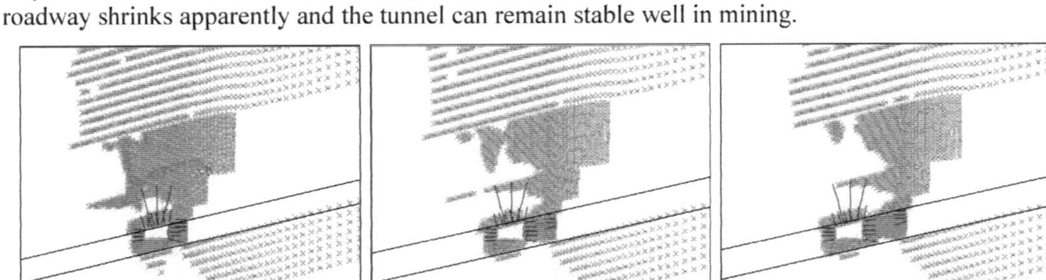

(a) 3m (b) 5m (c) 7m

(d) 10m (e) 15m (f) 20m

Figure 4. The failure field distribution of the wall rock with different coal pillar widths.

3.4. Comprehensive analysis of the reasonable coal pillar width

When the coal pillar width is 3m, 5m and 7m, the roof vertical stress peak located in the coal wall side, and when the coal pillar width is 3m, 5m, the values of maximum principal stress in the coal pillar is low. The coal pillar has suffered from the peak pressure during mining, leading to the plastic yielding and its failure zone running throughout, but the peak stress transferred to the inner of the coal wall side quickly with little high vertical pressure on the upper coal pillar and rocks remaining the residual strength after the peak stress. So for the coal pillar of 3 m, 5 m and 7 m wide, with little high vertical stress in the upper zone, they can keep itself stable through the residual strength of coal when adopting a certain reinforcement constraints on both sides of the coal pillar to keep the integrity of the yielded coal pillar. In addition, when the coal pillar width is 3 m, it cannot keep stability of itself even adopting a certain reinforcement constraints on both sides of the coal pillar. Consequently, the coal pillar will be deformed severely, and tends to cause air leakage for spontaneous combustion in the gob. Therefore, it is not conducive to make the roadway stable when the coal pillar width is 3 m.

Based on the analysis above, the coal pillar width can be divided into four regions to evaluate the stability of the roadway (figure 5). I zone and III zone embrace poor stability, otherwise II zone and IV zone are better stable. For the gob-side entry driving system of No.1232 (3) panel, I zone is below about 5 m; II zone is about 5 m to 10 m; III zone is about 10 m to 20 m; IV zone is over 20 m. Point A in Figure 5 represents a tipping point for leaping of the peak stress in the roof. When the coal pillar width is smaller than that in point A, the roof peak stress is in the coal wall side and gradually closes to the roadway with the increase of the coal pillar width, and this process is a gradual process; when the coal pillar width approaches point A, the roof peak stress transfers to the upper coal pillar from the entity coal seam, and this process is a mutation process; when the coal pillar width is larger than that in point A, the roof peak stress gradually stay away from the roadway, which is also a gradual process. Corresponding to the process above, the stability of the gob-side entry driving system also changes and its stability is poor when the coal pillar width near the roof peak stress mutation area, while the areas a certain distance from both sides of the mutation point are well stable. The coal pillar transits to a less stable region on the far area of the left mutation point because the width is too narrow. In this study, point A lies between 7 m to 10 m.

When choosing the reasonable width of the chain pillar of the gob-side entry driving system, not the wider coal pillar makes more stability. Because before and after mutation of the roof peak stress, the coal pillar width is in a chaotic region which is not conducive to the stability of the system. The coal pillar width in the range must be avoided in the production [7]. Outside of the chaotic area, the increase of the coal pillar width will make the gob-side entry driving system more stable. However, this is not conducive to improve the recovery rate of coal resources. Therefore, the reasonable width of chain pillars should choose II zone in Figure 5, and the reasonable pillar width should be 5 m.

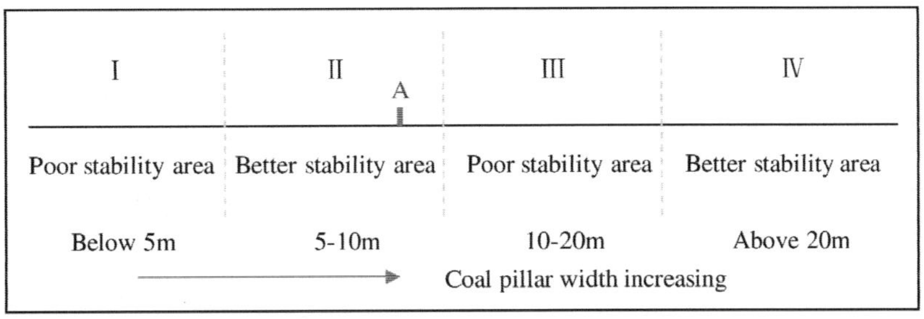

Figure 5. The stable area division of the rock mechanics system in gob-side entry driving.

4. Engineering Practice

4.1. The roadway support design

Given the complex situation of the coal seam, the weak roof layers formed, the joint grew well, and the regional tectonic stress was obvious. The soft and broken coal seam tended to form coal sliding and roof falling in the mining, and the two sides displacement was large and severe deformation of the surrounding rock took place. In order to solve the supporting problem, high-strength and high pre-stressed bolt and cable were applied for support experiment and research.

The roof was equipped with M5 steel belt and bolt supporting for cable reinforcement. The bolt featuring BHRB500 L-nonlongitudinal reinforced rebar, with 22 mm of diameter and 2.6 m of length, was anchored by total lengthened prestress, with 800 mm * 800 mm in row spacing. 6-hole M5 steel belt, with the length of 5.0 m and metal mesh canopy guard were applied. Anchor cable was 20 mm of diameter, 7.4 m of length, resin extended anchorage, with 1.8 m of length and 1200 x 800 mm in row spacing. Two-anchor bolt was with 2.0 m of length, while other forms were equal to the roof, and the metal mesh applied the diamond graticules woven with 8 # galvanized iron wires, with 1500 mm of web width, 30 * 30 mm in mesh specification and less than 100 mm of overlap length.

4.2. Roadway supporting performance

Figure 6 shows the displacement curves of the surface of the roadway surrounding rock. The roof subsidence was 60 mm, two sides of the roadway displacement was 207 mm and the floor heave was 305 mm. Among them, the lateral displacement of the coal pillar was 135 mm, and the strong deformation persists 7 d in pressure. Preserving 5 m of the chain pillar meet the safe production of mine and improved the recovery rate of coal resources greatly.

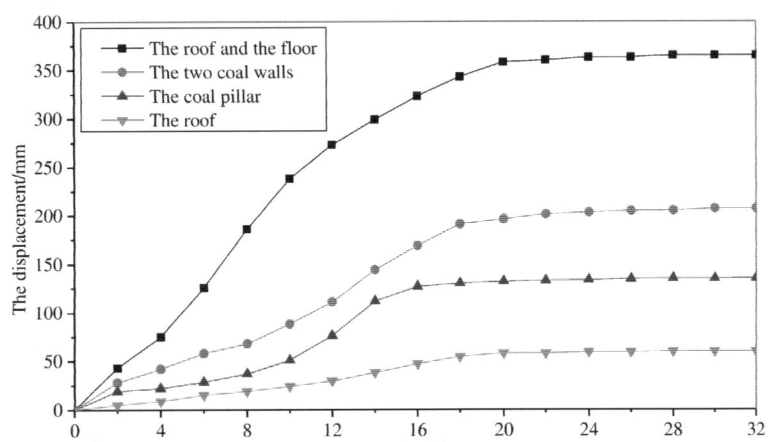

Figure 6. The displacement curves of surrounding rock of roadway.

5. Conclusions

(1) The control system was divided into 4 areas to evaluate the stability of the roadway. The tipping point A for leaping of the peak stress in the roof was located in zone II with poor stability and the point A was between 7 m and 10 m.

(2) When choosing the reasonable width of the chain pillar of the gob-side entry driving system, not the wider coal pillar made the roadway more stable. Because before and after mutation of the peak stress in the roof, the coal pillar width was in a chaotic region which was not conducive to the stability of the system.

(3) After preserving the narrow coal pillar, the displacement of the surrounding rock around roadway was monitored, with 135 mm of the lateral displacement of coal pillar. The strong deformation persisted about 7 d during the pressure. After preserving 5 m of the coal pillar width, the mine has made significant technical and economic benefits.

Acknowledgements

The research is supported by the Specialized Research Fund for the Doctoral Program of Higher Education (20120023110023) and the Fundamental Research Funds for the Central Universities (2014QZ03).

References

[1] Zhao Ming, Huang Kan, Lin Jia, Zhang Peng, Zhao Jian, Chen Lei and Ge Yong-yong 2016 Optimization research on weak rock control in coal roadway of large mining height *Rock and Soil Mechanics* vol. S2 pp 589-596.

[2] Kong De-zhong, Wang Zhao-hui and Li Xiao-meng 2014 Study of reasonable width of full-mechanized top-coal caving with large mining height *Rock and Soil Mechanics* vol. 35 pp 460-466.

[3] Xu Jin-hai, Miao Xie-xing and Zhang Xiao-chun 2005 Analysis of the time-dependence of the coal pillar stability *Journal of China Coal Society* vol. 30 pp 433-437.

[4] Zhang Ke-xue, Jiang Yao-dong and Zhang Zheng-bin 2014 Determining the reasonable width of narrow pillar of roadway in gob entry driving in the large pillar *Journal of Mining & Safety Engineering* vol. 31 pp 255-262.

[5] Peng Lin-jun, Zhang Dong-feng and Guo Zhi-biao 2013 Numerical analysis of thick coal seam small pillar along gob roadway and its application *Rock and Soil Mechanics* vol. 34 pp 3609-3616.

[6] Xie Guang-xiang, Yang Ke and Liu Quan-ming 2006 Study on distribution laws of stress in inclined coal pillar for fully-mechanized top-coal caving *Chinese Journal of Rock Mechanics and Engineering* vol. 25 pp 545-549.

[7] Wang Jin-an, Wei Wen-bing and Feng Jin-yan 2006 Stress evolution of an irregular pillar protected roadway system remained along goaf in fully mechanized top-coal caving mining field *Journal of University of Science and Technology Beijing* vol. 28 pp 317-323.

A novel self-embrittling strippable coating for radioactive decontamination based on silicone modified styrene-acrylic emulsion

Jing Wang[1,2], Jianhui Wang[1,2], Li Zheng[1,2], Jian Li[1], Can Cui[1] and Linmei Lv[1]

[1]Luoyang Hydraulic Engineering Technique Institute, Xishan Road, Luoyang, China
[2]Engineering Lab of Henan Province for Radioactive Decontamination Material and Equipment Technology, Wangcheng Street, Luoyang 471023, China

e-mail: jwang07@163.com

Abstract. Silicone modified styrene-acrylic emulsion and butyl acrylate were used as a main film-forming agent and an additive respectively to synthesize a self-embrittling strippable coating. The doping mass-ratio of butyl acrylate was adjusted at 0, 5%, 10%, 15%, 20%, and the results indicated the optimized doping ratio was 10%. $Ca(OH)_2$ was used to promote the coating film self-embrittling at a moderate doping mass-ratio of 20%. The synthesized coating's coefficients of α and β decontamination on concrete, marble, glass and stainless steel surfaces were both greater than 85%, which indicated the synthesized coating is a promising cleaner for radioactive decontamination.

1. Introduction

Strippable film coatings were highly focused in the last few decades for the application of radioactive decontamination. Numerous studies have shown that strippable film coatings were suitable for radioactive decontamination on flat surfaces [1-6]. However, continuously strippable films were more difficult for decontamination on complex geometrical surfaces or narrow space surfaces. Research have shown that self-embrittling strippable coating could be used for decontamination on those complex geometrical surfaces [7, 8]. This provided a new idea for the radioactive decontamination on the complex geometrical surfaces. The phenomenon that some silicone modified styrene-acrylic emulsion coatings presented embrittling glassy state after solidification was discovered in our previous studies. The cracking was irregular, nevertheless, the silicone modified styrene-acrylic emulsion was likely to become a new type of self-embrittling strippable coating for radioactive decontamination. Butyl acrylate (BA) was widely used as a typical soft functional monomer [9-11], which could regulate the flexibility and weaken the glassy state of coating films. The aim of this work was to prepare a novel self-embrittling strippable coating for radioactive decontamination based on silicone modified styrene-acrylic emulsion. In this paper, the formula of coating was optimized by evaluating its self-embrittling and strippable performance.

2. Experimental

2.1. Raw materials

Silicone modified styrene-acrylic emulsion was used for film-forming main agent, which was provided by Beijing Tongzhou Huyi Chemical Factory, and the material designation was HB-C. The main performance parameters of silicone modified styrene-acrylic emulsion were listed in Table 1. Butyl acrylate, provided by Hubei Qirui Chemical Factory, was used as an additive to regulate the flexibility and strippable performance of coating, and the material designation was BA-163. Calcium hydroxide was used to improve the self-fragmentation property of the coating.

Table 1. Performance parameters of silicone modified styrene-acrylic emulsion.

Property	Solid content/%	Viscosity/ mPa·S	PH	Glass transition temperature/℃
Parameter	48±1	1400~1600	7.5~8.5	25

2.2. Procedure

Different proportions of silicone modified styrene-acrylic emulsion and butyl acrylate as shown in table 2 were placed in a flask and stirred at 300r/min. Then calcium hydroxide was added and stirred at 1000r/min. The mixture was stirred for about 20 minutes. After the suspension was dispersed evenly, the thickener was added and the resulting mixture was stirred at 500r/min for about 20 minutes. Finally, the mixture was filtered to obtain a homogeneous coating sample.

Table2. Formulae of coatings.

No. of coating	Mass of silicone modified styrene-acrylic /g	Mass of butyl acrylate /g	Mass of $Ca(OH)_2$/g
C1	100	0	0
C2	100	5	0
C3	100	10	0
C4	100	15	0
C5	100	20	0
C6	100	10	10
C7	100	10	20
C8	100	10	30

3. Results and discussion

3.1. Content optimization of butyl acrylate

The mass-contents of BA were varied from 0 to 20%in the coating samples of C1~C5. The self-embrittling and strippable performances of these coating film were listed in table 3.

Table 3. Properties of coating films with different contents of BA.

No. of coating	Mass-content of BA/%	Self-embrittling performance	Strippable performance
C1	0	self-embrittling, hard fragment, uniform in fragment size	difficult to strip
C2	5	self-embrittling, hard fragment, uniform in fragment size	difficult to strip
C3	10	self-embrittling, soft fragment, uniform in fragment size	strippable
C4	15	self-embrittling, soft fragment, non-uniform in fragment size	strippable
C5	20	difficult to embrittle, soft coating film	strippable

As can be seen in table 3, silicone modified styrene-acrylic emulsion coating film could self-embrittling with hard fragment. But the adhesion to concrete surface was too strong to strip. After the introduction of BA, the coating film gradually became softer and more strippable. When the content of BA increased to 15%, the coating film of C4 represented a non-uniform fragment. And even when the content of BA increased to 20%, the coating film of C5 was difficult to embrittle. As a classical soft functional monomer, BA could decrease the hardness of coating film effectively. And also could modify the adhesion to concrete surface, which made the coating film strippable. However, excessive BA could make coating film too soft to self-embrittling. As can be seen from the formula experiment, the most appropriate doping mass-content of BA was 10%.

3.2. Content optimization of $Ca(OH)_2$

Addition of varied mass contents of $Ca(OH)_2$ from 10~30% has been achieved in the samples of C6~C8, and their properties and embrittling photos of the coating films were shown in table 4 and figure 1, respectively. The coating C3 that with no additional $Ca(OH)_2$ was also included as comparison.

Table 4. Properties of coating film with different contents of $Ca(OH)_2$.

No. of coating	Mass content of $Ca(OH)_2$/%	Fragment size/mm	Strippable performance
C3	0	20~30	strippable
C6	10	10~20	strippable
C7	20	5~10	strippable
C8	30	20~30	difficult to strip

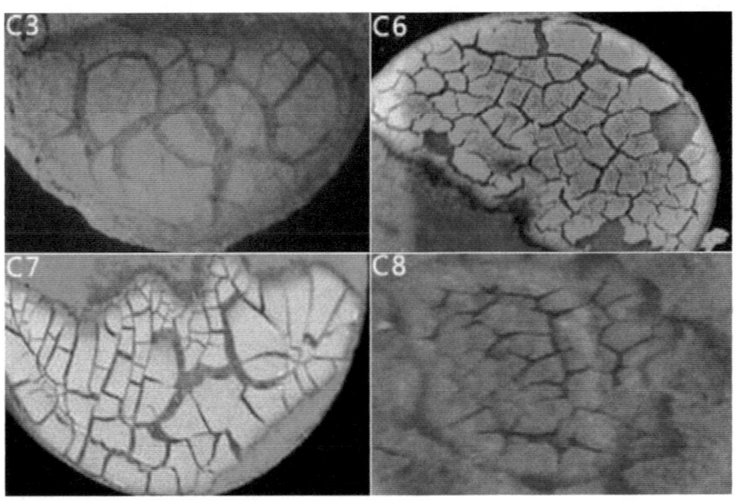

Figure 1. Embrittling photos of coatings with different $Ca(OH)_2$ contents.

As can be seen in table 4 and figure 1, moderate doping of $Ca(OH)_2$ could promote the coating films to self-embrittle, resulting in the decreased fragment size of coating films. When the mass-content of $Ca(OH)_2$ was 20%, the coating C7 self-embrittled well with a fragment size of 5~10mm. But when the doping content of $Ca(OH)_2$ increased to 30%, the coating film of C8 became hard and difficult to be stripped. Excessive doping of $Ca(OH)_2$ made the coating film harder and the adhesion to concrete surface stronger, mainly due to the transformation of Ca^{2+} to $CaCO_3$ during solidification. As can be seen from the formula experiments, the suitable doping mass-content of $Ca(OH)_2$ was 20%. Consequently, the optimized formula was obtained as coating C7 through formula experiments.

3.3. Viscosity of coating

Viscosity affected the application mode of the coating. Figure 2 was the viscosity-temperature chart between -10~30℃ of the coating C7. It showed that the coating had obvious viscosity-temperature characteristic. The coating' viscosity was below 2000 mPa·S when the temperature was above 5℃. Under this condition, the coating could be quickly sprayed by spraying machine. When the temperature was below 5℃, the coating was too adhesive to spray. Under this condition, it was just suitable for using as a brush-applied coating.

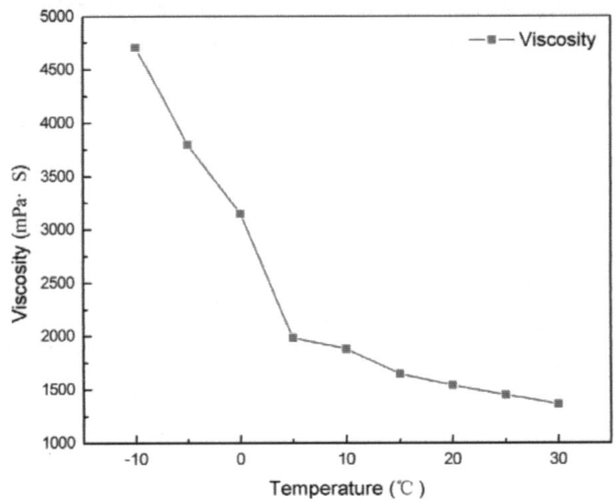

Figure 2. Viscosity-temperature chart of coating C7.

3.4. Coefficients of decontamination

Coefficients of decontamination on different surfaces of the coating were measured according to a previous literature [12], and the result was listed in table 5. The coefficients of α and β decontamination on glass and stainless steel surface were both above 90%, which were higher than those on concrete and marble surface. The above difference was ascribed to the porosity of different surfaces. Concrete had porous surface, therefore the coefficient of decontamination was smaller than those on other non-porous surface. To sum up, coefficients of α and β decontamination on four kinds of surface were both greater than 85%. It indicated the prepared coating was a promising cleaner for radioactive decontamination.

Table 5. Coefficients of decontamination on different surfaces.

Surface	Coefficient of α decontamination/%	Coefficient of β decontamination/%
Concrete	85.1	86.3
Marble	92.2	87.5
Glass	98.4	92.1
Stainless steel	97.6	91.3

4. Conclusions

In order to obtain a high performance coating for radioactive decontamination, the ratio of different components was optimized. The viscosity and coefficient of decontamination were also characterized. The conclusions can be drawn as follows.

(1) The ratio of the main film-forming agent and the additive was optimized. Butyl acrylate used as an additive could decrease the hardness of coating film effectively, and an optimized doping mass-ratio of BA was 10%.

(2) Ca(OH)2 used as a filler could promote the coating film to self-embrittling at a moderate doping mass-ratio of 20%.

(3) Coefficients of α and β decontamination on concrete, marble, glass and stainless steel surfaces were both greater than 85%. It indicated that the prepared coating is a promising cleaner for radioactive decontamination.

Acknowledgements
This work was financially supported by the Outstanding Talents Innovation Foundation of Henan Province (No. 154200510029).

References

[1] Bernaola O A, Filevich A 1970 *Health Physics* **19** 685
[2] Tan M, Whitaker J D and Schwartz D T 2009 *J Hazard Mater.* **162** 1111
[3] Guidi G, Cumo F, Santoli L D 2010 *Clean Techn Environ Policy* **12** 283
[4] Kohli R 2015 *Particle Adhesion and Removal* (Hoboken: John Wiley & Sons, Inc.)
[5] Moore R C, Tucker M D and Jones J A 2009 U.S. Patent 7,514,493
[6] Andre S, Andrieu X and Melin G 2008 U.S. Patent 0,159,706
[7] Liu R L, Li Y T and Zhou Y L 2016 *Chemistry Letters* **45** 1
[8] Wang T Y ang Li S 2014 C.N. Patent 103,695,205
[9] Sebastian G R and Axel H E M 1999 *Macromolecules* **32** 8331
[10] Wang J C, Chen Y H and Wang J H 2016 *e-Polymers* **9** 1079
[11] Bose R K, Hohlbein N and Garcia S J 2015 *Phys Chem Chem Phys* **17** 1697
[12] Wang X D, Wang X M and Xing G P 2001 *Chin J Radiol Health* **10** 167

A shear-mode magnetoelectric heterostructure for harvesting external magnetic field energy

Wei He[1], Jitao Zhang[2], Yueran Lu[1], Aichao Yang[3], Chiwen Qu[1], and Shuai Yuan[4]

[1] School of Information Engineering, Baise University, Baise 533000, China
[2] College of Electric and Information Engineering, Zhengzhou University of Light Industry, Zhengzhou 450002, China
[3] Jiangxi Electric Power Research Institute, Nanchang 330096, China
[4] College of Mechanical and Electric Engineering, Sichuan Agricultural University, Ya'an 625014, China
E-mail: weiheky@yeah.net

Abstract. In this paper, a magnetoelectric (ME) energy harvester is presented for scavenging external magnetic field energy. The proposed heterostructure consists of a Terfenol-D plate, a piezoelectric PZT5H plate, a NdFeB magnet, and two concentrators. The external magnetic field is concentrated to the Terfenol-D plate and the PZT5H plate working in shear-mode, which can potentially increase the magnetoelectric response. Experiments have been performed to verify the feasibility of the harvester. Under the magnetic field of 0.6 Oe, the device produces a RMS voltage of 0.53 V at the resonant frequency of 32.6 kHz. The corresponding output power reaches 44.96 μW across a 3.1 kΩ matching resistor.

1. Introduction

There are a variety of energy sources in our living environment for power generation [1-5]. The external magnetic field has attracted much attention due to its abundance [6-8]. Recently, research has been conducted on magnetoelectric (ME) laminated composites for their potential applications in magnetic energy harvesters. The ME laminated composites typically work in L-T mode, where the piezoelectric constant d_{31} is used. However, the shear piezoelectric constant d_{15} can attain a larger value for the typical piezoelectric materials of PZT5H and PMN-PT [9], and the magnetoelectric response might be further improved.

In this paper, we develop a ME heterostructure for energy harvesting from external alternating magnetic field, which operates in shear-mode. The shear vibration is achieved as a result of the magnetostriction of the magnetostrictive material and the retaining plates. Meanwhile, concentrators are employed to enhance the response of the device. Experiments have been carried out, which validates the feasibility of the proposed device.

2. Structure of the energy harvester

Figure 1 illustrates the structure of the proposed device. Two permalloy concentrators are bonded at both ends of the Terfenol-D plate, which can concentrate the external magnetic field. The NdFeB magnet provides bias magnet field on the Terfenol-D plate. The PZT5H is polarized along 3-direction. Magnetostriction of the Terfenol-D is induced under the external alternating magnetic field. The stress is transmitted to the PZT5H plate which works in shear-mode due to the stiffness of the retaining plates. The PZT5H then generates voltage output. The output power can potentially improve due to the concentration effect and the larger piezoelectric constant d_{15}.

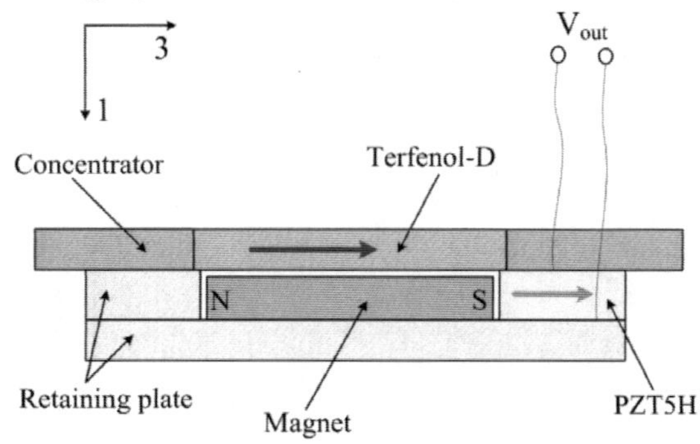

Figure 1. Structure of the proposed magnetic field energy harvester.

3. Analysis

The shear stress is transmitted to the PZT5H as a result of the magnetostriction of the Terfenol-D. At open-circuit condition, the piezoelectric constitutive equations are given by [10]

$$T_5 = c_{55}^{D} S_5 ,\ E_1 = -h_{15} S_5 , \tag{1}$$

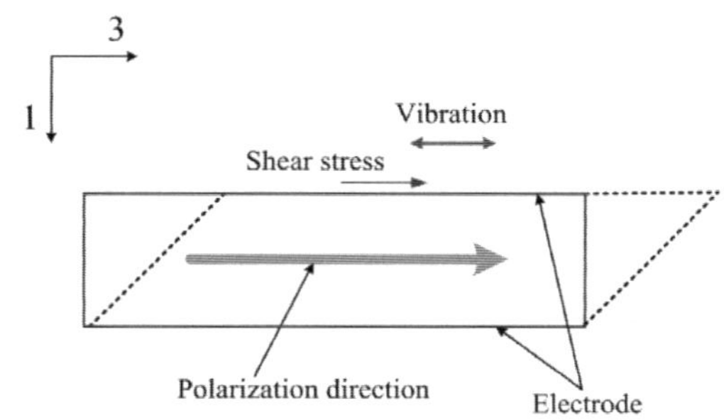

Figure 2. Shear vibration of the piezoelectric PZT5H.

where T_5 is the shear stress, c_{55}^{D} is the elastic stiffness coefficient (at constant D), S_5 is the shear stress, and h_{15} is the piezoelectric stiffness constant. According to equation (1), the electric field in 1-direction can be expressed as

$$E_1 = -h_{15} \frac{T_5}{c_{55}^{D}} . \tag{2}$$

The open-circuit voltage of the PZT5H can be given by

$$V_{open-circuit} = E_1 t = -h_{15} t \frac{T_5}{c_{55}^{D}} , \tag{3}$$

where t is the thickness of the PZT5H plate. Equation (3) indicates that the open-circuit voltage is proportional to the thickness of the PZT5H plate and the shear stress. For a given piezoelectric plate, the response of the device can be enhanced by selecting a suitable NdFeB magnet to attain optimal bias magnetic field.

4. Results and discussions

Figure 3 plots the output RMS voltage of the proposed device as a function of the frequency of the alternating magnetic field at H_{ac} = 0.6 Oe. It can be seen from figure 6 that the maximum output voltage attains 0.53 V at 32.6 kHz.

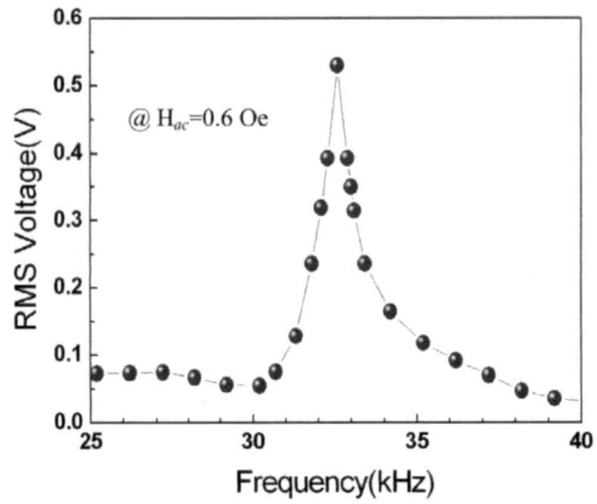

Figure 3. Output voltage as a function of the frequency of the external magnetic field at 0.6 Oe.

At the resonant frequency of 32.6 kHz, a resistance box is connected to the piezoelectric plate to investigate the output performance of the device. The results are shown in figure 4. It can be seen from figure 4 that, as the resistance increases, the output power increases first, and then reaches the maximum value. With the further increase of the load resistance, the power decreases. The maximum output power is 44.96 μW and the optimal load resistance is 3.1 kΩ.

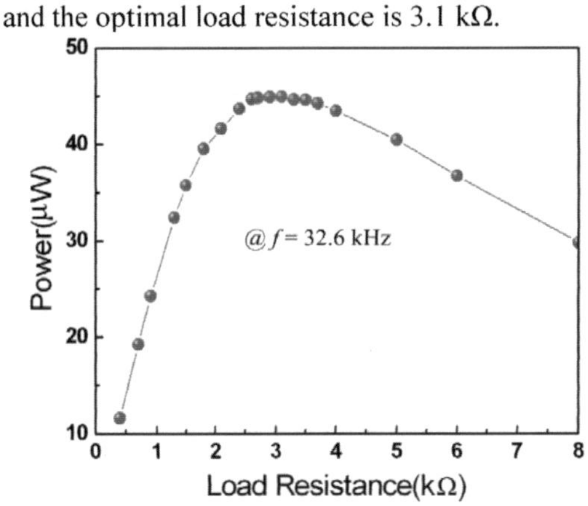

Figure 4. Output power versus load resistance at the resonant frequency of 32.6 kHz.

5. Conclusions

In conclusion, we have investigated a magnetoelectric energy harvester for scavenging external magnetic field energy. The proposed harvester uses permalloy concentrators to concentrate the alternating magnetic field and operates in shear-mode. Experimental results show that the device can feasibly work, with a maximum power of 44.96 μW on a 3.1 kΩ resistor.

Acknowledgments

This work is supported by the National Natural Science Foundation of China (Grant Nos. 61540039 and 61503344) and the Natural Science Foundation of Guangxi Province (Grant No. 2015GXNSFBA139263).

References

[1] Beeby S P, Tudor M J and White N M 2006 *Meas. Sci.Technol.* **17** R175
[2] Gu L and Livermore C 2012 *Smart Mater.Struct.* **21** 015002
[3] Wu H, Tang L H, Yang Y W and Soh C K 2014 *J. Intel. Mat. Syst. Str.* **25** 1875
[4] Wang X, Shi Z, Wang J and Xiang H 2016 *Smart Mater.Struct.* **25** 055005
[5] Palosaari J, Leinonen M, Hannu J, Juuti J and Jantunen H 2012 *J. Electroceram.* **28** 214
[6] Li P, Wen Y and Bian L 2007 *Appl. Phys. Lett.* **90** 022503
[7] Liu G, Ci P and Dong S 2014 *Appl. Phys. Lett.* **104** 032908
[8] He W, Lu Y, Zhang J, Qu C, Che G and Peng J 2016 *Eur. Phys. J. Appl. Phys.* **73** 30903
[9] Ren B, Or S W, Zhang Y, Zhang Q, Li X, Jiao J, Wang W, Liu D, Zhao X and Luo H 2010 *Appl. Phys. Lett.* **96** 083502
[10] Ikeda T 1990 *Fundamentals of Piezoelectricity* (Oxford University Press)

IOP Conf. Series: Earth and Environmental Science **59** (2017) 012028 doi:10.1088/1755-1315/59/1/012028

Determining an appropriate integrated assessment model of tourism safety risk within the Changbai Mountain Scenic Area

Lijun Zhou, Jisheng Liu

Geography Science College of Northeast Normal University
Email: zhoulj990@nenu.edu.cn

Abstract. Tourism safety is gradually gaining more attention due to the rapid development of the tourism industry in China. Changbai Mountain is one of the most famous mountainous scenic areas in Northeast Asia. Assessment on Changbai Mountain scenic area's tourism safety risk could do a favor in detecting influence factor of tourism safety risk and classifying tourism safety risk rank, thereby reducing and preventing associated tourism safety risks. This paper uses the Changbai Mountain scenic area as the study subject. By the means of experts scoring and analytic hierarchy process on quantified relevant evaluation indicator, the grid GIS method is used to vectorize the relevant data within a 1000m grid. It respectively analyzes main indicators associated tourism safety risk in Changbai Mountain scenic area, including hazard, exposure, vulnerability and ability to prevent and mitigate disasters. The integrated tourism safety risk model is used to comprehensively evaluate tourism safety risk in Changbai Mountain scenic area.

1. Introduction

Tourism safety is gradually gaining more attention due to the rapid development of the tourism industry in China. Understanding the influence factors and conducting risk assessments are the key topics of tourism safety risk management. Faulkner (2001)[1], Ritchie(2004)[2] constructed a framework for tourism disaster management. Prideauxa(2003)[3] explored the limits to formal tourism trends forecasting methods. Peter(2010)[4]、 (2014) [5]presented a model that used the principles of risk assessment to evaluate the environmental sustainability in the tourism and recreation sectors, and applied this model in Ireland. The tourism safety risk studies are very popular in Mountainous areas, because these areas are the common destinations of tourists gathering while the terrain is complex and accident-prone. Hu Wenhai(2000)[6] studied the forms of and preventive measures against disasters in mountain tourism; Li Xinjuan(2010)[7] explored the tourism safety evaluation method of mountain scenic; while Cen Qiao(2011)[8] conducted systematic studies on the perception, alerts, and rescues in mountain tourism safety. In general, the studies in mountain tourism safety are relatively fragmented and preliminary, with few quantitative examinations, as most focused only on the qualitative level.

Taking Changbai Mountain as the study area, based on the integrated tourism safety risk model, the authors identify the influence factors of tourism safety risk, quantify the relevant assessment indicators through expert scoring method, and gain the evaluation value, Then discuss the results of tourism safety risk assessment in the Changbai Mountain scenic area.

2. Research methods

2.1 Integrated tourism safety risk model

Tourism safety risk refers to the probability of accidents occurring within the tourism system. The magnitude of tourism safety risk is dependent on the combined effect of the hazard of risk source (H), the exposure (E) and vulnerability (V) of the hazard-bearing body, and the ability of disaster prevention and reduction (R). The relation can be expressed by the following equation:

Tourism safety risk indicator (*TSRI*) = hazard (H) × exposure (E) × vulnerability (V) / (1 + ability of disaster prevention and reduction (R)) (1)

$$H=W_{H1}X_{H1}+W_{H2}X_{H2}+\ldots+W_{Hi}X_{Hi} \quad (2)$$

$$E=W_{E1}X_{E1}+W_{E2}X_{E2}+\ldots+W_{Ei}X_{Ei} \quad (3)$$

$$V=W_{V1}X_{V1}+W_{V2}X_{V2}+\ldots+W_{Vi}X_{Vi} \quad (4)$$

$$R=W_{R1}X_{R1}+W_{R2}X_{R2}+\ldots+W_{Ri}X_{Ri} \quad (5)$$

In Equations (2) through (5), Xi is the weight of i after quantification and Wi is the weight of the indicator i, which represents the relative importance of various factors to tourism risk formation.

2.2 Grid GIS technology

Grid GIS is the application of geographic information system technology in the network environment. The authors vectorize the relevant data within a 1000m × 1000m grid and obtain the tourism safety risk value of the every grid, which is then further classified.

3. Determining an indicator system and a model of tourism safety risk assessment for the Changbai Mountain scenic area

Changbai Mountain is one of the most famous mountain destinations in Northeast Asia. Changbai Mountain started receiving visitors in the 1980s. At present, the tourism routes within China's boundary (as it separates the country from North Korea) are developed along the northern slope, western slope, and southern slope.

3.1 Determining an indicator system of tourism safety risk assessment

According to the integrated tourism safety risk model described above and based on the principles of representation and operability, the authors have further broken down the four factors—hazard, exposure, vulnerability, and the ability of disaster prevention and reduction—in an effort to select more specific assessment indicators for constructing an appropriate tourism safety risk assessment indicator system. This paper selected 18 performance indicators for the tourism safety risk assessment applicable to the Changbai mountain scenic area. The specific indicators are shown in Table 1.

Table 1. Tourism safety risk assessment indicators in the scenic area

Factor	Indicator	Weight
Hazard	Slope level	0.1742
	Altitude level	0.1334
	Precipitation level	0.1248
	Type of surface vegetation cover	0.1475
0.2600	Management standard indicator	0.2275
	Visitor behavior variability indicator	0.1925
Exposure	Visitor ratio	0.3021
0.3021	Facility ratio	0.2025
	Tourism income ratio	0.2353
	Tourism resource amount	0.2600
Vulnerability	Visitor vulnerability indicator	0.4484
0.2025	Tourism capacity	0.3213
	Tourism resource level	0.2302
Ability of	Visitor safety awareness	0.2711
Disaster	Safety management ability	0.1612

Prevention and	Medical rescue ability	0.1967
Reduction	General rescue ability	0.1817
0.2353	Convenience of transportation	0.1891

3.2 Weighted calculation of the tourism safety risk assessment indicators
The authors applied the Analytic Hierarchy Process (AHP) method to the tourism safety risk assessment indicators. During this process, experts in fields such as tourism management, natural disasters, and tourism geography were invited to score on the importance of various indicators to their corresponding hierarchy level, before final weights were derived.

3.3 Classification and scoring of the various indicators
After determining the weights of the various indicators using the above AHP method, we then used a five-level scoring method to assign values (i.e., 1 through 5) to each basic-level factor, and divided each indicator into five classes. The classification was done based on the actual status of the assessment indicators, followed by the assignment of different values.

4. Results of total tourism safety risk assessment for Changbai Mountain scenic area

4.1 Hazard assessment for Changbai Mountain scenic area
Based on the weights and scores of Changbai Mountain scenic area's tourism safety risk hazard indicators, the authors then calculated by the integrated tourism safety risk model. The result is spread in the GIS to obtain the hazard assessment map of the area (Figure 1).

Figure 1 shows that the hazards in the entire scenic area generally form a semi-circle around Tianchi Lake (reflected in red), indicating that the area has the highest degree of hazard, with its distribution range widest in the northern slope, followed by the southern slope, and the western slope. Grids with a moderate level of hazard are mainly distributed in the western and southern slopes, the northern slope having a slightly lower level than the others. Areas marked in yellow depict relatively low hazard areas, and their distribution is more fragmented, concentrated mostly along the edges of the southern and western scenic areas. Most of these areas are distanced away from tourism routes. Areas marked in green are located at the center of the volcanic cone and indicate a weak hazard level. The rest are areas with only minor hazards.

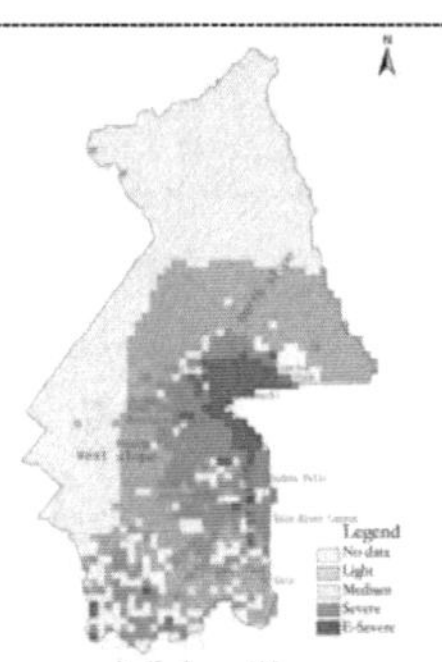

Figure 1. The hazard of tourism safety risk for Changbai Mountain scenic area

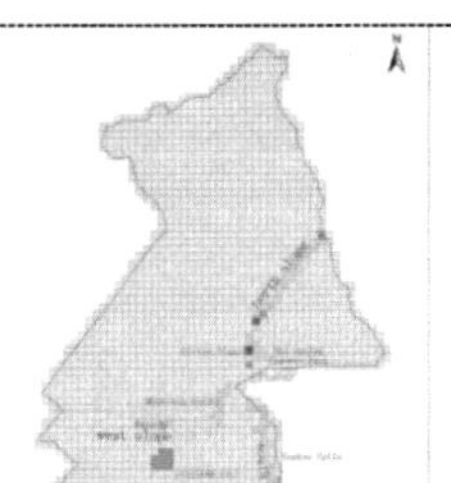

Figure 2. The exposure of tourism safety risk for Changbai Mountain scenic area

4.2 Safety exposure assessment for Changbai Mountain scenic area
Based on the weights and scoring of the exposure indicators for the Changbai Mountain scenic area's disaster occurrence, we then used the integrated tourism safety risk computation model to obtain the calculations shown in Figure 2. Two units—the car station and hot spring at the northern slope—are under extremely high exposures. These are also the areas with the highest tourist concentrations. Areas

under severe exposure, as a result of high tourism volume or concentration of resources, include the great canyon at the western slope, falls at the northern slope, Tianchi Lake, and entrances. Other destinations along tourism routes are under moderate exposure, while areas with slight exposure are largely tourism channels in the scenic area.

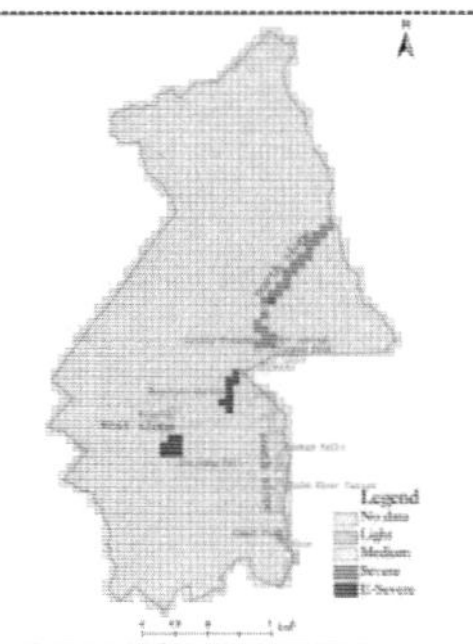

Figure 3. The vulnerability of tourism safety risk for Changbai Mountain scenic area

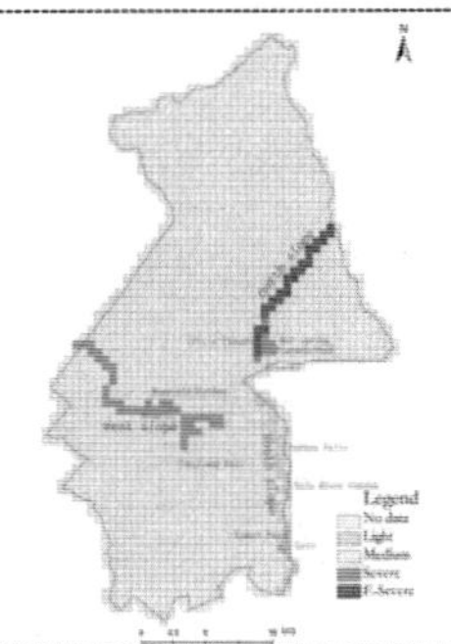

Figure 4. The ability of disaster prevention and reduction of tourism safety risk for Changbai Mountain scenic area

4.3 Safety vulnerability assessment for Changbai Mountain scenic area

Based on the weights and scoring of the vulnerability indicators for the Changbai Mountain scenic area, the authors then used the integrated tourism safety risk model for the calculations shown in Figure 3. In general, areas near the great canyon and mountain trails at the western slope are deemed extremely vulnerable; other areas along the northern slope and western slope routes are largely severe and moderate vulnerable areas. The southern slope routes, excluding Tianchi and the Camel Peak, are largely areas of slight vulnerability.

4.4 Disaster prevention and reduction ability assessment for Changbai Mountain scenic area

Based on the weights and scoring of the Changbai Mountain scenic area's ability of disaster prevention and reduction, we then used the integrated tourism safety risk computation model for the calculations shown in Figure 4. The northern slope has a greater ability of disaster prevention and reduction, which is largely attributable to its more advanced development and more comprehensive management organization and facilities. However, there are several areas with lower ability in the northern slope those are mainly inaccessible, as they are distanced from roads. The western slope's ability to prevent and reduce disasters is generally good, which have moderate ability in preventing and reducing disasters. The southern slope is furthest from the rescue center and has the weakest ability in the scenic area.

4.5 Results of total risk assessment for Changbai Mountain scenic area

Using the above studies on hazard, exposure, vulnerability, and the ability to prevent and reduce disasters, and the integrated tourism safety risk assessment equation, we obtained the total assessment value of Changbai Mountain scenic area's tourism safety, as illustrated in the following figure:

Figure 5 shows that the main tourism routes of individual slopes have relatively large differences in terms of their risk levels. Areas near Tianchi and the falls at the northern slope have the highest risk; some spots, such as Tianwen Peak, little Tianchi, hot springs, and the slope's entrance have high risks; the rest of the areas are moderate risk areas. As for the western slope, its highest risk level is lower than that of the northern slope, with destinations along the tourism route having relatively high risk levels. For the southern slope, areas with relatively high risks include the entrance, Yuehua Falls, Camel Peak , while the rest are moderate risk areas.

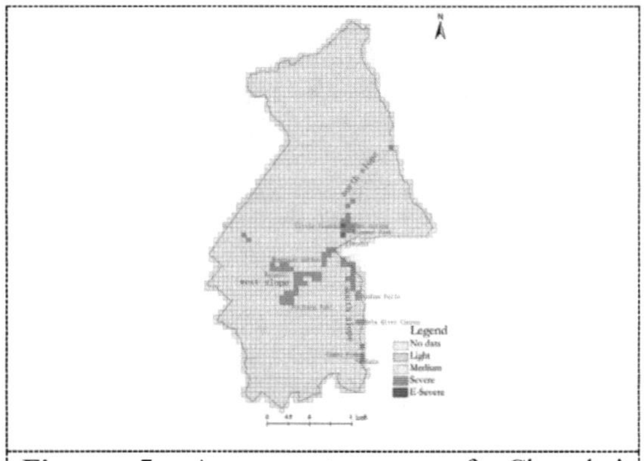

Figure 5. Assessment map of Changbai Mountain scenic area's tourism safety risk level.

5. Conclusions

This paper used the 1000m × 1000m scale standard grid as the basic unit for assessing tourism safety risks of the Changbai Mountain scenic area. With comprehensive considerations of the hazard, vulnerability, exposure, and the ability of disaster prevention and reduction for tourism safety accidents, using the integrated tourism safety risk indicator method and GIS techniques, we were able to derive a tourism safety risk assessment map for the area. Results show that areas with the greatest total risks are located around Tianchi Lake and the falls at the northern slope. Areas with relatively high risks include Tianwen Peak, little Tianchi, hot springs, car station, and the entrance at the northern slope, as well as the great canyon and major tourism routes at the western slope and its entrance, the Yuehua falls, Camel Peak, and the upstream tourism routes leading to the Tian at the southern slope. The rest of the areas show only moderate risk.

References

[1] Faulkner , B. Towards a framework for tourism disaster management[J] . Tourism Management. 2001，(22)：135-147.

[2] Ritchie B W.Chaos，Crises and Disasters: A Strategic Approach to Crisis Management in the Tourism Industry[J].Tourism Management，2004，(25):669-683.

[3] Prideauxa B，Lawsb E，Faulkner B. Events in Indonesia:Exploring the limits to Formal Tourism Trends Forecasting Methods in Complex Crisis Situations[J].Tourism Management，2003，(24):475-487.

[4] Peter Roe. A Risk Assessment Based Model for Assessing the Environmental Sustainability of Tourism and Recreation Areas. Dublin Institute of Technology .2010，(25):669-683

[5] Peter Roe , Victor Hrymak, Frederic Dimanche. Assessing Environmental Sustainability in Tourism and Recreation Areas: a Risk-assessment-based Model. Journal of Sustainable Tourism,2014,22(2):319-338.

[6] Wenhai Hu, Lailin Hu. Calamities in Tourism Mountain and their Prevention Strategy. Journal of Mountain Science.2000，18（6）：576-579.

[7] Xinjuan Li. Tourism Crisis Evaluation and Prevention for Mountainous Scenic Areas. Journal of Henan Polytechnic University. 2010,11(4)：158-162.

[8] Qiao Cen, Ying Huang. Research on Perception and Attitude toward Tourism Safety of Mountain Scenic Areas——Based on Investigation on Tourists and Staffs of Mountain Scenic Areas. Technology and Market.2011,18(6):347-350.

IOP Conf. Series: Earth and Environmental Science **59** (2017) 012029 doi:10.1088/1755-1315/59/1/012029

Study on highway transportation greenhouse effect external cost estimation in China

Chunchao Chu, Fengming Pan

China Academy of Transportation Science, Beijing, 100029, China

chucc79@163.com

Abstract. This paper focuses on estimating highway transportation greenhouse gas emission volume and greenhouse gas external cost in China. At first, composition and characteristics of greenhouse gases were analysed about highway transportation emissions. Secondly, an improved model of emission volume was presented on basis of highway transportation energy consumption, which may be calculated by virtue of main affecting factors such as the annual average operation miles of each type of the motor vehicles and the unit consumption level. the model of emission volume was constructed which considered not only the availability of energy consumption statistics of highway transportation but also the greenhouse gas emission factors of various fuel types issued by IPCC. Finally, the external cost estimation model was established about highway transportation greenhouse gas emission which combined emission volume with the unit external cost of CO_2 emissions. An example was executed to confirm presented model which ranged from 2011 to 2015 Year in China. The calculated result shows that the highway transportation total emission volume and greenhouse gas external cost are growing up, but the unit turnover external cost is steadily declining. On the whole overall, the situation is still grim about highway transportation greenhouse gas emission, and the green transportation strategy should be put into effect as soon as possible.

1. Introduction

Climate change is a crucial issue that concerns about the survival and development of human beings. It comes to a consensus gradually that carbon dioxide (CO_2), methane (CH_4), nitrous oxide (N_2O) and hydrogen fluoride (HFCs) eventually led to global climate change that we produced during all production and life activities. Carbon dioxide is the most important factors in greenhouse gases. From global perspective, highway transportation is one of the three main carbon sources of greenhouse gas emissions [1]. According to the 2007 EU Transport Ministers Meeting *Report of Carbon Dioxide Emissions Reducing in Transportation*, in 2003, greenhouse gases CO_2 accounted for 34% of the total emission which was produced by the fuel consumption in the national transportation industry (including the commercial and private transportation) of Organization for Economic Cooperation and Development (OECD). Among 34%, the highway transportation accounted for 23%, and other mode transportations accounted for 11%. It was shown that if the concentration of CO_2 in the atmosphere was doubled, the average temperature of the earth would be raised by 2.5°C [2]. The greenhouse effect causes global warming, polar ice and plateau glaciers melting, the sea level rise, the ecological system imbalance in mid-latitude region, frequent sudden climate disasters, diseases and epidemics spreading, and even serious impaction on human beings' survival and development. According to WHO's estimation, global climate change led to more than 100 thousand people losing their lives each year, which caused economic losses was equivalent to 5% of global GDP.

The external cost of highway transportation greenhouse effect mainly refers to the economic losses which were caused by the greenhouse gas emissions, such as CO_2 and so on, which brought about the global warming. Due to the large population and rapid economic development in China, the whole volume of energy consumption and greenhouse gas emissions was very large, and was increasing year by year in China. It was eventually replacing the United States as the largest major source of greenhouse gas emissions. With rapid growth of vehicles and huge consumption of fuel, it is more urgent to construct the estimated model on highway transportation greenhouse gas emissions and external cost, and moreover, put forward the management and technical measures of carbon reduction and energy saving. Based on the analysis of the composition and characteristics of highway transportation greenhouse emission gases, the model was presented which estimated the emission volume and external cost in highway transportation. Moreover, an example was concretely executed to confirm presented model which ranged from 2011 to 2015 Year in China. At last, some suggestion is putted forward for the further management and technical measures of energy saving and carbon reduction. The research results can provide decision-making references for the highway transportation management department, which includes allocating reasonably the transportation resources and implementing the low-carbon transportation project.

2. Composition and characteristics of greenhouse gases

Greenhouse gases include not only CO_2 but also CH_4, N_2O and CFCs, and so on. It was found that the greenhouse effect brought by various types of gases varied very large , which not only related to each type of gas content in the atmosphere, but also with retained time and life of each type of gas in the atmosphere, in other words, greenhouse effect was a function of greenhouse gases content and life. In table 1, it was listed about five type greenhouse gas content, life, sharing ratio and warming potential. Warming potential refers to a ratio of a certain greenhouse gas and CO_2 on the impacts of climate change, regarded as the relative effect index of each type of greenhouse gas [3].

Table 1: The eigen values of various greenhouse gases

greenhouse gases	Content (Volume fraction)(10^{-6})	Content growth rate (%)	Life (year)	Share ratio (%)	Warming potential
CO_2	354	0.5	50-200	55	1
CH_4	1.7	1.1	11	15	32
N_2O	0.3	0.3	150	6	150
CFC-11	0.0002	5	75	17	14000
CFC-12	0.00032	5	111		17000

The table reflects that the contribution rate of CO_2 was 55%, which is the most important greenhouse gas. CH_4, whose greenhouse effect is 32 times that of CO_2, has shorter life and the residence time is 11 years in the atmosphere. Chlorofluorocarbons (CFCs) are some synthetic chemicals that are widely used in air conditioning, foaming agent and solvent. Among CFCs, CFC-11 and CFC-12 play a major role in greenhouse effect, with extremely low concentration in the atmosphere, but in recent years they grow rapidly at an annual rate about 5%, with not only their warming potential as high as 14000 and 17000 but also long survival time.

3. Affecting factors of highway transportation greenhouse effect external cost

To calculate the external cost of highway transportation greenhouse effect, two type data were needed. For one thing, the energy consumption of highway transportation and its corresponding greenhouse gases emissions. For another, the amount of economic and social losses, which were caused by the unit of greenhouse gas emissions, known as the monetization of ecological damage, human health, agriculture, raw materials and ecosystems.

3.1 Energy consumption of highway transportation

In current statistical system in China, there is no direct statistical data of the energy consumption corresponding to highway transportation. In consideration of both the availability of data and using statistical data to improve data quality as possible, it was adopted which included transportation turnover and unit consumption to reckon the total highway transportation energy consumption. The calculated model is shown below in the formula (1).

$$FQ = \sum_{i=1}^{3} A_i \times TQ_i \tag{1}$$

Where FQ: Fuel consumption of highway transportation;

A_i: unit consumption corresponding to mode "i" of highway transportation;

TQ_i: transportation turnover corresponding to mode "i" of highway transportation;

i: mode "i" of highway transportation, including highway passenger transport, freight transport, urban public transport.

With the energy consumption data of highway transportation and presented greenhouse gas emission model, it can be calculated about greenhouse gas emission volume from highway transportation. Greenhouse gas emission model can be referred in Section 4.1.

3.2 Unit greenhouse gas external cost

United Nations Framework Convention on Climate Change estimated that each emission ton of CO_2 cost about \$5-25 (1990). National Research Council (1991) suggested that the emission cost per ton of CO_2 should be in the range of \$10-20 in the United States. According to the latest relevant research, the international market price of carbon emission trading was 17 euros per ton of CO_2 in general [4], while the domestic transaction price was range from 100 to 150 yuan. According to both these reference values and Chinese actual conditions, our research adopted 120 yuan per emission ton of CO_2 to estimate the external cost of Chinese highway transportation greenhouse gas emissions.

4. Estimated method of greenhouse effect external cost

In order to calculate the external cost of the greenhouse effect of highway transportation, two models need to be established. The first one was the greenhouse gas emission model, which could calculate the different greenhouse gas emission according to the energy consumption of highway transportation. The second one was the external cost estimated model of greenhouse effect, which would quantify and monetize the external effects of greenhouse gas emissions from highway transportation.

4.1 Greenhouse gas emission model

There are two kinds of quantification models about highway transportation greenhouse gas emissions: energy consumption statistics and vehicle test. The first model is widely used in the statistics of greenhouse gas emissions in the transportation department, such as the US Environmental Protection Agency greenhouse gas emission estimated methods, the WRI carbon emission model for motor vehicles, the *IPCC National Greenhouse Gas Inventories Guide* vehicle carbon emission model, and the NEMS model developed by the US Energy Information Administration (EIA). The methods and assumptions among these models varied widely, and the estimated results of greenhouse gas emissions distinguished obviously. In this research, based on these emission models, an improved greenhouse gas emission model was presented which considered not only the availability of energy consumption statistics of highway transportation but also the greenhouse gas emission factors of various fuel types issued by IPCC. The model was shown in equation (2).

$$GQ_j = \sum_{k=1}^{n} B_{kj} \times FQ_k \tag{2}$$

Where: GQ_j: emission of greenhouse gas "j"

B_{kj}: emission factor of greenhouse gas "j" corresponding to fuel "k"

FQ_k: consumption of fuel "k"

Where emission factor of greenhouse gas B_{kj} can be calculated as follows.

$$B_{kj} = \text{net calorific power} \times \text{carbon content} \times \text{oxygenation efficiency} \times \text{carbon conversion coefficient}$$

Where the net calorific power is the calorific value that per unit fossil fuel is completely combusted, known as the net heating value. The carbon content is the amount of carbon emitted by per unit calorific value of the fuel. The oxygenation efficiency is the carbon oxidized ratio. Carbon conversion coefficient is the carbon conversion coefficient of carbon dioxide, known as 44/12 = 3.67.
According to "the Report on Climate Change (2009)", the net calorific power, carbon content and oxygenation efficiency of various types of fuels were shown in Table 2. According to equation (3), the carbon emission factors of gasoline, diesel and natural gas were respectively 2.98 tCO_2/t, 3.16 tCO_2/t, 2.18tCO_2/t. Furthermore, according to the research [5], the relevant emission coefficients and factors CH_4 and N_2O corresponding to other fuels can be calculated and shown in Table 2.

Table 2: the relevant emission coefficients and factors of various types of fuels

	net calorific power (kJ/kg)	carbon content① ($kgCO_2/TJ$)	oxygenation efficiency②	emission factor (tCO_2/t)③	(tCH_4/t)	(tN_2O/t)
diesel	42650	20.20	1	3.16	0.55	0.09
gasoline	43070	18.90	1	2.98	3.83	0.09
natural gas	38931	15.30	1	2.18	0.28	0.04

Source: Report on Climate Change (2009). Note: ①emission factors of each fuel take the 95% confidence limit for each fuel emission factor, data from: *2006 IPCC Guidelines for National Greenhouse Gas Inventories Volume 2 Energy*; ② According to *2006 IPCC Guidelines for National Greenhouse Gas Inventories*; ③ The fuel calorific value is based on the low calorific value data.

4.2 Greenhouse effect external cost model
As global climate change has become one of the hot issues of the world, carbon emissions trading has become more popular. The model for estimating the external cost of the greenhouse effect from highway transportation is shown as follows:

$$EC = PC \times \sum_{j=1}^{m} C_j \times GQ_j \quad (j=1, 2, \ldots, m) \tag{3}$$

Where PC: The external cost of unit CO_2 emission;
C_j: The conversion coefficient of the greenhouse gas "j" to CO_2, known as the warming potential heating factor of the greenhouse gas "j", referred to Table 1;
GQ_j: Emission of the greenhouse gas "j".

5. Application Example

Since 2011, unit energy consumption of highway transportation began to sample statistics in Chinese statistical system. According to the China Statistical Bulletin of Transportation, transport turnover and unit energy consumption were shown in Table 3 about highway passenger transport, freight transport, urban public transport in China during 2011-2015.

Table 3: The energy consumption of highway transportation

	Year	2011	2012	2013	2014	2015
Freight	Transport turnover (100 million t·km)	51375	59535	55738	61017	57956
	Unit energy consumption (kg standard coal/100 ton·km)	2.2	1.7	1.9	2	1.9
Passenger	Transport turnover	51375	59535	55738	61017	57956

	(100 million t·km)					
	Unit energy consumption (kg standard coal/1000 person·km)	11.3	11.7	11.6	12.1	12.6
Urban public	Transport turnover (100 million person)	1094	1141	1174	1190	1163
	Unit energy consumption (ton standard coal/10000 person)	1.4	1.4	1.5	1.4	1.5
Total	ton standard coal (10 thousand ton)	14728	13879	13656	15332	14110

Note: 1) urban rail transit is deducted from urban public statistics.

It was classified transportation, warehousing and postal services as a whole in the energy consumption statistics system in China. With the total energy consumption of highway transportation, combined with the proportion of gasoline and diesel consumption in the whole energy statistics, the energy consumption of gasoline and diesel from highway transportation can be further calculated, and the results were shown in Table 4.

Table 4: The classified energy consumption of highway transportation

Year		2011	2012	2013	2014	2015
Transport, postal and warehousing as a whole	diesel(10000-ton)	9485	10727	10920	11746	12635
	gasoline(10000-ton)	3374	3753	4381	5035	5786
	natural gas(100 million m3)	138	154	176	200	228
Proportion of conversion standard coal	diesel	68%	68%	65%	63%	62%
	gasoline	24%	24%	26%	27%	29%
	natural gas	8%	8%	9%	9%	9%
Highway transportation energy consumption	ton standard coal (10000-ton)	14728	13879	13656	15332	14110
Classified energy consumption	diesel(10000-ton)	6827	6467	6088	6681	6004
	gasoline(10000-ton)	2429	2262	2442	2864	2750
	natural gas(100 million m3)	99	93	98	114	108

Note: 1) according to the energy ratio, 1 ton gasoline is equivalent to 1.4714 tons standard coal, 1 ton diesel is equivalent to 1.4571 tons standard coal, 10000 m^3 natural gas is equivalent to 12.143 tons standard coal.

2) The consumption fuels of highway transportation are mainly diesel, gasoline and natural gas. Kerosene and other fuels account for less, and are combined into the aforementioned three types during calculation.

3) In current energy consumption statistics, the proportions of energy consumption were released only to 2013, so the proportions after 2013 were calculated based on the existing proportion trend.

According to the model presented in the previous section and greenhouse gas emission factors, with the energy consumption of highway transportation, the greenhouse gas emissions could be estimated including CO_2, CH_4 and N_2O from commercial vehicles during 2011-2015, as is shown in Table 5. The warming potentials of CH_4 and N_2O are 32 times and 150 times of CO_2 respectively.

Table 5: The greenhouse gas emissions of highway transportation Unit: 10^4 ton

Year	Fuel type	Fuel consumption	CO_2 emission	CH_4 emission	N_2O emission	Conversation CO_2 emission
2011	diesel	6827	21575	3755	614	233908
	gasoline	2429	7237	9302	219	337680
	natural gas	99	1408	181	26	11067

	total		30220	13238	859	582654
2012	diesel	6467	20435	3557	582	221551
	gasoline	2262	6742	8665	204	314577
	natural gas	93	1316	169	24	10343
	total		28493	12391	810	546472
2013	diesel	6088	19238	3348	548	208578
	gasoline	2442	7279	9355	220	339603
	natural gas	98	1390	179	26	10932
	total		27907	12882	793	559113
2014	diesel	6681	21112	3675	601	228892
	gasoline	2864	8534	10968	258	398172
	natural gas	114	1614	207	30	12688
	total		31260	14850	889	639752
2015	diesel	6004	18973	3302	540	205697
	gasoline	2750	8195	10533	248	382360
	natural gas	108	1530	197	28	12032
	total		28698	14031	816	600089

Note: The measurement unit of natural gas is billion m^3, 10000 m^3 is equivalent to 6.5 tons in ordinary pressure.

At 120 yuan per ton of CO_2, both the total external cost of greenhouse gas emissions and the external cost of unit conversion transport turnover can be obtained from highway transportation in China. Further, the proportion may be calculated about the external cost of greenhouse effect from highway transportation in China GDP, as was shown in Table 6.

Table 6: Greenhouse gas emission external cost and the proportion in GDP

Year	Conversion transport turnover (100 million ton·km)	Greenhouse gas external cost (100 million RMB)	Unit turnover external cost (RMB / ton·km)	Proportion in GDP (%)
2011	53051	6992	0.13	1.44
2012	61382	6558	0.11	1.23
2013	56863	6709	0.12	1.14
2014	62225	7677	0.12	1.21
2015	59030	7201	0.12	1.06

Note: Conversion transport turnover = Freight turnover + Passenger turnover/10

As a whole of highway transportation, the external cost of greenhouse effect was growing up in 2011-2015, but the external cost of unit transport turnover steadily declined. The results indicated that the growth rate of greenhouse effect external cost caused by highway transportation was less than that of conversion transport turnover. So it can be seen that the highway transportation was gradually developing into low-carbon transport in China. At the same time, with the rapid development of national economy and the obvious growth of the economic scale, the proportion of the external cost of greenhouse effect from highway transportation in GDP will decrease year by year in the future.

6. Conclusions

According to the research above, the current greenhouse gas emissions from highway transportation are in a steady and good trend, but the overall situation is still grim. Thus the green transport strategy should be put into effect as soon as possible. First and foremost, carrying out the clean energy vehicle project, which will strengthen the access management of transport market, lay down strict vehicle emission standards, and accelerate the use of clean fuel transport. Secondly, adjusting highway transportation energy consumption structure, which may encourage to increase the proportion of diesel vehicles in large and medium vans, and improve continuously the ratio of clean energy consumption to reduce environmental pollution and greenhouse gas emissions from the origin. Thirdly, developing integrated transport, which will play comparative advantages of various transport modes. Compared with road transport, waterways and railways are more environmental-friendly and have unique advantages in per unit traffic energy utilization. More attentions should be paid attention to the joints among various modes of transport, and play combination advantages of transport modes. Finally, but not least, public transport should be developed actively. Non-commercial vehicles account for a large proportion in vehicle ownership in China. Along with the popularization of private cars, greenhouse gas emissions from non-commercial vehicles and their external costs are increasing year by year. Therefore, it should be actively promoted and encouraged about public transport.

References

[1] Yu, Y.C. & Yu, M.Y., Estimation and forecast of for-hire motor vehicle emission volume. *Journal of Highway and Transportation Research and Development*, 25, pp. 154-158, 2008.
[2] Li, Y., Study on Road Transport Sustainable Development Model in China. Beijing: China Communications Press, 2014.
[3] Mei, J., Fan, Q.H., Zhao, Y.C., *Greenhouse Gas Emission Reduction and Control Technology in Transportation*. Beijing: Chemical Industry Press, 2009
[4] International Energy Agency (IEA). 2012 Key World Energy Statistics:33. Paris, 2012
[5] Ou, X.M., Zhang, X.L., *Life-cycle Analysis of Automotive Energy Pathways in China*. Beijing: Tsinghua University Press, 2011

Demographic variables in coal miners' safety attitude

Wen-wen Yin, Xiang Wu*, Hui-Peng Ci, Shu-Qi Qin, Jia-Long Liu

(School of Engineering and Technology, China University of Geosciences, Beijing, 100083)
E-mail: wuxiang@cugb.edu.cn

Abstract: To change unsafe behavior through adjusting people's safety attitudes has become an important measure to prevent accidents. Demographic variables, as influential factors of safety attitude, are fundamental and essential for the research. This research– does a questionnaire survey among coal mine industry workers, and makes variance analysis and correlation analysis of the results in light of age, length of working years, educational level and experiences of accidents. The results show that the coal miners' age, length of working years and accident experiences correlate lowly with safety attitudes, and those older coal miners with longer working years have better safety attitude, as coal miners without experiences of accident do.However, educational level has nothing to do with the safety attitude. Therefore, during the process of safety management, coal miners with different demographic characteristics should be put more attention to.

1.Introduction

Mining industry is considered as one of the most dangerous industries in China[1]. Over the past one decade and more, coal mine accidents take a large proportion in industrial accidents. Although China has been improving the level of technology to reduce the accident rates, there is a big gap between China and developed countries in light of mortality of million tons of coal mine. At the same time, studies reveal that 88% of the accidents are caused by human factors [2],while human's unsafe behavior is the main factor leading to accidents in coal mining industry[3,4].Only improving the safety technology and facilities can't reduce the accident rate significantly[5]. How to reduce the occurrence of accidents caused by human factors through various effective ways in order to enhance the safety performance of enterprises has become an urgent problem for scholars in various countries. Lund suggests that accidents can be ultimately prevented by changing behavior through the change of attitudes [6]. Adjusting safety attitudes and thus changing unsafe behavior is an important measure to prevent accidents[7,8].

According to the accident proneness theory, some people are more possibly to experience accidents. For employees of different demographic characteristics, the possibility of experiencing accidents is different. Therefore, the research of demographic variables in the field of safety is important and fundamental. Kerlinger also believes that demographic variables have great importance in sociological researches[9]. However, there are few studies of demographic variables, and current studies mainly focus on the aspects such as work stresses [10], leadership behavior [11, 12], and so on. In view of this, we will do research from the perspective of the demographic variables, to study the influence of education, age, length of working years and accident experience on the safety attitudes of coal miners.

2.Research process

2.1. Literature Research and Hypothesis

The impact of demographic variables on safety in different industries has been discussed in the past. Guo Dong-chen finds the demographic characteristics impact on the safety management of highway bridge and tunnel construction teams, and thus established a safety management model[13]. Gao Guo-ping points out that the staff with higher education have more mature safety attitudes[14].Bu Meng-meng points out that age, driving years and educational level affect drivers' awareness of risk. And a man with longer driving years and higher educational level has higher level of risk awareness[15]. In manufacturing enterprises, the employees' safety attitudes are getting better with the increase of their ages. The safety attitudes of the employees with college or higher academic qualifications are obviously higher than that of employees who have the junior education. The longer their working years are, the stronger their safety awareness is, and thus they have better safety attitudes[16]. In construction industry, Siu suggests that age has an impact on the safety attitude of construction workers, and the results show that the older the construction workers are, the more positive attitudes of safety they have[17]. There are other researches about coal miners suggest that the employees' age, length of working years and other variables have impact on their intentions of unsafe behavior[18]. Therefore, this paper proposes the following hypotheses:

H1: Educational level influences the safety attitudes of coal miners, and those with higher academic qualifications have better safety attitudes than those less educated.

H2: Age influences the safety attitudes of coal miners, and older coal miner workers have better safety attitudes than younger ones.

H3: Length of working years influences the safety attitudes of coal miners, and senior workers have better safety attitudes.

The study of demographic variables reveals that variables such as age, working years and educational level are the demographic variables which are of great concern to the researchers, and few studies have been done of the accident experiences. Some scholars research the chemical industry, and point out that the safety attitudes can be affected by the frequency that accidents happen[19].Therefore, this study will also take into account the influence of the accident experiences on the safety attitude of coal miners, and propose the following hypothesis:

H4: Accident experiences influence the safety attitudes of coal miners, and those who have accident experiences have better safety attitudes.

2.2. Research methods

This research uses literature research, questionnaires, and data analysis to study the relationship between demographic variables and safety attitude. We analyze the demographic variables that affect the safety attitudes of different industries, and explore variables that may influence the safety attitudes of coal miners in demographic variables. The safety attitudes and demographic data of the workers are obtained from the questionnaires of safety attitude measurement. We employ SPSS22.0 to do single factor variance analysis and independent sample t-test to analyze the influence of these factors on the safety attitudes.

2.3. research tool

The demographic variables and safety attitudes of coal miners are studied in the form of questionnaires. The safety attitude measurement questionnaire used in this study are based on Cox[20] and Donald's[19] study, and the safety attitude questionnaire of coal miners initially developed by Liu Jia-long[21]. The questionnaire includes four dimensions: safety commitment of managers, work pressure, team safety atmosphere, fatalism, and has 17 items. As to demographic variables, this paper mainly investigates the four aspects: age, working years, educational level and accident experiences.

3.Data analysis

808 questionnaires were distributed and 808 questionnaires were given back. After eliminating the invalid questionnaires,593 effective questionnaires were left, and the effective rate is 73.4%. The demographic information is shown in Table 1.

Table 1. The demographic information of the sample

Statistics		Number	Percentage（%）
Age	21~30years old	337	58.6
	31~40 years old	149	25.9
	41~50 years old	78	13.6
	51~60 years old	11	1.9
Length of working years	1~2years	109	18.5
	3~4years	204	34.7
	5~6years	128	21.8
	Seven years and above	147	25.0
Educational level	Primary school	6	1.0
	Junior high school	114	19.2
	High school or secondary school	231	39.0
	University or college	216	36.4
Accident experiences	Without	102	17.2
	With	411	69.3

3.1. Reliability and validity analysis

The Cronbach's α is analyzed by doing exploratory factor analysis with the help of SPSS 22.0. The results are shown in Table 2.

Table 2. The reliability test of safety attitude scale

	Cronbach's α	Items number
fatalism	0.929	4
safety commitment of managers	0.778	5
work pressure,	0.764	4
team safety atmosphere,	0.642	4
Total questionnaire	0.872	21

Generally speaking, the internal consistency of the questionnaire reliability is insufficient when Cronbach's α is not more than 0.6.The results show that the value of total questionnaire's Cronbach's α is 0.872, and each dimension's Cronbach's α is 0.642~0.929, each one is not less than 0.6, which is acceptable. Based on the exploratory factor analysis, a formal questionnaire was developed to measure coal miners' safety attitudes. The next step was to perform a confirmatory factor analysis(CFA) to validate the questionnaire by analyzing $\chi 2$ / d.f., GFI, AGFI, TLI, CFI, RMSEA structural models. CFA is shown in Table 3, the measurement models all reach the requirements of the ideal model and are of a good degree of fit. In summary, the safety attitude questionnaire has good validity and reliability, and can effectively measure the safety attitudes of coal miners, which is the prerequisite for later rational research.

Table 3. Confirmatory factor analysis

Structural model	χ^2/d.f.	GFI	AGFI	TLI	CFI	RMSEA
Ideal model	<3	>0.90	>0.90	≥0.95	≥0.95	<0.08
Measurement model	2.86	0.93	0.91	0.95	0.96	0.06

3.2. An Analysis of the Relativity between Demographic Variables and Safety Attitudes

As the demographic variables in the age, working years, and educational level are divided into more than two different groups, and variance analysis should be used to analyze the differences in safety attitudes held by different groups, and then conduct Post hoc analysis on the dimensions where differences exist to find the differences between the groups. Finally, the correlation intensity is analyzed between the various dimensions and safety attitudes. The accident experiences are divided into two groups, workers with accident experiences and without accident experiences, and they are tested by independent sample t-test.

3.2.1. The Influence of Educational Level on Safety Attitude. From Table 4, we can see that the overall test F value of academic record is 2.387, P value is 0.061 (>0.05), not reaching the significant level. It means that there is no correlation between educational level and safety attitudes, i.e. there is no difference among coal miners of different educational level. It is different from the assumption in H1 that educational level influences the safety attitudes of coal miners, and those with higher educational level have better safety attitudes. As a result, H1 is not supported.

Table 4.ANOVA of sample' safety attitude

Variable	F-value	P-value
Age	4.286	0.03
Length of working years	2.914	0.00
Educational level	2.387	0.061

3.2.2. The Influence of Age on Safety Attitude. The P-value of age is 0.03,less than 0.05 (as shown in Table 4), reaching a significant level. However, the variance analysis F statistic belongs to the whole test, when the value of F is significant, it only means that there are at least two groups have difference between the mean value of the dependent variables. To identify the two groups that are different in the average of independent variable, post hoc analysis is still required. Using Tamhane's T2 method, we find that the mean difference in safety attitudes between group 1 and group 3 is -4.53032 *(* indicates significance).The results showed that the safety attitudes are different between age groups of 21 ~ 30 and 41 ~ 50.The safety attitude level of 41 ~ 50 age group is higher than 21 ~ 30 age group. The specific results are shown in Table 5.

Table 5.the post hoc analysis of the different age's safety attitude

	(I) age groups	(J) age groups	Average difference (I~J)	Significance
Safety attitude	21~30	31~40	-1.52120	0.596
		41~50	-4.53032*	0.016
		51~60	1.32749	1.000
	31~40	21~30	1.52120	0.596
		41~50	-3.00912	0.337
		51~60	2.84869	0.981
	41~50	21~30	4.53032*	0.016
		31~40	3.00912	0.337
		51~60	5.85781	0.684
	51~60	21~30	-1.32749	1.000
		31~40	-2.84869	0.981
		41~50	-5.85781	0.684

Wu Ming-Long[22] points out that the dependent variable significance generalized from variance analysis is statistical significance, to analyze the actual significance need to find the correlation strength ω^2.If $\omega^2<0.059$, then the independent variables and dependent variables are in low correlation; if $\omega^2>0.138$, then the independent variables and dependent variables are in high correlation; if ω^2 is between 0.059 and 0.138, then the independent variables and dependent variables are in moderate correlation.

The age group is selected as the dependent variable and the safety attitude as the dependent variable. The single factor procedure is used to determine the association degree of the age group with the safety attitude according to the correlation intensity ω^2. The results are shown in Table 6. The results show that the correlation intensity $\omega^2 = 0.017$(the adjusted R^2 value), which means that age and safety attitudes are in low correlation. The results show that the average scores of safety attitudes of the four age groups are

60.4184,61.9396,64.9487,59.0909, which mean that the safety attitude of employees aged 20-50 are improved with the increase of age. H2 is supported.

Table 6. The effect test of age and safety attitude

source	SS	MS	F-test	Significance	η^2	Observation of the inspection ability [b]
Corrected mode	1409.032[a]	469.677	4.286	0.005	0.022	0.864
Intercept	535337.797	535337.797	4885.127	0.000	0.895	1.000
Age	1409.032	469.677	4.286	0.005	0.022	0.864
Error	62573.166	109.585				
Sum	2231832					
The total number after correction	63982.198					

a.R^2=0.022（adjusted R^2=0.017）
b. alpha=0.05

3.2.3. The influence of Working Years on Safety Attitude. The P-value of the working years is 0.00, less than 0.05(as shown in Table 4), reaching a significant level. The post hoc analysis shows that only the first and fourth groups have difference in safety attitudes, the difference of safety attitudes is only 3.87899.It indicates that the safety attitudes of coal miners of 7 working years and more are higher than the safety attitudes of workers of 1-2 working years. The specific results are shown in Table 7.

Table 7.the post hoc analysis of different working years' safety attitude

	(I) age groups	(J) age groups	Average difference (I~J)
safety attitude	1~2	3~4	-2.54425
		5~6	-2.67661
		7 and above	-3.87899*
	3~4	1~2	2.54425
		5~6	-0.13235
		7 and above	-1.33473
	5~6	1~2	2.67661
		3~4	0.13235
		7 and above	-1.20238
	7 and above	1~2	3.87899*
		5~6	1.33473
		7 and above	1.20238

The degree of association of the age group with the safety attitudes is judged by the correlation intensity ω^2. The results of the analysis are shown in Table 8. $\omega^2 = 0.010$, the length of working years explains 1% variance of safety attitudes, and the length of working years and safety attitudes are in low correlation. At the same time the statistical results show that the average safety attitudes of the four groups of coal miners are: 59.0734,61.6176,61.7500,62.9524. This shows that the longer the working years are, the better the safety attitudes people have. Coal miners of 7 and more working years have the highest safety attitudes, and coal miners of 1-2 working years have the lowest. Therefore, the assumption in H3 that the length of working years has an impact on the safety attitude of coal miners, and the safety attitudes of long service coal miners are better than the short-term coal miners is supported.

Table 8. The effect test of working years and safety attitude

source	SS	df	MS	F-test	Signifi	η^2	Observation of the

					cance		inspection ability [b]
Corrected mode	962.702[a]	3	320.901	2.914	0.034	0.015	0.694
Intercept	2098807.829	1	2098807.829	19059.820	0.000	0.970	1.000
Length of working years	962.702	3	320.901	2.914	0.034	0.015	0.694
Error	64308.256	584	110.117				
Sum	2289849	588					
The total number after correction	65270.957	587					

a. R^2=0.015 (adjusted R^2=0.010)

b. alpha=0.05

3.2.4. The Influence of Accident Experiences on Safety Attitude. The coal miners' accident experiences of this research are divided into two groups: with and without accident experiences, using independent samples t test for analysis. As shown in Table 9, the safety attitude F test value is 0.106 (P = 0.745> 0.05), t = 2.049 (P = 0.041 <0.05). There is a difference in the safety attitudes of workers with accident experiences and workers without accident experiences.

Table 9.T-test of different accident experiences' safety attitude

		Leven test for equality of variance		T test for whether the mean values are equal						
		F	Significance	T	Df	Sig.(2-tailed)	Mean difference	Std. Error Difference	95% confidence interval of the difference Lower	Upper
Safety attitude	Equal variance	0.106	0.745	2.049	511	0.041	2.423	1.182	0.100	4.745
	Equal variance not assumed			2.101	159.950	0.037	2.423	1.153	0.146	4.700

To analyze the actual salience, we also need to find out the effect value. The effect value representing the total variance of the dependent variable can be explained by the grouping variable. If the effect value is less than or equal to 0.06, it means that there is a low correlation between the grouping variable and the test variable. If the value is more than or equal to 0.14, they are highly correlated. If the effect variable is between 0.06 and 0.14, they are moderate correlated. In this research, $\eta 2 = 0.008 < 0.06$ (as shown in Table 10). It indicates a low correlation between safety attitudes and accident experience. Therefore, the assumption in H4 that accident experiences have an impact on the safety attitudes of coal miners is supported. At the same time, the statistical results show that the average safety attitudes of coal miners with or without accident experiences are 61.16 and 63.58, which mean that coal miners who have not experienced accidents have better safety attitudes, so the assumption in H4 that workers with accident experiences have better safety attitudes than those without accident experiences is not supported.

Table 10. The effect test of accident experiences and safety attitude

source	Sum of class III squares	df	Mean square	F	Significance	Local Eta Square	Non-central parameter	Observation of the inspection ability [b]
Corrected mode	479.655[a]	1	479.655	4.200	0.041	0.008	4.200	0.534
Intercept	1271438.524	1	1271438.524	11133.684	0.000	0.956	11133.684	1.000
Accident experiences	479.655	1	479.655	4.200	0.041	0.008	4.200	0.534
Error	58354.907	511	114.197					
Sum	2007810.000	513						
The total number after correction	58834.561	512						

a. R^2 = .008（adjusted R^2 =0 .006）
b. alpha =0 .05

4.Conclusion and Outlook

The quantitative relationship between demographic variables and safety attitudes is examined in 593 samples of coal miners, and the results confirm that the correlations between coal miners' safety attitudes and demographic variables like age, length of working years and accidents experiences are consistent with the assumptions. In addition, this paper also finds a different conclusion from the research hypothesis: educational level has nothing to do with the safety attitudes of coal miners. The reason is that the middle-aged and experienced coal workers' educational level is generally low in the early career, but long-term work experience makes them have a stronger sense of safety, highly educated people have fluke mind or are lack of experience, which affect their safety attitudes. The accident experiences have a low correlation with the safety attitudes of the workers, but the employees who have accident experiences have lower safety attitudes than those who do not have accidents. The accidents have negative effects on employees' psychology and ideology and influence badly on their work. They doubt whether the safety of their work can be guaranteed, so the level of safety attitudes is reduced.

This paper also has a strong practical significance. Through the research conclusion, we can find that the safety attitudes of workers who are young, of short working years and have accident experiences are low. If managers are aware of this, then they should attach more importance to certain areas, to strengthen the guidance and training of these employees, and to reduce the loss due to the problems of the working places as much as possible. The research also finds that the level of education is not associated with safety attitude and therefore employees should not be discriminated against during the training in safety education.

At present, the studies of the impact of demographic variables on the safety attitude of coal miners are rather less. Future researches can expand the scope of data investigation, increase the number of people surveyed, use other scientific methods other than questionnaire method to analyze more demographic variables and enhance the universality of research results, which will have a significant impact on the safety development of the coal industry.

Acknowledgement

This work was supported financially by the National Undergraduate Training Programs for Innovation and Entrepreneurship of China (201611415064).

References

[1] Chen Da-wei, Fang Dong-ping, Zhang Yan-dong. Study on Countermeasures of Establishing Long Term Construction Safety Production Mechanism in China[J].Construction Economy,2009(2);46-49.

[2] Heinrich HW. Industrial Accident Prevention, A Scientific Approach[M].New York: McGraw-Hill;1931.

[3]Sun Cheng-kun,Fu Gui,Dong Ji-ye,Yang Chun.Application of Behavior Control Methods in Coal Minging Enterprise[J]. Safety and Environmental Engineering,2014,21(4):122-126.

[4]Tian Shui-cheng, Li Hong-xia,Wang Li, Chen Ting. Probe into the frequency of coalmine accidents from based on the theory of three types of hazards[J]. China Safety Science Journal, 2007, 17(1):10-15.

[5]TIAN Shui-cheng , PEI Zhuo-qin. On the Construction of Safety Culture in Coal Mine Enterprises[J]. Safety and Environmental Engineering, 2014， 21(4):122-126.

[6]Lund J, Aaro L E. Accident prevention. Presentation of a model placing emphasis on human, structural and cultural factors [J]. Safety Science, 2004, 42(4): 271-324.

[7]Nasab H S, Ghofranipour F, Kazemnejad A, et al. Evaluation of knowledge, attitude and behavior of workers towards occupational health and safety [J]. Iranian Journal of Public Health, 2009, 38(2): 125-129.

[8]Bergheim K, Nielsen MB, Mearns K, Eid J. The relationship between psychological capital, job satisfaction, and safety perceptions in the maritime industry[J].Safety Science, 2015, 74: 27-36.

[9]Fred N. Kerlinger. Foundations of Behavioral Research,(3rd Edition, Revised)[M]. New York: Harcourt Brace Jovanovich.

[10]Lu Jie-wen & Zhang Xi-chao. (2009). Related Research on Work Stress and Safety Attitude. Beijing social psychology Academy2009 Academic Annual Conference, 45.

[11] LI Nai-wen,HUANG Peng. The Relationship between Transformational Leadership， Safety Attitude and Safety Performance—An Empirical Research Based on Coal Mine Enterprise[J]. Soft Science,2012, 26(1): 68-71.

[12]Börjesson M, Österberg J, Enander A. Risk and Safety Attitudes Among Conscripts During Compulsory Military Training[J]. Military Psychology, 2011, 23(6): 659-684.

[13]Guo Dong-chen,Su Xin-guo,Huang Xue-wen,Zhang Wei-jun,Wu Lin-song. Safety Management Mode of Bridge & Tunnel Construction Team Based on Demographic Characteristics[J].Transport Research, 2015, 4(4): 66-74

[14]Gao Guo-ping. Research on Safety Production Control of Small and Medium - sized Enterprises in Wenzhou[D].Hubei: Huazhong University of Science and Technology.2011

[15]Bu Meng-meng,Cai Ya-qin,Sui Ying-nan. Analysis of driving risk difference based on personality and demographic variables[J]. Journal of East China Jiaotong University, 2016,33(3):86-92.

[16]Zhang Hai-ben.Research on Safety Attitude of Enterprise Employees and Its Determinants:Empirical Evidence from Manufacturing Enterprises in Ruian[D].Zhejiang: Zhejiang University of Technology.2015

[17]Siu O-l, Phillips DR, Leung T-w. Age differences in safety attitudes and safety performance in Hong Kong construction workers[J]. Journal of Safety Research, 2003, 34(2): 199-205.

[18]Liang Zhen-dong. Study on relationship of demographic characteristics with unsafe behavior and unsafe behavioral intention[J]. Journal of Safety Science and Technology, 2014,10(1):46-52.

[19]Donald I, Canter D. Employee attitudes and safety in the chemical industry[J]. Loss Prevent Proc, 1994, 7(3): 203-208.

[20]Cox S,Cox T.The structure of employee attitudes to safety: A European example[J]. Work & Stress, 1991,5(2):93-106.

[21]Liu Jia-long,Liu Bin-bin,Wu Xiang. Development of Coal Mine Safety Attitude Measurement Scale[J]. Journal of Wuhan University of Technology,2016,38(2):164-167.

[22]Wu Ming-long. Questionnaire Statistical Analysis Practice - SPSS Operation and Application[M]Chongqing: Chongqing University Press; 2010.

IOP Conf. Series: Earth and Environmental Science **59** (2017) 012031 doi:10.1088/1755-1315/59/1/012031

Prospect and technology of unconventional gas exploration and development in China

Wang Nan[1,2], Zhao Qun[1,2], Zang Huanrong[1,2], Guo Wei[1,2], Liu Dexun[1,2]

[1]Research Institute of Petroleum Exploration & Development-LangFang, Hebei 065007,China
[2]National Energy Research Institute of Petroleum Exploration and Development, PetroChina, 065007,China.

Abstract. The rapid development of natural gas lays the foundation for the unconventional gas development. In 2015, unconventional natural gas production is 44.9Bcm, accounting for 34.06% of total natural gas production. Increasing demand of natural gas provides a wide developing space for unconventional natural gas. China has deployed unconventional natural gas resource evaluation to different extent, which mainly includes tight gas, CBM and shale gas. Unconventional gas E&P in China is still in its initial stage. By 2015, 1.6 Tcm proved reserves and 36 Bcm production has made China tight gas entering a large scale and commercializing stage. CBM development is still in the initial stage of E&P, with less reserve and production, in the early stage for commercialized development. Shale gas has a significant breakthrough since its proved reserves reached to 544.1Bcm, and production reached to 4.5Bcm. At present, conventional gas develops rapidly in China. During the next 5-10 years, it will also take the dominant position in the natural gas development. Unconventional gas will play a more and more important role in the long-term development of natural gas.

1. Introduction

Natural gas resource is abundant in China, with great potential for reserve and production increase. The total gas resource in China is 56 Tcm. By the end of 2015, the proved reserve is 12.47 Tcm. The resource proved ratio is 22%, which shows great potential. The gas production in China increased from 22.3 Bcm in 1998 to 131.8 Bcm in 2015, with an annual average increase rate of 12%. The rapid development of natural gas lays the foundation for the unconventional gas. The top 14 large gas fields with more than 1 Bcm productive capability have established all over the country, such as Kela2, Sulige, Yulin, Jingbian, etc. Changqing, Tarim, and Xinan gas field are the three largest gas areas which secure the safe gas supply to West-East and Shanjing pipelines.

The development of Unconventional gas is still in the early stage. By 2015, 1.6 Tcm of proved reserve and 36 Bcm Production have made tight gas entering large scale and commercializing stage. By 2015, Proved CBM reserves accounts for 418.5 Bcm and accumulative production accounts for 4.5 Bcm, which was showing CBM industry still in the initial stage of commercialization. Now, Shale gas has a significant breakthrough, as its production accounted for 4.5 Bcm. In the year of 2015, unconventional natural gas production is 44.9 Bcm, accounting for 34.1% of total natural gas production (fig 1). By 2020, it is expected that natural gas demand would reach 300 Bcm with shortage of 120 Bcm in China.

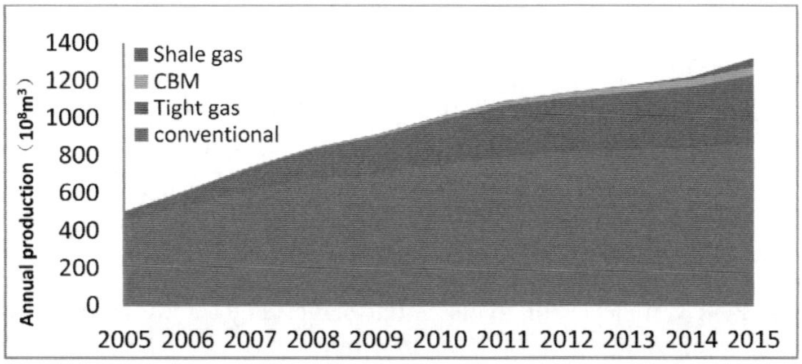

Fig 1 China Unconventional Gas Annual Production

2. China Unconventional Gas Resource and its Distribution

China has deployed unconventional natural gas resource evaluation to different extent, which mainly includes tight gas, CBM and shale gas. Based on the analyses of RIPED-Langfang, PetroChina, tight gas resource accounts for 16 Tcm. CBM resource shallow than 2000m accounts for 36.8 Tcm. In 2015, resource survey and evaluation of shale gas is showing that shale gas in China is accounting for 30 Tcm, and recoverable resource accounted for 1.28 Tcm.

2.1 Tight Gas

The favorable exploration area of tight gas is about 13× 10^4 km^2; total resource volume is about 16～25 Tcm (RIPED-Langfang, PetroChina, 2014). Tight gas is extensively distributed in more than 10 basins in China, such as Ordos basin, Sichuan basin, Songliao basin, Bohai bay and Tarim basin

2.2 Coalbed Methane

The overall CBM resource is 36.8 Tcm (shallow than 2000m) which assessed by National Land & Resource Ministry. There are 8 basins with resource more than 1 Tcm which are mainly distributed in central and western China. The total resource in 8 large basins is 28 Tcm, accounting for 76% of the total resource. In north-eastern and north-western areas, small scale coal-bearing layers mainly occurred in continental facies, with low rank coal, lower gas content. In southern areas, limited scale and scatter-distributed coal layers are intensively transformed and severely damaged due to the late structural movement, and scattered distribution of resource. The total resource is 7 Tcm which regarded as the main exploring target.

2.3 Shale gas

Shale gas mainly occurred in the marine facies and China marine facie depositional area amounts to 2.8 million km^2. In Huabei area, shale gas mainly occurred in Cambrian, Ordovician, Carboniferous and Permian which has lower thickness and lower resource abundance. In South China, shale gas mainly occurred in Cambrian, Ordovician and Silurian system. Longmaxi formation of Silurian and Qiongzhusi formation of Cambrian are the favourable targets with single layer thickness ranging from 30 to 50 meters, high gamma ray and high TOC in bottom.

Two sets of the marine shale have been developed in the Longmaxi formation of Silurian and Qiongzhusi formation of Cambrian in Chuanexiangqian, Qiannan-Guizhong, Qiandong-Qianxi, Suzhewan, Chuandongnan-Chuandongbei, which are the main exploration target for shale gas. To process the resource assessment, different organizations and experts have gotten different evaluation results (Tab.1).

Tab.1 China shale gas resource assessments by different organization

Area	Resource Volume (Tcm)	Organizations	Year
China	100	Rogner	1997

IOP Conf. Series: Earth and Environmental Science **59** (2017) 012031 doi:10.1088/1755-1315/59/1/012031

China	15-30	Colorado School of Mines	2002
China	100	IHS	2008
China	31(recoverable)	China Ministry of Land & Resource	2011
China	10-15(recoverable)	RIPED-Langfang	2011

3. Current Status of Unconventional Gas E&P in China

Unconventional gas E&P in China is still in its initial stage. Tight gas has abundant reserve and relatively large scale production. CBM is still in the initial stage of E&P, with less reserve and production. Shale gas has significant production breakthrough with huge resource.

3.1 Tight gas

Tight gas manly distributed in Ordos and Sichuan basins. Since 2000, economic development has been realized in Sulige. In 2010, the annual tight gas production is 16Bcm, accounting for 17% of the total gas production nationwide, in which Ordos basin takes 84%.Ordos basin is about 380,000 km^2. The tight gas area of this basin is about 80,000 km^2; resource volume is about 9Tcm and tight gas reserve is about 1.1Tcm, Layer mainly distribute in the Carboniferous and Permian.

Sulige gas field is situated in Ordos City, the main part of the Ordos Basin. Upper Paleozoic strata of Sulige gas field is divided into Benxi Formation of Carboniferous; Taiyuan, Shanxi, Lower Shihezi, Upper Shihezi formation and Shiqianfeng formation of Permian. Benxi No. 8 coal seam is widely distributed in the Sulige region, constituting a stable regional mark. The lenticular sand is controlled by the Upper Paleozoic braided river. The main layer of Permian Shanxi Formation include Shan No.1 and Shihezi No. 8 is delta deposit, the average depth is about 3295m; The sand thickness is 15 ~ 48m and normally thicker than 20m, porosity ranges from 5 to 12%, the pressure coefficient is 0.82 ~ 0.93, recoverable reserves of abundance varies from 27 million to 65 million cubic meter per square kilometres.

3.2 Coal bed methane

China CBM has entered the initial stage of commercialization. By 2011, the total national CBM wells accounts for 8765, which mainly distributed in the East of Ordos Basin and Qinshui Basin. The annual CBM productivity is 6 Bcm, Which includes 2.3 Bcm of PetroChina, 0.8 Bcm of CUCBM, 2.3 Bcm of Jinmei Mine and 0.6 Bcm from other companies. The annual total CBM production is 2.3 Bcm and it has commercialized in Qinshui, eastern Ordos, Fuxin and Tiefa area.

There is rich CBM resource in Shanxi province with resource about 7 Tcm shallow than 2000m. E&P activities from China 3 major Petroleum companies and more than 15 domestic and overseas enterprises mainly focus on their CBM activities in Shanxi Province. Fanzhuang area of Qinshui coalbed methane field located in the south of Qinshui Basin in Shanxi Province, which has 182 km^2 areas and 40.2 Bcm proved reserve. The target coal is No.3 coal of Shanxi formation in Permian. The buried depth of coal ranges from 185 to 1300 meter. Reflectance ratio ranges from 2.0 to 4.0%； gas content varies from 10 to 26 m^3/t； thickness ranges from 5 to 6 meter， Permeability of coal is smaller than 0.5 mD. Yearly production of 0.7Bcm has been achieved in Fanzhuang project. For Multi-lateral horizontal well, 2 trunks and 6 branches with open-hole completion are adopted, average single well gas rate is around 7000 m^3 per day.

3.3 shale gas

Till the end of 2015, about 600 wells were drilled in total, including 400 horizontal wells, and more than 500 Bcm reserves were proved in Sichuan Basins. In 2015, the shale gas production was 4.5 Bcm. More than 10 companies had drilling activities in exploration, but only 2 major companies have real production, respectively 1.2 Bcm of CNPC and 3.2 Bcm of Sinopec. It is estimated that will exceed 7 Bcm in 2016

Compare with U.S.A. (tab 2) China is characterized by marine shale which deposited in Cambrian and Silurian of early Paleozoic. Shale is mature and over-matures with Ro of 2.5-4.6%. Most shale reservoir is more than 5000 meters and few of them are shallow than 3000 meter. It has been proved that China shale gas is greatly influenced by the tectonic movement and induced fault and transformation will intensively influence the reservoir pressure and gas content quality. The tectonic activity should be considered as one of the important elements during the sweet spot screening.

Tab2 Comparison of Reservoir Forming Conditions between marine shale of late Paleozoic of South China and USA Shale

Area	Age	Deposition Facies	Geochemical Character		Fragile mineral content (%)	Depth (m)	Preserve condition	topography
			TOC (%)	Ro(%)				
USA	D,C	Deep water and continental shelf	2-14	1.1-3.5	40-70	1000 -3500	Stable distribution structure Flat formation	flat
China	∈,S	Deep water and continental shelf	2-8	2.5-4.0	37-80	1500 - 4000	Well-development Faults due to Multi-cycle structure movements	mountain

4. Conclusions

In China natural gas is developing rapidly. Although the unconventional resource is abundant and has a bright prospect, it is still in early development stage and facing many challenges. It is good for all developer to cooperate with from energy industry to face these challenges. It will speed up unconventional gas E&P and meet natural gas demand in China, which will contribute to global gas industry as well.

References

[1] Baihly, Raphael Altman, Raj Malpani, Shale Gas Production Decline Trend Comparison Over Time and Basins, SPE135555

[2] Curtis. J B. Fractured Shale To Gas Systems [J] . AAPG,Bulletin ,2002 ,86 (11) :192121938.

[3] Douglas J. Lehr, David D. Cramer, Best practice For Multi-zone Stimulation Using Composite Bridge Plugs, SPE141456

[4] Hasan A. AL-Ahmadi, Anas M. Almarzooq, 2010. Application of Linear Flow analysis to shale gas wells .–Field Cases. SPE 130370.IEA.

[5] World energy outlook 2015[EB/OL]. (2015-04)[2015-11-20]. http://www.worldenergyoutlook.org/media/weowebsite/2015/WEO2015_ToC.pdf.Jason

[6] M. Nobakh, D.M. Anderson and others Modeling Well Performance Data From Overpressured Shale Gas Reservoirs, SPE 137693

[7] Schepers K.C. and others, Reservoir Modeling In Support Of Shale Gas Exploration, SPE 123057

[8] Vassilellis G.D, C. Li, R. Seager and others, Investigating The Expected Long-Term Production Performance of ShaleReservoirs. SPE138134

[9] Wang Nan, Liu Xinyuan, et al., 2012. Lessons from U.S. and Canadian shale gas industry policies . International Petroleum Economics, 20(9): 69-73.

IOP Conf. Series: Earth and Environmental Science **59** (2017) 012032 doi:10.1088/1755-1315/59/1/012032

Research on Ore-controlling factors and Metallogenic Prognosis of Dongchuan Copper Mining Area, Yunnan, China

Tianguo Wang[1, 2], Jishun Liu[1, 2], Chaowen Huang[1, 2]

[1]Key Laboratory of Metallogenic Prediction of Nonferrous Metals and Geological Environment Monitoring, Ministry of Education, Central South University, Changsha 410083, China

[2]School of Geosciences and Info-Physics, Central South University, Changsha 410083, China

wangtianguo@csu.edu.cn

Abstract. Dongchuan is a famous copper producing area in China, where has good geological condition to form many copper deposits with local characteristics. This article summarizes the metallogenic regularities of Dongchuan copper deposits based on the previous works, research findings and the latest research advances. The study suggests that the Dongchuan copper-ore area has a good metallogenic background and immense prospecting potential. Copper ore is mainly found in Yinmin Formation, Luoxue Formation, Heishan Formation, Daqiaodi Formation, and in the Sinian-aged basement, and its occurrence is closely related to volcanic magmatism. The discussions regarding prospecting directions of Dongchuan copper ore were conducted based on the metallogenic regularity research, the results shows that further expanding of the exploration activities from the existing ore-bodies would be carried out; the breakthrough points focus on the Pingdingshan Formation, Qinglongshan Formation, and Daqiaodi Formation strata; at the same tine, more attention should be paid to the new types of deposits, such as Yanziya-type or hydrothermal deposits.

1. Introduction

The Dongchuan copper production site is located in Dongchuan, Kunming city, north Central Yunnan, whicn is one of the most famous copper production sites in China. There are good metallogeny conditions and complicated mineralization environment in Dongchuan area. Several deposit types and multiple ore-bearing strata have been found in this area, including Dongchuan-type, Xikuangshan-type, Lanniping-type, and Taoyuan-type deposits [1-4]. A series of geological prospecting and scientific research projects were launched in the Dongchuan area, especially in the late 1980s to 1990s. Significant results were obtained from the ore prospecting activities conducted in Dongchuan area by applying the latest metallogeny theories and exploration techniques, revealing it as a good copper-ore prospecting targets in China with immense potential [5-7]. Hence, it is meaningful to study the ore controlling factors and the prospecting direction in this area to further increase the availability of copper resources from deeper or outer Dongchuan production area.

2. Metallogenic geological conditions and ore controlling factor analysis

The Dongchuan area is in the middle of the western margin of the Yangzi Plate, which has a complicated crustal movement evolution history [8-10].

1) Ore–bearing strata and paleogeography

Dongchuan copper ore is mainly found in Yinmin Formation, Luoxue Formation, Heishan Formation, and in the Sinian-aged basement, although copper mineralization can also be observed in Qinglongshan and Pingdingshan Formations. The Yinmin Formation contains Xikuangshan-type Cu–Fe deposits, and the edifice should be closely related to the copper-ore mineralization position and regional distribution of the deposit. A Dongchuan-type copper deposit is also found within the dolomite strata of Luoxue Formation, and it is the most important industrial deposit in this area. Most ore-bodies are distributed within the lower Luoxue Formation and the bottom impure dolomite; copper metallogeny increases with increasing thickness of the ore-bearing strata. The copper deposit in Heishan Formation is located in the bottom carbo-argillaceous dolomitic slate under redox conditions. At other locations, only pyrite is observed in Heishan Formation without the existence of copper mineralization. The weathering-deposited copper ore in the Sinian-aged basement, which is a Lanniping-type copper deposit, is a special type of deposit found in the Dongchuan area. Its deposition is related to the weathering degree, preservation conditions, and paleotopography of the ancient weathered crust possibly formed in the depressions of an unconformable surface.

Previous researchers generally considered the Qinglongshan Formation as non-Cu-bearing, although a 20-m-thick Cu-bearing bed was found in the Tuobuka old-house village. The Pingdingshan Formation contains high-potential Cu-bearing strata because of the presence of Na-rich volcanic rocks (mafic-intermediate intrusive rocks) that favor large-scale copper mineralization and also the presence of a mineable Cu-rich ore-body in the same strata as that in the Yanziya area.

Daqiaodi Formation is a volcanism–magmatism-related copper deposit, occurring in volcanic rocks, cryptoexplosion breccia, and intrusive rocks. It is also the Cu-bearing ore bed.

2) Volcanism–magmatism and metallogeny

Kunyang Formation in the Dongchuan area underwent relatively aggressive volcanism–magmatism that created highly favorable metallogeny conditions. Deep-seated, rifting-controlled basic magma is very closely related to copper ore and is beneficial to its mineralization. Hence, areas with vigorous volcanism–magmatism activities exhibit higher mineralization potential. Volcanic rocks and intrusive rocks can also contain ore-bearing strata.

3) Distribution pattern of geological structures and ore-bearing lithology

The distribution of the ore-bearing lithology in the Dongchuan copper-ore area is structurally controlled, especially by folds. This is obvious in the Cu-bearing strata of Yinmin and Luoxue Formations; the EW and SN-strike strata are controlled by EW and SN-trend folds. Faults of smaller scales would cause displacement of the ore-body, whereas larger-scale faults would destroy or overstep the ore-bearing strata, for example, the Luoyindongyi ore-belt is deeply overstepped, possibly because of the presence of faults.

4) Uplifting–subsidence of fault blocks and deepening of the ore–bearing lithology

The Dongchuan area was separated into several fault blocks by few regional faults and differentially uplifted or subsided because of post-depositional tectonic movements. These tectonic events produced fault blocks with different degrees of denudation, directly affecting deep-seated ore prospecting. The Shanmuqing fault block exhibits the highest degree of subsidence, whereas the ore-bearing strata are well preserved. It has great prospecting potential but is deeply buried under the Dayingpan Formation. The Tuobuka–Boka fault block exhibits the highest uplifting and significant denudation, thereby indicating that the current ore-bearing strata might be remnants of the denudation, resulting in lower prospecting potential. However, there are some relatively well-preserved localities in Tuobuka–Boka fault block, especially the syncline core and the relatively subsided area, which exhibit significant prospecting potential.

The Tangdan–Lanniping fault block is slightly uplifted but the ore-bearing strata are still preserved. Therefore, it has some prospecting potential in the deeper parts. The top cover of the Luoyin fault block's ore-bearing strata was recently denudated but the strata itself are slightly damaged. Yet, this area also has high deep-seated ore-prospecting potential.

Theoretically, the chance of discovering a copper deposit or a deep-seated ore-body increases when the prospect target area satisfies more prospecting criteria. All criteria may not be simultaneously met during the actual prospecting in a target area; however, this does not affect the evaluation of the target area, which depends on the main prospecting information during the process.

3. Metallogeny prognosis

There are two major directions of metallogeny prognosis in the Dongchuan copper-deposit area: prospecting prognosis of the known ore-body (belt); and prediction of "edge," "deep," and "new" areas.

1) Prospecting direction of the known ore-body (belt): This is the focus of classical prospecting. The continuous progress of geological exploration, mining, and scientific research in this direction is expected to increase the reserves and scales of the known ore-body (belt) and also improve the prospecting and exploration methods. Although the Dongchuan copper-deposit area is highly researched, it has immense prospecting potential as the area is big and contains many ore-bodies. Hence, this is an important direction for further progress.

2) "Edge," "deep," and "new" areas: Research and exploration in the Dongchuan area has only been concentrated until now on the known ore-body (belt); however, the surrounding areas have not been investigated. Some important geological problems and observations related to prospecting have remained unsolved for a long time because of lack of investment. Thus, strategic issues regarding the exploration of these areas outside the known ore-body (belt) have not been considered. Hence, the "edge," "deep," and "new" prospecting direction has been realized.

(1) Edge: It refers to the areas that are relatively away from the current exploration and mining sites. Although the exploration degree is low and the working conditions are hard, these areas exhibit immense prospecting potential, and ore-bodies have been found in some cases till now as well. Hence, more prospecting efforts should be put into these "edge" areas, including the area from the west of Lanniping to the Jiulong area, the area from the west of Yinming to the Xiasikeshu area, the area from Renzhanshi to Wagangzhai, and the area to the north of Xintang.

(2) Deep: This refers to the deep-seated and hidden ore-bodies. There are some areas where huge undiscovered ore-bodies may be present, such as the east-wing of the Luoyin syncline, both wings of the semi-concealed Shanmuqing syncline, the Xiaoliukou–Yinmin hidden ore-body that is covered by mixed breccia of the Daqiaodi Formation (Dashuigou- or Zhangkoudong-type), and the Potou–Buwei area located east of Tuobuka, which is also covered by mixed breccia of the Daqiaodi Formation.

(3) New: refers to the new theories, new strata, new deposit types, and new methods. Some units have shown remarkable contributions in these aspects, such as the (previously known as) China National Non-ferrous Metal Company, the Nonferrous Geological Exploration Bureau of South-western China, and the Dongchuan Copper Mines Administration. We should look for prospecting breakthrough in new strata such as those in Pingdingshan Formation, Qinglongshan Formation, and Daqiaodi Formation other than the ore-bearing strata of the known industrial ore-bodies. We should also look for new deposit types such as Yanziya-type and hydrothermal-type deposits.

Although a significant amount of time has passed since the discovery of the Yanziya-type copper deposit, nothing has been done regarding the prospect, ore-bearing strata, type of deposit, ore-body scale, and geological characteristics of this ore-body. A private company has currently set more focus on the Yanziya copper deposit, which is a large-scale deposit with thick and rich ore-bodies. It is clear that this Yanziya-type deposit has immense prospecting potential, requiring further prospecting to be conducted in the Dongchuan area even if the ore-body may not be located in continuous strata and the metallogeny may be complex. The author also thinks that the ore-bearing Pingdingshan Formation is parallel to the non-mineralized Pingdingshan Formation. Some copper-mineralized Pingdingshan Formation outcrops were found in Qinjialiangzi Formation and Hongmenlougou Formation in Tuobuka area. Hence, exploring the Yanziya-type copper deposit in the Dongchuan area is an important prospecting direction.

Volcanism–magmatism activities have been vigorous in the Dongchuan area. Not only massive mafic rocks but also a huge Jinningian-aged granitic body has been found in Xinchun (Dongchuan area). These magmatic activities bring along ore-forming materials and induce important mineralization. In fact, a lot of rich and hydrothermal-related ore-bodies were found in the Dongchuan area. However, owing to their small scale, they were considered as independent deposit types and did not attract much attention. The discoveries of Yanziya-type copper deposit and mixed breccia copper deposit in the Pingdingshan Formation indicate that some intrusive-related copper deposits showing hydrothermal activity may be related to copper mineralization. This should be considered during further mineralization studies and prospecting.

4. Conclusion

The Dongchuan area is an important copper-ore deposit in China and has immense prospecting potential. Several geological studies have been conducted by previous researchers. The known ore-body (belt) still remains the major target area for existing levels of prospecting; however, the exploration of remote areas needs to be reinforced. In the near future, it will be possible to achieve some breakthroughs in the strata of Pingdingshan Formation, Qinglongshan Formation, and Daqiaodi Formation. Furthermore, the exploration of new deposit types, such as Yanziya-type and hydrothermal-type deposits, should be reinforced.

References

[1] Gao H, Pei R F, Wang A J, Chao D H, Fan S J and Mei Y X 2012 *Geological Bulletin of China* **08** 1332–51
[2] Liu J S Wu, Y Z and Duan J R 1996 *Journal of Central South University of Technology* **27** 8–12
[3] Qiu H N, Wijbrans J R, Li X H, Zhu B Q, Zhu C L and Zeng B C 2001 *Bulletin of Mineralogy,Petrology and Geochemistry* **04** 358–9
[4] Zhang Y F 2003 *Yunnan Geology* **04** 360–70
[5] Nie T, Fang W X and Du Y L 2014 Z W, Qian X G and Tian M 2014 *Geotectonica et Metallogenia* **04** 813–21
[6] Jiang J S, Li T F, Chen X S 1996 *Sedimentary Geology and Tethyan Geology* **02** 205–19
[7] Zhao X F, Zhou M F, Li J W, Sun M, Gao J F, Sun W H and Yang J H 2010 *Precambrian Research* **182** 57–69
[8] Huang R Q, Wu J M and Wang Z P 1999 *Geology and Exploration* **04** 15–8
[9] Yin F G, Sun Z M, Ren G M and Wang D B 2012 *Acta Geologica Sinica* **12** 1917–1932
[10] Wang T R, Fang W X and Zhao J J 2012 *Yunnan Geology* **S1** 272–4

Source apportionment of groundwater pollution in a city's eastern part using multivariate statistical techniques

Peng Jiang[1,2], Zhenmin Ma[1,2*], Ming Wen[1,2]

[1]School of resource and environment, Jinan University, Jinan 250022, China
[2]Research center of groundwater numerical simulation and pollution control engineering in Shandong province, Jinan 250022, China
*corresponding author's e-mail: stu_mazm@ujn.edu.cn

Abstract: This study was carried out to assess the overall water quality and identify major chlorinated hydrocarbon variables affecting the groundwater quality. The source apportionment of groundwater pollution is important for the efficient management of groundwater resources.Based on 13 variables surveyed at 43 monitoring sites,the comprehensive application of different multivariate methods were used for determining source apportionment of groundwater chlorinated hydrocarbon pollutants in study area. Factor analysis and cluster analysis were applied to the identification of pollution sources and four potential pollution sources that explained 92.810% of the total variance were identified.The absolute principal component score-multiple linear regression was adopted to calculate the contribution of each pollution source. Regression results revealed that most variables were primarily influenced by chemical industry,electrical manufacturing,chemical fiber and agricultural source.The contributions of each pollution source to the entire study area were 43%, 32%, 14% and 11% respectively.

1.Introduction

As an important natural resource which supports our life system,groundwater is extensively utilized for drinking, irrigation, and industrial purposes. The quality of groundwater is declining due to increasing population, urbanization, industrialization, and agriculture activities[1-4]. Due to the diversity of sources of groundwater pollution, to acquire the knowledge of sources is important for effective pollution abatement [5]. Source apportionment of groundwater pollutants can enhance the comprehensive understanding to the environmental conditions and help researchers establish priorities in the sustainable groundwater management[6,7].

Based on the ambient data registered at monitoring sites, the purpose of source apportion is to identify qualitatively the type of pollutant sources, moreover, to calculate the contribution of various sources, which consequently provide a certain basis forgroundwater pollution management.[8,9]. In recent years, multivariate statistics (such as cluster analysis (CA) and factor analysis (FA)) have been effectively used globally multivariate statistical method study, and used to understand clustering of sampling stations, discharge sources and their origin[10]. The study illustrates the availability of multivariate statistical techniques in the analysis and interpretation of complex data sets, the water quality assessment,and the identification of possible factors sources that influence water systems offering a valuable tool for reliable management of water resources, both in quantity and quality level [11-13]. This study indicates the necessity and usefulness of multivariate statistical techniques for the evaluation and interpretation of the data. It facilitates better information about the water quality and designs some remedial techniques to prevent future contamination[14].

Given all the above considerations, the main objectives of this work are:(1) analyzing the pollutants data and understanding the groundwater environments clearly.(2) exploring the main sources of groundwater pollutants with the assistant of factor analysis (FA) and cluster analysis(CA).(3) quantitatively calculating the contributions of each source type to each pollutant and the entire study area respectively by the absolute principal component score-multiple linear regression (APCS-MLR).

2.Materials And Methods

2.1. Sample collection and analysis

The monitoring data for this study was compiled from the groundwater pollution investigation of the typical study area. In view of the characteristics of shallow groundwater pollution system in the study area, 43 civilian wells and drilling wells were chosen as sample sites. The sample locations are shown in Fig.1.Due to the concentration of chlorinated hydrocarbon in groundwater is low, mostly in ppb level, the test requirements of concentration of chlorinated hydrocarbon in groundwater should be controlled strictly, like test method, test condition, and test equipment. Based on China Standard Method and the U.S. Environment Protection Agency (EPA) method as the standard the concentration of chlorinated hydrocarbon in groundwater are analyzed by GC/MS (gas chromatography/mass spectrum) method. The data description was presented in table 1.

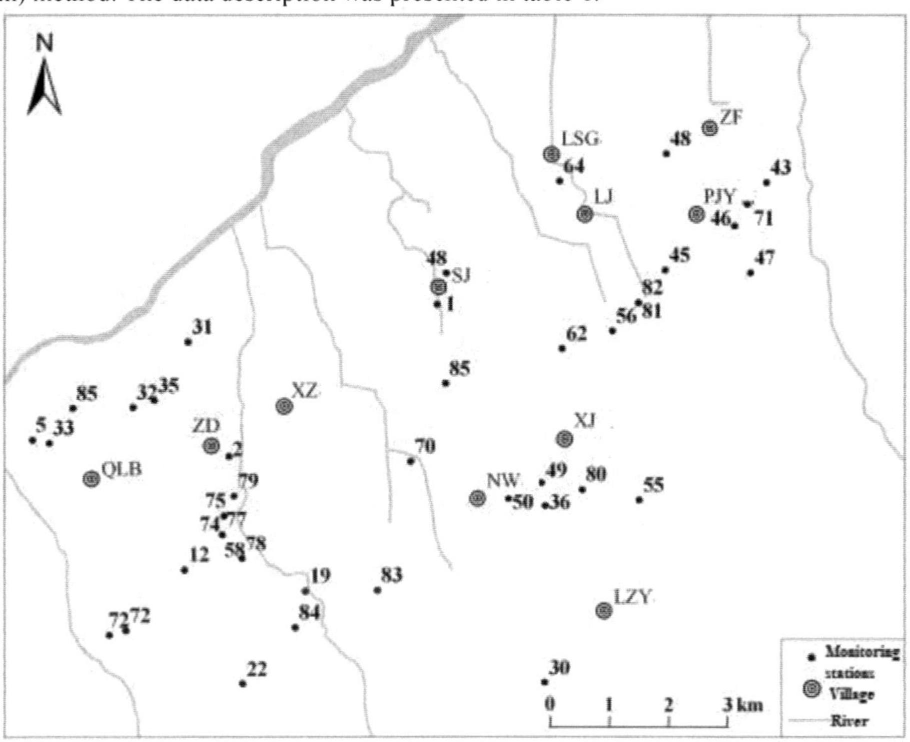

Fig 1 The location map of study area and spatial pattern of the 43 sampling sites.

Table 1 The statistical description of measured variables in study area and "*"represents national quality standards for drinking water, China(GB 5749-2006),other's is the reference of the United States EPA drinking water quality standards.

Parameters	φ/%	ρ/(μg/L)

	detection rate	Exceeding rate	Min	Max	Mean	quality standards
Vinyl chloride	93.1	60.34	0.47	8.53	3.87	2
Methylene chloride	94.83	94.83	5.21	16.85	10.04	5
1, 2 - dichloroethane	94.83	67.24	2.03	12.56	6.6	5
1, 1-2 vinyl chloride	96.55	72.41	1.11	10.46	5.34	7
cis-1,2-Dichloroethene	50	0	1.12	2.42	1.95	50.00*
trans-1,2-Dichloroethylene	32.76	0	0.44	1.31	1.03	50.00*
Chloroform	100	0	0.2	2.4	1.5	60.00*
1,1,1 - trichloroethane	94.83	0	1	2.05	1.59	200
Trichloroethylene	100	94.83	2.1	73.64	48.87	5
Tetrachloroethylene	94.83	70.69	3.01	32.25	10.07	5
Carbon tetrachloride	100	94.83	1.5	23.31	10.28	2.00*
HCH	15.52	0	0.1	0.52	0.37	5.00*
DDT	8.62	0	0.1	0.12	0.11	1.00*

2.2. Statistical analysis

Since the elemental concentrations varied greatly among the major and trace elements, the raw data were standardized (with a mean of 0 and a standard variation of 1) before the statistical analysis.

Factor analysis (FA) is a common and convenient multivariate statistical method for data dimension reduction based on the principal components analysis (PCA) which could identify the unobservable and latent pollution sources[15]. It has gained widely application in the pollutants source apportionment. The principle of factor analysis is to interpret the internal correlations of complicated multiple variables by extracting some principal components that represent a cluster of interrelated variable[16]. Readers who arr interesting to the mathematical and steps details of factor analysis can take it as reference[16,17].

Clustering analysis(CA) is a progress based on quantitative or qualitative characteristics of a large amount of experimental data for packet classification so as to understand the internal structure of a data set, and to describe each data set. The main principle of CA is that samples similar to each other in the same class,while, there has enormous diversity in different kinds of samples. Clustering analysis is normally used to verify the results of the conclusions of the factor analysi[18,19]. In this research, adopting the R cluster analysis method, and according to the similarity of variables to classify. The first step of CA is to calculate the correlation coefficient between each variable in the sample , then variables will be clustered based on the calculation results.

In the process of source apportionment, absolute principal component score-multiple linear regression(APCS-MLR) is one of the most commonly used methods in source apportionment to calculate the contribution rate of each pollution source[20]. It is important that characterizing groundwater quality and apportionment of pollution sources to groundwater pollution for efficient water management. In order to recognize and quantify the pollution sources receptor modeling by Multi-Linear Regression of the Absolute Principal Component Scores (APCS-MLR) has been used to evaluate the source apportionment of groundwater pollution [21].After determining the number and variety of sources, APCS-MLR can quantify efficiently the contribution rate of each pollution source type [9]. In this study, APCS-MLR was employed to estimate the contribution of each pollution source to each pollutant and the entire study area.

3.Result and Discussion

3.1. Identification of pollution sources

The results of the KMO and Bartlett's Test are showed in the table 2. In general,when the KMO value is closer to 1 and significance level is closer to 0,then the correlation between variables is more stronger, which means the original variables are more suitable for factor analysis. In this test the KMO value was 0.826 and the Bartlett's Test reached extremely significant level (0.000) which express factor analysis in this article was adaptive.

Table 2 Results of the KMO and Bartlett's Test.

Kaiser-Meyer-Olkin value of sampling adequacy.		0.826
Bartlett's test of sphericity	Test value (χ^2)	1146.832
	Degree of freedom (df)	78
	Significance level (Sig.)	0.000

The results of the FA are reported in the table 3. Based on the principle of eigenvalues is greater than one, the study extracted four common factors which accounted for 92.810% of the total variance. The first factor explaining 42.384% of the total variance, was strongly and positively related to Vinyl chloride,Methylene chloride, cis-1,2-Dichloroethene trans-1,2-Dichloroethylene,Chloroform,Carbon tetrachloride.The second factor accounting for 27.464%, showed strong positive factor loadings on 1,1,1-trichloroethane, Trichloroethylene, Tetrachloroethylene. The third factor which explained 12.852% of the total variance had high loadings on 1,2- dichloroethane and 1,1-2 vinyl chloride. The fourth factor explaining 9.928% of the total variance, had positive factor loadings on HCH and DDT.

Comparing with different clustering methods, the Ward's Method was found to be a suitable method to classify the variables. Based on the correlation efficients calculated by FA, the dendrogram of a cluster analysis of the 13 variables is illustrated in Fig. 2, and four distinct clusters are identified. The results of cluster analysis (CA) was consistent with the factor analysis (FA), indicating that the identification results were suitable and reliable.

Table 3 Total variance explained and component matrixes (four common factors extracted).

Total variance explained						
Component	Initial eigenvalues			Extraction sums of squared loadings		
	Total	% of Variance	Cumulative (%)	Total	% of Varianc e	Cumulative (%)
1	5.510	42.384	42.384	5.510	42.384	42.384
2	3.594	27.646	70.030	3.594	27.646	70.030
3	1.671	12.852	82.882	1.671	12.852	82.882

4	1.291	9.928	92.810	1.291	9.928	92.810
5	0.394	3.029	95.840			
6	0.220	1.690	97.529			
7	0.122	0.935	98.464			
8	0.071	0.544	99.008			
9	0.039	0.298	99.306			
10	0.032	0.248	99.555			
11	0.028	0.217	99.771			
12	0.016	0.121	99.893			
13	0.014	0.107	100.000			

Component matrixes

Element	Component Matrix				Rotated Component Matrix			
	Factor1	Factor2	Factor3	Factor4	Factor1	Factor2	Factor3	Factor4
Vinyl chloride	0.933	0.177	0.150	0.197	**0.956**	-0.054	-0.210	-0.046
Methylene chloride	0.821	0.467	0.196	0.131	**0.924**	0.243	-0.133	-0.133
1, 2 - dichloroethane	-0.691	0.416	-0.006	0.567	-0.270	0.255	**0.911**	-0.067
1, 1-2 vinyl chloride	-0.630	0.388	-0.024	0.647	-0.199	0.182	**0.943**	-0.070
cis-1,2-Dichloroethene	0.936	0.197	-0.039	-0.034	**0.835**	-0.034	-0.401	-0.240
trans-1,2-Dichloroethylene	0.790	-0.259	0.197	0.424	**0.811**	-0.437	-0.075	0.236
Chloroform	0.610	0.565	0.225	-0.303	**0.787**	0.383	-0.342	-0.182
1,1,1 - trichloroethane	-0.294	0.831	0.377	-0.036	0.043	**0.896**	0.338	-0.025
Trichloroethylene	-0.569	0.719	0.283	-0.065	-0.255	**0.833**	0.408	-0.025
Tetrachloroethylene	-0.169	0.768	0.260	-0.328	-0.003	**0.880**	0.016	-0.139
Carbon tetrachloride	0.877	0.266	0.033	0.326	**0.954**	-0.071	-0.051	-0.172
HCH	-0.195	-0.581	0.775	0.049	-0.133	-0.105	-0.031	**0.975**
DDT	-0.163	-0.580	0.791	0.020	-0.112	-0.093	-0.070	**0.981**

Extraction method: Principal Component Analysis; Rotation method: Varimax with Kaiser Normalization; Rotation converged in 5 iterations.

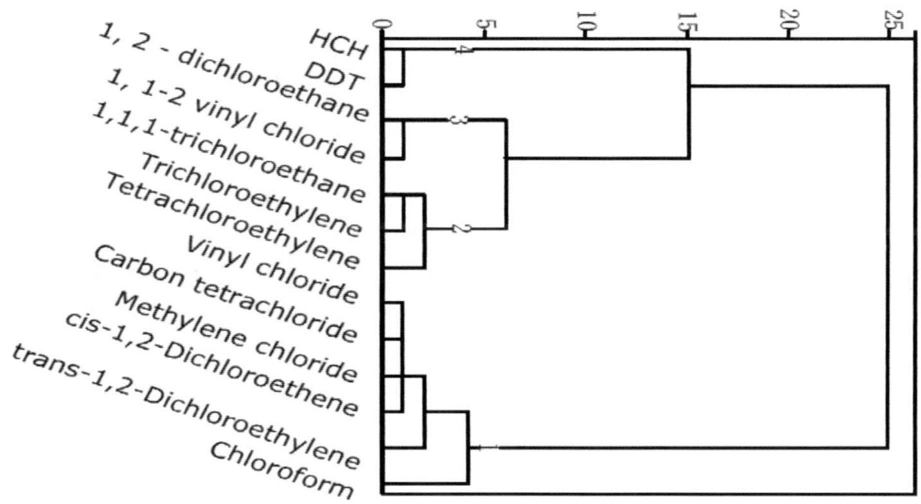

Fig 2 Dendrogram from the R-cluster analysis showing the four distinct clusters.

The variables that have high factor loading are vinyl chloride, methylene chloride, cis-1,2-Dichloroethen, trans-1,2-Dichloroethylene, chloroform and carbon tetrachloride in the first common factor（F1）. These chlorinated hydrocarbons are important chemical raw materials, and they are widely used in chemical production. The vinyl chloride is an important monomer used in polymer chemistry. There is a large plastics factory in the northeast of the researched area where the groundwater has been seriously polluted. The chloroform is an important raw material in organic synthesis process. It is often used as solvents in the manufacturing process of chemical products, including freon, resin, rubber, paint, phosphorus and iodine. At the same time, it is a critical raw material for the manufacturing of dry cleaners, pesticides, floor wax, etc. As an essential chemical raw materials, the chlorinated carbon is usually used as organic solvents, fire extinguishing agent, organic chloride agent, leaching agent of spice, degreasing agent, drug extraction agent, lubricant, the fabric dry cleaners and so on. From what has been discussed above, the first common factor containeing chlorinated hydrocarbon pollutants originates from chemical industry enterprises. Therefore the first common factor can be defined as chemical pollution.

The variables that have high factor loading are 1,1,1-trichloroethane, Trichloroethylene, Tetrachloroethylene in the second common factor（F2）.1,1,1-trichloroethane is a kind of incombustible solvent.It is often used as cleaning agent and metal degreasing cleaning agent.As a kind of good solvent. Trichloroethylene is commonly used as metal cleaners, metal degreasing agent and metal parts processing and surface treatment agent.Tetrachloroethylene is recyclable and it is widely used as a dry cleaning solvent for all countries in the world. From what has been discussed above 1,1,1-trichloroethane, trichloroethylene and tetrachloroethylene are widely used in metal cleaning agent and metal degreasing agent.The second common factor can be defined as the mechanical and electrical manufacturing sources.

The variables that have high factor loading are 1, 2 - dichloroethane, 1, 1-2 vinyl chloride in the third common factor（F3）.The 1, 2- dichloroethane is more widely employed as organic solvents in industrial application.It is not only used as the raw materials of organic synthesis, but also can be used as binder, wetting agent, soaking agent, manufacturing acetyl fiberd in chemical fiber industry.The 1,1-2 vinyl chloride is not uased as solvent in the industrial production because its volatile is bigger.The 1,1-2vinyl chloride is used to manufacture all kinds of copolymer and compounding fiber.By comparing the profiles of 1,2-dichloroethane, with 1,1-2 vinyl chloride, the researcher found that these two kinds of chlorinated hydrocarbon pollutant concentration are more higher. So the pollution of the 1, 2 - dichloroethane and 1, 1-2 vinyl chloride mainly come from the chemical fiber enterprises. The third common factor can be defined as chemical fiber contamination.

The variables that have high factor loading are HCH and DDT in the fourth common factor（F4）. The HCH and DDT are the most typical of organochlorine pesticides that have been used in domestic production. The HCH and DDT belong to persistent organic pollutants. Properties are stable, and hard to break down even in the environment that is contaminated. What's more, they will accumulate in the environment, which can worsen the environmental pollution. Even though the HCH and DDT have not been used in the study area, the residues of HCH and DDT in groundwater still have an certain impact on groundwater in the study area. From what has been discussed above, it is clear that the source of chlorinated hydrocarbon pollutant is pesticide pollution in the fourth common factor, so the fourth common factor can be defined as the pesticide pollution.

3.2. Calculation of the contributions of pollution sources

Common factor scores obtained from the factor analysis as independent can be took as independent variable, each variable's standardized concentration can be seen respectively as a dependent variable , the stepwise regression method was adopted to calculate the contributions of pollution sources to each pollutant with the assistance of SPSS. The results of the APCS-MLR are presented in the table 4. Based on correlation coefficients (R2) of each APCS-MLR, the conclusion of all the APCS-MLRS were relatively accurate could be reached. The quantitative information regarding the contributions of each source type to each pollutant was calculated accurately. Most variables in the study area were primarily influenced by chemical industry (73.86%, 64.48%, 56.57%, 54.65%, 34.65% and 79.70% of vinyl chloride, methylene chloride, cis-1,2-Dichloroethene,trans-1,2-Dichloroethylene, Chloroform and Carbon tetrachloride), electrical manufacturing(72.61% ,55.68% and 86.36% of 1,1,1 trichloroethane, trichloroethylene, tetrachloroethylene),chemical fiber (60.61% of 1, 2 - dichloroethane and 67.65% of 1, 1-2 vinyl chloride) and agricultural pollution (80.38% of HCH and 78.11% of DDT). In addition, according to the results, It could be find other unidentified sources excepting for the four pollution source types. The types of the unknown sources might need more field investigation or other feasible methods to identify.

Taking common factor scores obtained from the factor analysis as independent variables and the aggregate variables' standardization concentrations of each sampling point as dependent variables, the same APCS-MLR method above was applied to estimate the contributions of pollution sources to the entire study area. The results are illustrated in table 5. Under the confidence level of 95%, Sig values were equal to 0.000 represented the equation was significant. R2 value was equal to 0.98, indicating that the regression model had a good fitting effect for the primary data. Consequently, the APCS-MLR was applicable and reliable. From the table the regression equation was calculated as following: $Z = 4.41F1 + 3.08F2 + 1.30F3 + 1.09F4$ .This formula showed that contributions of chemical industry,electrical manufacturing,chemical fiber and agricultural source to pollutions of the entire study area were 43%, 32%, 14% and 11% respectively. The contribution of the chemical industry was the highest, theelectrical manufacturing took the second place, the chemical fiber and agricultural sources were following subsequently.

Table 4 Source contribution to each variable.

Parameters	Potential pollution source					R^2
	F1	F2	F3	F4	US[a]	
Vinyl chloride	78.36	4.43	17.21			0.96
Methylene chloride	64.48	16.96	9.28		9.28	0.95
1, 2 - dichloroethane	17.96	16.97	60.61	4.46		0.97

1, 1-2 vinyl chloride	14.28	13.06	67.65	5.02		0.97
cis-1,2-Dichloroethene	56.57		27.17		16.26	0.92
trans-1,2-Dichloroethylene	54.65	29.45			15.90	0.90
Chloroform	34.65	34.42	20.19		10.74	0.83
1,1,1 - trichloroethane		72.61	27.39			0.92
Trichloroethylene	17.05	55.68	27.27			0.92
Tetrachloroethylene		86.36		13.64		0.80
Carbon tetrachloride	79.70	5.93		14.37		0.95
HCH	10.96			80.38	8.66	0.98
DDT	8.92			78.11	12.97	0.99

[a] US means unidentified sources.

Table 5 Regression coefficients.

Model	Unstandardized Coefficients		Standardized Coefficients	t	Sig.	R^2
	B	Std. Error	Beta			
Constant	$9.310E^{-6}$	0.03		0.00	1.00	
F1	4.14	0.03	0.76	124.12	0.00	
F2	3.08	0.03	0.57	92.31	0.00	0.98
F3	1.30	0.03	0.24	39.04	0.00	
F4	1.09	0.03	0.20	32.79	0.00	

Sig.: Significant level. R2: coefficient of determinatio.

4.Conclusions

In this study, multivariate statistical methods like factor analysis (FA), cluster analysis (CA) and absolute principal component score-multiple linear regression (APCS-MLR) were adopted to identify the different groundwater pollution source types and calculate the contribution of each type of pollution source in study area, using 43 sampling sites data. Results showed that the comprehensive application of FA, CA and APCS-MLR was effective for groundwater pollution source apportionment.

Through the analysis of data of 43 sampling sites, we could draw a conclusion that the groundwater in the study area was polluted by chlorinated hydrocarbon. According to the results of the FA and CA, the pollution sources could be grouped into four categories,namely chemical industry,electrical manufacturing,chemical fiber and agricultural source , explaining 92.810% of the total variances.

The calculation of each pollution source based on APCS-MLR revealed the contribution rates of these four pollution sources to the entire study area were respectively 43%, 32%, 14% and 11%. However, it exists another latent sources in the study area. Therefore, more field investigations or other feasible methods need be required to have a further identification of the latent pollution sources.

Acknowledgments
This work was financially supported by Shandong provincial key research project (no.2015GSF117025) and Shandong provincial natural science foundation, China (no. ZR2014DM011).

References

[1] Hani, H (1990) .The analysis of inorganic and organic pollutants in soil with special regard to their bioavailability. *Int J Environ Anal Chem* ,**39**(2):197–208.

[2] Kumar, N., (2010) .Evaluation of Groundwater quality in shallow and deep aquifers: a case study. *Report and Opinion* . **2**(9):75–87.

[3] Pangarkar, B.L.,Sane, M.G.and Mahendra, G.,(2011) .Reverse osmosis and membrane distillation for desalination of groundwater: a review. *Isrn Materials Science*, **2011**(2090-6080).

[4] Chidambaram, S., Prasad, M,B,K.and Prasanna,M.V(2014).Evaluation of Metal Pollution in Groundwater in the Industrialized Environs in and Around Dindigul, Tamilnadu, India.*Water Quality Exposure & Health*,**7**(3):1-11.

[5] Xue, L., Lang, Y.and Liu, A., et al (2010). Application of CMB model for source apportionment of polycyclic aromatic hydrocarbons (PAHs) in coastal surface sediments from Rizhao offshore area, China. *Environmental monitoring and assessment*, **163**(1-4), 57-65..

[6] Kolovos, A., Christakos, G.,and Serre, M.L., et al (2002). Computational BME solution of a stochastic advection reaction equation in the light of site-specific information. *Water Resources Research* ,**38** (12), 1318–1334.

[7] Huang, F., Wang, X. and Lou, L.,et al (2010). Spatial variation and source apportionment of water pollution in Qiantang River (China) using statistical techniques. *Water Research*, **44**(5), 1562-1572.

[8] Tauler, R., Viana, M.,and Querol, X., et al(2009). Comparison of the results obtained by four receptor modelling methods in aerosol source apportionment studies. *Atmospheric Environment*, **43**(26), 3989-3997.

[9] Cai,W.L.,Luo,G.and Xiao-Yi,X.U.,et al(2012).Contamination Characteristics of Polycyclic Aromatic Hydrocarbons(PAHs) in Surface Water from Jialing River in Chongqing. *Environmental Science*, **33**(7):2341-6.

[10] Varol,M.,Gökot, B.and Bekleyen,A .,et al(2012).WATER QUALITY ASSESSMENT AND APPORTIONMENT OF POLLUTION SOURCES OF TIGRIS RIVER (TURKEY) USING MULTIVARIATE STATISTICAL TECHNIQUES—A CASE STUDY. *River Research & Applications*, **28**(9):1428–1438.

[11] Muangthong, S.and Shrestha,S.(2015).Assessment of surface water quality using multivariate statistical techniques: case study of the Nampong River and Songkhram River, Thailand.*Environmental Monitoring & Assessment,***187**(9):1-12.

[12] Khound, N.J.and Bhattacharyya,K.G.,(2016).Multivariate statistical evaluation of heavy metals in the surface water sources of Jia Bharali river basin, North Brahmaputra plain, India.*Applied Water Science*, 1-10.

[13] Hema, S., Subramani, T.and Elango,L., (2014).Assessment of Surface Water Quality Using Multivariate Statistical Techniques in a Part of River Cauvery, Tamil Nadu, India.*Journal of Environmental Science & Engineering*,**56**(3):277-282.

[14] Su, S., Zhi, J., and Lou, L., Huang, F., et al (2011). Spatio-temporal patterns and source apportionment of pollution in Qiantang River (China) using neural-based modeling and multivariate statistical techniques.*Physics and Chemistry of the Earth, Parts A/B/C*, **36**(9-11), 379-386.

[15] Li, X., Lee, S. L.and Wong, S. C., et al (2004). The study of metal contamination in urban soils of Hong Kong using a GIS-based approach. *Environmental Pollution*, **129**(1), 113-124.

[16] Pekey, H., Karakas, D.and Bakoglu, M., (2004). Source apportionment of trace metals in surface waters of a polluted stream using multivariate statistical analyses. *Marine Pollution*

Bulletin ,**49**(9-10), 809-818.

[17] Yidana, S. M., Banoeng-Yakubo, B.and Akabzaa, T. M(2010). Analysis of groundwater quality using multivariate and spatial analyses in the Keta basin, Ghana. *Journal of African Earth Sciences*, **58**(2), 220-234.

[18] Wang, X. S.and Qin, Y.,(2006) .Environmental rise and sources of heavy metals in Xuzhou urban topsoil. *Geochimica*, **35**(1), 88-94.

[19] Facchinelli, A., Sacchi, E.and Mallen, L.,(2001). Multivariate statistical and GIS-based approach to identify heavy metal sources in soils. *Environmental pollution*, **114**(3), 313-324.

[20] Zhou, F., Huang, G. H.and Guo, H.,et al (2007). Spatio-temporal patterns and source apportionment of coastal water pollution in eastern Hong Kong. *Water research*, **41**(15), 3429-3439.

[21] Gulgundi,M.S.and Shetty,A(2016).Identification and Apportionment of Pollution Sources to Groundwater Quality.*Environmental Processes*, **3**(2):1-11.

IOP Conf. Series: Earth and Environmental Science **59** (2017) 012034 doi:10.1088/1755-1315/59/1/012034

Numerical investigation of nanofluid convection performance in the fully developed flow regime of the pipe with constant wall temperature

Yongfu Liu[1] **and Hanping Hu**[1, 2]

[1]Department of Thermal Science and Energy Engineering, University of Science and Technology of China, Hefei, Anhui 230027, China

[2]Author to whom any correspondence should be addressed. E-mail: hphu@ustc.edu.cn

Abstract. In this work, single-phase dispersion model (SPD) is used to numerically study laminar force convection performance of Al_2O_3/water nanofluid in the fully developed flow regime of a constant wall temperature pipe, and a new average Nusselt number expression for Al_2O_3/water nanofluid flow in such a condition is put forward by modifying the classical average Nusselt number formula. In the numerical calculation, two-dimension equation is solved by using finite volume method, dispersion effects causing the increase in thermal conductivity and viscosity of nanofluid have been considered. It is shown that the numerical simulation results are in agreement with experimental data, the deviations are less than 7% in terms of Nusselt number and convective heat transfer coefficient. In addition, the modified average Nusselt number formula can predict heat transfer performance well for Al_2O_3/water nanofluid laminar flow in a constant wall temperature pipe.

1. Introduction

Nanofluid refers to a uniform and stable working fluid filled with the 10~100nm metal or non-metallic oxide nano powder [1]. As an innovation in heat transfer fluid medium, nanofluid has aroused the interest of many researchers.

For the laminar flow of nanofluid, some similar conclusions of intensifying heat transfer were gotten from experiments [2-5]. Numerical simulations were also conducted, many of which focused on comparision and modification of single- and two-phase models of nanofluid flow under the constant heat flux condition [6-9].

There were few numerical investigations on nanofluid laminar full developed flow in a constant wall temperature pipe. Single-phase models in some existing literatures [10] are incomplete, because they don't take the effect of dispersion or Brownian into consideration on the thermal conductivity and viscosity of nanofluid, besides, the properties of nanofluid obtained was not based on actual experimental measurement. So it's necessary to predict nanofluid heat transfer performace more accurately by using more perfect model [8] for nanofluid laminar flow in a constant wall temperature pipe. Besides, the classical average Nusselt number formula also needs to be modified for Al_2O_3/water nanofluid flow in such a condition in order to calculate heat transfer performance better.

Present study aims to numerical investigation of Al_2O_3/water nanofluid laminar flow in the fully developed flow regime of a constant wall temperature pipe. For the sake of accuracy, the SPD model considering the impacts of dispersion on thermal conductivity and viscosity of nanofluid has been adopted in such a condition, instead of currently used sing-phase model(SPH) without any

IOP Conf. Series: Earth and Environmental Science **59** (2017) 012034 doi:10.1088/1755-1315/59/1/012034

modification of thermal conductivity and viscosity [10]. The simulation results are compared with experimental data available in literature [11] and that of SPH model.

2. Mathematical models

In single-phase dispersion model, the difference of velocity and temperature between nanoparticle and base fluid is ignored, nanofluid can be regarded as classical Newton fluid and nanofluid flow is assumed to be a steady incompressible flow. The continuity, momentum and energy equations therefore are expressed as follows:

$$\nabla\cdot(\rho_{nf}\vec{V})=0 \tag{1}$$

$$\rho_{nf}(\vec{V}\cdot\nabla\vec{V})=-\nabla P+\nabla\cdot(\mu_{eff}\nabla\vec{V}) \tag{2}$$

$$\nabla\cdot(\rho_{nf}\vec{V}C_{P,nf}T)=\nabla\cdot(k_{eff}\nabla T) \tag{3}$$

where ρ_{nf}, μ_{eff}, $C_{p,nf}$ and k_{eff} are the density, effective viscosity, specific heat, and effective thermal conductivity of nanofluid, respectively.

$$k_{eff}=k_{nf}+k_{disp} \tag{4}$$

$$k_{disp}=C(\rho C_P)_{nf}ud_pR\varphi \tag{5}$$

$$\mu_{eff}=\mu_{nf}+\mu_{disp} \tag{6}$$

$$\mu_{disp}=\frac{k_{disp}}{C_{P,bf}}\mathrm{Pr}_{nf} \tag{7}$$

where k_{nf} is the thermal conductivity of nanofluid, k_{disp} is called the dispersion thermal conductivity [12], C is a constant which is determined by matching experimental data, u is mean velocity of nanofluid, d_p is particle diameter (40nm), R is the inner radius of pipe (5mm), φ is the particle volume fraction, μ_{nf} is the viscosity of nanofluid, μ_{disp} is the dispersion viscosity, and Pr_{nf} is the Prandtl number of nanofluid.

3. Problem statement and relevant parameters

A copper tube with 2m length and 5mm inner diameter is heated by water vapor and Al_2O_3/water nanofluid with 40 nm average nanoparticle size flow passes through a horizontal isolated copper tube with 50 cm length before it enters the heating zone, which is the same set-up as that in the experiment conducted by Heyhat [11].

Two dimensional axisymmetric formulation is considered in numerical simulation. Control volume technique is used to solve continuity, momentum and energy equations, and the residual is less than 10^{-5}.

Inlet velocity is determined by the Reynolds number adopted in experiment. The tube wall temperature is maintained at a constant temperature of 373K.

Relevant parameters such as average heat transfer coefficient, average Nusselt number, are calculated from results of simulation by following equations:

$$\overline{h_{nf}}=\frac{C_{P,nf}\rho_{nf}UA(T_{out}-T_{in})}{\pi DL(T_w-T_i)_{LM}} \tag{8}$$

$$\overline{Nu_{nf}} = \frac{\overline{h_{nf}} D}{k_{nf}} \tag{9}$$

where $(T_W - T_i)_{LM}$ is the logarithmic mean temperature difference, U is the mean velocity of nanofluid, A is the cross-sectional area of the pipeline, T_{in}, T_{out}, and T_w are the inlet temperature, outlet temperature and wall temperature, respectively, D is the inner diameter and L is the length of tube.

Grid independence have to be checked, and 50×5000 grid is selected as a suitable grid for present simulation calculation, the grid is concentrated near the wall. In order to verify the validity of current model and grid, the predicted Nusselt number for water in the same condition has been compared with experiment and theory (equation (10)) and good agreements have been seen, as is shown in figure 1.

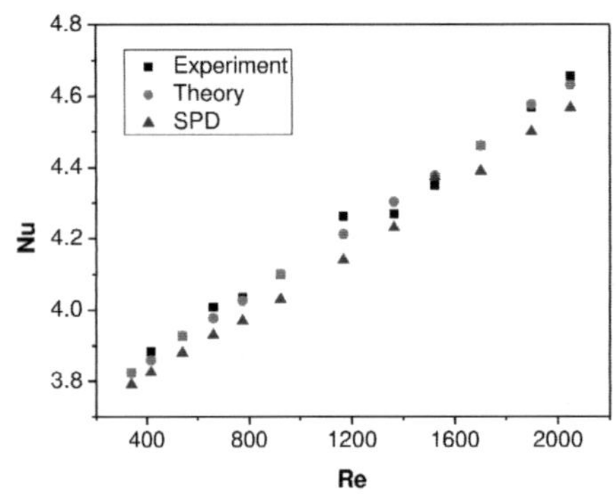

Figure 1. Average Nusselt number changing with Reynolds number for water

4. Results and discussion

The purpose of this work is to simulate nanofluid flow in the fully developed flow regime of a constant wall temperature pipe in terms of heat transfer by using SPD model. Dispersion coefficient C is calibrated by experimental data at Reynolds number 810 and 0.01 volume fraction.

As shown in figure 2, the results of SPD model are compared with the experiment and that of SPH model in terms of Nusselt number. The Nusselt number obtained from SPD model increases with the increase of Reynolds number, which is consistent with the experimental results and the largest deviation gotten is 7% at $\varphi = 0.001$, Re = 1900. With the increase of nanoparticle volume fraction, the accuracy of simulation results of SPD is rapidly improved and much better than that of SPH, which shows that SPD model is more effective in predicting Nusselt number of nanofluid.

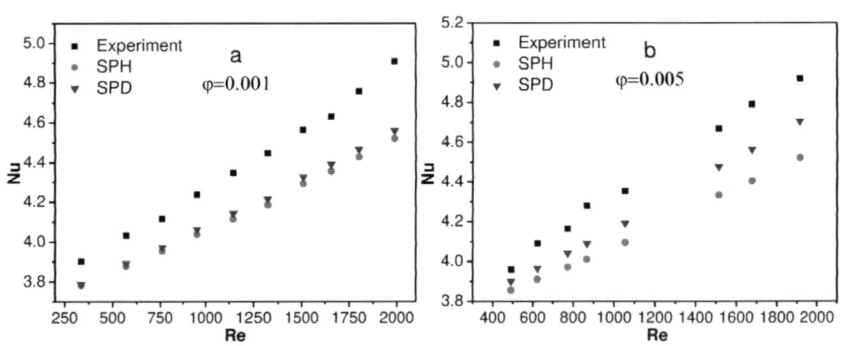

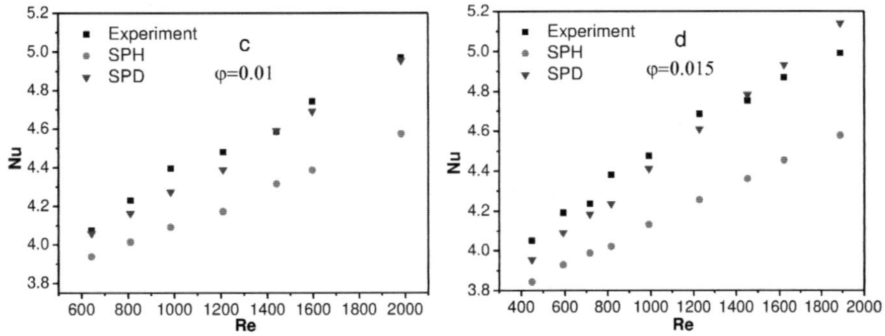

Figure 2. Average Nusselt number changing with Reynolds number at four kinds of nanoparticle volume fraction for Al_2O_3/water nanofluid

Figure 3 illustrates the change of average convective heat transfer coefficient with nanoparticle volume fraction at different Reynolds number. It is obvious that the results obtained from simulation are in good agreement with that of experiment. The convective heat transfer coefficient increases with the increase of nanoparticle concentration and the maximum deviation is 6.2% occurred at the φ=0.001, Re=1900, the deviations of other results are less than 4.2%. The validity of SPD model is further proved.

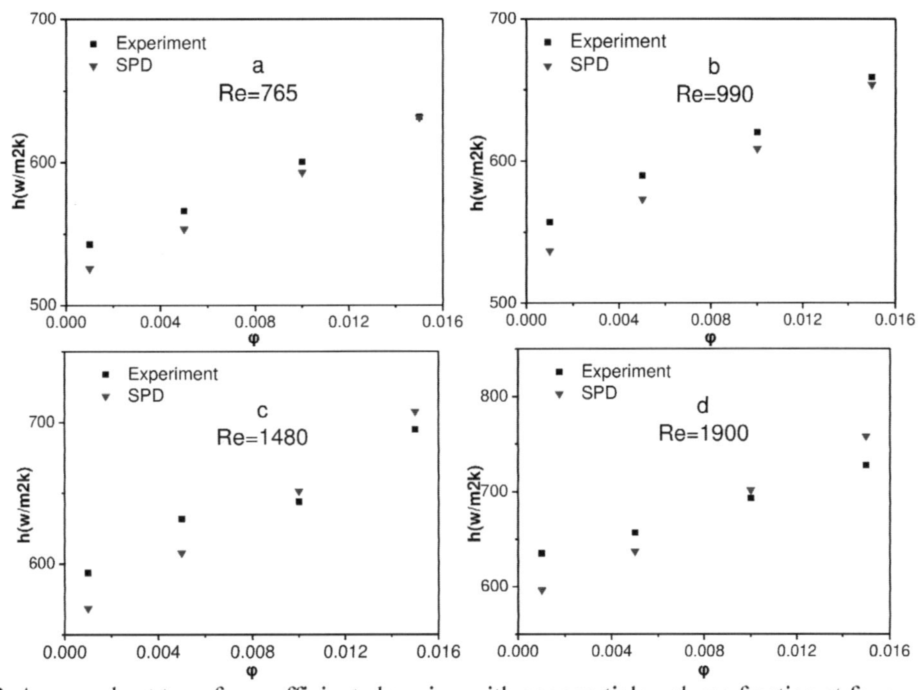

Figure 3. Average heat transfer coefficient changing with nanoparticle volume fraction at four Reynolds number for Al_2O_3/water nanofluid

It was pointed out in literature [11] that the classical average Nusselt number formula of fluid flow in a pipe with constant wall temperature (equation (10)) failed to predict the average Nusselt number of Al_2O_3/water nanofluid flow in such a condition.

$$Nu = 3.66 + \frac{0.0668(D/L)\mathrm{Re}^{(14/13)}\mathrm{Pr}_{nf}}{1+0.04[(D/L)\mathrm{Re}\,\mathrm{Pr}_{nf}]^{2/3}} \tag{10}$$

So here, we propose a modified average Nusselt number correlation formula for Al_2O_3/water nanofluid laminar full developed flow in a constant wall temperature pipe by fitting of experiment data obtained.

$$Nu = 3.66 + \frac{0.0668(D/L)\mathrm{Re}^{(14/13)}\mathrm{Pr}_{nf}}{1+0.04[(D/L)\mathrm{Re}\,\mathrm{Pr}_{nf}]^{2/3}}[1+3.5335\mathrm{Re}^{-0.3171}(1+\varphi)^{0.9258}] \tag{11}$$

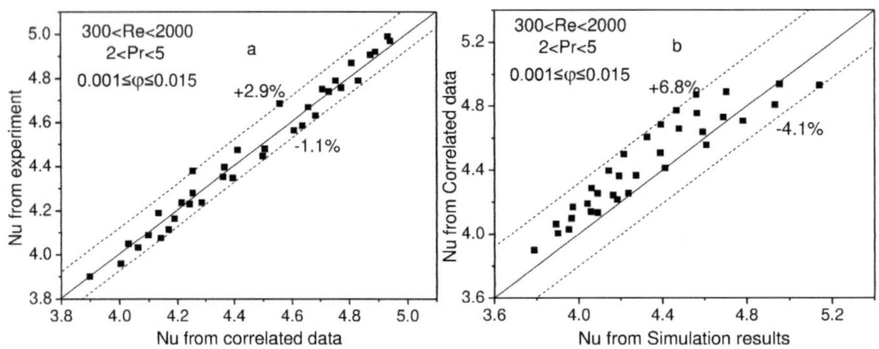

Figure 4. Comparison of correlated Nu with experiment and simulation

The average Nusselt number deviation between equation (11) and the experiment is depicted in figure 4(a) and it is about -1.1% ~ +2.9% in the range of 300 < Re < 2000, 2 < Pr_{nf} < 5, $0.001 \le \varphi \le 0.015$, which validates the effectiveness of this correlation formula.

The average Nusselt numbers obtained from equation (11) and SPD model are also compared. SPD model is an appropriate model in predicting heat transfer characteristics of nanofluid flow in a pipe, according to the analysis above, and, as is shown in figure 4(b), the average Nusselt number deviation between equation (11) and SPD model is about -4.1% ~ +6.8% , so, a comparatively good conformity is also seen.

5. Conclusion

Laminar full developed flow of Al_2O_3/water nanofluid in a pipe with constant wall temperature is simulated by using SPD model in terms of heat transfer and the results are compared with experimental data, which shows that the numerical simulation results are in agreement with that of experiment, heat transfer performance of nanofluid increases with the increase of Reynolds number and nanoparticle volume fraction, and the deviations of simulation are less than 7% in terms of Nusselt number and convective heat transfer coefficient. In addition, the classical average Nusselt number correlation formula for the convective heat transfer of fluid in a pipe has been modified for predicting Al_2O_3/water nanofluid laminar full developed flow in a pipe with constant wall temperature and its validity and good practicability are illustrated.

Acknowledgments

This research is partly supported by NSFC under the grant No. 51276178.

References

[1] Hussein A M, Sharma K V, Bakar R A and Kadirgama K 2014 A review of forced convection heat transfer enhancement and hydrodynamic characteristics of a nanofluid *Renewable and Sustainable Energy Reviews.* **29** 734-743

[2] Wen D S and Ding Y L 2004 Experimental investigation into convective heat transfer of nanofluids at the entrance region under laminar flow conditions *International Journal of Heat and Mass Transfer.* **47** 5181-88

[3] Hwang K S, Jang S P and Choi S U S 2009 Flow and convective heat transfer characteristics of water-based Al_2O_3 nanofluids in the fully developed laminar flow regime *International Journal of Heat and Mass Transfer.* **52** 193-199

[4] Duangthongsuk W and Wongwises S 2009 Heat transfer enhancement and pressure drop characteristics of TiO_2-water nanofluid in the double-tube counter flow heat exchanger *International Journal of Heat and Mass Transfer.* **52** 2059-67

[5] Suresh S, Venkitaraj K P, Selvakumar P and Chandrasekar M 2012 Effect of Al2O3-Cu/water hybrid nanofluid in heat transfer *Experimental Thermal and Fluid Science.* **38** 54-60

[6] Moraveji M K and Ardehali R M 2013 CFD modeling (comparing single and two-phase approach) on thermal performance of Al_2O_3/water nanofluid in mini-channel heat sink *International Communications in Heat and Mass Transfer.* **44** 157-164

[7] Goktepe S, Atalik K and Erturk H 2014 Comparison of single and two-phase models for nanofluid convection at the entrance of a uniformly heat tube *International Journal of Thermal Sciences.* **80** 83-92

[8] Heshmati F and Erturk H 2016 Single-phase models for improved estimation of friction factor for laminar nanofluid flow in pipes *International Journal of Heat and Mass transfer.* **95** 416-425

[9] Labib N N, Nine M J, Arianto H, Chung H and Jeong H 2013 Numerical investigation on effect of base fluids and hybrid nanofluid in forced convective heat transfer *International Journal of Thermal Sciences.* **71** 163-171

[10] Fard M H, Esfahany M N and Talaie M R 2010 Numerical study of convective heat transfer of nanofluids in a circular tube two-phase model versus single-phase model *International Communications in Heat and Mass Transfer.* **37** 91-97

[11] Heyhat M M, Kowsary F, Rashidi A M, Momenpour M H and Amrollahi A 2013 Experiment investigation of laminar convective heat transfer and pressure drop of water-based Al_2O_3 nanofluids in fully developed flow regime *Experiment Thermal and Fluid Science.* **44** 483-489

[12] Xuan Y M and Roetzel W 2000 Conception for heat transfer and flow features of nanofluids *International Journal of Heat and Mass Transfer.* **43** 3701-07

Numerical simulation of carbon dioxide removal from natural gas using supersonic nozzles

Wenjuan Sun[1], Xuewen Cao[1,3], Wen Yang[2] and Xuetang Jin[1]

[1] College of Pipeline and Civil Engineering, China University of Petroleum, 66 Changjiang West Road, Qingdao 266580, China
[2] South China Branch, SINOPEC Sales Co., Ltd., 191 Tiyu West Road, Guangzhou 510620, China

[3] Corresponding author. E-mail address: caoxw@upc.edu.cn

Abstract. Supersonic separation is a technology potentially applicable to natural gas decarbonation process. Preliminary research on the performance of supersonic nozzle in the removal of carbon dioxide from natural gas is presented in this study. Computational Fluid Dynamics (CFD) technique is used to simulate the flow behavior inside the supersonic nozzle. The CFD model is validated successfully by comparing its results to the data borrowed from the literature. The results indicate that the liquefaction of carbon dioxide can be achieved in the properly designed nozzle. Shock wave occurs in the divergent section of the nozzle with the increase of the back pressure, destroying the liquefaction process. In the supersonic separator, the shock wave should be kept outside of the nozzle.

1. Introduction

Natural gas plays a more and more important role in the worldwide energy supply network. When burned as fuel for energy, natural gas emits less harmful pollutants to the environment than other fossil fuels. Natural gas is a gaseous mixture of hydrocarbons and various contaminants when produced from the reservoir. The impurities such as water vapor, carbon dioxide, hydrogen sulphide, and heavier hydrocarbons must be removed from the natural gas before it is brought to the market. Conventional technologies such as absorption and adsorption may have good separation performance, but they need relatively large equipment, complicated system and consequently high investment [1]. Meanwhile, they may have a negative influence on the environment due to the use of chemicals [2].

Supersonic separator has emerged as a promising technology to eliminate some of the disadvantages mentioned above [3]. It is a compact device capable of separating condensable components from gaseous mixture by combining the principles of adiabatic cooling and cyclonic separation [4]. The Laval nozzle is a key part of the supersonic separator. Supersonic flow is obtained in properly designed converging-diverging nozzle, resulting in low pressure and temperature.

Researches on the flow behavior in the supersonic nozzles have been reported in voluminous literature [5-9]. However, the investigation on the performance of supersonic nozzles in separating carbon dioxide from natural gas is deficient at present. The flow behavior of methane-carbon dioxide mixture inside the supersonic nozzle is investigated and the influence of back pressure on the performance of the supersonic nozzle is discussed in this work.

2. Nozzle structure

The Laval nozzle contains three sections: convergent section, throat and divergent section. The converging contour is calculated by a cubic polynomial equation, which can be expressed by equation (1). The structure parameters of the nozzle employed in this work are shown in table 1.

$$\frac{D-D_t}{D_l-D_t}=\begin{cases}1-\dfrac{1}{X_m^2}\left(\dfrac{x}{L}\right)^3 & \dfrac{x}{L}\le X_m \\ \dfrac{1}{\left(1-X_m\right)^2}\left(1-\dfrac{x}{L}\right)^3 & \dfrac{x}{L}> X_m\end{cases} \tag{1}$$

where x is the distance from the inlet. D, D_l and D_t are the diameters at x, inlet and throat, respectively. L is the length of the convergent section. X_m=0.52 is the relative coordinate of the junction point between the front section and the back section.

Table 1. Geometry size of supersonic nozzle.

Parameter	Value (mm)
Inlet diameter	50.00
Throat diameter	11.00
Outlet diameter	12.00
Converging length	33.80
Diverging length	17.86

3. Mathematical model

3.1. Governing equations

In a supersonic nozzle, the gas flow characteristics can be depicted by partial differential equations, including mass, momentum, and energy equations, as described in equations (2), (3) and (4).

$$\frac{\partial\rho}{\partial t}+\frac{\partial}{\partial x_j}\left(\rho u_j\right)=0 \tag{2}$$

$$\frac{\partial}{\partial t}\left(\rho u_i\right)+\frac{\partial}{\partial x_j}\left(\rho u_i u_j\right)=-\frac{\partial p}{\partial x_i}+\frac{\partial}{\partial x_j}\left[\mu\left(\frac{\partial u_j}{\partial x_i}+\frac{\partial u_i}{\partial x_j}-\frac{2}{3}\delta_{ij}\frac{\partial u_j}{\partial x_j}\right)\right]+\frac{\partial}{\partial x_j}\left(-\rho\overline{u_i' u_j'}\right) \tag{3}$$

$$\frac{\partial}{\partial t}\left(\rho E\right)+\frac{\partial}{\partial x_j}\left(\rho u_j E+u_j p\right)=\frac{\partial}{\partial x_j}\left(k_{\mathrm{eff}}\frac{\partial T}{\partial x_j}+u_i\tau_{\mathrm{eff}}\right) \tag{4}$$

where ρ is the density, t is the time, u is the velocity, p is the pressure, μ is the viscosity, δ_{ij} is the Kronecker delta, E is the total energy, k_{eff} is the effective thermal conductivity, τ_{eff} is the effective viscous stress.

3.2. Numerical method

The flow behavior in the Laval nozzle is simulated by computational fluid dynamics (CFD) modeling using Fluent software. The density based solver and k-ω turbulence model are employed in this study as they are applicable to compressible flow.

According to the flow characteristic of the supersonic compressible flow, pressure boundary conditions are imposed for the inlet and outlet. The total pressure, static pressure, total temperature, turbulence parameters, and species are fixed for the inlet and outlet. No-slip and adiabatic boundary conditions are posted to the wall.

The unstructured meshes are generated and a fine gird scheme is employed in the boundary layer. Gird independence is investigated for the designed nozzle before the simulation is carried out. The second order upwind scheme is employed for the flow equations, and the first order upwind scheme is applied to the turbulent kinetic energy and specific dissipation rate. The solution is considered as converged when the residuals drop below 10^{-6} for the energy and 10^{-3} for the others, while the mass flow rate relative error between the inlet and outlet is less than 0.05%.

4. Results and discussion

4.1. Model validation

The model is validated by the work of Arina [10]. The working fluid is air. The inlet temperature and pressure are 288 K and 0.1 MPa, respectively. The outlet pressure is 83% of the inlet pressure. The same nozzle geometry, inlet and outlet conditions are used in the validation study.

The pressure and density distribution along the nozzle axis are presented in figure 1. As can be seen form figure 1, the simulation results are in excellent agreement with the data in Arina's work and the discontinuous changes of fluid properties caused by shock wave are captured accurately by the simulation, which illustrates that the model discussed above is suitable to simulate the flow behavior inside the Laval nozzle.

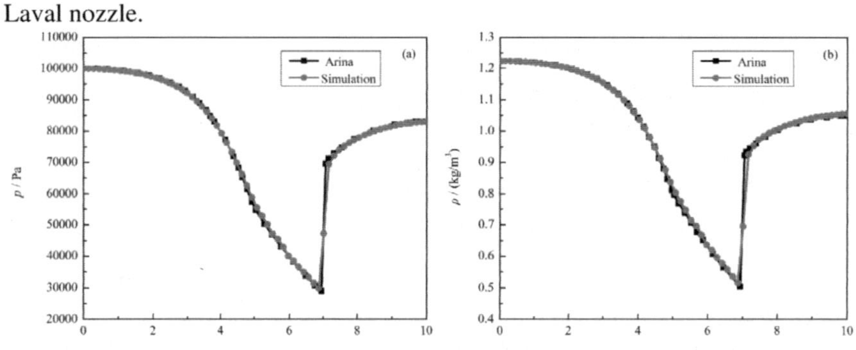

Figure 1. Comparison of pressure and density distribution along the nozzle axis.

4.2. Nozzle flow behavior

The flow field in the supersonic nozzle is simulated based on the above numerical method. A binary mixture of methane (70 mole %) and carbon dioxide (30 mole %) is chosen as the working fluid in this study. Total pressure and total temperature at the inlet are set to be 8 MPa and 303 K, respectively. The pressure at the outlet is not specified as the flow will be supersonic at the exit.

The pressure-temperature variation is shown in figure 2, where the phase envelope of the gas mixture is also plotted. As figure 2 shows, the pressure-temperature curve enters the two-phase region as the stream flows through the supersonic nozzle, which illustrates that the liquefaction of carbon dioxide can be achieved.

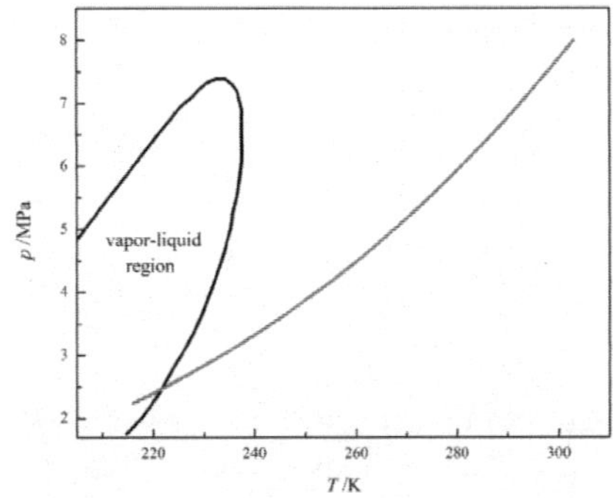

Figure 2. Phase diagram and pressure-temperature distribution along the nozzle axis.

4.3. *Effect of back pressure*

The back pressure at the exit has a great influence on the flow behavior of the gas stream in the nozzle. A shock wave may occur in the divergent section of the nozzle under certain conditions, which will change the flow from supersonic to subsonic. The effect of back pressure on the flow behavior is simulated, under the conditions that the inlet parameters remain constant (8 MPa and 303 K).

Figure 3-4 show the phase diagram and the pressure-temperature variations at different back pressure. It is clear that the presence of the shock wave prevents the gas from expanding and leads to the abrupt change in the pressure and temperature, which will cause the evaporation of the condensed carbon dioxide. It also can be seen that with the decrease of the back pressure, the gas flow expands more and the temperature and pressure are lower before the occurrence of the shock wave. As indicated in figure 4, when the back pressure is 5.6 MPa, the pressure-temperature curve is kept outside of the two-phase region and the flow will remain in a single phase all though the nozzle. In the supersonic separator, the shock wave should be enforced into the diffuser and the condensed liquid should be separated from the natural gas before the occurrence of the shock wave.

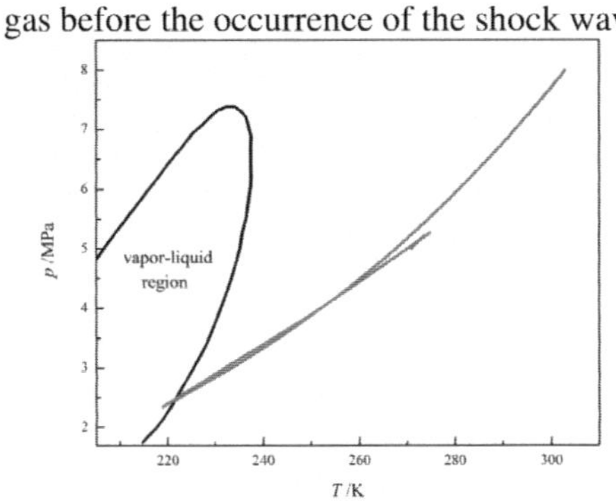

Figure 3. Phase diagram and pressure-temperature distribution with back pressure 5.0 MPa.

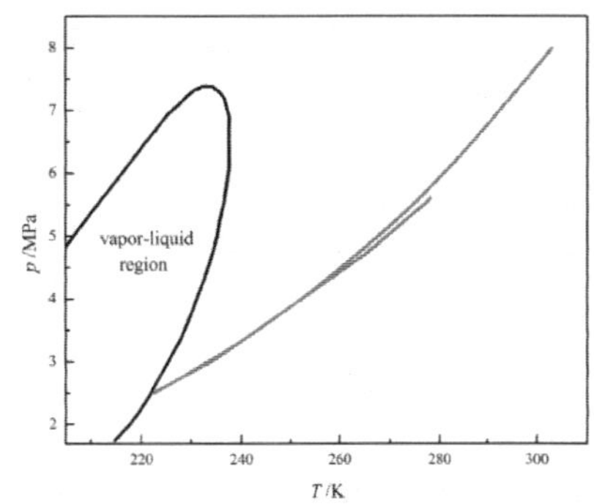

Figure 4. Phase diagram and pressure-temperature distribution with back pressure 5.6 MPa.

5. Conclusion

The performance of the Laval nozzle in removing carbon dioxide from the natural gas is simulated by Fluent. Due to the expansion in the Laval nozzle of supersonic separator, it is possible to cool the gas to a sufficient extent that the liquefaction of carbon dioxide can occur. Shock wave occurs in the divergent section of the supersonic nozzle with the increase of the back pressure. The emergence of the shock wave significantly impairs the expansion process, destroying the liquefaction condition.

Therefore, the back pressure should be reasonably controlled to avoid the emergence of the shock wave in the nozzle.

Acknowledgments
This study is supported by the National Natural Science Foundation of China (grant number 51274232) and the Fundamental Research Funds for the Central Universities (grant number 16CX06004A).

References
[1] Netusil M and Ditl P 2011 *J. Nat. Gas Chem.* **20** 471–6
[2] Yang Y, Wen C, Wang S L and Feng Y Q 2014 *Appl. Energ.* **132** 248-53
[3] Alfyorov V, Bagirov L, Dmitriev L, Feygin V, Imayev S and Lacey J R 2005 *Oil Gas J* **103** 53–8
[4] Haghighi M, Hawboldt K A and Abdi M A 2015 *J. Nat.Gas Sci. Eng.* **27** 109–21
[5] Jassim E, Abdi M A and Muzychka Y 2008 *Pet. Sci. Technol.* **26** 1757–72
[6] Jassim E, Abdi M A and Muzychka Y 2008 *Pet. Sci. Technol.* **26** 1773–85
[7] Karimi A and Abdi M A 2009 *Chem. Eng. Process.* **48** 560–8
[8] Wen C, Cao X W, Yang Y and Li W L 2012 *Energy Educ. Sci. Technol. Part A Energy Sci. Res.* **30** 651–60
[9] Castier M 2014 *J. Nat.Gas Sci. Eng.* **18** 304–11
[10] Arina R 2004 *Appl. Numer. Math.* **51** 409–26

Control of a selective catalytic reduction system based on NARMA-L2 model

W M Han[1], Y J Wang[1], T X Zheng[1,3], T L Zhou[1], Y Zhang[2] and R Tan[2]

[1]School of Automation, Chongqing University of Posts and Telecommunications, Chongqing 400065, China
[2]School of Advanced Manufacturing Engineering, Chongqing University of Posts and Telecommunications, Chongqing 400065, China
[3]zhengtx@cqupt.edu.cn

Abstract. The plant of the selective catalytic reduction (SCR) system is characterized by significant nonlinearity, time delay and temperature sensitivity. In order to control the urea injection accurately, the (nonlinear auto regressive moving average) NARMA-L2 model based control is applied to the SCR system. In this paper, a data-based technique is taken and a model of the plant is identified on the basis of input-output data. Then the identified model is used to the design of a NARMA-L2 controller. Simulation of the NARMA-L2 model based control for the SCR system is presented to demonstrate the effectiveness and superiority. The comparison results show better performance over the traditional PID control.

1. Introduction

Diesel engines have attracted increasing attention in the past decades because of the well-known performance in fuel economy, power capability, durability and reliability. However, due to their special combustion process, diesel engines typically generate more NO_x emissions than gasoline engines [1-3]. One of the promising aftertreatment methods is combining the SCR system with other filters such as diesel oxidation catalyst (DOC) and diesel particulate filter (DPF) [3].

The performance of the SCR system can be influenced by many factors such as SCR catalyst size, SCR gas space velocity, temperature, NH_3/NO_x ratio and NO/NO_2 ratio. To simultaneously minimize the NO_x emissions and limit the ammonia slip, plenty of researches have been conducted [4-7]. The NARMA model is an exact representation of the input - output behavior of a finite-dimensional and nonlinear discrete time dynamic system in a neighborhood of the equilibrium state [9]. However, due to the dynamic nonlinearities, it is relatively difficult to implement it for real-time control systems. Fortunately, NARMA-L2 model is convenient to be practically implemented using multi-layer neural networks. The main concept of a NARMA-L2 controller is to transform a nonlinear dynamic system into a linear dynamic system. Therefore, the NARMA-L2 model can be a good choice for the control of a nonlinear and complex system.

In this paper, a NARMA-L2 model based control strategy is applied to a nonlinear selective catalytic reduction system. Because the plant is characterized by significant nonlinearity, time delay and temperature sensitivity, we utilize NARMA-L2 model to identify the plant model and then design the input controller. We compared the proposed method with the traditional PID control strategy. Simulation results show that the NARMA-L2 model based control gives better results.

2. Controlled process description

A schematic diagram of SCR system is given in Fig.1. As can be seen from the figure, temperature, NO_x and NH_3 sensors are placed upstream and downstream of the SCR catalyst. A DOC is typically installed upstream of a SCR catalyst to convert part of NO to NO_2 . In the meantime, a DPF is generally installed between a DOC and a SCR for reducing PM emissions. The objective of this work is to achieve high NO_x conversion efficiency and keep the ammonia slip to a desired set point by accurately controlling the injection of Adblue.

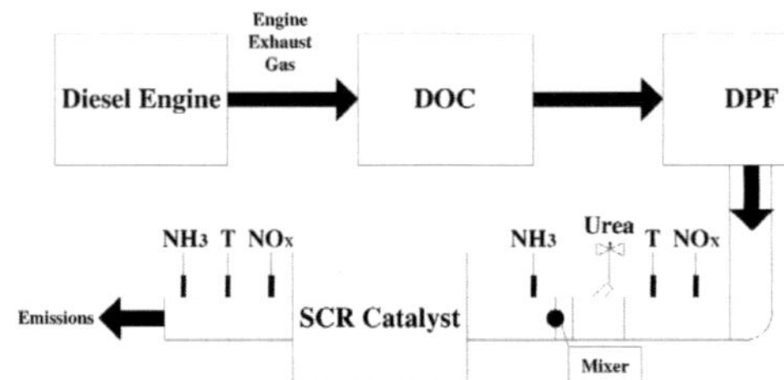

Figure 1. Schematic diagram of SCR system

3. NARMA-L2 control

Due to the high approximation accuracy and fast convergence speed, NARMA model can be used to capture the dynamics of nonlinear and complex systems. The main objective of the control method is to transform the nonlinear system into approximate linear system. There are two basic steps in NARMA: system identification and nonlinear controller design. Generally, one standard model of discrete-time nonlinear systems is represented as

$$y(k+d)=N[y(k),\cdots,y(k-n+1),u(k),\cdots,u(k-m+1)] \tag{1}$$

where $u(k)$ is the system input, $y(k)$ is the system output an d is the relative degree. The positive integers m and n are respectively the number of measured values of inputs and outputs. Three-layer neural networks can be trained to approximate the function N. In order to let the system outputs follow a reference trajectory $y(k+d)=y_r(k+d)$. The control input can be expressed as

$$u(k)=G[y(k),\cdots,y(k-n+1),u(k),\cdots,u(k-m+1)] \tag{2}$$

It is hard to achieve parameters of the neural network function G during plant real time control using static back propagation. One of the effective solutions is to use a NARMA-L2-based model

$$\begin{aligned} y(k+d)= & f[y(k),\cdots,y(k-n+1),u(k),\cdots,u(k-m+1)]+ \\ & g[y(k),\cdots,y(k-n+1),u(k),\cdots,u(k-m+1)]u(k) \end{aligned} \tag{3}$$

where f and g are the neural networks to get prediction of $y(k+d)$.

Considering the realization, then the resulting theoretical controller is

$$u(k+1)=\frac{y_r(k+d)-f[y(k),\cdots,y(k-n+1),u(k),\cdots,u(k-m+1)]}{g[y(k),\cdots,y(k-n+1),u(k),\cdots,u(k-m+1)]} \tag{4}$$

which is realizable for $d \geq 2$. Block diagram of NARMA-L2 controller is shown in Fig. 2.

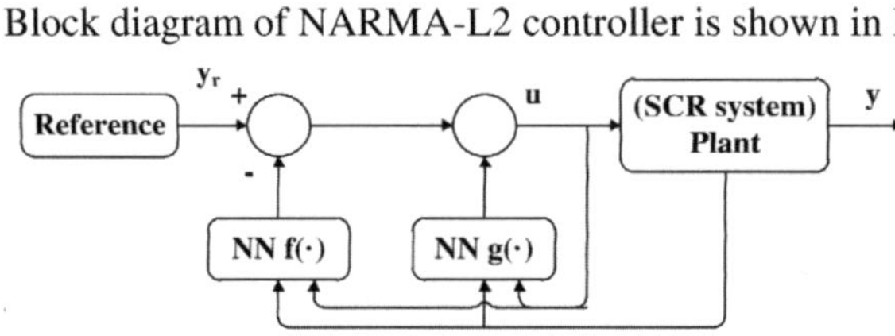

Figure 2. Block diagram of NARMA-L2 controller

4. Results and discussion

To illustrate the validity for the after-treatment process, NO_x conversion efficiency and ammonia slip are taken as the output and the injection of urea (concentration of the inlet ammonia) is taken as the input. To show the effectiveness of the NARMA-L2 controller, we compare it with the traditional PID control strategy. Based on a certain type of diesel engine, the SCR system is established. The detail parameters of the parts are listed in Table 1 and Table 2. The simulations are operated in the case of set-point tracking and the two control strategies are shown as follows.

Table 1. The detail parameters of the Engine

Parameter	Value and Unit
Engine Type	4-cylinder
Engine Capacity	3.0 L
Bore	100 mm
Stroke	110 mm
Connecting Rod Length	145 mm

Table 2. The detail parameters of the SCR

Parameter	Value and Unit
Cell Density	400 $inch^{-2}$
Length	250 mm
Discretization Length	25 mm
Active Surface Density	125 $mole/m^3$

4.1 The traditional PID controller

The simulation results obtained by the PID controller are shown in Fig. 3 and Fig. 4. As can be seen from the graphs, the controller reaches an optimal operating point in which about 96.2% of NO_x is reduced while allowing about 25 ppm NO_x slip past the catalyst. And the control response obtained using PID controller has a large overshoot and a long settling time.

4.2 The NARMA-L2 controller

There are two neural networks to be trained and the neural network model has 3 delayed plant inputs, 2 delayed plant outputs and 6 neurons in the hidden layer. As can be seen from the graphs, the controller reaches an optimal operating point in which about 96.4% of NO_x is reduced while allowing about 25 ppm NO_x slip past the catalyst. And the control response obtained using NARMA-L2 controller has a smaller overshoot and a relatively shorter settling time.

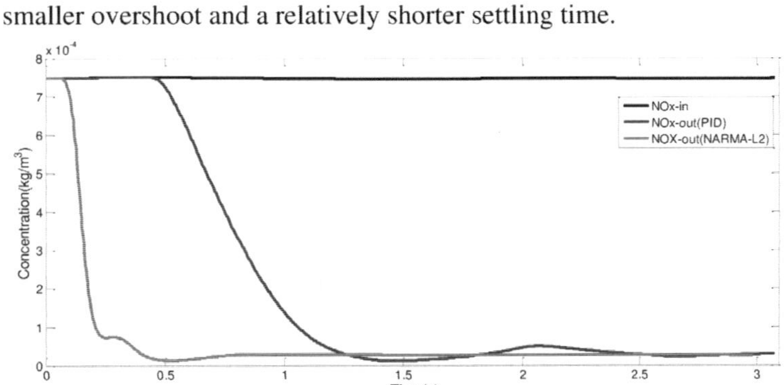

Figure 3. NO_x concentrations before and after the SCR system for the two controllers

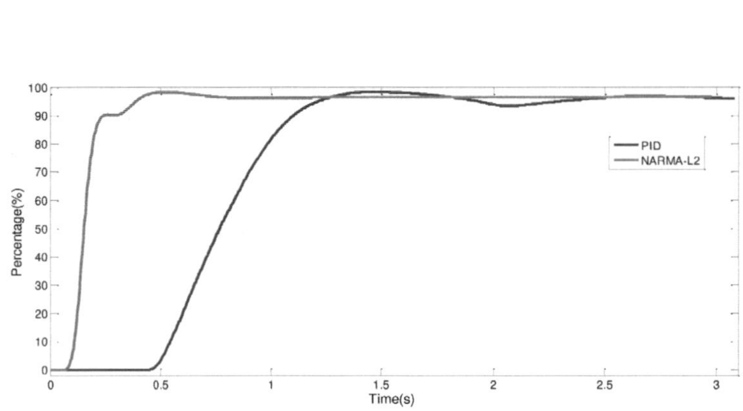

Figure 4. NO_x conversion efficiency for the two controller

5. Conclusions

In this paper, a NARMA-L2 controller is applied to a nonlinear selective catalytic reduction system. The main concept of a NARMA-L2 controller is to transform a nonlinear dynamic system into a linear dynamic system.NARMA-L2 includes two neural networks to approximate the dynamics of the system.. The simulations showed the better tracking capability and response speed of the NARMA-L2 controller. The controlled SCR system achieved a higher NO_x conversion efficiency within a shorter settling time.

Acknowledgments

This work is jointly supported by Scientific and Technological Innovation Program of Chongqing (Grant No. CYS16168), Chongqing Natural Science Foundation (Grant No. CSTC2014JCYJA60003), Chongqing Natural Science Foundation of Education Commission (Grant No. KJ120511).

Reference

[1] Zheng M, Reader G T, Hawley J G. 2004. Diesel engine exhaust gas recirculation—a review on advanced and novel concepts[J]. *Energy Conversion and Management*, 45(6): pp 883–900

[2] Johnson, Timothy V. 2012. Vehicular Emissions in Review[J]. *SAE International Journal of Engines*, 5(2): pp 216-234

[3] Yan F, Wang J. 2013. Control of diesel engine dual-loop EGR air-path systems by a singular perturbation method[J]. *Control Engineering Practice*, 21(7): pp 981-988

[4] Lee S I, Park S Y. 2014. Numerical analysis of internal flow characteristics of urea injectors for SCR dosing system[J]. *Fuel*, 129: pp 54-60

[5] Chen P, Wang J. 2015. A novel cost-effective robust approach for selective catalytic reduction state estimations using dual nitrogen oxide sensors[J]. *Proceedings of the Institution of Mechanical Engineers, Part D: Journal of Automobile Engineering*, 229(1): pp 83-96

[6] Yun B K, Kim M Y. 2013. Modeling the selective catalytic reduction of NOx by ammonia over a Vanadia-based catalyst from heavy duty diesel exhaust gases[J]. *Applied Thermal Engineering*, 50(1): pp 152-158

[7] Hsieh M, Wang J. 2011. A two-cell backstepping-based control strategy for diesel engine selective catalytic reduction systems[J]. *Control Systems Technology*, 19(6): pp 1504-1515

[8] Bonfils A, Creffy Y, Lepreux O, Petit N. 2014. Closed-loop control of a SCR system using a NOx sensor cross-sensitive to NH3[J]. *Journal of Process Control*, 24(2): pp 368-378

[9] Mandic D P, Chambers J A. 2000. Advanced RNN based NARMA predictors[J]. *Journal of VLSI signal processing systems for signal, image and video technology*, 26(1-2): pp 105-117

IOP Conf. Series: Earth and Environmental Science **59** (2017) 012037 doi:10.1088/1755-1315/59/1/012037

Development situation and prospecting division of geothermal resources in Yangshan county, Guangdong Province

Lijun Yin[1], Yizhen Luo[1], Huiying Ma[2], Yanxiao Xie[1], Zhenzhen Liu[1]

[1] Guangdong nonferrous mining geological disaster prevention and control center, Guangzhou 510080, China

[2] Hunan institute of geological survey, Changsha 410116, Hunan, China

455078211@qq.com

Abstract. Yangshan County has abundant low-geothermal resources in its northeast, southwest, and midwest regions. These low-temperature geothermal resources in Yangshan County can prove to be beneficial for different purposes such as tourism, recuperation, sauna, and agriculture. Thirteen geothermal hot springs (spots) and seven geothermal anomalies have been discovered till now in this area. The geothermal resources are grouped on the basis of their conditions as follows: The Chengjia–Dianzhan and Dongguan–Jietan geothermal areas are classified as priority development zones, the Huangben–Mazishui and Qigongyuntankeng areas as sub-priority development zones, the Jiangying geothermal area as a general development zone, the Yangshan geothermal area as a potential development zone, and the Chengjia and Longfeng geothermal areas as restricted development zones.

1. Introduction

Geothermal energy is a pollution-free energy source. Few industries such as agriculture, cultivation, and tourism, have taken advantage of this energy source [1-4]. The geothermal resources in the Yangsan County follow the distribution of regional geothermal sources, which is related to the regional structure and batholiths. Yangshan County has relatively better geological conditions and geothermal resource distribution in its northeast, southwest, and midwest regions than in the remaining parts of the County. Thirteen hot springs (spots) and seven geothermally anomalous areas have been discovered till now in these areas. The Chengjia, Longfeng, Qigongyuntangkeng, and Huangben hot springs are presently under development, whereas the rest are not in use or possess deep-seated geothermal resources, e.g., the Mazishui, Jiangyingyankou, Leigongkeng, Dongguan, and Jietan hot springs. Overall, the geothermal exploration in the Yangshan County has been limited to the discovery of only two sites till now. The development and utilization of these resources is limited to tourism and does not extend to industrial and agricultural activities or residential heating [5-8]. Environmental protection measures are not efficiently followed and there is a lack of environmental monitoring and protection measures.

2. Geothermal resource development and utilization

The utilization of geothermal resources depends on the temperature and the quantity and quality of the source water. Based on the temperature classification in the "Geological Exploration Standard of Geothermal Resources" (GB/T 11615-2010), the geothermal resources in the Yangshan County can be primarily classified as low-temperature geothermal resources. These resources can be used directly for tourism, recuperation, and agriculture after accounting for the environment, industrial structure, and economic development of the County.

1) Tourism

Tourism is a major industry, pursued by the County, where the geothermal resources can be utilized. The 13th Five-Year Plan for the economic and social Development of the Yangshan County emphasizes upon the development of ecotourism and modern agriculture as well as the need to promote industrial development activities.

2) Agriculture

The low-temperature geothermal resources in the County are suitable for the cultivation of vegetables, fungi, and flora and also for the hatching of bird eggs in greenhouses.

3) Geothermal water drilling, intake, and heating

Geothermal water drilling and heating is a possible use of the geothermal water resources distributed in the County. The Chengjia Village in the Yangshan County has cold winters; hence, geothermal energy can be used for heating residences and industrial facilities. Further investigation of the geothermal resources and drilling for geothermal water intake and heating will improve the utilization of hot springs in the surrounding towns and villages.

3. Geothermal resources

Multiple factors such as the scale, temperature, depth of geothermal reservoir, geothermal water yield, location of the field, and natural topography and environment are taken into consideration for evaluating the development and utility value of a geothermal field [9-10]. In this study, the geothermally anomalous areas of Chengjia, Dongguan–Shiluo, Huangben–Mazishui, Qigong, Jiangying, Leigongkeng, and Chengjiadi–Longfengdi were evaluated and divided into different categories.

1) Chengjia geothermal zone

It is located along the secondary EW-striking fault within the inner contact zone of the Dadongshan rock massif, which is a part of the northern extension of the NE-striking Chengjia–Qigong fault zone. The Paleozoic outcrop majorly consists of limestone and quartz-sandstone. Massive intermediate felsic intrusive rocks from the Yanshanian age are also observed. After the Palaeozoic outcrop, the sedimentation-cover layer is limited and represents the deposits of a long-lasting uplift. The major structures include the NE- striking structures, the contact zone of the rock massif, the EW-striking structures, and the Neo-Cathaysian structures. These structures form a complicated structural condition beneficial to the generation of geothermal fluids. In short, this zone has excellent geothermal conditions and shows good potential for further geothermal development. The hydrochemistry is of the HCO3–Na•Ca type with relatively high levels of fluorine that reach the concentration levels similar to that in mineral water used for physiotherapy. Thus, it can be named as fluorine water, with high physiotherapy-utilization value. There are two geothermal areas in this zone. The first is the Chengjia hot spring, which is developed and utilized, and the second is the recently discovered Dianzhan hot spring. This zone has a very high tourism potential as it is well-connected to the Shaoguan area in the east. There are many scenic areas such as the Chengjia natural reserve park and the Top Mountain of Guangdong. The geothermal resources of the Chengjia hot spring area are limited; hence, the existing scenic areas and parks need to be well-protected while undertaking the developmental activities by accounting for the natural flow of the geothermal waters in the design. Further geothermal exploration needs to be done to investigate the occurrence and characteristics of the geothermal resources in this zone without compromising on the protection measures for the existing Chengjia hot spring. A third industrial chain can be created with the hot spring at the opening by combining the development of saunas and outdoor adventure activities and using the residual geothermal water for agriculture.

2) Dongguan–Shiluo geothermal zone

It is located at the upstream of the Xiaobeijiang River, in the middle of the NW-striking Shuikou–Shiluo fault and in the northern part of the NE-striking fault. The geothermal reservoir consists of Carboniferous limestone. The zone is located at the intersection of the NE- and NW-striking structures and has favorable geological conditions. The hydrothermal activity within the zone is vigorous.

There are four or five geothermal hot springs (spots) with warm geothermal waters having temperatures ranging between 40°C and 57°C. Only the Longfeng and Shiluo hot springs have been explored till now. The waters are of calcium-sulfate-type and rich in trace elements such as metasilicic acid, fluorine, and strontium. It can be named as fluorine water or strontium-rich spring. The water quality and degree of utilization are high.

The geothermal anomaly is 20–30 km away from the Yangshan area with relatively good accessibility. The Longfeng spring in the Reshui Village of the Yangshan County along the bank of the Xiaobeijiang River is a scenic spot and has a high tourism potential.

The other hot springs are not as well-explored and are in the deeper parts of the area. Overall, this zone can be developed for leisure, recuperation, and business tourism.

3) Huangben–Mazishui geothermal zone

It is located on the NW-trending fault zone and along the northern and southern parts of the NE-striking Chengjia–Qigong Fault. The outcrop of this zone, which is a part of a collapsed basin filled with Quaternary sediments, contains Late Paleozoic limestone and quartz-sandstone. The major structures in the area are the NW-trending Qinglian–Dadongshan Fault and the NE-trending Daao–Ouwu structure belt. These relatively large-scale faults provide the necessary conditions for the occurrence and formation of geothermal fluids. The NW-trending Huangben–Lingbei–Mazishui belt intersects with the NE-trending Chengjia belt in the Lingbei area. It was initially thought that there were undiscovered geothermal resources in the NW direction, especially around the Lingbei area.

Three geothermal spots are found in this zone despite the lack of extensive geothermal exploration; the low-temperature (33–35°C) Huangben hot spring and the warm (43°C) Mazishui hot spring. The hydrochemistry of the Huangben and Mazishui hot springs is of the HCO3–Ca type. The fluorine content of the Huangben hot spring is 0.664 mg/L, whereas that of the Mazishui hot spring is 0.89 mg/L. The low concentration of fluorine is attributed to the mixing with shallow cold water during surface water sampling. The fluorine concentration might change if water samples are obtained from deeper parts of the spring. The metasilicic acid content is higher in the Mazishui hot spring (38 mg/L).

The Huangben and Mazushui hot springs have no direct access to roads but trails are available. The geothermal potential of the hot springs needs to be investigated. The Huangben Gaolan and Huangben Linchang hot springs can be used for aquaculture, greenhouse cultivation, and irrigation activities. The Mazishui hot spring can be developed for recuperation and tourism with some modification of the existing facilities; however, new heat sources need to be identified.

4) Qigong geothermal zone

It is located in the inner contact zone of the Lianyang rock massif in the southern part of the Yangshan County. The NE-striking Chengjia–Qigong Fault is the largest fault in this area that controls the distribution of the geothermal hot springs. There are two geothermal hot springs (spots) in this zone. The Yuntangkeng hot spring with water temperatures of 46–47°C and the Huangshakeng hot spring, a seasonal hot spring located northeast of the Yuntangkeng hot spring and with better geothermal conditions. There is no exploration activity conducted in this geothermal zone till now. The hydrochemistry of the zone is of the HCO3•SO4–Na•Ca-type, and the concentrations of trace elements such as fluorine, metasilicic acid, and radon are similar to those of the thermal mineral water for physiotherapy; the waters thus have high physiotherapy value and are well utilized. The Yuntangkeng and Huangshakeng hot springs in the Qigong geothermal zone are 6 km from the town of Qigong and are connected by a country road.

5) Jiangying geothermal zone

It is located in the southern limb of the Yankou Syncline in the northeastern parts of the Jiangying Town. The bedrock consists of Carboniferous carbonates. A low-temperature (29–35 °C) hot spring is found in this zone. The permitted heat withdrawal and geothermal fluid withdrawal capacities are 6.10 × 1010 kJ/a and 132 × 104 m3/a, respectively. The hydrochemistry of the hot spring is of the calcium–magnesium carbonate type and it is suitable for aquaculture and irrigation activities.

6) Leigongkeng geothermal zone

It is located on the secondary EW-striking fault zone and on the southwestern side of the NE-striking fault belt in the Yangshan Basin. The EW-striking fault contains sandstone and quartz-sandstone in the north and limestone in the south. One geothermal hot spring is found in this zone, with water temperature in the range of 27–29°C. This zone has not been officially explored yet. The current potential is low as it might be a deep-seated resource. There is an investment risk involved because of the unfavorable geological conditions; therefore, it has been suggested to locally explore the area and investigate the current economic conditions before undertaking any developmental activities. Overall, the Leigongkeng geothermal zone has some potential.

7) Chengjia and Longfeng geothermal fields

The geological exploration of these geothermal fields has been concluded and the fields are already in use. They belong to a resource-protected mining area but are within the monitoring range of the groundwater and surface water environments.

In summary, the six geothermal anomalous zones were categorized into five groups, ranging from priority development zone to restricted development zone, on the basis of their hydrothermal fluid characteristics and according to the local economic status.

1) Priority development zone: The Chengjia geothermal anomalous zone (except the Chengjia hot spring) and the Dongguan–Shiluo anomalous zone (except the Longfeng hot spring) are rich in geothermal resources, have thick geothermal reservoirs, and large water withdrawal capacities. Moreover, they are evenly distributed and shallow, which facilitates their development. These two fields have huge geothermal potential, are accessible and scenic, and can be further developed. The priority exploration and development at the outbound areas is suggested to follow the heat-controlling structures and should proceed without affecting the Chengjia and Longfeng hot springs.

2) Subpriority development zone: The Huangben–Mazishui geothermal anomalous zone and the Qigong–Yuntangkeng geothermal field are relatively rich in geothermal resources, evenly distributed, and have a thick geothermal reservoir and large water withdrawal capacity. The shallow resources are easy to develop but they are placed in this group because they are not easily accessible.

3) General development zone: The Jiangying–Yankou geothermal anomalous zone has low water temperature and is inaccessible; however, it can be developed and utilized.

4) Potential development zone: The deep-seated Leigongkeng geothermal anomalous zone needs to be further studied prior to its development and utilization.

5) Restricted development zone: The exploration of the Longfeng and Chengjia geothermal fields has been concluded and they are already in the development and utilization stage. The exploration of the surrounding fields needs to be stopped to prevent duplication of activities.

4. Conclusion

The Yangshan County is rich in geothermal resources that can be utilized for tourism, recuperation, sauna, agriculture, and aquaculture activities. The geothermal resources are distributed in the northeastern, southwestern, and midwestern areas, with thirteen geothermal hot springs (spots) and seven geothermal anomalous zones discovered till now. The Chengjia and Dongguan–Shiluo geothermal fields are grouped into the priority development zone; the Huangben–Mazishui geothermal anomalous zone and Qigong Yuntankeng geothermal field belong to the subpriority development zone; the Jiangying–Yankou geothermal anomalous zone belongs to the general development zone; the Leigongkeng geothermal anomalous zone belongs to the potential development zone; and the Longfeng and Chengjia geothermal fields belong to the restricted development zone.

References

[1] Liu J, Song M Y and Tian G H 2012 *Geological Survey and Research* **35** 67–73
[2] Cheng B Y 2007 *Resources Environment & Engineering* **21** 107–10
[3] Liao Y, Li W, Chen J G Wang L Tan W and Lv X J 2014 *Resources Environment & Engineering* **28** 884–7
[4] Liang H B, Qian Z, Xin S L, Zhao K J and Zhu L R 2010 *China Petroleum Exploration* **15** 63–8
[5] Li A N and Lv J B 2001 *Geology in China* **28** 41–5

[6] You W J, Liu Y F and Guo M J 2013 *Safety and Environmental Engineering* **20** 24−8, 34
[7] Wang B Y 1999 *Geology and Exploration* **35** 47−8
[8] Zhang D Z and Wang Y Y 2000 *Acta Geoscientica Sinica* **21** 151−4
[9] Zhu Z D and Zhen X C 1999 *Geological Review* **45** 627−30
[10] Wu L J 2016 *Geology and Exploration* **52** 300−6

IOP Conf. Series: Earth and Environmental Science **59** (2017) 012038 doi:10.1088/1755-1315/59/1/012038

Research on the industry environmental total factor productivity in Jiangsu Province based on the SBM-SML

Sun Lingfang[1,2] Wang Han[2] and Gong Jian[2]

[1]College of Economics and Management, Taizhou University, Taizhou, Jiangsu, 225300,China
[2] School of Economics and Management, Jiangsu University of Science and Technology, Zhenjiang, Jiangsu,212003,China

E-mail: 310649254@qq.com

Abstract. This paper uses the SBM-SML to measure the industry environmental total factor productivity in Jiangsu province of its 13 cities during 2005-2014 with SO^2 emissions as the undesirable output, and discomposes the total factor productivity into the pure technical efficiency, the scale efficiency change, the pure technical change and the scale technical change. The research shows that the overall trend of the industry environmental total factor productivity is increasing in Jiangsu province during 2005-2014, the technical change is a main reason pushing up growth rates of economy, and the pure technical change is the intrinsic motivation of the technical change.Introduction.

1. Introduction

Since the reform and opening up, Jiangsu, as an economically developed province, has made tremendous contributions to the rapid development of China's economy, among which the vigorous industrial development of Jiangsu is outstanding. In 1979 to 2014, the average growth rate of industrial added value in Jiangsu was 13.9%, and industrial GDP accounted for 41.4% of the total GDP of Jiangsu in 2014. However, the excellent achievements made by industrial development are also accompanied with increasingly serious problems of environmental pollution. "Haze" has become a popular word in recent years. The large-scale haze in Jiangsu was obvious since 2011 and became more serious after 2012. The orange alerts of air quality have been issued in 13 cities repeatedly, and masks have become the essential goods. It was proposed in the Twelfth Five-Year Plan of Jiangsu that "environmental priorities should be insisted, and the ecological construction should be further promoted; to solve the problems threatening people's health and affecting sustainable development, effectively control the environmental pollution and ecological destruction, and fully complete the pollutant emission reduction task assigned by the country." Therefore, under the environmental factors, the study on the industry total factor productivity of Jiangsu is conducive to the comprehensive understanding of the changes in industrial economy in Jiangsu, and providing help for the government decision-making.

In terms of the research on environmental total factor productivity, domestic and foreign scholars tend to use the data envelopment analysis method with the environmental pollutants as non-expected output. Pittman (1983) treated the non-expected output as shadow price [1]; Fare (1989) proposed the weak disposability of input-output via the non-linear programming method[2]; Hailu et al. (2003) treated the non-expected output variable as input[3]; Scheel (2001) and Zhu (2003) took the reciprocal of non-expected output as the input, and analyzed via the traditional DEA model[4][5]; Fare (2003) proposed a method of directional distance function[6]; in addition, Tone (2003) proposed a non-radial non-angular SBM model to solve the relaxation of input-output caused by the selection of angle and radial direction[7]. Since then, this method has been widely used in the study on environmental total factor productivity. The domestic studies based on this method included: Zhou et al. (2006) measured the environmental regulation and impact of carbon dioxide emissions of 30 organizations for economic cooperation in 1998-2002 using the SBM efficiency method[8]. Liu et al. (2009) selected SBM-NS environmental efficiency model to evaluate the environmental efficiency of 43 enterprises in Anhui in 2007, and proved the effectiveness of SBM model[9]; Tu Zhengge and Liu Leike (2011) analyzed China's industrial efficiency with energy and environmental factors based on the provincial data of SBM model[10]; Li Ling (2011) analyzed the green total factor productivity of pollution-intensive industries based on the empirical analysis of SBM model, and made a conclusion that environmental regulation did not really play a substantive role, and China's pollution-intensive industries are still high-energy consumption and high-pollution[11]; Liu Hui and Li Zhicui (2013) compared the difference of industry environmental efficiency and economic efficiency in western China from the SBM model and the traditional CCR model and confirmed that the SBM model could avoid the choice problem of radial direction and angle, and better deal with the problem of non-expected output, which reflected the objective and real industry environmental efficiency[12]; Hu Biao et al. (2015) studied the urban ecological civilization construction efficiency based on the SBM model, illustrated by Tianjin, and made a conclusion that the fine overall efficiency of ecological civilization construction in Tianjin in the past 12 years and the irrational industrial structure are the root causes of ineffective comprehensive technical efficiency[13].

In conclusion, although the study on industry environmental total factor productivity based on SBM model has made a lot of achievements, there are few decomposition studies on the industry environmental total factor. Therefore, this paper analyzes the industry environmental efficiency of Jiangsu based on SBM-SML, and then analyzes it as four aspects, namely pure technical efficiency change, scale efficiency change, pure technical change and scale technical change, so as to better investigate the industrial economic development thoughts of Jiangsu and provide a theoretical basis for the common development of environment and economy.

2. Method

2.1. SBM model of non-expected output

Traditional CCR model does not consider the slack variable of input and output, and also ignores the inevitable non-expected output problems in production, so it overestimates the technical efficiency in the process of actual production with the distortion. But SBM model based on the non-expected output makes up the defects of the traditional model well.

Assuming that there are n *DMUs* with three factors for each: input, good output and bad output, presented as $x \in R^m$ $y^g \in R^{s_1}$ and $y^b \in R^{s_2}$, respectively. In order to make the formula more concisely, it is expressed by matrix. $X = [x_1, x_2, \cdots, x_n] \in R^{m \times n}$ $Y^g = \left[y_1^g, y_2^g, \cdots, y_n^g \right] \in R^{s_1 \times n}$ $Y^b = \left[y_1^b, y_2^b, \cdots, y_n^b \right] \in R^{s_2 \times n}$ and $X > 0$ $Y^g > 0$ $Y^b > 0$. The production possibility set is presented as:

$$P = \left\{ \left(x, y^g, g^b \right) \middle| x \ge X\lambda, y^g \le Y^g \lambda, y^b \ge Y^b \lambda, \lambda \ge 0 \right\} \quad (1)$$

SBM model of non-expected output is obtained:

$$\begin{cases} \rho^* = \min\left(1-\frac{1}{m}\sum_{i=1}^{m}\frac{s_i^-}{X_{i0}}\right)\left[1+\frac{1}{s_1+s_2}\left(\sum_{r=1}^{S_1}\frac{s_r^g}{y_{r_0}^g}+\sum_{r=1}^{S_2}\frac{s_r^b}{y_{r_0}^b}\right)\right]^{-1} \\ s.t.\ \ x_0 = X\lambda + s^- \\ \quad y_0^g = Y^g\lambda - s^g \\ \quad y_0^b = Y^b\lambda + s^b \\ \quad s^- \geq 0, s^g \geq 0, s^b \geq 0, \lambda \geq 0 \end{cases} \tag{2}$$

In the formula (2), s^-, s^g and s^b is the input slack variable, expected output slack variable and non-expected output slack variable, respectively; λ is the weight vector, the objective function is strictly decreased and the target value $\rho^* \in [0,1]$. When $\rho^* = 1$, s^{-*}, s^{g*} and s^{b*} are 0, DMU is effective.

2.2. SML index and its decomposition

SML index (Sequential Malmquist) is based on DEA sequence; in other words, the reference set in each phase contains all reference sets t in previous periods, and the reference set in t phase is:

$$s^{s(t)} = s^1 \cup s^2 \cup \cdots \cup s^t = \{(x_j^1, y_j^1)\} \cup \{(x_j^2, y_j^2)\} \cup \cdots \cup \{(x_j^t, y_j^t)\} \tag{3}$$

So the technical regression or "sunken" status in production boundary can be avoided in the analysis. The specific calculation method is as follows:

$$MI(x^{t+1}, y^{t+1}, x^t, y^t) = \left(\frac{E^{s(t)}(x^{t+1}, y^{t+1})}{E^{s(t)}(x^t, y^t)}\frac{E^{s(t+1)}(x^{t+1}, y^{t+1})}{E^{s(t+1)}(x^t, y^t)}\right)^{1/2} \tag{4}$$

$$EC = \frac{E^{s(t+1)}(x^{t+1}, y^{t+1})}{E^{s(t)}(x^t, y^t)}$$

$$TC = \left(\frac{E^{s(t)}(x^t, y^t)}{E^{s(t+1)}(x^t, y^t)}\frac{E^{s(t)}(x^{t+1}, y^{t+1})}{E^{s(t+1)}(x^{t+1}, y^{t+1})}\right)^{1/2}$$

$$MI = EC \times TC \tag{5}$$

According to the decomposition method of Zofio, EC is broken down into PEC and SEC, TC is broken into PTC and STC, namely

$$MI = EC \times TC = PEC \times SEC \times PTC \times STC \tag{6}$$

In the formula, EC is the technical efficiency change, TC is the technical change, PEC is the pure technical efficiency change, SEC is the scale efficiency change, PTC is the pure technical change, and STC is the scale technical change.

3. Data selection and processing

13 cities in Jiangsu Province in 2005-2014 were selected as the decision making units, and the input and output data were obtained according to the *Statistical Yearbook of Jiangsu, Statistical Yearbook of China's Regional Economic, China City Statistical Yearbook* and statistical yearbooks in each city. And according to related literature, as well as the availability and integrity of data, 2 input indexes, 1 expected output index and 1 non-expected output index were selected here.

Capital stock input index: the fixed capital investment of each city over the years were selected, and the perpetual inventory method was used to calculate the capital stock of each city over the years.

Labor input index: the average number of employees in each industry was selected.

Expected output index: the industrial added value of each city over the years was selected, and made into the constant price with the year 2005 as base year.

Non-expected output index: due to the increasingly serious haze, sulfur dioxide emission was selected as the non-expected output.

The descriptive statistics of each index are as shown in Table 1.

Table 1. descriptive statistics of each index

Index	Unit	Min.	Max.	Mean	S.D.
Capital stock	Hundred million yuan	838.09	16346.52	4548.50	3522.32
Labor	Ten thousand people	11.09	345.20	77.85	72.32
Industrial added value	Hundred million yuan	125.96	6516.82	1385.45	1246.83
SO_2 emmission	Ton	18251	242362	78923.46	48695.52

4. Empirical analysis

According to the panel data of 13 cities in Jiangsu Province, *maxdea6.8pro* software was used to calculate the industry environmental total factor productivity index of each city based on SBM-SML, and it was decomposed into the pure technical progress, scale technical change, pure technical efficiency change and scale efficiency change. The industry environmental total factor productivity and internal factors of Jiangsu Province will be analyzed below from the whole industry and city.

4.1. Change and decomposition analysis of industry environmental total factor productivity of Jiangsu Province

Table 2 shows the industry environmental total factor productivity and values of each decomposition index of Jiangsu Province in 2005-2014.

Table 2. the industry environmental TFP and decomposition index of Jiangsu Province

Year	SMI	EC	PEC	SEC	TC	PTC	STC
2005-2006	1.1107	1.0139	1.0242	0.9900	1.0955	1.0600	1.0335
2006-2007	1.1231	1.0061	1.0046	1.0015	1.1162	1.0795	1.0340
2007-2008	1.0022	0.9660	0.9664	0.9995	1.0375	1.0196	1.0176
2008-2009	1.1746	1.0061	1.0380	0.9692	1.1675	1.0909	1.0702
2009-2010	1.0943	1.0339	1.0056	1.0281	1.0584	1.0535	1.0046
2010-2011	1.2042	1.0214	1.0137	1.0075	1.1790	1.1104	1.0618
2011-2012	1.1560	1.0036	1.0081	0.9956	1.1518	1.0966	1.0504
2012-2013	1.1520	0.9926	0.9813	1.0116	1.1606	1.1197	1.0365
2013-2014	1.1447	1.0023	1.0042	0.9981	1.1421	1.1078	1.0309
Geometric mean	1.1277	1.0049	1.0049	1.0000	1.1222	1.0816	1.0375

According to Table 2, the variation trend map about the industry environmental total factor productivity and decomposition index of Jiangsu Province was drawn, as shown in Figure 1, Figure 2 and Figure 3.

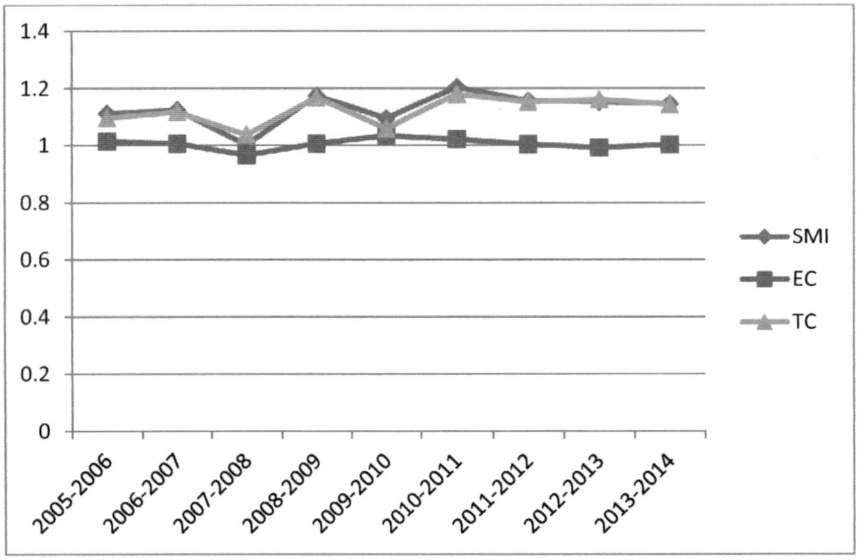

Figure 1. variation trend of SML and decomposition index

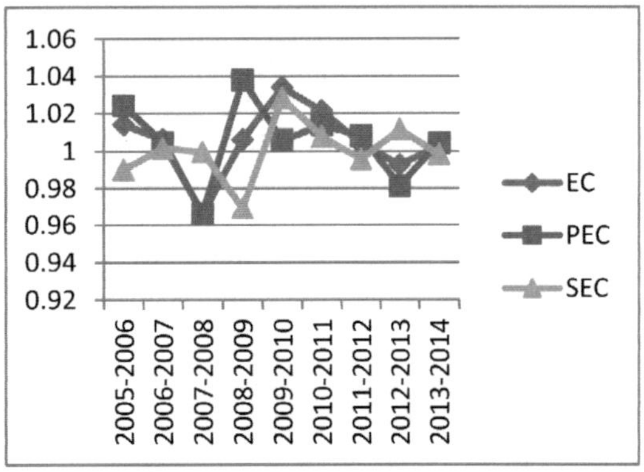

Figure 2. variation trend of EC and decomposition index

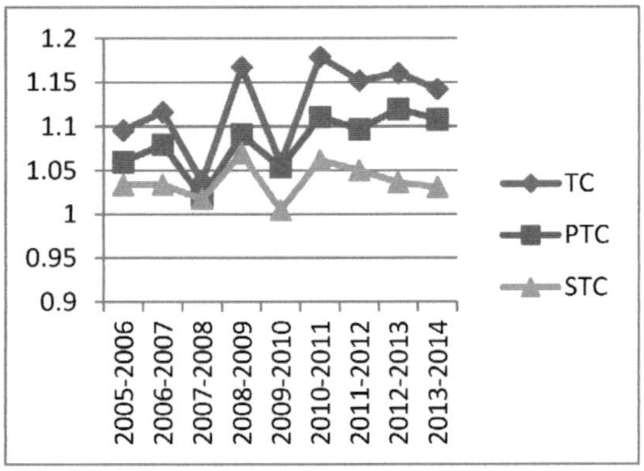

Figure 3. variation trend of TC and decomposition index

The geometric mean of industry environmental total factor productivity of Jiangsu Province in 2005-2014 was 1.1277, indicating that the industry environmental total factor productivity of Jiangsu Province was on the rise in this decade. But it can be seen from Figure 1 that the growth rate of total factor productivity was not stable. In 2008, the global economic crisis leads to a decline in growth with the growth rate of 0.22%; subsequently, Jiangsu government conducted the large-scale government investment plans, and accelerated technological innovation, promoting the growth. However, Jiangsu Province was in the industrialization of heavy industry in the following years, so the sulfur dioxide emissions sharply rose in 2011 by 12.17% compared with that last year, sharply reducing the industrial total factor productivity growth included into the environmental impact. Figure 1 intuitively shows that the changing curve of SML index is consistent with that of technical progress, illustrating the technical progress change is the main factor of promoting the change in productivity growth.

Secondly, from the perspective of specific factor decomposition, in terms of the decomposition of technical efficiency change, Table 2 shows that the geometric mean of technical efficiency change in Jiangsu Province during the past decade was 1.0049, and that of the pure technical efficiency and scale efficiency change was 1.0049 and 1, suggesting that the pure technical efficiency change is the main internal factor of technical efficiency change. Technical efficiency change had the negative growth during 2007-2008 and 2012-2008, the former of which was because both pure technical efficiency and scale efficiency change were less than 1, and the latter of which was because the scale efficiency change growth was not enough to offset the negative growth of pure technical efficiency change. Combined with Figure 2, it can be seen that the scale efficiency change had negative growth many times, which shows that the industry in Jiangsu Province is optimized without blind pursuit of mass production. From the perspective of decomposition of technical changes, the geometric average of pure technical change was 1.0816, while the geometric average of scale technical change was 1.0375, so the pure technical change was the main internal cause of technical change. Combined with Figure 3, it can be seen that the fluctuations in the three curves are almost the same, a fluctuant increasing trend, showing that the industrial technology of Jiangsu Province was continuously improved.

4.2. Industry environmental total factor productivity and decomposition analysis of each city in Jiangsu Province

Table 3 shows the industry environmental total factor productivity and mean of each decomposition index of each city in Jiangsu Province in 2005-2014.

Table 3. the industry environmental TFP and its decomposition index of each city

DMU	SML	EC	SEC	PEC	TC	STC	PTC	Ranking
Nanjing	1.2597	1.0000	1.0000	1.0000	1.2597	1.0425	1.2083	1
Changzhou	1.2048	1.0398	1.0047	1.0350	1.1586	1.0390	1.1151	2
Suzhou	1.1777	0.9984	0.9984	1.0000	1.1795	1.0495	1.1239	3
Yangzhou	1.1752	1.0323	1.0033	1.0289	1.1385	1.0496	1.0846	4
Yancheng	1.1334	1.0185	0.9918	1.0269	1.1128	1.0261	1.0845	5
Zhenjiang	1.1230	1.0000	1.0000	1.0000	1.1230	1.0521	1.0673	6
Taizhou	1.1154	0.9819	0.9900	0.9917	1.1360	1.0790	1.0528	7
Nantong	1.1114	1.0157	0.9931	1.0227	1.0943	1.0171	1.0759	8
Wuxi	1.1060	1.0000	1.0000	1.0000	1.1060	1.0073	1.0979	9
Xuzhou	1.0711	1.0029	1.0012	1.0018	1.0680	0.9987	1.0694	10
Lianyungang	1.0705	0.9917	1.0193	0.9730	1.0794	1.0318	1.0461	11
Suqian	1.0673	0.9930	0.9930	1.0000	1.0748	1.0621	1.0119	12
Huaian	1.0638	0.9918	1.0056	0.9862	1.0726	1.0360	1.0354	13

From the perspective of the ranking of environmental total factor productivity in each city, SML index of Nanjing was 1.2597, namely the growth rate of environmental total factor productivity of 25.97%, ranking in the first place. And Suzhou ranked in the third place with the growth rate of 2.71% lower than that of Changzhou, which was due to the negative growth of technical efficiency change, and caused by the scale efficiency change. But we can see that technical change growth in Suzhou was high, which was related to the continuous development of new technology and construction of high and new technology zones of Suzhou in recent years. The environmental total factor productivity of Huaian was the lowest, 1.0638, and it can be seen that the negative growth of pure technical efficiency change of Huaian was the chief culprit hindering the improvement of its total factor productivity, so Huaian should strengthen the effective utilization of resources, and speed up the optimization of resource allocation. Suqian was the last but one; although its technical efficiency change was the same as that in Huaian with a negative growth, the internal cause was different; the scale efficiency affected the growth of total factor productivity. Overall, the environmental total factor productivity in each city of Jiangsu Province increased in different degrees.

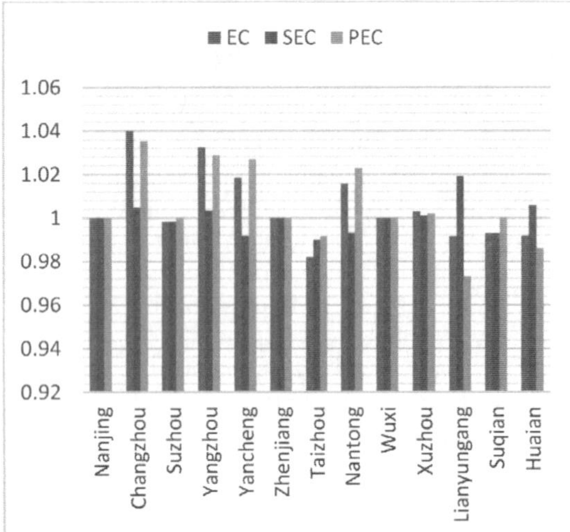

Figure 4. EC and its decomposition index of each city

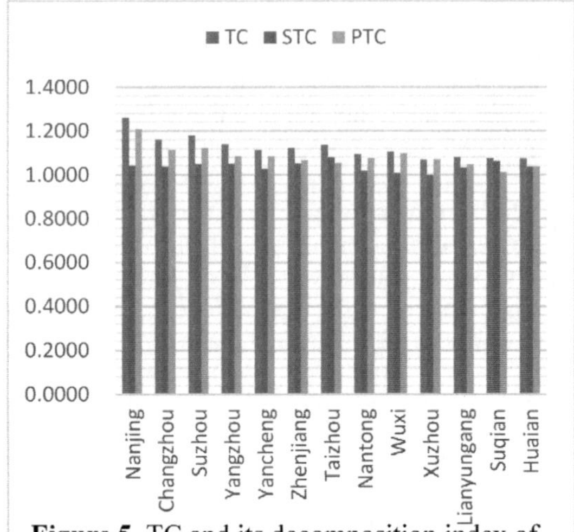

Figure 5. TC and its decomposition index of each city

It can be seen from Figure 4 that the technical efficiency of 5 cities had a negative growth, among which the technical efficiency of Taizhou fell by 1.81% on average during the past decade. When it came to the reasons, the pure technical efficiency and scale efficiency had a negative growth at the same time. From the perspective of regional division, it can be found that Lianyungang and Huaian in north Jiangsu had the negative growth of high technical efficiency caused by the low resource utilization rate; Suzhou in south Jiangsu had the negative growth in technical efficiency because of the decreased scale efficiency. There were 5 cities whose mean technical efficiency was greater than 1, among which that of Changzhou was the highest with the growth of 3.98%, and large growth of pure technical efficiency and scale efficiency advantage were the intrinsic motivation.

It can be seen from Figure 5 that the technical change in Nanjing was the highest, 1.2597, and the pure technical change was the main factor of technical change compared with scale technique. The technical change in Xuzhou was the lowest, increasing by 6.8%. Overall, the pure technical change in each city was more stable, and the fluctuations in pure technical change and technical change tended to be consistent, suggesting that pure technical change was the main factor of technical change. From the perspective of regional division, it can be seen clearly that the average technical change in cities in south Jiangsu was significantly higher than that in middle Jiangsu, and that middle Jiangsu was also significantly higher than that in north Jiangsu. This is consistent with the reality, because there are many high-tech zones in south Jiangsu, such as Xinbei District in Changzhou, Jiangning District in Nanjing and Suzhou Industrial Park with strong technological innovation ability; but although north Jiangsu has rapidly developed in recent years, the industrial base is weak and the technology innovation ability is lagged compared to that of south Jiangsu.

5. Conclusions

Through the concrete analysis of the environmental total factor productivity and its internal factors from the perspective of Jiangsu Province and each city, the following conclusions can be made:

(1) During the study, the growth rate of industry environmental total factor productivity of Jiangsu Province was fluctuant, but it was on the rise with the average geometric growth rate of 12.77%, which made a great contribution to the industrial economic growth in Jiangsu, and was one of the important factors to promote economic development in Jiangsu. After decomposition, it was found that technical progress is the main factor relative to technical efficiency, and the pure technical efficiency, pure technical change and scale technical change were the intrinsic motivation of growth of industry environmental total factor productivity.

(2) During the study, the industry environmental total factor productivity of each city in Jiangsu Province had the growth in different degrees, among which Nanjing and Changzhou ranked the first and second place, and Huaian was the last one. From the perspective of decomposition, technical progress was the main factor of the growth of industry environmental total factor productivity, followed by technical efficiency.
(3) From the perspective of administrative division of Jiangsu Province, the industry environmental total factor productivity in cities in south Jiangsu was generally higher than that in middle Jiangsu, and that in middle Jiangsu was generally higher than that in north Jiangsu. From the perspective of decomposition, technical progress was the reason why industry environmental total factor productivity was higher in cities in south Jiangsu than others, and the pure technical change was the main factor of technical progress. Pure technical efficiency of cities in north Jiangsu had a negative growth generally, hindering the improvement of industry environmental total factor productivity in each city.

Acknowledgments
(1) Fund project: Key philosophy and social science research project for universities in Jiangsu province, "dynamics analysis and structure optimization research on industry and energy consumption of atmospheric pollution in Jiangsu province" (2015ZDIXM039)
(2) Fund project: A fund project for the professor(Dr.) in Taizhou University, "research on optimization of manufactural competitive environment and development strategy in Jiangsu Province based on "made in China2025""(TZXY2015JBJJ004)

References
[1] Pittman R W. Multilateral Productivity Comparisons with Undesirable Outputs[J]. Economic Journal, 1983, 93(372): 883-91.
[2] Faere R, Pasurka C. Multilateral Productivity Comparisons When Some Outputs Are Undesirable: A Nonparametric Approach[J]. Review of Economics & Statistics, 1989, 71(1): 90-98.
[3] Hailu A, Veeman T S. Non-parametric Productivity Analysis with Undesirable Outputs: An Application to the Canadian Pulp and Paper Industry[J]. American Journal of Agricultural Economics, 2003, 85(85): 1075-1077.
[4] Scheel H. Undesirable outputs in efficiency valuations[J]. European Journal of Operational Research, 2001, 132(2): 400-410.
[5] Maleyeff J. Quantitative Models for Performance Evaluation and Benchmarking: DEA with Spreadsheets and DEA Excel Solver[M]. Springer, 2009: Pages.
[6] Färe R, Grosskopf S, Pasurka C A. Environmental Production Functions and Environmental Directional Distance Functions: A Joint Production Comparison[J]. Social Science Electronic Publishing, 2004.
[7] Tone K. Dealing with Undesirable Outputs in DEA : A Slacks-based Measure (SBM) Approach(DEA(1))[J]. abstract book of Japan's operational research societies in spring, 2004, 2004: 44-45.
[8] Zhou P, Ang B W, Poh K L. Slacks-based efficiency measures for modeling environmental performance[J]. Ecological Economics, 2006, 60(1): 111-118.
[9] Liu Yong, Li Zhixiang, Li Jing, Environmental Efficiency Evaluation Based on SBM-NS [J], Mathematics in Practice and Theory, 2009, 39 (24): 25-30.
[10] Tu Zheng-Ge, Liu LeiKe, Evaluation of China's Industrial Efficiency with Energy and Environmental Factors: A Provincial Data Analysis Based on SBM [J], Economic Review, 2011 (2): 55-65.
[11] Li Ling, Tao Feng, Green Total Factor Productivity and Its Influencing Factors of Pollution-Intensive Industries: An Empirical Analysis Based on SBM Directional Distance Function [J], Economists, 2011 (12): 32-39.
[12] Liu Hui, Li Zhicui, Studies on the Differences of Industrial Environmental Efficiency and Economic Efficiency in Western China: A Comparative Analysis Based on DEA-SBM and

DEA-CCR Model [J], West China Forum, 2013 (6): 90-96.

[13] Hu Biao, Wang Feng, Li Jianyi, et al., An Empirical Study on the Efficiency Evaluation of Urban Ecological Civilization Construction Based on Non-expected Output SBM: Illustrated by Tianjin [J], Journal of Arid Land Resources and Environment, 2015, 29 (4): 13-18.

IOP Conf. Series: Earth and Environmental Science **59** (2017) 012039 doi:10.1088/1755-1315/59/1/012039

The study on the movement law of overlying strata in fully mechanized caving face under alluvium with huge thickness in deep mine

CHEN Yong [1], GONG Feng-gang [2], ZHAO Ming[1]

1.College of Resources and Safety Engineering, China University of Mning and Technology(Beijing) ,Beijing 100083 China; 2. Inner Mongolia Fu cheng Mining Company, Inner Mongolia 017000 China

Email:815220255@qq.com

Abstract. Xin ju-long Mine has characteristics of deep buried depth and large alluvium's thickness, and the ground pressure is obvious. In order to realize the effective control of the overlying strata in the fully mechanized caving face under alluvium with huge thickness in deep mine, the movement law of overlying strata is studied. The relationship between the key stratum's fracture and the movement of overlying strata is analyzed based on 'the key stratum' theory, and the movement law is obtained by numerical simulation. The results show that:1)after the bedrock is completely broken, the strength of the overlying alluvium becomes low, and its supporting capacity is gradually lost and it will gradually sink with the breaking of the bedrock;2)with the increase of the alluvium's thickness, the roof subsidence is gradually increasing, and with the continuous advancement of the working face, the influence range of the support pressure and the amount of roof subsidence also show a trend of increasing; 3)the study on the movement law of overlying strata under this condition provides a reliable and theoretical guide for the selection of the support and the control of ground pressure.

1.Introduction

The depth of Xin ju-long mine is more than 800 meters, the bedrock is relatively thin, and the alluvium's thickness is huge,so the mine is faced with strong pressure appearance in the process of mining.Safe and efficient mining under alluvium with huge thickness in deep mine has become an important research topic in recent years. In combination with the production practice of coal mine and the observation of overlying strata movement, 'the voussoir beam' theory is put forward by Qian Ming-gao(Qian et al.,1989; Qian et al.,2010). Academician Qian Ming-gao and Professor Miao Xie-Xing put forward the 'key stratum' theory based on 'the voussoir beam' theory, who believe that the key stratum plays a major role in the activities of rock mass, and the fracture of the key stratum will make the overlying strata have obvious movement and lead to the strata behaviours (Qian et al.,1996). Based on a large number of field measurements and theoretical analysis, Academician Song Zhen-qi put forward 'the transferring rock beam' theory,which is centered on the overlying strata movement(Tan et al.,2007). Professor Jiang Fu-xing put forward 'the spatial structure of overlying strata' theory, which is a systematic analysis on the mechanical mechanism of mine dynamic disasters such as rock burst(Jiang et al.,2006).On the basis of previous studies(Xu et al.,2008; Xu et al.,2009; Ma et al.,2010; Liu et al.,2011; Ju et al.,2011; Ma et al.,2013), using the methods of theoretical analysis and numerical simulation, the movement law of overlying strata is studied and the law of the support pressure's distribution and the amount of roof subsidence is analyzed.

2. Calculation of the Key Stratum's Thickness

There is a great relationship between the movement of overlying strata and the fracture of the basic roof. Once the fracture of the basic roof occurs, the subsidence of the overlying strata will occur, which will lead to the subsidence of the ground surface. So the experts gives the concept of the key stratum which plays a major role in the activities of rock mass. The key stratum's thickness is usually determined by the method of field measurement. In addition, it can be obtained according to the weighting average of the first weighting interval and the period weighting interval.
The thickness of the key stratum obtained from the first weighting interval is as follows.

$$m_{E0}=\frac{C_0^2\gamma_E}{2\times 1000\sigma_t} \tag{1}$$

Where: C_0 - the key stratum of the first weighting interval, m; γ_E - the key stratum of the average bulk density, 25kN/m^3; σ_t - the tensile strength of the key stratum, kN/m^2.
The key stratum's thickness of the first and the second periodic weighting is obtained according to the periodic weighting interval.

$$m_{E1}=\frac{3\gamma_E\left[\left(C_1+\frac{C_0}{4}\right)^2-\left(\frac{C_0}{4}\right)^2\right]}{1000\sigma_t} \tag{2}$$

$$m_{E1}=\frac{3\gamma_E\left[\left(C_2+\frac{C_1}{4}\right)^2-\left(\frac{C_1}{4}\right)^2\right]}{1000\sigma_t} \tag{3}$$

Where: C_1 -the first period of weighting interval,m; C_2 - the second period of weighting interval, m.
According to the weighted average of the periodic weighting interval, the key stratum's thickness is obtained.

$$m_E=\frac{m_{E0}C_0+m_{E1}C_1+m_{E2}C_2}{C_0+C_1+C_2} \tag{4}$$

3.Mechanical Analysis on Movement Law of Overlying Strata

3.1 Shear Stress Analysis of the Composite Strata

The combined action of the composite strata results in the compounded effect of the overlying strata. Suppose that the cross section of the composite strata is rectangular, the middle is soft rock, the upper and the lower strata are hard rock.The bending elastic modulus is smaller if the middle strata is thinner and it is greater if the upper and the lower strata are thicker.
The diagram of composite strata is shown in Figure 1.

h^1, E^1- the thickness, elastic modulus of the upper strata,; h^2, E^2- the thickness, elastic modulus of the lower strata; h^3, E^3- the thickness , elastic modulus of the middle strata;

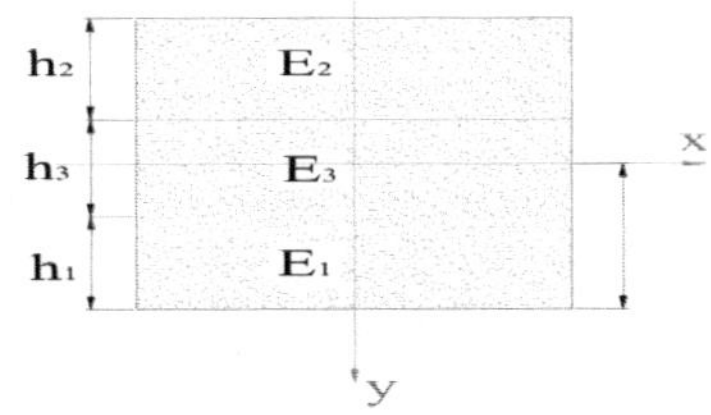

Fig.1 Cross section of the composite strata

The following equations are obtained from the knowledge of the mechanics of materials.

$$\left.\begin{aligned} \sigma_1 &= \frac{E_1 y}{\rho} \\ \sigma_2 &= \frac{E_2 y}{\rho} \\ \sigma_3 &= \frac{E_3 y}{\rho} \end{aligned}\right\} \tag{5}$$

Where: ρ -The curvature radius of the rock beam's neutral layer.
In order to facilitate the calculation, we take the width of the rock beam as one, then the inertia moment of each cross section is as follows.

$$\left.\begin{aligned} I_1 &= \frac{h_1^3}{12} + h_1(a-h_1)^2 \\ I_2 &= \frac{h_2^3}{12} + h_2\left(h_1 + \frac{h_2}{2} + h_3 - a\right)^2 \\ I_3 &= \frac{h_3^3}{12} + h_3\left(h_1 + \frac{h_3}{2} - a\right)^2 \end{aligned}\right\} \tag{6}$$

Where:a-The distance from the neutral axis to the lower cross section.
The dislocation of the rock beam is caused by Q (the shear force). According to the knowledge of the mechanics of materials, the maximum shear failure occurs at the neutral plane. Based on the above analysis, τ_{13} (the shear stress between the upper and middle strata), τ_{23} (the shear stress between the middle and lower strata) and τ_3 (the maximum shear stress on the neutral plane)are obtained as follows.

$$\left.\begin{aligned} \tau_{23} &= \frac{QE_1(2h_2 + h_1 + 2h_3 - 2a)h_1}{2(E_1I_1 + E_2I_2 + E_3I_3)} \\ \tau_{13} &= \frac{QE_2(2a - h_2)h_2}{2(E_1I_1 + E_2I_2 + E_3I_3)} \\ \tau_3 &= \frac{Q[E_2(2a-h_2)h_2 + E_3(a-h_2)^2]}{2(E_1I_1 + E_2I_2 + E_3I_3)} \end{aligned}\right\} \tag{7}$$

3.2 Discrimination of the Composite Strata

Suppose that the composite fracture of the overlying strata occurs , then:

$$\left.\begin{array}{l}\tau_{13\max} \le \tau_{13e} \\ \tau_{23\max} \le \tau_{23e} \\ \tau_{3\max} \le \tau_{3e}\end{array}\right\} \quad (8)$$

Where: $\tau_{13\max}$, τ_{13e} -the maximum shear stress, the shear strength between the middle and upper strata; $\tau_{23\max}$, τ_{23e} - the maximum shear stress, the shear strength between the lower and middle strata is; $\tau_{3\max}$, τ_{3e} - the maximum shear stress, the shear strength on the neutral plane.

Once the strata is in accordance with the formula (8), it will form the composite strata which become the main key stratum.The lower part of the strata and the relatively soft strata will be damaged and the separation of the overlying strata will occur. With the advancement of the working face the exposed area of the lower strata is increasing continuously.When the formula ‘ $\tau_{13\max} \le \tau_{13e}$ ’ is satisfied,the shear failure will occur in the lower strata. The broken strata acts on the soft strata and then acts on the supports,the working face shows a state of pressure at this time.

4 Numerical Simulation of Overlying Strata movement

4.1 Establishment of FLAC3D Numerical Model

According to the geological column of No.2301 fully mechanized caving face, the $FLAC^{3D}$ numerical model is founded. The length in the direction of inclination is 200 m, and the strike length is 250 m, and the height respectively are 200 m, 400 m and 600 m. The transition area on the left is set to 20 m, and the transition area on the right is set to 40 m. The thickness of coal seam is 9 m. The above values are conducive to the calculation and analysis.

The normal displacement and velocity around the model is zero.In the model, and the top of the model is affected by the uniformly distributed load whose size is the weight of the overlying strata. The physical and mechanical parameters in the numerical model are shown in Table1.

Tab. 1 The physical and mechanical parameters in the numerical model

Lithology	Thickness /m	Density /kg/m^3	Bulk modulus /GPa	Shear modulus /GPa	Cohesion /Pa	Internal friction angle/°	Tensile strength /MPa
Fine sandstone	2.4	2879	9.81	5.46	4.2e6	42	8
Medium sand	7.2	2851	9.96	6.52	4e6	40	9
Siltstone	3.5	2877	8.13	4.87	3.4e6	41	7
Fine sandstone	5.0	2879	9.81	5.46	4.2e6	42	8
Fine sandstone	12.32	2879	9.81	5.46	4.2e6	42	8
Medium sand	11.6	2851	9.96	6.52	4e6	40	9
Grit stone	27.6	2480	8.24	5.87	6.1e6	38	7
Fine sandstone	4.0	2879	9.81	5.46	4.2e6	42	8

Siltstone	2.63	2876	8.13	4.87	3.4e6	41	7
3#Coal seam	9.0	1339	5.68	3.42	5.2e6	18	1.2
Siltstone	22.4	2877	8.13	4.87	3.4e6	41	7

4.2 Simulation Results of Roof Subsidence under Different Alluvium's Thickness

When the alluvium's thickness is 200 m,choose 20m, 40m, 60m, 80m, 100m and 120m respectively as the advancing distance, then the simulation results are shown in figure 2.

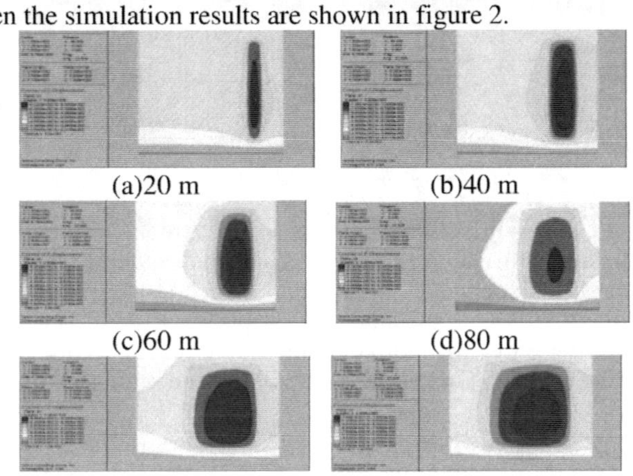

(a)20 m (b)40 m
(c)60 m (d)80 m
(e)100 m (f)120 m

Fig. 2 Roof subsidence of different advancing distance under 200m-alluvium's thickness

When the alluvium's thickness is 400 m,choose 20m, 40m, 60m, 80m, 100m and 120m respectively as the advancing distance, then the simulation results are shown in figure 3.

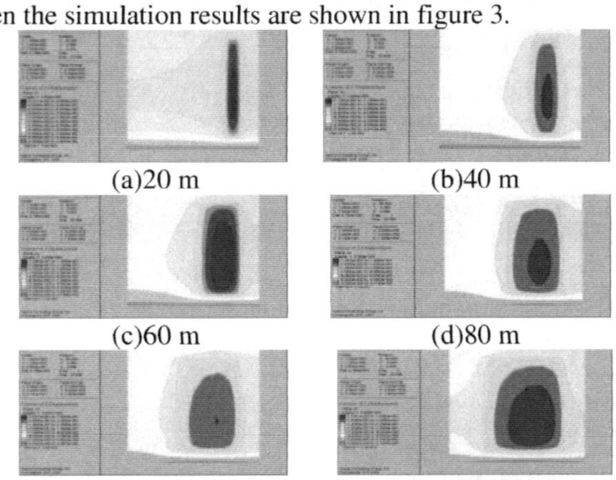

(a)20 m (b)40 m
(c)60 m (d)80 m
(e)100 m (f)120 m

Fig. 3 Roof subsidence of different advancing distance under 400m-alluvium's thickness

When the alluvium's thickness is 600 m,choose 20m, 40m, 60m, 80m, 100m and 120m respectively as the advancing distance, then the simulation results are shown in figure 4.

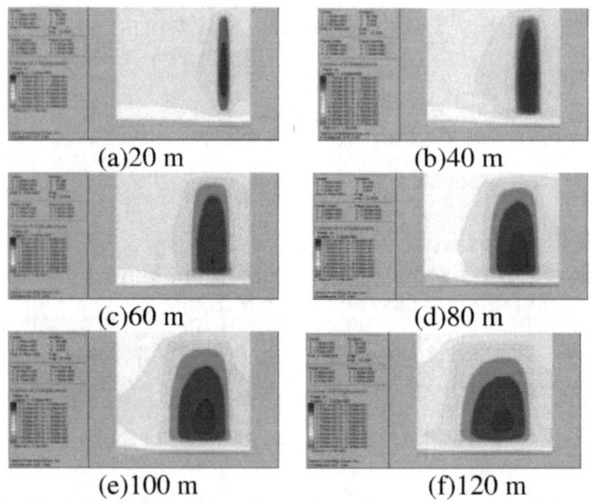

Fig. 4 Roof subsidence of different advancing distance under 600m-alluvium's thickness

According to the amount of roof subsidence at 23 metres from the top of roof, the law of roof subsidence with different advancing distance under different alluvium's thickness is existed as follows.

(1) Under a certain alluvium's thickness, the amount of subsidence at the open-off cut and the coal wall is minimum in early stage of caving mining and the amount is maximum at the middle of the working face. The vertical displacement of the roof gradually decreases from the middle of the working face to the open-off cut and the coal wall.

(2) Under a certain thickness of the alluvium, with the advancement of the working face the amount of roof subsidence gradually increases and the influence range gradually enlarges.The influence range is the rectangular area centered on the middle part of the stope.

(3) Because the 3# coal seam is nearly horizontal, the roof strata are in the same direction.With the advancement of the working face, the amount of roof subsidence gradually decreases.

(4) Compared with the general conditions, the roof of the working face under alluvium with huge thickness is more active, and the strata behaviour is more obvious, and the supporting work is more difficult.

4.3 Change of the Roof's Vertical Displacement under 600m-alluvium's Thickness

When the alluvium thickness is 600 m and the advancement distance is 120 m, the fracture and evolution of overlying strata at different locations above the roof is shown in Figure 5.

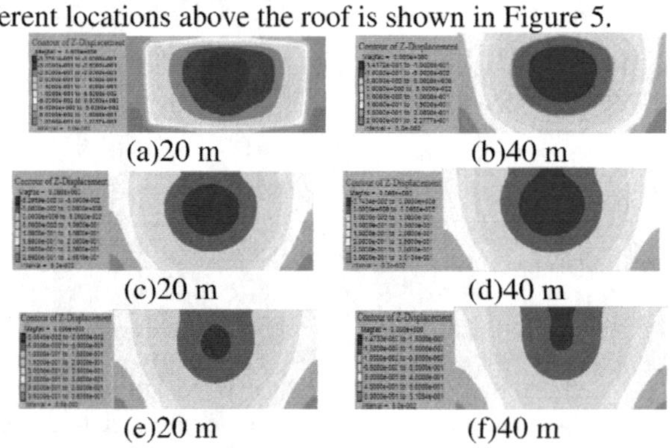

Fig. 5 Change of roof's vertical displacement at different locations above the roof

Figure 5 reveals the following law.

(1) The farther away from the roof of the working face, the more the roof subsidence is not obvious;

(2) The amount of roof subsidence gradually decreases from the middle part of working face to both sides.
(3) The amount of roof subsidence at the upper and the lower end shows asymmetrical distribution, and the amount at upper end is obviously smaller than that at the lower end.

5 Conclusions

(1) There is a close connection between the key stratum and the movement of overlying strata, the fracture of the key stratum plays an important role in the movement of the overlying strata., especially under the condition of large depth and huge thickness of the alluvium. After the bedrock are all broken, the strength of the overlying alluvium became lower and the supporting capacity was lost,then the overlying alluvium gradually subsides. Because the alluvium's thickness is very large, the bench convergence does not occur on the ground surface.
(2) With the increase of the alluvium's thickness,the amount of roof subsidence in fully mechanized caving face increases gradually. With the continuous advancement of the working face,the influence range of bearing pressure and the amount of roof subsidence also increase gradually.
(3) To find out the movement law of overlying strata in fully mechanized caving face under alluvium with huge thickness in deep mine can achieve effective control of the overlying strata, and further provide theoretical guidance for strata control and support selection of working face.

References

[1]QIANMing-gao,HeFulian,1989. The Behaviotir of the Main Roof in Longwall Mining. Wighting Span, Fracture and Disturbance.Journal of Mine, Metals&Fuels,6-7:240-246.
[2]QIAN Ming-gao ,SHI Ping-gao,XU Jia-lin,2010. Ground Pressure and its Control. China University of Mining and Technology press,65-93.
[3]QIAN Ming-gao,MIAO Xie-xing,1996.Research on the theory of key stratum in strata control.Journal of China Coal Society,21(3):225-230.
[4]TAN Yun-liang,WU Shi-liang,YIN Zeng-de,et al.,2007.Ground Pressure and Strata Control. China Coal Industry Publishing House, 7-16.
[5]JIANG Fu-xing, 2006. Viewpoint of spatial structures of overlying strata and its application in coal mine. Journal of Mining & Safety Engineering,23(1):30-33.
[6]XU Nai-zhong,DAI Hua-yang,2008. Research status of mining subsidence rule and control under the condition of thick loose layer. Safety in Coal Mines,39(11):53-55
[7]XU Jia-lin , ZHU Wei-bing, WANG Xiao-zhen, et al.,2009. Classification of key stratum structure of overlying strata in shallow coal seam. Journal of China Coal Society, 7:865-870.
[8]MA Shu-ming,LV Wen-yu,SUN Bing-yu,2010. Current situation and trend of the research on the ground pressure's theory in deep mining field. Coal engineering,10:87-89.
[9]LIU Fei,MA Hua,DING Yan-lu,2011. Principle of Ground Pressure and Strata Control.Coal Mine Modernization,1:98-100.
[10]JU Jin-feng,XU Jia-lin,WANG Qing-xiong,2011. Cantilever structure moving type of key stratum and its influence on ground pressure in large mining height workface. Journal of China Coal Society, 36(12):2115-2120.
[11]MA Li-qiang,ZHANG Dong-sheng, SUN Guang-jing,et al.,2013. Thick alluvium full-mechanized caving mining with large mining height face roof control mechanism and practice. Journal of China Coal Society, 2:199-203.

Analysis of the Correlation between Environmental Performance in Enterprise Green Management and the Competitive Edge

Xing Li[1, a], [*]**Zihan Yang**[1, b], **Peiyao Zhang**[1, c],

[1]College of Economics and Management,LongDong University ,QingYang City,Gansu

[a]lixing@ldxy.edu.cn, *[b]455248638@qq.com,[c]annaree@163.com

Abstract: In recent years, there has been an increasing concern over the topics of enterprise green management and environmental performance's influence on a company's competitive edge. This paper mainly analyzed the effect of enterprise green management on a company's environmental performance and its competitive advantages, as well as environmental performance's influence over the competitive edge. Furthermore, this paper discussed the role of environmental performance between enterprise green management and competitive edge. Objects of this study are manufactures in northwestern China which have acquired the ISO 9001 certification before December, 2014. Research discoveries shows that enterprise green management would exert positive influence on the environmental performance and its competitive edge; the environmental performance could also add significant competitive edge to a company; last but not the least, environmental performance also plays an intermediary role between enterprise green management and its competitive advantages

1. Research Background

Over the past few years, there have been haggles between environmentalists and business sector over topics of climate change, air, water and soil pollution, and decreasing biodiversity. Under such circumstance, environmental protection NPOs have been set up one after another, and the international communities have raised their awareness of environmental protection and called for a healthier global environment. Therefore, several international conventions were signed among countries, such as the Washington Convention (1973), the Montreal Protocol (1987), the Framework Convention on Climate Change (1992), the Basel Convention and other declarative conventions, which afterwards were proved to be inefficient in shielding the environment from deteriorating. Consequently, the Kyoto Protocol which has legal constraints was signed among 128 countries in 2005, showing the resolution of the world community to reduce the green gas emission. This global movement indicates that the global intention to protect environment is getting more proactive and measures taken are of increasing legal binding forces (Kirchgeorg, 1990; Peattie, 1992; Porter and van der Linde, 1995; Ramus and Steger, 2001; Shrivastava, 1994; Steger, 1993).

The previous research regarding enterprise green management mainly focused on the countermeasures that company should take, so as to wall off the increasingly strict rules set by environmental organizations and pressures from the stakeholders. However, the industrial circle takes green innovation and management as unnecessary investments, and even thinks they would increase the operational costs and retard company's development. As a result, this research bridges the academic gap of a comprehensive study on the influence of green management over company's competitive advantages, so as to examine the hypothesis of environmental performance would exert positive effects on company's competitive edge.

Consequently, there are three main objectives for this research:

(1) To explore the influence of enterprise green management on its competitive edge and environmental performance;
(2) To discuss the influence of environmental performance on the competitive edge;
(3) To study the role of environmental performance between enterprise green management and competitive edge.
What's more, the research objects are manufactures in northwestern China which have acquired the ISO 9001 certification before December, 2014.

2. Literature Review and Hypotheses

2.1 Environmental Performance

Performance is the benchmark for a company to evaluate the result of its resource utilization, while different scholars hold distinctive definitions of organization performance, it is mainly because they have different views about the nature of an organization (Goodman and Pennings, 1977). Financial performance and environmental performance are two indicators generally used to evaluate an organization's performance (Hart, 1995; Judge and Douglas, 1998; Russo and Fouts, 1997). The corporate environmental performance can be classified into four types: Organization system;Stakeholder relationship;Environmental impact;Obedience to the rule of law.

Pollution reduction measures can be divided into ex-post control and ex-ante prevention (Barret and Segerson, 1997; Nehrt, 1998; Shnvastava, 1995c). Ex-post control refers to the disposing of pollutants, wasted water, air and other materials via pollution control equipment, so as to meet the regulatory standards of environmental protection laws; while ex-ante prevention are measures taken to put down the pollution at the source, so as to ameliorate environmental performance through clean production, green marketing and proactive administration (Berry and Rondinelli, 1998; Hart, 1995; Nehrt, 1998). Therefore, the paper raises several hypotheses as follows:

H1: Enterprise green management has positive influence on the environmental performance.
H1-1: Clean production has positive influence on the environment operational performance.
H1-2: Proactive administration has positive influence on the environment operational performance.
H1-3: Green marketing has positive influence on the environment operational performance.
H1-4: Clean production has positive influence on the environment managerial performance.
H1-5: Proactive administration has positive influence on the environment managerial performance.
H1-6: Green marketing has positive influence on the environment managerial performance.

2.2 Environmental Performance's Influence on the Competitive Edge

Hutchmoson (1992) pointed out that a proactive attitude taken by companies towards environmental problems could invite many potential benefits, such as improving company images and popularity; attracting more green-loving consumers; passing the investment review successfully; increasing production efficiency and lower the costs by recycling, energy-conservation and control over contamination; avoiding heavy premium fee; keeping good terms with surrounding neighbors and producing green products with high value. The Enterprise's innovative green performance would positively influence its competitive edge, and the innovative green performance covers several areas, such as energy-conservation technology, pollution prevention technology, waste recycling and reusing technology, green products design and green management etc. The innovative green performance will have greater positive influence on the company's competitive edge as its competitor's ability decreases or the industrial growth rate increases. Hence, hypotheses are made as follows:

H2: the environmental performance has positive influence on the competitive edge.
H2-1: the environment operational performance has positive influence on the competitive edge.
H2-2: the environment managerial performance has positive influence on the competitive edge.

2.3 The Intermediary Role of the Environmental Performance

According to the above literature reviews, a conclusion can be drawn that enterprise green management could not only increase its competitive edge (Berry and Rondrnelli, 1998; Porter and van der Linde, 1995), but also bring about better environmental performance (Berry and Rondinelli, 1998; Hart, 1995; Nehrt,

IOP Conf. Series: Earth and Environmental Science **59** (2017) 012040 doi:10.1088/1755-1315/59/1/012040

1998; Porter and van der Linde, 1995). Besides, improvements in the environmental performance could also add competitive advantages to a company (Hutchinson, 1992; Stock da/., 1997; Maxwell da/., 1997; Chen a/., 2006). While modern theories think efficient use of energy and waste reduction could save cost (Berry and Rondmelli, 1998; Hart, 1995; Porter and van der Lind, 1995; Shrivastava, 1995c) and improve company's competitive edge (Hutchinson, 1992; Stock d a/., 1997; Maxwella/., 1997; Chena/., 2006). Therefore, based on the above literature reviews, this paper deducts that the environmental performance plays an intermediary role between enterprise green management and competitive edge, and the following are the hypotheses proposed:
H3: Green management can influence company's competitive edge through environmental performance.
H3-1: Clean production can influence company's competitive edge through environmental performance.
H3-2: Proactive administration can influence company's competitive edge through environmental performance.
H3-3: Green marketing can influence company's competitive edge through environmental performance.

3. Framework of the Research

This study will focus on the influence of enterprise green management on the environmental performance and the competitive edge, and will further analyze environmental performance's effect on competitive edge. In addition, this study will take environment performance as the intermediate variable to testify whether it would play an intermediary role between green management and competitive edge. And the relation can be illustrated as follows Figure1:

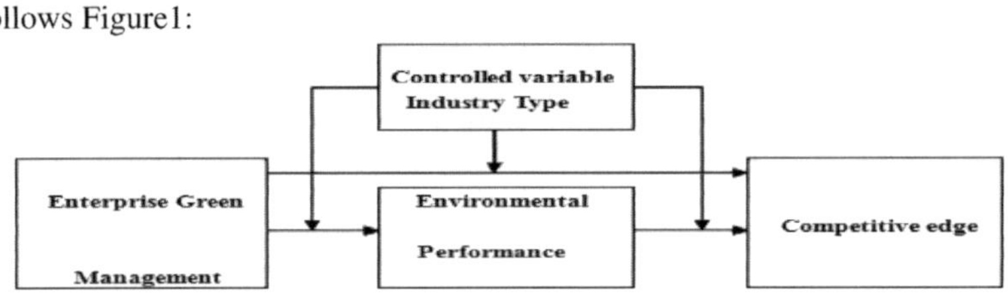

Figure 1

4. Sample Study

Objects of this study are manufactures in northwestern China which have acquired the ISO 9001 certification before December, 2014, and questionnaire surveys have been conducted among these manufactures. In April, 2014, 1261 questionnaires were sent to them and in order to assure the reply rate, self-addressed envelopes have also been enclosed within questionnaires. The investigated were required to return the questionnaire within two weeks. Two weeks later, phone calls were made to recollect questionnaires from the manufactures, and there were some follow-up mailings as well. People who answered the questionnaires were principles of companies, directors or deputies of environmental protection departments.

Among the 1261 questionnaires, 202 were recollected and the recollection rate of effective questionnaires was 16.01%. The study has adopted the Cronbach's alpha to do reliability analysis of each dimension, and the results has been listed in chart 3. Among the three dimensions of enterprise green management, the reliability of clean production, proactive management and green marketing are 0.875, 0.847, and 0.852 respectively, and they are all relatively high. In terms of the two dimensions of environmental performance, the reliability of environment operational performance and environment managerial performance are 0.938 and 0.902 accordingly, which are both high. And the reliability of environmental performance is 0.910, which is also relatively high. Generally speaking, the reliability analysis of this study is quite reasonable.

Table 1 Multiple Regression Analysis of the Environmental Performance and the Competitive Advantages

Controlled Variable				
Industry Type	-0.028	1.000	0.020	1.000
Dependent Variable				
Environment Operational Performance	0.486**	1.000		

Environment Managerial Performance			0.517**	1.000
F	30.787**	36.335**		
Adjusted R-squared	0.229		0.260	

* P < 0.05 , P < 0.01

Table 2 Analysis of the Intermediary Role of Environmental Performance

Dependent Variable	competitive edge		
	Mode 1	Mode 2	Mode 3
Controlled Variable			
Industry Type	-0.033	-0.025	-0.028
Dependent Variable			
Clean Production	0.284**		0.209*
Proactive Administration	0.325**		0.188*
Environmental Performance		0.530**	0.262**
F	29.802**	39.029**	25.500**
Adjusted R-squared	0.301	0.275	0.328

* P < 0.05 , P < 0.01

5. Conclusions

The discovery of this study shows that the three dimensions of enterprise green management, which are clean production, proactive administration, and green marketing accordingly, could levy positive influence on a company's competitive edge. In terms of green management and environmental performance, clean production and proactive administration have obvious positive influence on the environment operational and managerial performances; however, subtle influence of green marketing can be seen on the environment operational and managerial performances, it might due to the fact that green marketing doesn't have any significant influence on the environmental performance of an organization, while it has conspicuous effect on the external stakeholders and could improve the company's competitive edge consequently. In the aspect of environmental performance and competitive advantages, discoveries show that both environment operational performances and environment managerial performances can positively influence the competitive edge. That is to say, environmental performance is not the operational cost nor the hindrance to a company's growth, instead, it is the impetus of a company's competitive edge.

The research also find that environmental performance plays an intermediary role between enterprise green management and competitive edge, that is to say, clean production and proactive administration can positively affect competitive advantage through environmental performance. Hence, enterprise green management could bring benefits to companies, it could not only improve the environmental performance and competitive edge, but also could further strengthen the competitive edge via environmental performance. This also indicates that enterprise's undertaking of green innovation and green management could not only meet the requirements of environmental protection regulations and of the stakeholders, but also could decrease environmental damages caused by human productive activities, therefore improves the economic performance.

Acknowledgements
Longdong University doctoral research start-up funding project (No.: XYBE1605)

References

[1] Aragon-Correa, J. A., "Strategic Proactivity and Firm Approach to the Natural Environment," Academy of Management Journal, Vol. 41, No. 5, 1998, pp. 556-567.
[2] Bansal, P., Business Strategy and the Environment, Business and the Natural Environment, Butterworth-Heinneman, Oxford, U.K, 1997.
[3] Barney, J. B., "Firm Resources and Sustained Competitive Advantage," Journal of Management,

Vol. 17, No. 1, 1991, pp. 99-120.

[4] Baron, R. M. and Kenny, D. A., "The Moderator-Mediator Variable Distinction in Social Psychological Research： Conceptual, Strategic, and Statistical Considerations," Journal o^f Personality and Social Psychology, Vol. 51, 1986, pp. 1173-1182.

[5] Barret, J. and Segerson, K., "Prevention and Treatment in Environmental Policy Design," Journal of Environmental Economics and Management, Vol. 33, 1997, pp. 196-213.

[6] Berry, M. A. and Rondinelli, D. A., "Proactive Corporate Environmental Management: A New Industrial Revolution," Academy of Management E^ecuti^e, Vol. 12, No. 2, 1998, pp. 38-50.

[7] Bragdon, J. H. and Marlin, J. A. T., "Is Pollution Profitable?" Risk Management, Vol. 19, No. 4, 1972, pp. 157-169.

[8] Callan, S. J. and Thomas, J. M., Environmental Economics and Management: Theory, Policy, and Applications, Irwin, Chicago, 1996

Removal efficiency of polycyclic aromatic hydrocarbons and phthalate esters by surface flow wetland in Shunyi district, Beijing

Binghua Li*, Licai Liu, Li Han and Yong Yang

Beijing Engineering Research Center for Non-conventional Water Resources Utilization and Water Saving, Beijing Water Science and Technology Institute, Chegongzhuang west road, Haidian District, Beijing, 100048, China

*Corresponding author e-mail: libinghua75@163.com

Abstract. The surface flow wetland (SFW) system was located on Shunyi district, Beijing. It was built to treat industrial wastewater and domestic sewage, which were looked as its influent. Here sixteen polycyclic aromatic hydrocarbons (PAHs) and six phthalate esters (PAEs) were detected by gas chromatography-mass spectrometry (GC-MS).To determine treatment effect of SFW system, concentrations of targeted compounds in the influent were compared with those in the effluent. Results showed typical compounds of industrial wastewater were naphthalene (NAP), phenanthrene (PHE), dibutyl phthalate (DBP), di-(2-ethylhexyl) phthalate (DEHP), and their concentrations were ranged from 122.6ng.L^{-1} to 760.6 ng.L^{-1}.However typical compounds of domestic sewage were NAP, anthracene (ANT), PHE, DBP, diethyl phthalate (DEP), DEHP, and their concentrations were ranged from 280 ng.L^{-1} to 7998.1 ng.L^{-1}. Typical compounds of effluent were NAP, PHE, DBP, DEHP, and their concentrations changed between 4.2 ng.L^{-1} and 1430.74 ng.L^{-1}. The removal rate of those compounds were 10% ~ 99%, and nineteen compounds removal rate reached above 70%.Therefore, it can be concluded that SFW system had a strong effect on the removal of these compounds.

1. Introduction

Surface flow wetland (SFW) is an ecological treatment technique for slightly polluted wastewater. It utilizes the synergistic reaction of soil, plant and microorganisms for purification wastewater; especially the interaction of microorganisms with other organisms and plants plays an important role for the absorption and degradation of pollutants [1, 2]. This technique is widely used because of high efficiency, low operation cost and convenient maintenance.

Here the SFW system was located in Shunyi district, Beijing, China. Its area was about 17.3 hm^2, and consisted of two stabilization ponds, eight grade series SFWs and one water storage pond. It was built to treat industrial wastewater and domestic sewage nearby with total 7500m^3 per day. Effluent of this SFW system was mainly used for forest irrigation, and some was discharged into Chaobai River.

Polycyclic aromatic hydrocarbons (PAHs) and phthalate esters (PAEs) are two kinds of trace organic compounds with mutagenic effect. They belong to typical persistent organic compounds (POPs), and are ubiquitous in aquatic environment. Moreover, they can be bioaccumulation and threaten the ecological safety and human health [3, 4]. Thus they should be paid more attention.

This study aimed to determine the removal efficiency of SFW system in Shunyi district, Beijing by comparing concentrations of compounds in the influent and effluent of that SWF system.

2. Methods

Water samples were filtered by 0.45 μm thickness glass fiber film. Then the filtered water were added with substitute and evenly passed through the activated solid phase extraction (SPE) for elution. Finally, the elution liquid was dewatered and stored in refrigerator. Quantitative of PAHs and PAEs was performed by GC-MS. All samples were added indicators to control the recovery rate of the whole analysis process, thus ensure accuracy of this detection. Detection limit and recovery rate of the method was ranged from 0.10 ng.L^{-1} to 0.40 ng.L^{-1}, and 85.2% to 96.3%, respectively.

3. Results and discussions

3.1. Detection of PAHs and PAEs in influent water and effluent water

3.1.1. Detection of PAHs. The detection of PAHs in industrial wastewater, domestic sewage and wetland water were shown in Table 1.

Table 1. Concentration of PAHs in influent and effluent of SWF system (unit: ng.L^{-1})

Compounds	influent		effluent	Standard①
	Industrial wastewater	Domestic sewage		
NAP	182.3	1 168.7	76.5	
ANY	5.0	21.1	19.0	
ANA	26.7	31.4	14.0	
FLU	73.2	125.2	33.4	
PHE	122.6	280.2	42.00	
ANT	11.7	395.8	5.5	
FLT	36.8	88.9	9.2	
PYR	27.7	81.4	7.6	
BaA	13.0	22.6	2.5	
CHR	14.4	17.5	10.0	
BbF	11.8	22.2	3.0	
BkF	1.7	2.8	0.7	
BaP	2.9	ND②	0.5	2.8
IPY	9.1	2.4	3.5	
DBA	3.4	3.8	1.4	
BPR	5.2	6.5	2.5	
∑PAHs	547.5	2 270.6	231.3	2 000

①Chinese standard of surface water environment quality(GB3838-2002), ②ND means not detected.

As can be seen all sixteen PAHs were detected in industrial wastewater, and concentrations of PHE, NAP were relatively higher, which were 182.3 ng.L^{-1} and 122.6 ng.L^{-1}. The other compounds concentration was between 1.7 ng.L^{-1}and 73.2 ng.L^{-1}. Total PAHs was 547.5 ng.L^{-1}, which did not exceed Chinese standard of surface water environment quality (GB3838-2002). However, it should be noted that the concentration of BaP in industrial wastewater slightly exceeded that standard limit. Except for BaP, the other fifteen PAHs were detected in domestic sewage, and concentrations of NAP, ANT, PHE were relatively larger, which were 1168.7 ng.L^{-1}, 395.8 ng.L^{-1}, 280.2 ng.L^{-1}, respectively. The other PAHs concentration was changed between 2.4 ng.L^{-1} and 125.2 ng.L^{-1}. Total PAHs was 2270.6 ng.L^{-1}. All sixteen PAHs in wetland effluent were detected. The concentrations of PHE and NAP were 76.5 ng.L^{-1}, 42.0 ng.L^{-1}, and the others were ranged from 0.5 ng.L^{-1}to 33.4 ng.L^{-1}. Total PAHs was 231.3 ng.L^{-1}, and did not exceed Chinese standard of surface water environment quality.

3.1.2. Detection of PAEs. The detection of PAEs in industrial wastewater, domestic sewage and wetland water were shown in Table 2. As can be seen all six PAEs were detected in industrial wastewater. Concentration of DEHP was the highest, which reached to 760.6 ng.L^{-1}, and concentration of DnOP was the lowest, which was only 4.7 ng.L^{-1}.Those six PAEs were also detected in domestic sewage. The concentrations of DBP, DEHP and DEP were relatively higher, which were 7998.0 ng.L^{-1}, 2889.0 ng.L^{-1}, 1584.4 ng.L^{-1}, respectively. Moreover, the concentration of DBP was 2.7 times of that Chinese standard of surface water environment quality. The concentration of BBP was the lowest, which was only 29.85 ng.L^{-1}. Compared with the concentrations of PAEs in industrial wastewater, those in domestic sewage were one order higher. Six PAEs were all detected in the wetland effluent. The concentrations of DBP and DEHP were 1430.7 ng.L^{-1}, 410.9 ng.L^{-1}, and the concentration of BBP was the lowest, which was only 0.7ng.L^{-1}.

Table 2. Concentrations of PAEs in influent and effluent of SWF system (unit: ng.L^{-1})

Compounds	influent		effluent	Standard*
	Industrial wastewater	Domestic sewage		
DMP	73.3	671.4	53.5	
DEP	256.4	1584.4	56.8	
DBP	392.0	7998.1	1430.7	3000
BBP	6.9	29.8	0.7	
DEHP	760.6	2889.0	410.9	8000
DnOP	4.7	31.6	2.2	
∑PAEs	1493.8	13204.2	1954.8	

*: Chinese standard of surface water environment quality (GB3838-2002)

3.2. Difference of PAHs and PAEs in influent water and effluent water

Concentrations of ten PAHs including NAP, ANY, FLU, PHE and ANT in domestic sewage was higher than those in industrial wastewater, and especially the concentration of ANT in domestic sewage was 34 times of that in industrial waste. However, the other six PAHs concentrations were roughly similar. In addition, concentration of individual PAH and total PAHs in effluent water was significantly lower than that of industrial wastewater and domestic sewage.

As for six PAEs, concentrations in domestic sewage were all higher than those in industrial wastewater and effluent, especially that of DBP, which was 20 times of that in industrial wastewater. One reason of this difference was that their origin was different. The other reason maybe that domestic sewage was directly discharged without any treatment, but industrial wastewater was sedimentation, filtration and other processes before flowing into the SWF system.

3.3. Removal rate of PAHs and PAEs by the SWF system

The removal rates of targeted compounds by the SWF system were shown in Table 3. It revealed that the wetland system had a strong effect on the removal of PAHs and PAEs. Removal rate of nineteen compounds including NAP, ANT, DMP and BBP were over 70%, especially those of ANT and BBP, which reached 98%. The main mechanism for the removal of trace organic compounds in wetland system was the degradation of organic compounds by heterotrophic microorganisms, which transformed those compounds to CO_2 and H_2O. Insoluble trace organic compounds after precipitation, filtration and assimilation can be used by microorganisms, and those soluble organic compounds were easily metabolism. One reason for higher removal of ANT was its strong volatilization, the other maybe that it was small molecular weight and apt to be degradation [3]. As for BBP, its higher removal rate maybe closely relate to degradation under suitable environment. There were reports that BBP were reduced to 10% by activated sludge after one day.

Table 3. Removal rate of PAHs and PAEs by the SWF system

Compound	Removal rate (%)	Compound	Removal rate (%)
NAP	93	BaP	74
ANY	12	IPY	60
ANA	70	DBA	76
FLU	79	BPR	74
PHE	87	Σ PAHs	90
ANT	98	DMP	92
FLT	91	DEP	96
PYR	92	DBP	80
BaA	91	BBP	98
CHR	61	DEHP	86
BbF	89	DNOP	93
BkF	81	Σ PAEs	84

4. Conclusions

(1) Most of target PAHs and PAEs were detected in influent and effluent of the SWF system. Typical organic compounds of SWF system were NAP, PHE, DBP and DEHP, which were 76.5 ng.L^{-1}, 42.0 ng.L^{-1}, 1430.7 ng.L^{-1}, 410.9 ng.L^{-1}, respectively.

(2) Generally concentrations of PAHs in industrial wastewater were the highest, and concentrations of PAEs in domestic sewage were higher than those in industrial wastewater and effluent of SWF system.

(3)The SWF system had good removal effect on PAHs and PAEs, most removal rate of target compounds was over 70%.

Acknowledgments

This work were support by Beijing Natural Science Foundation (NO:8162021) and Major National Science & Technology Projects(NO:2016YFC0401405).

References

[1] J.G. Allen, M.W. Beutel, D.R. Call, et.al, Effects of oxygenation on ammonia oxidation potential and microbial diversity in sediment from surface-flow wetland mesocosms, Biores. Techno, 101, 4(2009).

[2] A.B. IA, S.A SR, A. N, et.al, Performance assessment of pilot horizontal sub-surface flow constructed wetlands for removal of diesel from wastewater by Scirpus grossus. Water Sci Technol, 68,10(2013).

[3] Y.J. Lien, H.Y. Ku, P.H. Su, et.al, Prenatal Exposure to Phthalate Esters and Behavioral Syndromes in Children at Eight Years of Age: Taiwan Maternal and Infant Cohort Study, Environ Hea Perspec, 123, 1(2014).

[4] K.H. Kim, S.A. Jahan, E. Kabir, et.al, A review of airborne polycyclic aromatic hydrocarbons (PAHs) and their human health effects. Environ Internat, 60, 5(2013).

[5] A.K. Haritash and C.P. Kaushik, Biodegradation aspects of Polycyclic Aromatic Hydrocarbons (PAHs): A review, J. Hazar. Mater, 169, 1(2009).

IOP Conf. Series: Earth and Environmental Science **59** (2017) 012042 doi:10.1088/1755-1315/59/1/012042

The evolution analysis of seasonal drought in the upper and middle reaches of Huai River basin based on two different types of drought index

D Huang[1], J F Liu, W Z Wang, A B Peng

Hydrology & Water Resources Department, Nanjing Hydraulic Research Institute, 225 Guangzhou Road, Nanjing 210029, P. R. China

Email: [1]huangdui2004520@163.com

Abstract. The Huai River Basin is in the transitional zone between the climates of northern and southern China, and is liable to seasonal drought. This study is based on two types of drought index, one is the traditional meteorological drought index including precipitation anomaly percentage (P_a) and Standardized Precipitation Index (SPI), and the other is the remote sensing drought index (water deficit index, WDI), analyzing the serious winter and spring drought development and evolution of the Huai River Basin. The results showed that: (1) the drought development was similar based on P_a and SPI, but the indicated drought intensity was different, SPI was closer to the actual drought degree. (2) WDI reflect drought and spatiotemporal characteristics well, and respond quickly to soil wet conditions caused by short-term rainfall. (3) WDI had significant correlation with P_a and SPI, and the correlation coefficients were 0.93 and 0.92, respectively, which means WDI can supplement meteorological drought index, but WDI is better than meteorological drought index on the whole, which is helpful to improve the ability of assessing regional drought.

1. Introduction

Drought is the first major meteorological disaster affecting agricultural production in China. The complex climatic condition, the uneven distribution of the intra-annual precipitation and the large change of inter-annual precipitation, resulted in frequent droughts in Huai River basin [1]. The causes of drought are complex and diverse, it's difficult to monitor drought than other natural disasters, especially its occurrence and conclusion. At present, there has been a lot of research on drought indicators. Drought is usually divided into four types including meteorological or climatologically, agricultural and hydrological droughts; meteorological drought is the basis of other types of drought research. From the standpoint of meteorological drought, precipitation anomaly percentage (P_a), standardized precipitation index (SPI) [2] and other indicators mainly supplemented by the meteorological observation data are widespread in drought monitoring. Remote sensing acquires the distribution of drought in space and time changes with advantages of large converge and long-term of dynamic monitoring, which makes up the deficiency of traditional drought monitoring. There are a series of drought indices based on remote sensing data, among them the TVDI index [3] is famous, but it has the shortcomings of small application range and neglecting meteorological factors. Wang et al [4] proposed a water deficit index (WDI) applied in a semi-arid region of United States using remote sensing and meteorological data, which was based on the trapezoidal relationship of surface temperature and vegetation index. In this paper, the meteorological drought index of P_a and SPI (one-month time scale) and the remote-sensing index of WDI were used for analysing the temporal and spatial evolution of seasonal drought in the upper and middle reaches of the Huai River basin, which could provide basis for drought monitoring and utilization of water resources.

2. Study area

The Huai River Basin lies between the Yangtze and Yellow River basins with a catchment area of 270,000 km^2, is divided into three parts: the upper reaches, the middle reaches and the lower reaches. The upper and middle reaches of the Huai River were selected as the study area and it's located at 111°55'-119°E and 30°55'-34°55'N, with an area of 190,000 km^2. Annually, the average air temperature is 11 ~ 16 °C, increasing from north to south, from east to west; the average precipitation is 920mm, decreasing from south to north; the water surface evaporation is 900-1500mm, decreasing from north to south. The study area is shown in Figure 1.

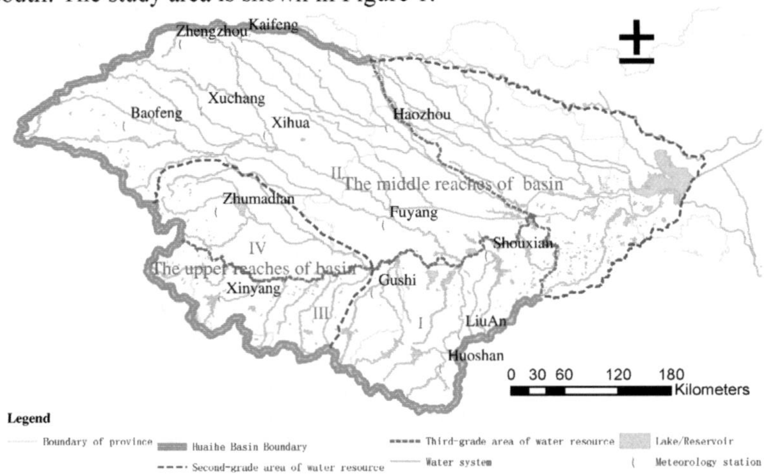

Figure 1. Overview of the upper and middle reaches of the Huai River Basin.

3. Data

3.1. Meteorology data

The meteorological data used in the paper include daily air temperature, wind speed, relative humidity, and rainfall provided by the National Meteorological Information Centre (NMIC) of China Meteorological Administration. There are 13 basic meteorological stations (Figure 1), data length was from 1958 to 2009 and the outliers and missing values were excluded.

3.2. Remote sensing data

Moderate Resolution Imaging Spectroradiometer (MODIS) has a high spectral resolution (36 channels) and a spatial resolution of 250-1000 m. NASA provides many kinds of free long time series surface products. The main remote sensing data are shown in Table 1. A series of operations include remote sensing data selection, re-projection, re-sampling to ensure data quality and space-time consistency.

Table 1. MODIS data used in the study.

Data type	Contents	Spatial resolution	Temporal resolution
MOD09A1	Surface Reflectance	500 m	8 days
MOD11A1	Surface Temperature and Emissivity	1 Km	daily
MOD13A1	Vegetation Indices	250 m	8 days
MOD15A2	Leaf Area Index	1 km	8 days

4. Methodology of Drought monitoring

4.1. Precipitation-based drought index

P_a indicates the rainfall is more or less than normal value at a certain period, which can directly reflect the drought caused by precipitation anomaly, is used to evaluate the monthly, seasonal and annual drought events. P_a (%) is calculated by dividing the difference between the precipitation of a certain period (P, mm) and the multi-year mean precipitation of the corresponding time period ($\overline{p}$, mm) by the latter.

SPI is a commonly used precipitation-based drought index recommended by the World Meteorological Organization (WMO) in 2009. It is a probability index proposed by McKee [2] indicates the scarcity of drought for any rainfall station with historical data at a given time scale. A long-term daily or monthly precipitation database with 30 years or more length is usually required. The classification of drought grade of monthly-scale P_a and SPI is seen in Table 2 according to China's first published national standard for monitoring drought disasters [5].

Table 2. Grade classifications of drought based on monthly-scale P_a and SPI.

Type	Normal	Light drought	Middle drought	Severe drought	Extreme drought
P_a	(-40,+∞)	(-60,-40]	(-80,-60]	(-95,-80]	[-95,-∞)
SPI	(-0.5,+∞)	(-1.0,-0.5]	(-1.5,-1.0]	(-2.0,-1.5]	[-2,-∞)

4.2. Remote sensing-based drought index

Water Deficit Index (WDI) for evaluating relative water shortage of both full-cover and partially vegetated sites, which was first proposed by Moran [6]. Wang et al [4]proposed the modified WDI based on the trapezoid relationship between land surface temperature (Ts) and enhanced vegetation index (EVI) using both remotely sensing and meteorological observation data. This method constructs a Ts-EVI trapezoidal space for each image pixel based on the principle of energy balance, and considers the influence of underlying surfaces and meteorological factors on the four theoretical extreme points of each trapezoid space.

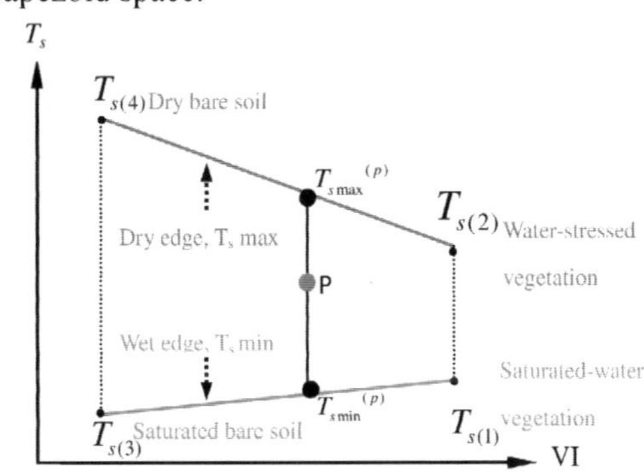

Figure 2. The hypothetical trapezoidal space of each pixel.

As shown in Figure 2, for each pixel P $(T_s^{(p)}, EVI^{(p)})$, a trapezoidal space with four extreme vertices $T_{s(i)}$ ($T_{s(1)}$, $T_{s(2)}$, $T_{s(3)}$, $T_{s(4)}$) is constructed. The four extreme vertexes indicate that the pixel P is theoretical covered by different land surface (bare soil, full coverage) and water supply conditions (extremely dry, extreme wetness). According to the location of the pixel P in the trapezoid space, the wet and dry state of the pixel P is calculated with the following equation:

$$WDI^{(p)} = \frac{T_s^{(p)} - T_{s\min}^{(p)}}{T_{s\max}^{(p)} - T_{s\min}^{(p)}} \quad (1)$$

Where, $T_{s\max}^{(p)}$ 、 $T_{s\min}^{(p)}$ refer to minimum and maximum values, respectively are interpolated linearly on the dry edge ($T_{s(2)}$ 、 $T_{s(4)}$) and wet edge ($T_{s(1)}$ 、 $T_{s(3)}$) of the *Ts*~EVI trapezoid. A WDI close to 1 indicate extreme dry, whereas 0 indicates soil moisture saturation. The method is mainly used in semi-arid areas, however the impact of the climatic conditions on the trapezoidal space is still need studied in the warm temperate-subtropical zone of Huai River basin.

A simplified method is proposed to calculate WDI by estimating the saturated bare soil and saturated-water vegetation point. We assume that the surface temperature of bare soil in the extreme wet condition is the same as the surface temperature of saturated-water vegetation. For each image, the minimum surface temperature of the regional water body (rivers, lakes and reservoirs) is selected as wet edge. The dry edge calculation is still performed by the existing iteration algorithm.

As an important component of energy balance, soil heat flux (G) is usually considered to be linearly related to net radiation (Rn), that is, G/Rn is a certain value. There are very different conclusions about G / Rn under different land cover and climate conditions. In this study, G/Rn to be 0.05 for both vegetated points and 0.2 for both bare soil points.

5. Results and discussion

According to the historical meteorological data and drought conditions, the study period is from November 2008 to April 2009, and winter is from Nov. to Jan. and spring is from Feb. to Apr.

5.1. Application of P_a and SPI

In accordance with WMO regulations, the latest 30-year (1973-2002) average is used as the climate baseline average for computing P_a. The P_a and SPI values of 13 meteorological stations and water resources zoning are calculated, as shown in Figure 3 and Figure 4. It indicates from November 2008 to January 2009 all sites suffered different degrees of drought except area I; Regions II and IV suffered drought in March and April 2009, respectively. The degree of drought indicated by the two indices is different: in December and January, P_a indicates that the areas II, III, and IV have suffered heavy drought (P_a <-80), while SPI indicates that the regions II, III and IV suffered middle drought (-1.5 <SPI <-1) in December.

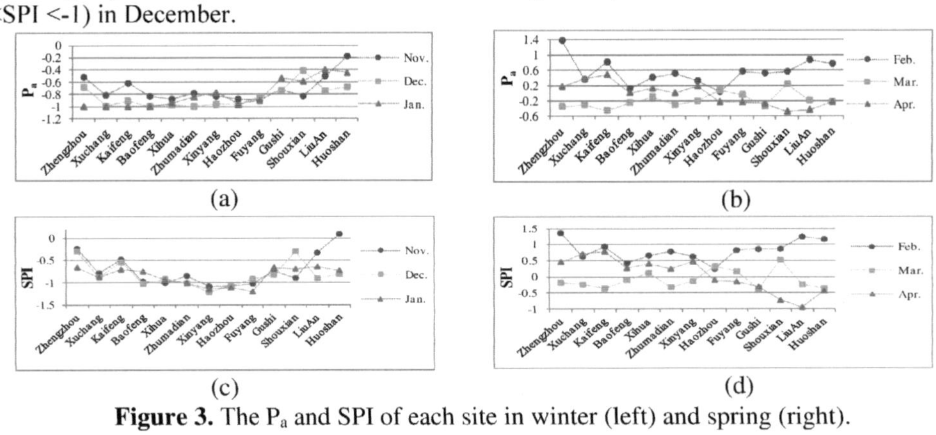

Figure 3. The P_a and SPI of each site in winter (left) and spring (right).

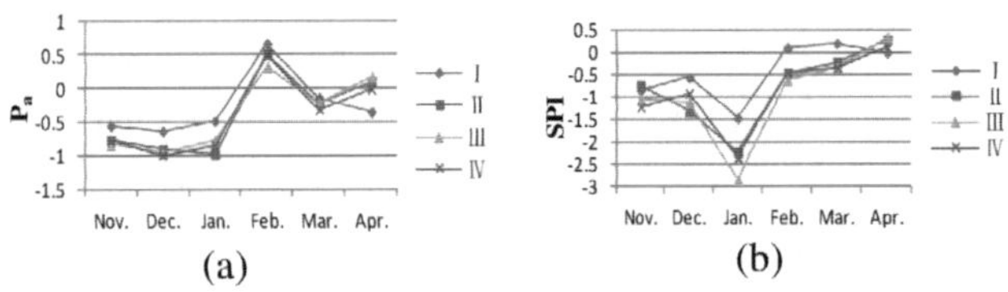

Figure 4. P_a and SPI of water resources zoning.

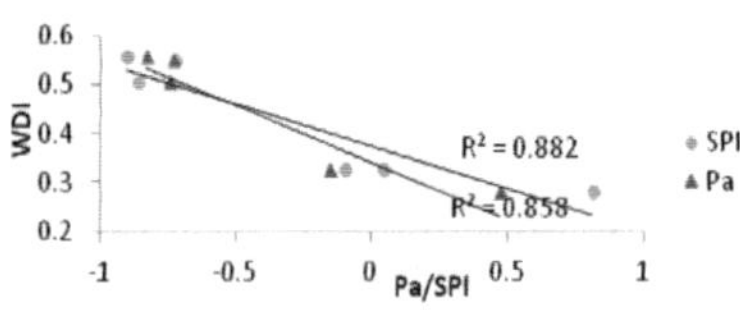

Figure 5. The correlation coefficient between WDI and SPI, P_a.

5.2. Application of WDI

WDI images of different time are obtained as shown in Figure 6. It showed the study area suffered different degrees of drought during the period. In terms of space, the degree of drought decreased from north to south, which is consistent with the spatial distribution of average annual precipitation; In terms of time, the most severe drought happened in December (WDI>0.5), and relieved in March 14 (WDI<0.3). The reason maybe there were different degrees of rainfall in the whole area for two consecutive days before March 14. The precipitation of Baofeng Station on March 12 was 14.7mm, which indicated that WDI respond quickly to soil wet and dry conditions of short-term rainfall.

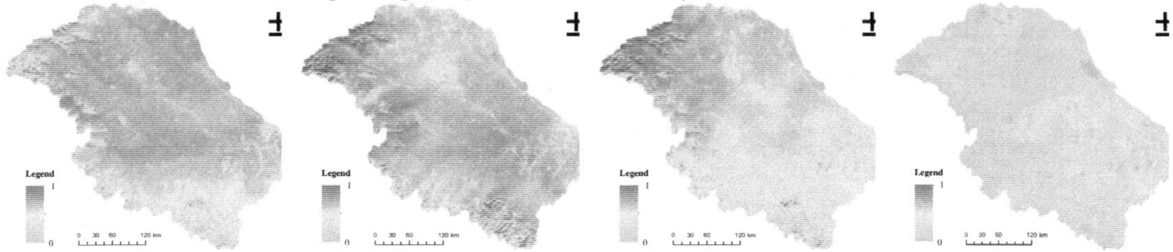

(a) 2008-11-19 (b) 2008-12-9 (c) 2009-1-10 (d) 2009-3-14

Figure 6. Some WDI maps during the period from November, 2008 to April, 2009.

5.3. Comparison of drought monitoring results with different index

The corresponding drought index of 13 sites in the whole area was extracted and compared, as shown in figure 5. The correlation coefficients of WDI and P_a, WDI and SPI were 0.93 and 0.92 respectively, which indicates WDI has obvious correlation with the precipitation index. WDI is obviously superior to P_a and SPI in terms of reflecting the spatial distribution and development of drought.

6. Conclusions

The seasonal drought evolution characteristics were analyzed using the traditional meteorological drought indexes (P_a and SPI) and remote sensing drought index (WDI). P_a and SPI showed the same drought development, while the drought intensity indicated by SPI was closer to the actual drought. WDI, which combines the meteorological and remote sensing data with obvious physical basis, reflect the temporal and spatial of drought during the study period well, is sensitive to short-term rainfall.

Acknowledgements:

Financial supports from the Basal Research Fund of Central Public-interest Scientific Institution of Nanjing Hydraulic Research Institute (Y516028, Y516029) are gratefully acknowledged.

References

[1] He J, Zhang M, Wang P et al. 2011.Climate characteristics of the extreme drought events in southwest China during recent 50 years. *Acta Geographica Sinica* 6(9):1179–1190.

[2] McKee TB, Doesken NJ ,Kleist J.1993.The relationship of drought frequency and duration to time scales. *In Proceeding of the 8th Conference on Applied Climatology*, Anaheim: California; 179–184.

[3] Price J C .1990. Using spatial context in satellite data to infer regional scale evapotranspiration. *IEEE Transactions on Geoscience and Remote Sensing*. 28: 940- 948.

[4] Wang W, Huang D, Wang X G, Liu Y R, and Zhou F.2011.Estimation of soil moisture using trapezoidal relationship between remotely sensed land surface temperature and vegetation index, *Hydrol. Earth Syst. Sci.*, 15(5), 1699-1712.

[5] GB/T 20481-2006, *Classification of meteorological drought* .Beijing,China Standard Press.

[6] Moran M S, Clarke T R, Inoue Y, Vidal A. 1994. Estimating crop water deficit using the relation between surface-air temperature and spectral vegetation index. *Remote Sensing of Environment* 49,246–263.

Observation and forecast on urban heat island effect based on Landsat data

YOU Cheng-zeng[1,2], SHU An[2,3], YANG Xin-yuan, PENG Ling[1], LI Xiang[1,2], YAO Xiao-jing[1]

[1] Institute of Remote Sensing and Digital Earth, CAS, Beijing 100094, China;
[2] University of Chinese Academy of Sciences, Beijing 100049, China
[3] Shanghai Institute of Technical Physics of the Chinese Academy of Sciences, Shanghai, 200083, China

Corresponding author: Ling Peng ,Email:pengling@radi.ac.cn

Abstract. This paper studies the urban heat island (UHI) effect of Jinan city with remote sensing image based on qualitive analysis. By inversing temperature from Landsat image of Jinan city at 2002, 2009 and 2015, Jinan city's UHI effect distribution and change tendency were analysed. And then, by using the Markov chain model, the UHI effect in 2021 and the steady state distribution in Jinan were predicted.

1. Introduction

The urban heat island(UHI) effect is the phenomenon that when the city develops to a certain scale, the temperature of urban places is significantly higher than that of suburbs, because of the change of city underlying surface, air pollution and the discharge of human-made exhaust heat [1]. There are lots of scholars using the brightness temperature data or land surface temperature (LST) data of images (such as NOAA/AVHRR, Terra (Aqua)/MODIS, Aster, etc.) to study the spatial structure and seasonal variation of the UHI in Beijing, Nanjing, Changsha, Shenyang, etc. And they have made some achievements [2]-[4].However, most researches focused on monitoring, analysing and abstracting UHI effect, but few applied a certain model to forecast the future state and steady state of the UHI, which is also very important for macro decision-making and city environment improving. Therefore, we used the temperature inversed from the thermal infrared band of Landsat data, and used the statistic of October 2009 and October 2015 to compute transition matrix and the matrix of Initial state, and then forecasted the UHI of Jinan (36 ° 40 'N, 117 ° 00 'E) in 2021. Then, we applied the theory of Markov Chain to forecast the distribution of UHI in steady state.

2. Data Processing and LST Retrieval

In this study, we chose Landsat images at October 6, 2002, October 17, 2009 and October 15, 2015 to do LST Retrieval (the details of the temperature retrieval are shown in the Reference [5], so we won't repeat the process in this article), before which we did radiometric calibration, atmospheric correction, geometric rectification, image mosaic and image cropping.

3. Results and Analysis

To show inversion results of the UHI better, we chose mean-standard deviation to classify the inversion results to six classifications[6] , which were extra high temperature region (Ts≥μ+std), high temperature region (μ+0.5std≤Ts<μ+std), sub high temperature region (μ≤Ts< μ+0.5std), medium temperature region (μ-0.5std ≤Ts<μ), sub medium temperature region (μ-std≤Ts <μ-0.5std) and low temperature region (Ts< μ-std).The classification maps are shown as fig.1.

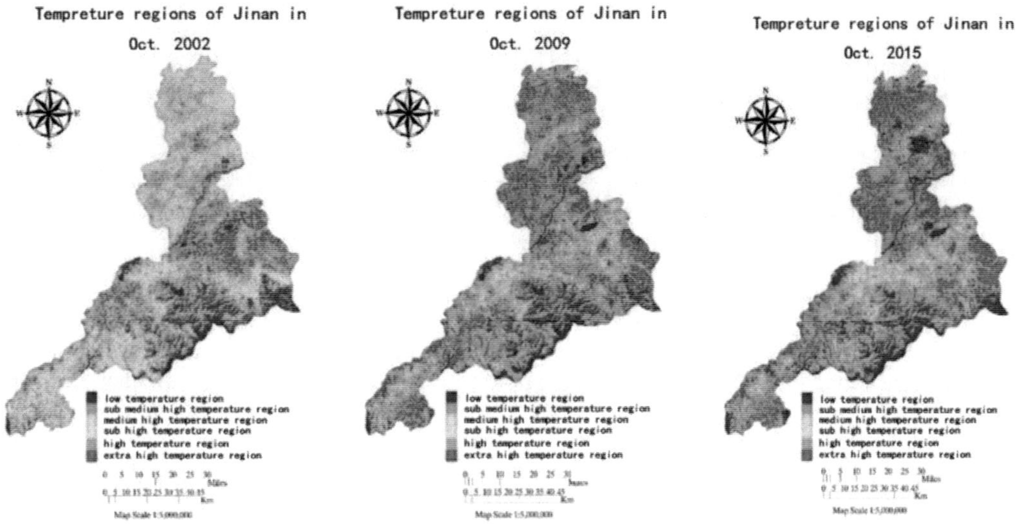

Fig. 1 temperature regions of the UHI of Jinan in Aug

From the results of classification, we can figure that:

(1) From 2002 to 2015, the UHI tended to move from the south to the north, with an increasing UHI in the north of Yellow river and a decreasing UHI in the south year by year.

(2) The UHI tended to move to the eastern area (Licheng, Zhangqiu), where the UHI strengthening year by year.

(3) Weak UHI appeared in medium and low temperature region such as streams, pools, lakes and southern coteau.

To further analyse the change of the UHI of Jinan in time series, we counted the area of each temperature region (details are shown in table.2). It shows that: since 2002, the area of middle region (medium temperature region, sub high temperature region) changed from 54.92% to 39.64% ,and then to 37.02%, showing apparent decreasing trend; Left side region (low temperature region, sub medium temperature region) changed from 19.47% to 30.76% ,and then to 31.99%, showing apparent increasing trend; Right side region (high temperature region, extra high temperature region) changed from 25.61% to 29.60%, and then to 30.99%, showing apparent increasing trend. In short, the UHI of Jinan (polarized into two opposite trend) showed apparent trend of Polarization.

4. The Trend Forecast of the UHI

The basic thought, K-step transition probability and stationary state of Markov Chain are shown in the reference [7],[8]. In this study, Stationary Markov chain combined with the attributes of the UHI has numerous solutions, and it is not easy to get the proper solution by conventional methods. Therefore, we solved the Stationary Markov Chain by programming, and got satisfactory results after the gradual optimization and validation.

4.1. The performing of matrixes of state transition about each temperature region

According to the area of difference from October 2009 to October 2015, we could calculate the matrix of the transformations, as shown in table.1:

Table.1 Transition matrix of probability

K year	K+1 year low temperature region	sub medium temperature region	medium temperature region	sub high temperature region	high temperature region	extra high temperature region
low temperature region	0.7003	0.2051	0.0599	0.0167	0.0087	0.0094
sub medium temperature region	0.2091	0.3894	0.2656	0.0904	0.0298	0.0158
medium temperature region	0.0701	0.2339	0.3588	0.2266	0.0801	0.0306
sub high temperature region	0.0237	0.0910	0.2395	0.3105	0.2150	0.1203
high temperature region	0.0121	0.0401	0.1378	0.2495	0.2846	0.2759
extra high temperature region	0.0071	0.0151	0.0455	0.1098	0.2245	0.5979

4.2. Analysis of the prediction result of Markov Chain

According to the Markov chain prediction model, we could get the prediction results of the UHI in 2021. And then by solving the nature of Stationary Markov chain, we obtained the temperature regions of Jinan city when the proportion of the forecast results tended to be stable. After testing, each predicted value was greater than 0, and the sum of each predicted value was 1, which meant the prediction results were reasonable.

We can get the percentages of all temperature regions in October 2009 and in October 2015, as well as the prediction results of the 2021 and the percentages of temperature regions of the steady state, as shown in table.2:

Table.2 Measured and predicted values of each temperature region（%）

	low temperature region	sub medium temperature region	medium temperature region	sub high temperature region	high temperature region	extra high temperature region
2009	13.60	17.16	20.36	19.28	11.81	17.79
2015	15.26	16.74	19.74	17.28	13.76	17.23
2021	16.27	16.65	19.26	16.93	13.71	17.19
steady state	18.53	17.11	18.82	16.19	13.03	16.32

(1) As is shown in the table 2, from the beginning of 2009, the trend of middle temperature region (including medium temperature region and sub high temperature region) is from 39.64% to 37.02% and to 36.19%, significantly decreasing year by year; The trend of left temperature region(including low temperature region and sub medium temperature region) is from 30.76% to 31.99% and to 32.92%, significantly increasing year by year; The trend of right temperature region(high temperature region and extra high temperature region) is from 29.60% to 30.99% and to 30.90 %, which has significantly increasing trend, too. To sum up, from 2009 to 2015 and then to the prediction of 2021, the UHI effect of Jinan has obvious phenomenon of polarization, which is consistent with the results of the analysis in chapter 3.

(2) From the predicted results of the stable state, the left temperature region continued to maintain a growth of 35.64%, and there was a growth of 2.72% compared with 2019; The medium temperature region and sub high temperature region continued to decrease, having a total reduction of 1.78%. It is noteworthy that, from the beginning of 2002, the strong heat island region (the right temperature region) in 2015 (30.99%) was the cut-off point, and by contract, there were different reductions for 2021 (30.90%) and the stable state (29.35%).

By analysis above, it is concluded that the strengthening trend of the UHI effect has somewhat eased after 2009. And the proportion of each temperature region tends to be stable. Which indicates that the Jinan municipal government's policies (Eastern Expansion, Western development, Southern regulation, northern span and central relief) is effective. While maintaining the current artificial and natural disturbing condition, after several years of Jinan city, the strong heat island tends to be stable, which maintained at around 29.35%.However, although there is no worsening trend, the ratio of the extra high temperature region still reached 16.32%, and the heat island effect is still serious, too.

5. Conclusions

The main conclusions of this study are as follows:

(1) According to the analysis of interannual variation, the UHI effect of Jinan has obvious moving trend from the south to the north, and the regional UHI effect becomes more and more serious in the areas of the north of the Yellow River from 2002 to 2015.

(2) According to the analysis of the UHI of Jinan from 2002 to 2015, it is found that the UHI effect is not simply increasing year by year, but from the middle temperature region (medium temperature region and sub high temperature region) to two other temperature regions (low temperature region and sub medium temperature region, high temperature region and extra high temperature region). In other words, it is polarizing obviously.

(3) As is shown in the spatial distribution, the UHI effect is closely related to the underlying surface, which shows that the urban land and the bare land are obvious high temperature areas. And water and green land are obvious low temperature regions, that is to say cooling effect is remarkable.

(4) According to the prediction results, while maintaining the current artificial and natural disturbing condition, it shows that the UHI effect has polarized phenomenon in 2021, which keeps the characteristics of the UHI effect of Jinan in time series, and this shows that the Markov chain has a certain reliability for the research of the UHI. Through the steady-state prediction results, it can be found that the UHI effect worsening trend has significantly eased, but still at a high level. Therefore, the government departments should make more effective policies to strengthen law enforcement to ease the impact and the harm brought by the UHI.

Acknowledgments

This research was sponsored by the National Science-technology Support Plan Project of China (Grant Nos. 2015BAJ02B00) and the Special Foundation of the Chief of the Institute of the Remote Sensing and Digital Earth, Chinese Academy of Sciences (Grant Nos. Y6SJ2800CX)

References

[1] PENG Shao-linl,ZHOU Kai,YE You-hual,SU Juan,et al. Research progress in urban heat island [J]. Ecology and Environment,2005,14(4):574-279.

[2] YAN Hao,DENG Lian-tang. Character Analysis Of Heat Island In Shanghai City Using Remote Sensing Parameters [J]. Journal of Tropical Meteorology,2004,20(5): 579-585.

[3] Li Le, Xu Hanqiu. Urban Expansion and Thermal Environment Changes in Hangzhou City of EastChina[J]. Remote Sensing Technology and Application,2014,29(2):264-272.

[4] ZHANG Xinyi. RS and GIS-based research on Shanghai land use/cover change with respect to thermal Envirnnmental effect[D], East China Normal Unversity,2006.

[5] You Chengzeng, Yang Xinyuan Shu An, Peng Ling， Li Xiang. Analysis and estimation on Urban Heat Island Effect Based on Landsat Data [J]. Environmental Engineering ,2016,S1:1145-1150+1158.

[6] ZHANG Dianjiang,ZHOU Bin,LI Hongwe,WANG Xiao-tang,SI Min. Prediction of Urban Heat Island Expansion Based on Markov Chain Theory [J]. China Population,Resources and Environment,2013,S2:321-325.

[7] Qian Minping, Qian Min. THE DECOMPOSITION INTO A DETAILED BALANCE PART AND A CIRCULATION PART OF AN IRREVERSIBLE STATIONARY MARKOV CHAIN[J]. Science in China,Ser.A,1979,S1:69-79.

[8] The Stationary Distribution of Markov Chains in Random Environments [J]. Chinese Journal of Applied Probability and Statistics,2012,04:337-344.

Effect of inert gas injection on gas explosion in the process of sealing fire zone

ZHANG Li-li, LI Cheng-yu*

Xinjiang Institute of Engineering, Urumqi 830091, Xinjiang, China

15099605118@126.com

Abstract. In order to study the influence of inert gas injection to gas explosion in the process of fire zone sealing, tunnel fire model was established by COMSOL software, and numerical simulation is carried out to study change laws of flow field and concentration field in the process of fire zone sealing. Through comparing the distribution law of flow field and concentration field whether the inert gas injection and different flows, obtained the influence of inert gas injection to gas explosion in the process of fire zone sealing. As the decrease of entrance average wind speed, the maximum wind speed in the centre vertical section of tunnel gradually reduces. when the entrance average wind speed is 1m/s, counter current layer is produced near the fire resource, length and thickness of counter current layer and maximum current speed rises with the declining of the entrance average wind speed, and swirl will be formed when the entrance average wind speed is less than 0.01m/s. In the process of fire zone sealing, the maximum gas concentration in the fire area and on the side of leeward goes up with the declining of the entrance average wind speed, the nitrogen concentration will gradually increase with the declining of entrance average wind speed.

1. Introduction

When the mine fire to a certain extent can't be directly extinguished, the fire zone must be closed. However, it is very easy to cause gas explosion in the process of fire zone sealing. How to close the fire zone quickly, safely and efficiently is the difficult point of fire area management [1]. In order to ensure the safe and efficient closure of the fire zone, it is necessary to study the law of gas migration and the explosion hazard in the process of fire zone sealing. In recent years, some scholars have studied the law of heat and mass transfer, gas accumulation and danger during the process of fire zone sealing, and the law of spontaneous combustion induced gas explosion in the process of fire zone sealing [1-8]. In this paper, combined with the existing research results, using numerical simulation method, to study the influence of inert gas injection to gas explosion in the process of fire zone sealing. Because numerical simulation of fire zone involves the coupling of multiple physical fields, this paper uses COMSOL Multiphysics software for numerical simulation.

2. To establish models and conditions

The calculating model is a horizontal rectangular tunnel designed with 4 m width, 3 m high and 50 m long. The roof, floor and sides of tunnel are designed as 0.5m thick rock. Fire source is coal heap, the center of coal heap is 16m from the entrance, the coal heap with basal diameter of 1m, height of 1m and top face diameter of 0.5m is shown in Fig.1. The model shown in Fig.2 uses free tetrahedron mesh, coal heap surface shown in Fig.3 is encrypted separately with free triangle, the whole model is divided into 434533 meshes.

There are three physical fields of flow field, heat transfer field and mass transfer field in fire zone. Flow field uses turbulence model, as inflow is in the left and outflow is in the right, forced convection exists because of volume force effect on airflow after being heated, entrance initial normal speed is 2m/s, outlet is pressure boundary. The model uses solid-gas coupling heat transfer model, sets coal heap as heat source, and supposes that there are enough combustibles in the whole combustion process, this paper estimates heat output with oxygen consumption method because of different coal quality with different generating heat, the energy generated by oxygen consumption per unit mass is 13.1MJ/kg, coal heap, roof and floor, and sides are solid heat transfer, because external is formation rock, the external boundary is set as convection cooling, airflow is fluid heat transfer, initial temperature is 293.15K(20℃);Mass transfer uses convection diffusion model, each gas concentration in airflow entrance is constant, the combustion reaction of coal occurred in coal heap surface, methane emits from tunnel wall. The initial value of the entrance: O_2 concentration is 21%, CH_4 concentration is 0.1%, CO_2 concentration is 0.03%, CO concentration is zero. At the initial stage of fire is oxygen-enriched combustion. Oxygen concentration is more than 15%, the oxygen consumption of combustion reaction on the coal heap surface has been calculated by that. The gases produced mainly consist of CH_4, CO and CO_2, the ratio of gases produced is 0.05:1:1 when fire zone temperature is before 1300℃, the ratio of gases produced changes to 0.05:1:2 after fire zone temperature reaches 1300℃, gas emission of tunnel wall is 0.048 m^3/s.

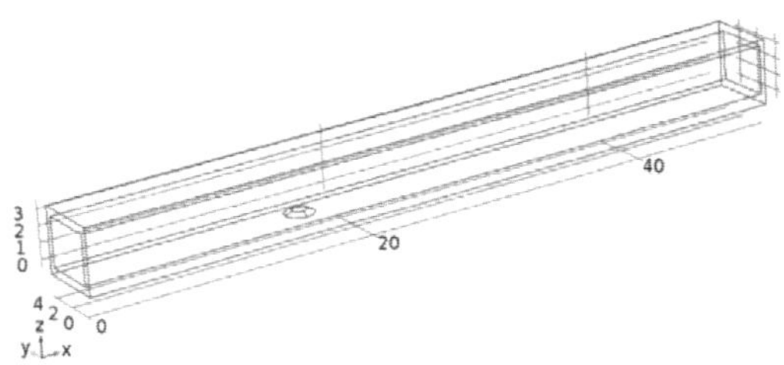

Fig.1 Calculation model

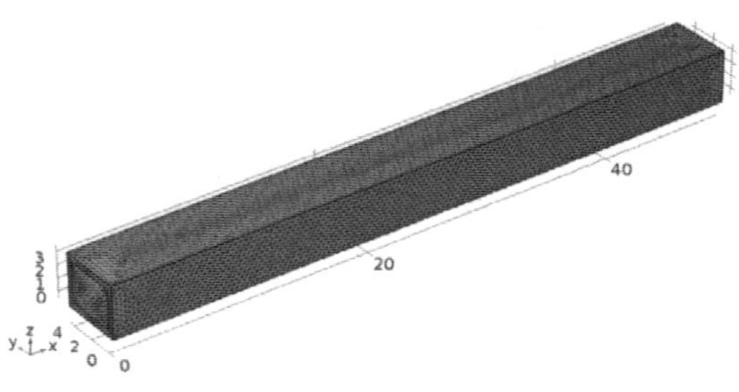

Fig.2 Mesh generation

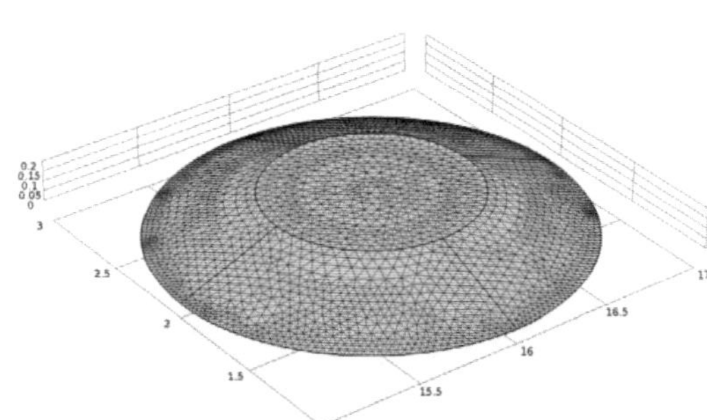

Fig.3 Local grid refinement

As entrance wind speed gradually decreases in the process of sealing fire zone, flow field is inevitably in change. The effects of reducing air supply volume on heat and gases produced in combustion are ignored, supposing that air supply volume decreases linearly, closure process continues for 60 minutes, a little air leakage exits in fire zone ($0.12m^3/s$), entrance wind speed boundary is changed to piecewise function about time.

$$\begin{cases} v = 0.01 & t > 3582 \\ v = 2 \times \dfrac{3600 - t}{3600} & t \le 3582 \end{cases}$$

The entrance of the left side is the closed construction site, and a 20cm diameter injection inert gas pipeline outlet is arranged at the upper side of the seal-wall, and the inert gas is injected, the inert gas flow is $300m^3$ / h, and the air leakage is $0.12m^3$ / S, the other boundary conditions remain unchanged.

3. Simulation of injection inert gas flow field in closed

The maximum wind speed change trend of centre vertical section of tunnel in the process of fire zone sealing is shown in Figure 4. The thickness change trend of counter current layer is shown in Figure 5, and the length change trend of counter current layer is shown in Figure 6. The maximum counter current speed change trend is shown in Fig.7. The flow field change of centre vertical section during the sealing process is shown in Fig. 8 ~ Fig.12

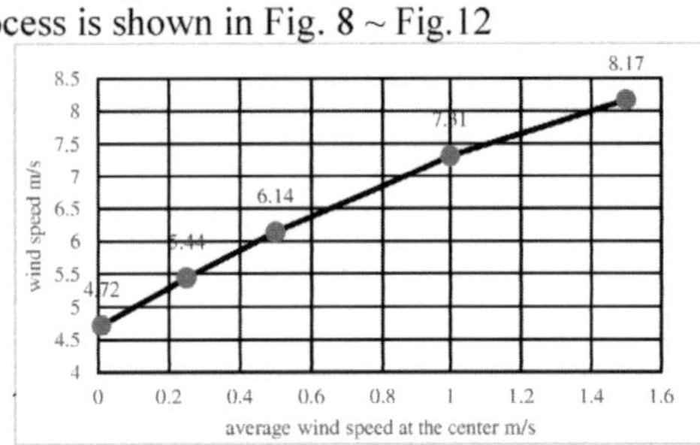

Fig4 The maximum wind speed of centre vertical section of tunnel

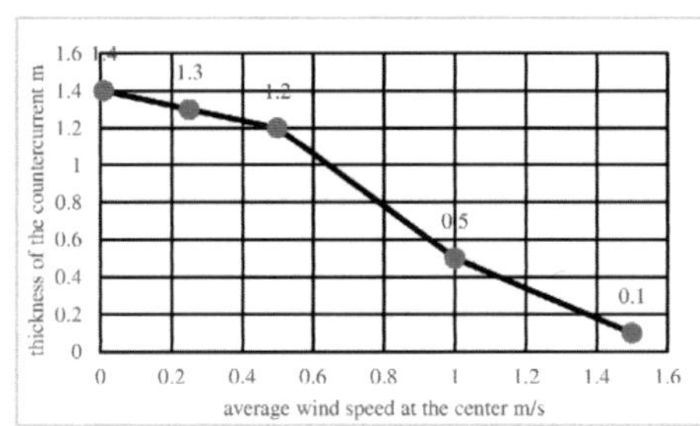

Fig5 Thickness of counter current layer

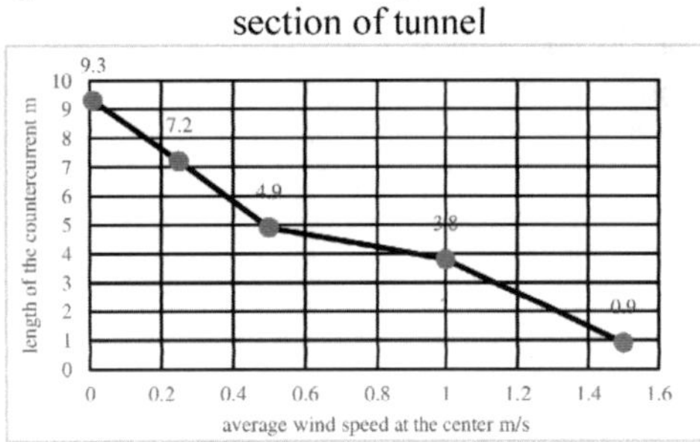

Fig6 length of counter current layer

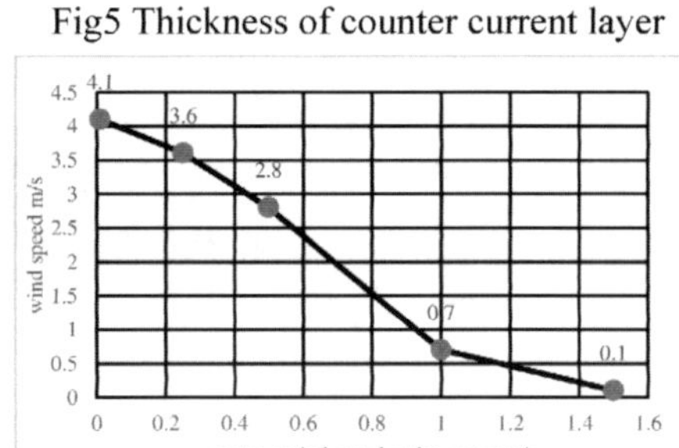

Fig7 The maximum speed of counter current flow

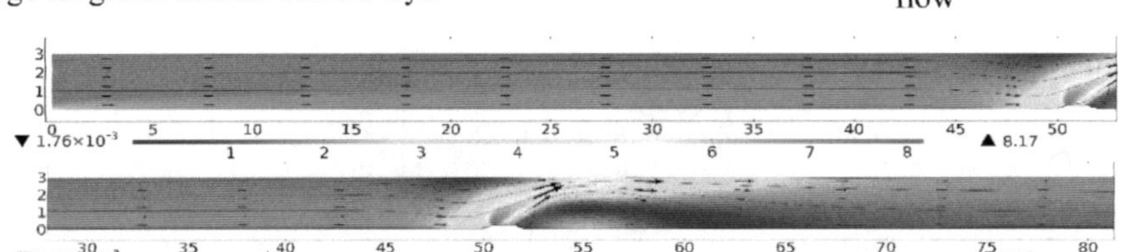

Fig8 Distribution map of tunnel wind speed as entrance average wind speed is 1.5m/s

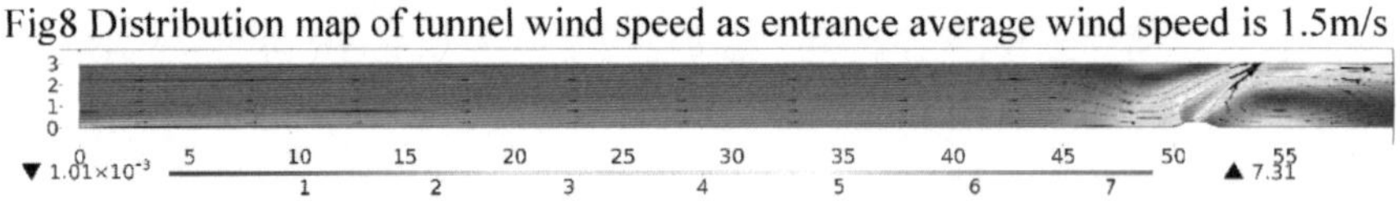

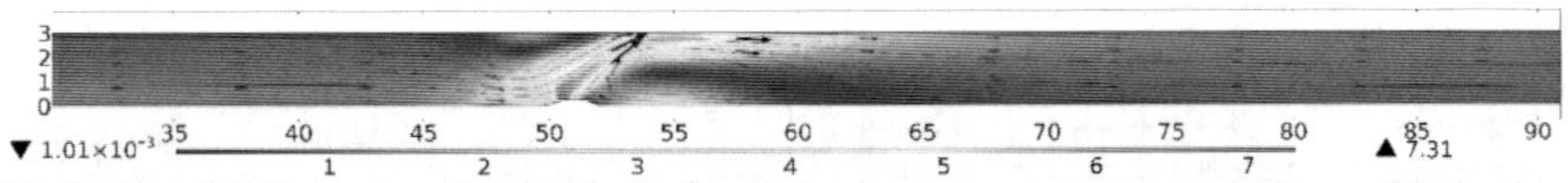

Fig9 Distribution map of tunnel wind speed as entrance average wind speed is 1m/s

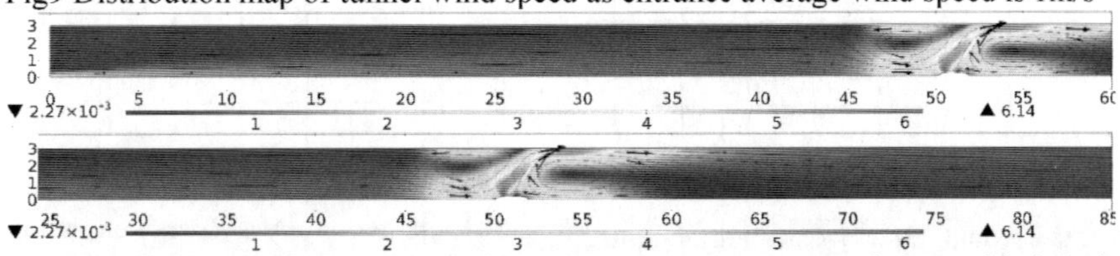

Fig10 Distribution map of tunnel wind speed as entrance average wind speed is 0.5m/s

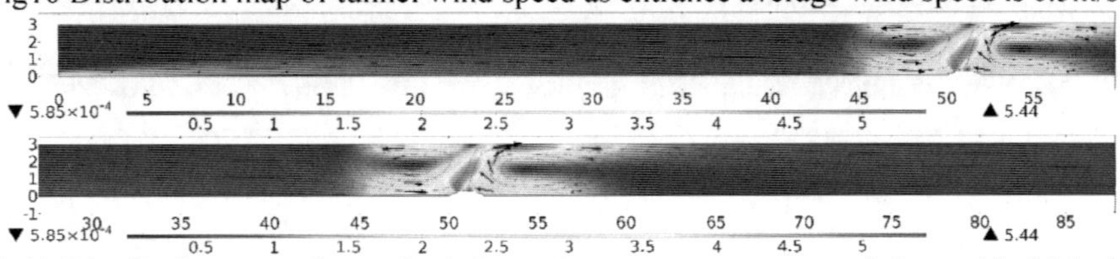

Fig11 Distribution map of tunnel wind speed as entrance average wind speed is 0.25m/s

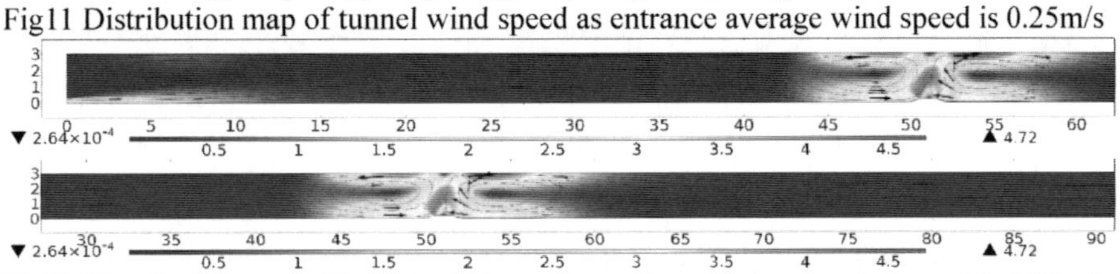

Fig12 Distribution map of tunnel wind speed as entrance average wind speed is 0.01m/s

It can be seen from Fig.4 ~ Fig.7 and Fig.8 ~ Fig. 12 that the maximum wind speed at the centre vertical section of the tunnel decreases with the decrease of the entrance average wind speed in the process of fire zone sealing. According to the arrows of streamline and velocity vector, When the average entrance wind speed is 1m / s, the counter current layer is formed near the fire source. The length and thickness of the counter current layer, the maximum counter flow speed all increase with the decrease of the entrance average wind speed. As the sealing process progresses, the entrance wind speed decreases gradually, and when it reaches 0.01m / s, eddy current is formed at the entrance.

When the closure is completed, the maximum wind speed of centre vertical section of tunnel is 4.72m / s, the counter current layer thickness is 1.4m, the counter current length is 9.3m and the maximum counter current speed is 4.1m / s. The maximum wind speed, counter current length, counter current layer thickness and counter current speed of the central vertical section of the tunnel are smaller than those of the non-injected inert gas.

4. Simulation of Gas Concentration Variation during Closed with Injection of Inert Gas

The trend of the maximum gas concentration in the vertical section of the centre of the tunnel is shown in Figure 13, and the trend of the maximum gas concentration on the downwind side of the fire source is shown in Figure 14, the gas concentration distribution of the centre vertical section of tunnel is shown in Figure 15 to Figure 19, The Nitrogen concentration distribution of the centre vertical section of tunnel is shown in Figs. 20 to 24.

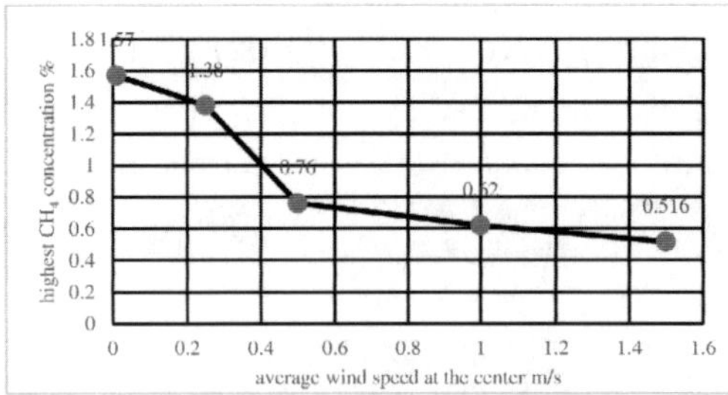

Fig13 The maximum CH_4 concentration of centre vertical section of tunnel

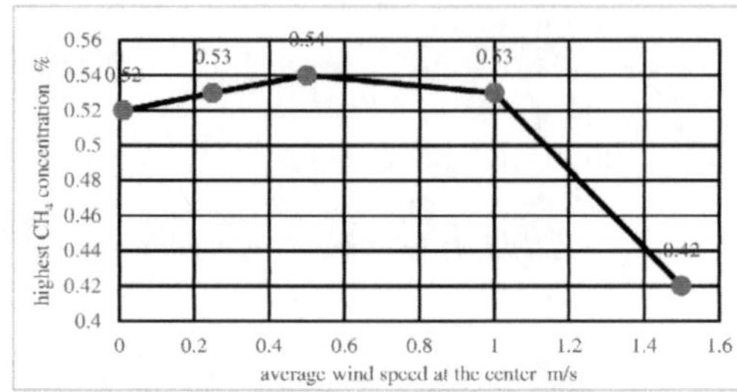

Fig14 The maximum CH_4 concentration below fire source

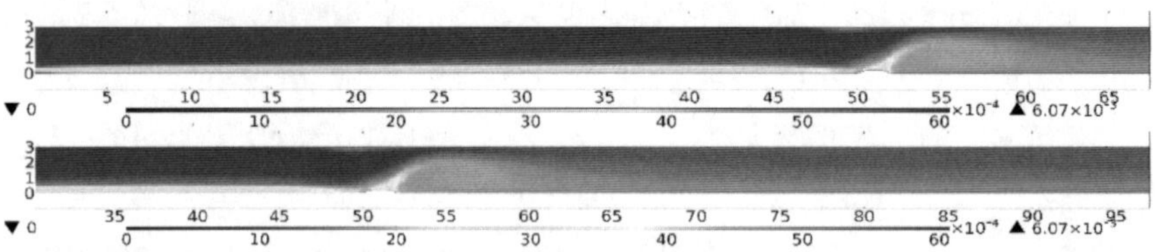

Fig15 Distribution map of CH_4 concentration as entrance average wind speed is 1.5m/s

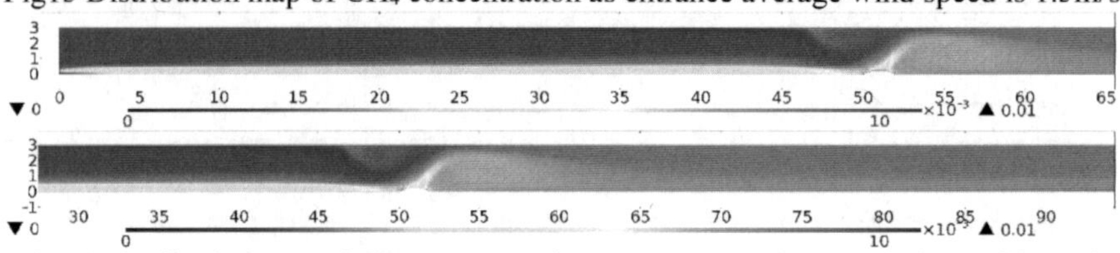

Fig16 Distribution map of CH_4 concentration as entrance average wind speed is 1m/s

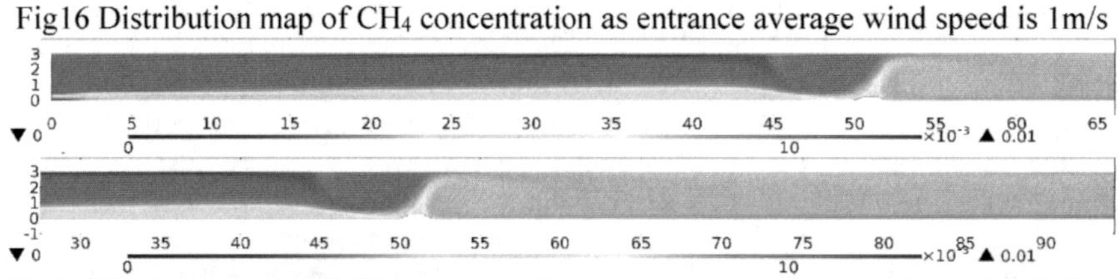

Fig17 Distribution map of CH_4 concentration as entrance average wind speed is 0.5m/s

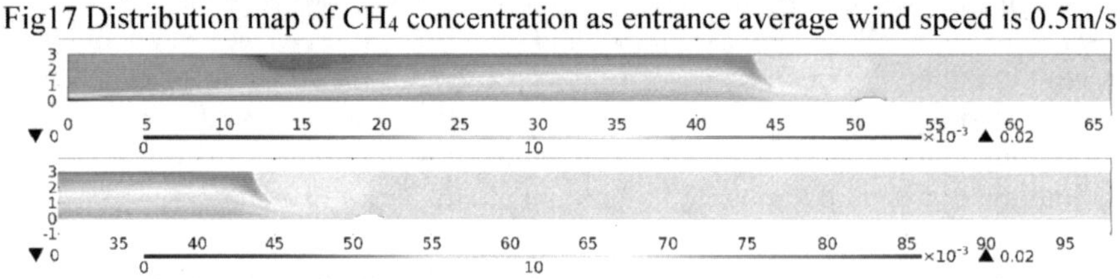

Fig18 Distribution map of CH_4 concentration as entrance average wind speed is 0.25m/s

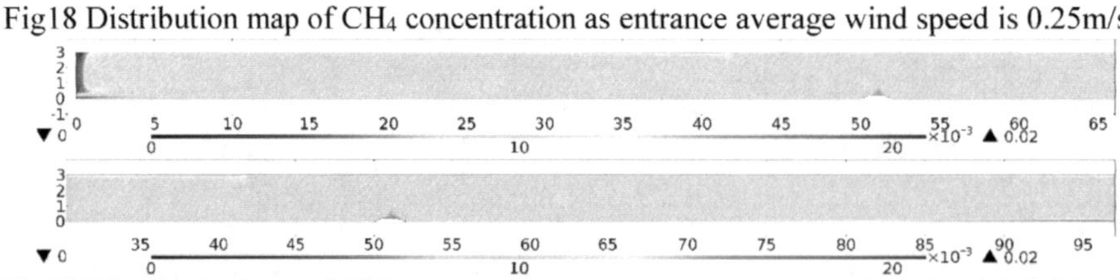

Fig19 Distribution map of CH_4 concentration as entrance average wind speed is 0.01m/s

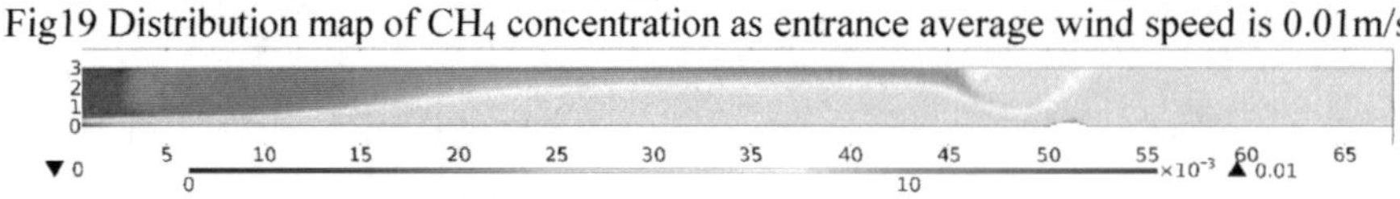

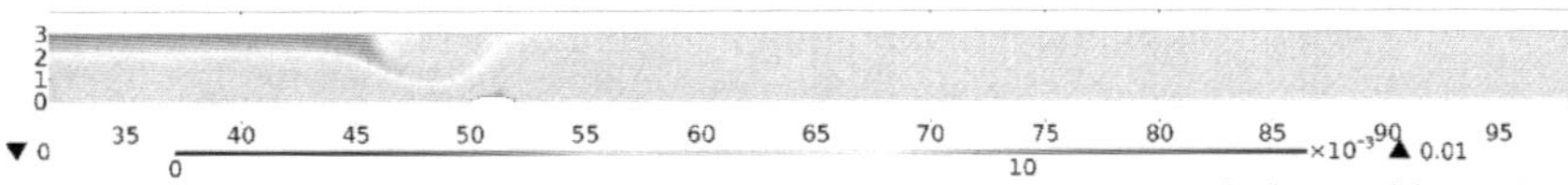

Fig20 Distribution map of CH_4 concentration as entrance average wind speed is 0.25m/s

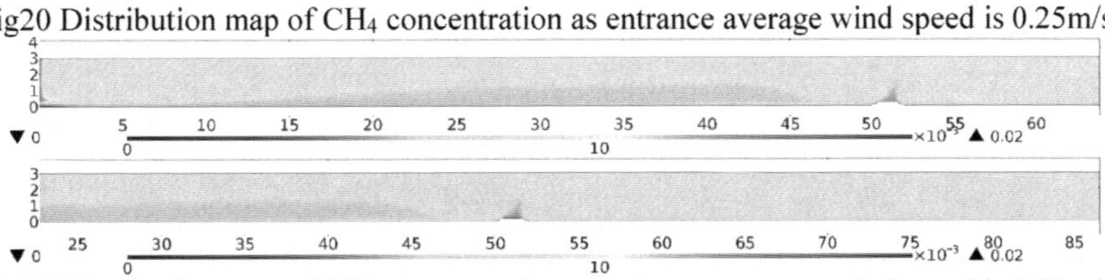

Fig21 Distribution map of CH_4 concentration as entrance average wind speed is 0.01m/s

It can be seen from Figure 15 to Figure 19 that the maximum gas concentration in the fire zone gradually increased to a maximum of 2.11%, and the maximum gas concentration at the downwind side increases slowly with the decrease of the entrance average wind speed, the maximum gas concentration at the vertical section of the middle of the tunnel appears at the edge of the counter current layer. When the wind speed is reduced to 0.25m / s, the maximum gas concentration at the vertical section of the tunnel is 1.92%, present to 36m ahead of the fire source near the roof, the maximum gas concentration in the vertical section of the tunnel is 2.11%, after the seal-wall, the gas concentration at the edge of the counter current layer is 1.48%. It can be seen that when the wind speed is less than 0.25m / s, the injected gas flow rate is relatively small, the flow rate is relatively small, inert gas mixed with the gas in front of the fire slowly, formation squeeze on fire area, due to interaction with the counter current layer, the lighter gas moves and converges toward the entrance side near the top plate.

Try to change the parameters of inert gas injection, inert gas flow change to 1000 m^3 / h, wind speed change to 0.25m / s and 0.01m / s, gas concentration distribution is shown in Figure 20, Figure 21.

The maximum concentration of gas is 1.36%, 1.63%, the average gas concentration of the whole fire zone is less than 1%, and the dangerous of gas emission eliminated. So field application should increase injection inert gas flow rate and injection inert gas speed as far as possible.

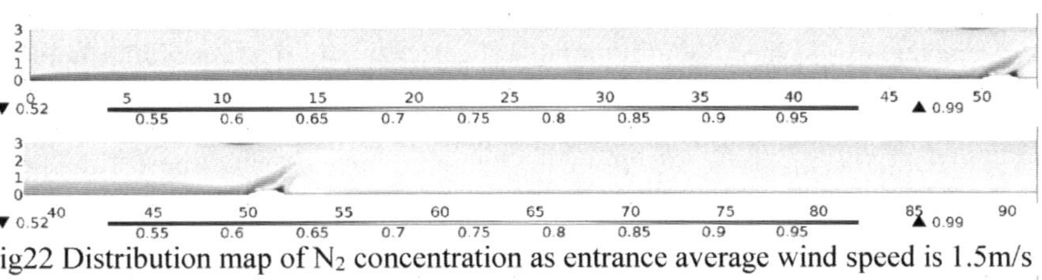

Fig22 Distribution map of N_2 concentration as entrance average wind speed is 1.5m/s

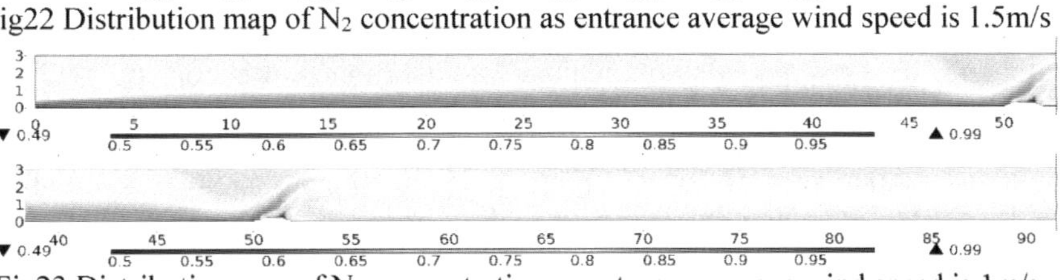

Fig23 Distribution map of N_2 concentration as entrance average wind speed is 1m/s

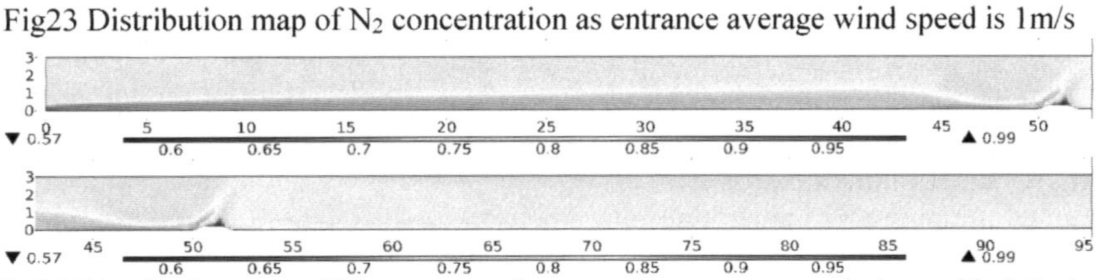

Fig24 Distribution map of N_2 concentration as entrance average wind speed is 0.5m/s

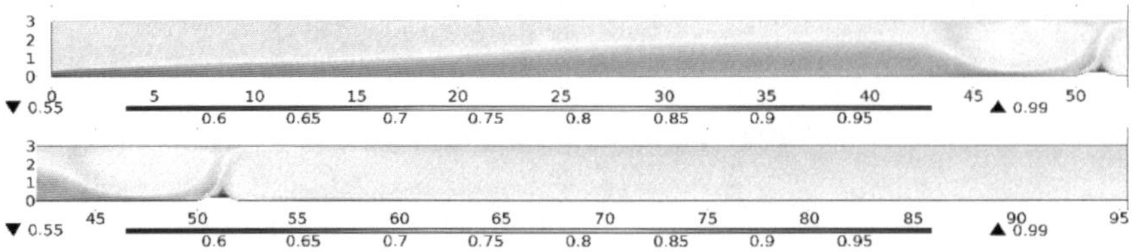

Fig25 Distribution map of N_2 concentration as entrance average wind speed is 0.25m/s

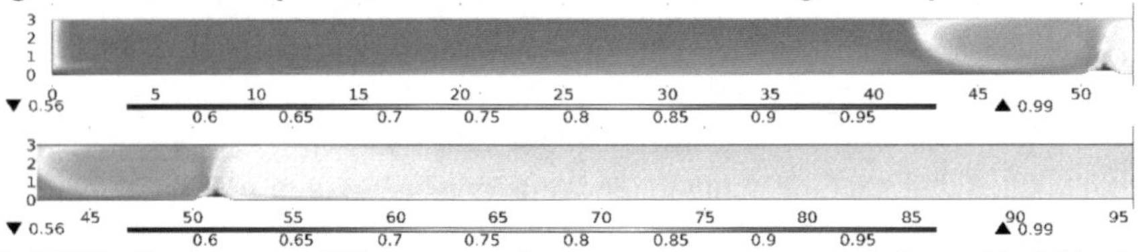

Fig26 Distribution map of N_2 concentration as entrance average wind speed is 0.01m/s

From Fig.22 to Fig.26, it can be seen that with the decrease of the wind speed in the fire zone, the nitrogen concentration in the fire zone increases gradually, and the injected nitrogen gas before the fire source is not mixed uniformly with the air in the fire area. Affected by thermodynamic nitrogen gas well blended with air in the fire area near the fire source.

5. Summary

(1) In the process of fire zone sealing, when inert gas flow is $300m^3$ / h, the maximum wind speed at the central section of the tunnel decreases with the decrease of the entrance average wind speed. When the entrance average wind speed is less than 1 m / s, the counter current layer is formed in the vicinity of the fire source. The thickness and length of the counter current, the maximum counter flow speed all increase with the decrease of the entrance average wind speed.

(2) When inert gas flow is $300m^3$ / h, and the wind speed of the fire zone reduce to 0.01m / s, swirl will be formed at the entrance, the maximum wind speed of centre vertical section of tunnel is 4.72m / s, the counter current layer thickness is 1.4m, the counter current length is 9.3m and the maximum counter current speed is 4.1m / s. The maximum wind speed, counter current length, counter current layer thickness and counter current speed of the central vertical section of the tunnel are smaller than those of the non-injected inert gas.

(3) When inert gas flow is $300m^3$ / h, the maximum gas concentration in the fire zone gradually increased to a maximum of 2.11% in the process of fire zone sealing, when the wind speed is less than 0.25m / s, the injected gas flow rate is relatively small, the flow rate is relatively small, inert gas mixed with the gas in front of the fire slowly, formation squeeze on fire area, due to interaction with the counter current layer, The lighter gas moves and converges toward the entrance side near the top plate, the maximum gas concentration at the vertical section of the tunnel is 1.92%, present to 36m ahead of the fire source near the roof.

(4) When inert gas flow change to 1000 m^3 / h, nitrogen gas well blended with air in the fire area near the fire source. the average gas concentration of the whole fire zone is less than 1%, and the dangerous of gas emission eliminated. So field application should increase injection inert gas flow rate and injection inert gas speed as far as possible.

References

[1] Wang Deming. Mine fire science [M]. China University of Mining and Technology Press .2008: 327-328

[2] DENG Cunbao, WANG Jiren, HONG Lin.Gas migration law in closed fire area of mine [J]. JOURNAL OF LIAONING UNIVERSITY OF TECHNOLOGY (NATURAL SCIENCE EDITION). 2003,23 (3): 295-298

[3] Wang Jiren, Deng Cunbao, Ding Baichuan.Study on Heat Exchange and Unsealing Time of Mine Enclosed Fire Area [J]. Journal of Liaoning Technical University (Natural Science Edition) .2003,22 (4): 452-454

[4] DUAN Yu-long, ZHOU Xin-quan, DING Xiao-lei, et al.Discussion on gas distribution law of gas injection in enclosed fire zone [J], 2010, 35 (1): 81-85[J]. Journal of China University of Mining & Technology (Natural Science Edition)

[6] Wang Zhongwen. Mine fire-induced explosion dynamic evolution and prevention techniques [D]. China University of Mining and Technology (Beijing) .2013

[7] NIU Hui-yong, DENG Jun, ZHOU Xin-quan, et al.Study on the Law of Gas Accumulation in the Process of Closed Coal Fire Area and Its Risk Analysis [J]. Journal of Central South University (Science & Technology), 2013,44 (9): 3918- 3924

[8] JIAO Yu, DUAN Yu-long, ZHOU Xin-quan, et al.Study on the law of spontaneous combustion-induced gas explosion in the closed process of coal mine fire zone [J] .Acta Coal Science, 2012,37 (5): 850-856.

Inorganic plugs removal using ultrasonic waves

Nasir Khan, Chunsheng Pu*, Li Xu and Zhang Lei

School of Petroleum Engineering, China University of Petroleum (East China), Qingdao, 266555, China

* Corresponding author Email: n.kh55@yahoo.com, Tel# 00 86 132 1084 6217

Abstract. It is essential to recover the lost productivity caused by formation damage in the proximity of the wellbore during different well operations. In comparison to conventionally used methods, the efficiency, reliability, environment friendly, and simple and convenient technique of ultrasonic waves make it more attractive in petroleum industries. In current study, ultrasonic waves were applied to mitigate the formation damage caused by deposition of calcium carbonate (CaCO3) nearby well bore. Results showed that 100 minutes exposure time could efficiently recover 38.1% of original productivity but further increase in irradiation time (120mins) would decrease the recovery to 37.1%. This aberration can be attributed to the particle-bridge formation formed by larger particles at later stages and tendency of acoustic wave to push back the fluid flow. Moreover, ultrasonic waves transducer#2 (Frequency 20KHz and Power 1000W) could recovery maximum recovery of 36.3%, however, high frequency transducer was not effective in this recovery. This inorganic removal can be attributed to the cavitation and thermal energy produced through three different ways including cavitation, boundary friction and transformation upon hitting the medium.

1. Introduction

Oil industries invest significant amount of total revenue to mitigate formation damage which happens during different well operations. Formation damage can be defined as the descent in initial productivity loss of the reservoir rock nearby wellbore [1,2]. This undesirable phenomenon can be owing to either pore obstruction, inorganic scaling deposition or others [1,2]. However, formation is mainly caused by inorganic plugs. Since $CaCO_3$ as the main culprit be formed by mixing two different kinds of water sources. Where one source has calcium ions abundance while other riches in carbonate at the basic medium. In addition, it can also be formed by either change in reservoir pressure or pH value of the production fluid [3,4] as shown by the reaction (1).

$$Ca^{2+}(aq) + 2HCO^{3-}(aq) \rightarrow CaCO_3(s) + H_2O + CO_2 \quad (1)$$

In order to prevent expensive and complex treatment, it's wise to prevent inorganic deposition as first line defense. In spite of preventive measures, inorganic scale usually develops and damages the initial permeability then chemical treatment, thermal treatment and hydraulic fracturing are usually suggest to mitigate the damages. However, extensive laboratory work and tendency of bottom hole assemblies' deterioration in conventionally applied technique make ultrasonic aided recovery more attractive [5,6].

This current study studies the application of efficient, reliable, relatively adaptive, environment friendly and convenient technique termed as ultrasonic wave technique for inorganic plugs removal. In addition, optimum frequency and power along with suitable irradiation time were investigated for effective and promising results. The valid reasons behind optimum parameters were also vividly enlightened in this study.

2. Materials and methods

2.1. Materials and brine Preparations

Brine solution was prepared by dissolving NaCl, $CaCl_2$ and $MgCl_2.6H_2O$ in corresponding proportion of 7, 0.6 and 0.4.Whereas, aqueous solution of Na_2CO_3 (200,000mg/l) and $CaCl_2$ (210,000mg/l) were separately prepared to acquire exact 1:1 stoichiometric ratio. Analytical-grade chemical reagents were obtained from Sinopharm Chemical Reagent Co., Ltd and was used as received. Artificial core samples were obtained from Haian Petroleum Research Instrument Co., Ltd and their characteristics are shown in Table 1.

Table 1 Characteristics of core samples.

Constitutes	Permeability (μm^2)	Length (cm)	Diameter (cm)	Porosity (%)	Pore Volume (cm^3)	ρ (g/cm^3)
Quartz	30×10^{-3}	6.0	2.5	20	5.88	1.67
	80×10^{-3}	6.0	2.5	22	6.47	
	150×10^{-3}	6.0	2.5	24	7.06	

2.2 Methods

2.2.1. Critical velocity and threshold salinity determination. Core samples were initially vacuumed and brine solution (80,000mg/l) was pumped through it by maintaining annular pressure at least 3MPa greater than the inlet pressure. Afterwards, lines were primed with minimum flow velocity (<1ml/min) using vent valve. Subsequently, flooding velocity was increased gradually to 5ml/min through 0.5ml/min incremental step.

Different brine solutions with the concentration of 40,000mg/l, 30,000mg/l, 20,000mg/l, 10,000mg/l and 5,000mg/l were termed as C1, C2, C3, C4 and C5. Firstly, core sample was flooded with 0.8times of critical velocity of C1 solution and reading was noted down when flow became constant. Subsequently, C2 brine solution of 10-15 times the pore volume was injected through the core sample at 0.1ml/min - 0.2ml/min velocity and was remained in the same brine solution for 12hours. Afterward, C3 and C4 were injected through the core in similar manner prior to pumping the identical volume and flowrate of distilled water.

2.2.2. Dynamic experiment. Before saturating the core samples for 8hours, dry weight of core samples and gas permeabilities were measured to be $30\times10^{-3}\mu m^2$, $80\times10^{-3}\mu m^2$ and $150\times10^{-3}\mu m^2$. Experimental setup for ultrasonic aided plug removal as shown in figure 1. Original permeability (K_o) of all three sets of core samples were measured by injecting brine (40,000mg/l) at 0.8times rate of the critical velocity (0.8*2ml/min).

In order to plugs the core sample, Na2CO3 (2*pore volume) was pumped at 0.8times the critical velocity to prevent channeling effect. In next stage, core samples were remained in CaCl2 for 2hours by injecting CaCl2 solution for complete synthesis of CaCO3 as shown in following reaction (2). Similar procedure was repeated for other two sets of core sample. Afterwards, damaged permeability (Kd) of the core sample was measured.

$$CaCl_3 + Na_2CO_3 \rightarrow CaCO_3 + 2NaCl \quad (2)$$

In next step, damaged core sample was irradiated for 6minutes and treated permeability (Kt) was measured. Eventually, percentage permeability recovery was calculated by applying equation (3) for the selection of best-performing transducer.

$$((Kt - Kd)/Ko)* 100\% \quad (3)$$

Thereafter, optimum exposure time was determined using transducer#2 for variant time spans ranging from 60min to 120min and treated permeability (Kt) was measured.

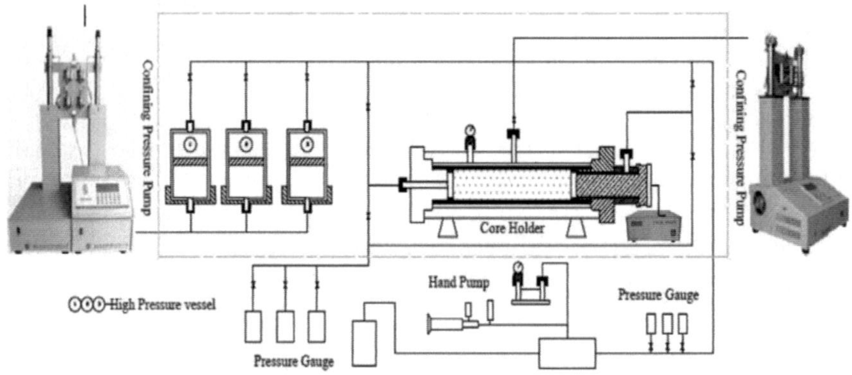

Figure 1. Experimental schematic of ultrasonic plug removal.

3. Results and Discussion

3.1 Critical velocity and threshold salinity of the core

Critical flow velocity and threshold salinity were indispensably measured before recovering the damaged core samples. Critical velocities were investigated to be 2ml/min for Core 3-12 and each 2.5ml/min for Core 8-38 and Core 15-32. Whereas, threshold salinity for Core 3-11, Core 8-37 and Core 15-31 with the gas permeability of $30\times10^{-3}\mu m^2$, $80\times10^{-3}\mu m^2$ and $150\times10^{-3}\mu m^2$ were measured to be 15,000mg/l and 10,000mg/l respectively.

3.2. Dynamic experiment of ultrasonic waves for the inorganic plug mitigation

3.2.1 Frequency and power optimization. Acoustic wave with appropriate frequency and power has undoubtable potential to mitigate formation damaged caused by inorganic plugs near the wellbore proximity. For this purpose, six core samples with $30\times10^{-3}\mu m^2$ gas permeability were exposed to six different kinds of transducers with variant frequency and power in order to choose the optimum one as depicted in Table 2. First three transducers #1, #2 and #3 with the same power of 1000W but different corresponding frequencies of 18KHz, 20KHz and 25KHz irradiated the core samples. Consequently, #2 transducer recovered 34.1% of the natural permeability and led among the six given transducers while 31.1% and 33.3% proportions of the original permeability were acquired using #1 and #3 transducers respectively. The principal reason of relatively less recovery of the former is due to falling of 18KHz in the audible range which led to transformation in undesirable sound energy. While in the latter case, relatively large gap between core sample and transducer resulted into inefficient transformation of acoustic energy into heat and other unwanted forms which can adversely influence the effect of ultrasonic waves.

Afterward, same power but different frequencies of 30KHz and 40KHz were applied using #4 and #5 transducers and their recoveries' proportion were witnessed to be 9.8% and 8.6% repectively. However, #6 (10.9%) had high potential to remove inorganic plugs in comparison to #4 and #5 owing to its high power.

Likewise, comparable results were obtained upon performing identical experiment on remaining two kinds of core samples with 80×10-3μm2 and 150×10-3μm2 gas permeabilities as depicted in Table 2 and Table 3. As far as Table 3 is concern, the proportion of damaged permeability recoveries were 31.7%, 36.3%, 32.4%, 9.7%, 7.6% and 10.9% by the application of #1, #2, #3, #4, #5 and #6 transducers. Considering Table 4, transducer #2 (20 KHz and 1000W) and #5 (40KHz and 60W) produced maximum and minimum recovery of 35.7% and 9.9% respectively. While #6 having maximum frequency (50KHz) and power (200W) improved only 12.9% of the original permeability. In addition, Transducer #1 and #3 with the same power (1000W) and the respective different frequencies of 18KHz and 25KHz recovered 31.5% and 34.4% proportions of the natural permeability of the core samples. In spite of identical power, acoustic wave with suitable frequency (20KHz) likely to mitigated the damaged core sample owing to matching of acoustic frequency with natural frequency of the particles and led to amplitude enhancement. Consequently, pore structure tried to follow the same pattern of the incident wave which resulted into peristaltic transport of the resident fluid when trough and crest of the two adjacent layers would coincide. Damaged permeability recovery can be attributed to drain down of loosely suspended core constituents due to occurrence of high local pressure caused by cavitation during wave propagation.

Cavitation mechanism can be referred to the bubble formation in low pressure zone due to existence of tiny nuclei containing undissolved gas [7,8]. Subsequently, prior to imploding of bubble with 5075K±156K temperature [9] and 1000atm pressure [9], it grows to certain size due to energy absorption. Furthermore, cavitation is broken up into two categories on the basis of cavity's life. In transient cavitation, bubble can survive less than one acoustic cycle, however, those bubbles which can exist beyond one acoustic cycle come under the heading of stable cavitation. Transient cavitation controls cavitation intensity which can be produced by keeping driving acoustic frequency lower than the natural frequency [10]. The produced shocking wave radially propagates and causes improvement in the recovery of damaged permeability which can further be influenced by fluid nature and site nucleation as shown in figure 2. From the above results, in a nutshell, high power of ultrasonic wave transducer likely to improve lost natural productivity due to thermal energy generation via three different means. Heat energy is probably to produce when acoustic wave transforms upon hitting the targeted porous media. Another enormous source of heat is the implosion of bubbles during cavitation process and the third reason could be the existence of boundary friction. In addition, ultrasonic wave is generally proportional to the boundary friction and absorption. Consequently, high frequency tends to absorb more and causes greater boundary friction. Likewise, sound wave intensity is directly related to the heat energy generation and vicious of the cavitation [7].

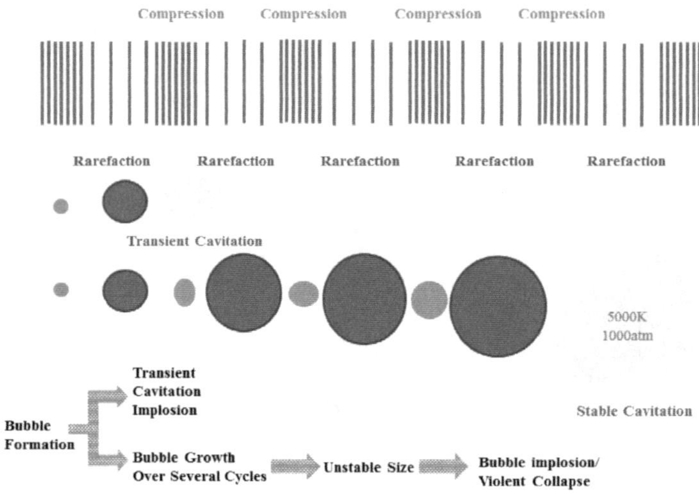

Figure 2. Cavitation Process.

Table 2. Experiment results of ultrasonic-removed inorganic scale with cores of air permeability around $30\times10^{-3}\mu m^2$.

Core #	Transducer	Frequency (KHz)	Power (W)	K_o ($10^{-3}\mu m^2$)	K_d ($10^{-3}\mu m^2$)	K_t ($10^{-3}\mu m^2$)	$(K_t-K_d)/K_o$ (%)
3-14	1	18	1000	20.77	11.91	18.37	31.1
3-33	2	20	1000	21.10	9.98	17.17	34.1
3-34	3	25	1000	20.03	11.49	18.15	33.3
3-30	4	30	60	19.84	10.83	12.78	9.8
3-32	5	40	60	21.60	10.74	12.61	8.6
3-29	6	50	200	21.03	10.15	12.31	10.2

Table 3. Experiment results of ultrasonic-removed inorganic scale with cores of air permeability around $80\times10^{-3}\mu m^2$.

Core #	Transducer	Frequency (KHz)	Power (W)	K_o ($10^{-3}\mu m^2$)	K_d ($10^{-3}\mu m^2$)	K_t ($10^{-3}\mu m^2$)	$(K_t-K_d)/K_o$ (%)
8-143	1	18	1000	62.37	36.28	56.07	31.7
8-26	2	20	1000	56.93	32.38	53.02	36.3
8-106	3	25	1000	61.82	32.52	52.57	32.4
8-144	4	30	60	66.71	35.49	41.95	9.7
8-146	5	40	60	59.20	35.22	39.75	7.6
8-10	6	50	200	64.42	34.44	41.48	10.9

Table 4. Experiment results of ultrasonic-removed inorganic scale with cores of air permeability around $150\times10^{-3}\mu m^2$.

Core #	Transducer	Frequency (KHz)	Power(W)	K_o ($10^{-3}\mu m^2$)	K_d ($10^{-3}\mu m^2$)	K_t ($10^{-3}\mu m^2$)	$(K_t-K_d)/K_o$ (%)
15-80	1	18	1000	108.16	60.76	94.88	31.5
15-37	2	20	1000	111.81	55.30	97.04	35.7

15-74	3	25	1000	111.76	62.94	101.41	34.4
15-129	4	30	60	122.13	58.29	71.24	10.6
15-89	5	40	60	125.69	64.31	76.81	9.9
15-11	6	50	200	123.33	59.72	75.64	12.9

3.2.2 Irradiation Time.

Prior to measuring the optimum ultrasonic waves' irradiation time, best-performing transducer #2 with 20KHz frequency and 1000W power was already determined. Table 5 and Table 6 illustrate permeability recovery proportions using transducer#2 when core samples were irradiated for 60mins, 80mins, 100mins and 120mins. Considering Table 5, 60mins irradiation time to the ultrasonic wave brought 34.1% reduction in the damaged permeability while this proportion increased by 0.06% when exposure time increased to 80mins. In addition, maximum recovery (35.9%) was achieved by 100mins irradiation time and considered as the optimum time span. Despite further exposure, it dropped to 35.2% when exposure time was increased to 120mins. It can be concluded that irradiation beyond optimum exposure time may detach relatively larger particles which can provide resistance to the flow by forming particle-bridge at the later stages. Moreover, over exposure of the ultrasonic waves such as beyond 100mins irradiation has greater tendency to push back the fluid flow as well as waste of useful acoustic energy into undesired forms of energy upon transformation in the fluid present between the gap as demonstrated in figure 3.

Similar trend is shown in Table 6 for the core sample with 80×10-3μm2 gas permeability, ultrasonic wave began to recover the damaged permeability by 36.3% and this recovery raised to 37.5% for 60mins and 80mins exposure time respectively. Furthermore, damaged permeability recovery jumped to the peak of 37.6% (100mins exposure time) but excess irradiation dipped the recovery to 35.5%. Likewise, ultrasonic waves showed identical trend in Table 7 for the core sample with 150×10-3μm2 gas permeability. Whereas, recoveries for irradiation time of 60mins, 80mins, 100mins and 120mins were 34.4%, 36.6%, 38.1% and 37.1% respectively. Although effective irradiation time span is still in developmental stage and need more research work.

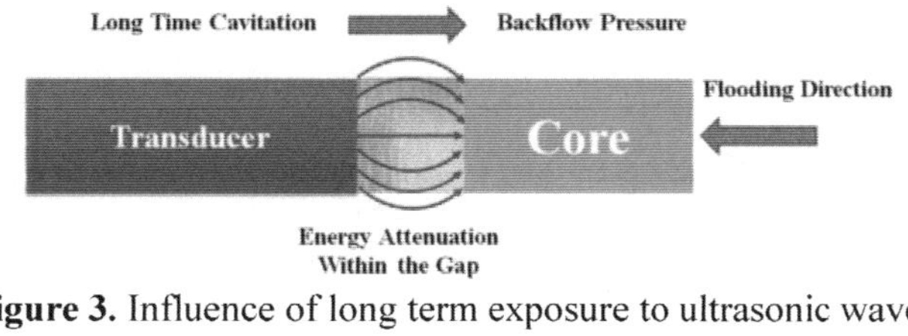

Figure 3. Influence of long term exposure to ultrasonic waves.

Table 5. Experiment results of optimizing ultrasonic integrating treating time with cores of gas permeability around $30\times10^{-3}\mu m^2$.

Core #	Time duration (min)	K_o ($10^{-3}\mu m^2$)	K_d ($10^{-3}\mu m^2$)	K_t ($10^{-3}\mu m^2$)	$(K_t-K_d)/K_o$ (%)
3-33	60	21.10	9.98	17.17	34.1
3-28	80	18.89	10.89	17.45	34.7
3-27	100	19.94	10.48	17.64	35.9
3-26	120	22.34	11.64	19.49	35.2

Table 6. Experiment results of optimizing ultrasonic integrating treating time with cores of gas permeability around $80\times10^{-3}\mu m^2$.

Core #	Time duration (min)	K_o ($10^{-3}\mu m^2$)	K_d ($10^{-3}\mu m^2$)	K_t ($10^{-3}\mu m^2$)	$(K_t-K_d)/K_o$ (%)
8-26	60	56.93	32.38	53.02	36.3
8-38	80	60.70	36.03	58.79	37.5
8-39	100	63.28	32.56	56.32	37.6
8-42	140	61.45	34.09	55.92	35.5

Table 7. Experiment results of optimizing ultrasonic integrating treating time with cores of gas permeability around $150\times10^{-3}\mu m^2$.

Core #	Time duration (min)	K_o ($10^{-3}\mu m^2$)	K_d ($10^{-3}\mu m^2$)	K_t ($10^{-3}\mu m^2$)	$(K_t-K_d)/K_o$ (%)
15-74	60	111.76	62.94	101.41	34.4
15-54	80	117.34	63.28	106.27	36.6
15-2	100	121.43	65.60	111.90	38.1
15-93	120	127.35	62.23	109.52	37.1

4. Conclusions

Following conclusions are drawn;

1. #2 transducer was considered as the best-performing transducer with frequency 20KHz and power 1000W on the basis of extent of inorganic plugs mitigation. The produced maximum permeability recoveries were 34.1%, 36.3% and 35.7% in the core samples with initial gas permeability of $30\times10^{-3}\mu m^2$, $80\times10^{-3}\mu m^2$ and $150\times10^{-3}\mu m^2$ respectively.
2. 100mins exposure time was witnessed to be optimum which led to maximum inorganic plugs removal of 35.9%, 37.5% and 38.1% in the core samples of $30\times10^{-3}\mu m^2$, $80\times10^{-3}\mu m^2$ and $150\times10^{-3}\mu m^2$ initial gas permeability respectively.

Acknowledgement

We appreciate Fundamental Research Funds for the Central Universities of Chinese government for their financial support (Grant No. 16CX02020A) of this project. The authors also extend their gratitude to Higher Education Commission (HEC) of Pakistan for awarding scholarship to carry out research work.

References

[1] Ezenweichu C L and Laditan O D 2015 the Causes , Effects and Minimization of Formation Damage

in *Pet. Coal* **57** 169–84

[2] Faruk civan 2014 *Reservoir Formation damage* ed oxford (Elsevier B.V.)

[3] Al-Taq A A, Muhaish S A, Nakhli M M and Alrustum A A 2015 Organic/Inorganic Deposition in Oil Producing Wells from Carbonate Reservoirs: Mechanisms, Removal and Mitigation. SPE 177447. *Abu Dhabi Int. Pet. Exhib. Conf.* 1–15

[4] Ohen H, Williams L, Lynn J and Ali L 2004 Assessment and Diagnosis of Inorganic-Scaling Potential Using Near-Infrared Technology for Effective Treatment *SPE Prod. Facil.* **19** 18–20

[5] Krueger R F 1988 SPE 17459 An Overview of Formation Damage and Well Productivity in Oilfield Operations: An Update *Soc. Pet. Eng.* 535–53

[6] Zekri A Y, Shedid S A and Alkashef H 2007 A new technique for treatment of permeability damage due to asphaltene deposition using laser technology *J. Pet. Sci. Eng.* **59** 300–8

[7] Guo X, Du Z, Li G and Shu Z 2004 High Frequency Vibration Recovery Enhancement Technology in the Heavy Oil Fields of China *SPE Int. Therm. Oper. Heavy Oil Symp.* 1–18

[8] Duhon R D and Campbell J M 2013 The Effect Of Ultrasonic Energy On The Flow Of Fluids In Porous Media *SPE East. Reg. Meet.*

[9] Flint E B and Suslick K S 1991 The temperature of cavitation. *Science* **253** 1397–9

[10] Mujumdar S, Senthil Kumar P and Pandit A B 1997 Emulsification by ultrasound : Relation between intensity and emulsion quality *Indian J. Chem. Technol.* **4** 277–84

Preparation of TiO_2 nanoparticle and photocatalytic properties on the degradation of phenol

Guanyu Wang[1], Deping Xu[1,*], Weijie Guo[1], Xiaohui Wei[1], Zhe Sheng[1] and Zhi Li[2]

[1]School of Chemical & Environmental Engineering, China University of Mining & Technology, Beijing, Beijing 100083, China
[2]Institute of Geology and Geophysics. CAS, Beijing 100083, China

Corresponding e-mail: siyueming@126.com

Abstract. Titanium dioxide (TiO_2) nanoparticle in the diameter of 30-50nm in this study was prepared by sol-gel method. X-ray diffraction (XRD) and scanning electron microscope (SEM) were employed to investigate the characteristics of TiO_2 nanoparticle. The effect of amount of distilled water and absolute ethyl alcohol, calcination temperature on photocatalytic activity were explored on the degradation of phenol. The results showed that the optimal amount of distilled water and absolute ethyl alcohol and calcination temperature was 5mL, 90mL and 575°C, respectively. Moreover, in the photocatalytic activity test, the effect of TiO_2 nanoparticle amount and the pH value of phenol solution on the degradation of phenol were discussed. The results showed that the optimal amount of TiO_2 nanoparticle was 75mg in 150mL of phenol solution (100mg/L) and the degradation reached 92.4% when the pH value was 1. The utilization of TiO_2 nanoparticle on the degradation of phenol is promising for the coal-chemical wastewater.

1. Introduction

TiO_2 photocatalyst is suitable for coal-chemical wastewater due to its inexpensiveness, nontoxicity and high photochemical stability [1-4]. Generally, the photocatalysis is initiated when a photon excited by UV-light has equal or greater energy than its band gap, it would escape from valence band to conduction band and form electron (e^-) and positive holes (h^+). The generated positive holes (h^+) and electrons (e^-) transfer to different position on the photocatalyst surface and react with water into hydroxyl radicals(•OH) which have high oxidation activity and mineralize the organic contaminants into water and carbon dioxide [5-9]. In the recent years, many researches about TiO_2 have been reported. Antonopoulou et al. [10] prepared char/N-TiO_2 and char/N-F-TiO_2 composite photocatalysts to degrade the phenol (5mg/L) under solar light in 4 hours, although phenol was degraded all, the concentration of phenol was low. Zang et al [11] utilized TiO_2 to degrade phenol (20mg/L) in different pulsed discharge systems. TOC mineralization efficiency of phenol was enhanced to 65%, however, the concentration of phenol was low and the energy consumption was high. The researches mentioned above were not suitable for industrialization.

In this work, TiO_2 nanoparticle was prepared by sol-gel method and the factors of TiO_2 nanoparticle preparation, such as amount of distilled water and absolute ethyl alcohol, calcination temperature were investigated. It provide a theoretical reference about the preparation of TiO_2 nanoparticle by sol-gel method. Meantime, the effect of TiO_2 nanoparticle amount and the pH value on photocatalytic activity were also studied. It is suitable for industrialization due to the degradation of high concentration phenol solution and low cost. It also provides a novel way for the treatment of coal-chemical wastewater.

2. Experimental

2.1 Preparation of TiO_2 nanoparticle

TiO_2 nanoparticle was prepared by sol-gel method. Tetrabutyl titanate (10mL) was mixed with absolute ethyl alcohol (50mL) under vigorous stirring for 30min, which obtained solution A. Subsequently, distilled water (5mL) was mixed with absolute ethyl alcohol (50mL) and glacial acetic acid (6mL) under rapidly stirring for 30min, which obtained solution B. At last solution A was added into solution B drop by drop under rapidly stirring. Thus, TiO_2 gel was obtained. The gel was aged at room temperature for 24 hours, followed by drying at 80 °C for 24 hours. After calcination in air at 575 °C for 4 hours, the obtained solid was grinded into particles, which was TiO_2 nanoparticle.

2.2 Photocatalytic Experiment

The photocatalytic experiments were carried out in a water-jacket reactor at a constant temperature of 10°C and initiated by irradiating with a 24W high-pressure mercury lamp (λ= 240~270 nm), which was inserted in the phenol solution. In every experiment, 75mg of test sample and 150mL of 1×10^{-4} mg/L phenol solution were introduced in the reactor with magnetic stirring. 4mL of the irradiated solution was extracted from the reactor at certain intervals. After centrifugal separation, the concentration of phenol was analyzed by a UV-vis spectrophotometer at λ=270 nm.

3. Results and Discussion

3.1 Effect of factors of TiO2 nanoparticle preparation.

In the same condition, the totally amount of absolute ethyl alcohol was 10mL, 30mL, 50mL, 70mL, 90mL, 110mL and 130ml, respectively. Obtained TiO_2 nanoparticle was employed to photodegrade the phenol solution. The results were showed in Figure 1. With the increasing amount of absolute ethyl alcohol, the degradation was kept sustainable growth until 90mL of ethyl alcohol added, the degradation reached the top of 73.8%. This is mainly due to the hydrolysis reaction of tetrabutyl titanate is rapid. Absolute ethyl alcohol slows down the hydrolysis reaction and narrows the diameter size of TiO_2 nanoparticle. Thus, the specific surface area is improved and activity sites per unite mass becomes more, which cause the photodegradation efficiency better. However, with the continuing increasing amount of absolute ethyl alcohol, the degradation declined. This is mainly due to excess absolute ethyl alcohol makes $Ti(OH)_x(OBu)_y$ do not touch each other. Thus, excess absolute ethyl alcohol do not generate gel and makes photodegradation efficiency worse.

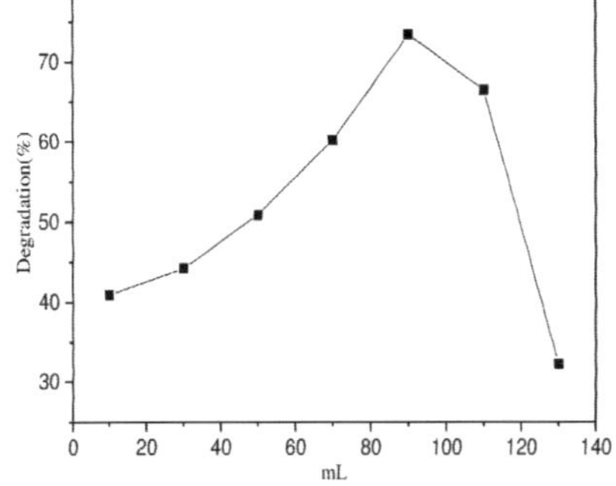

Figure 1. Effect of amount of absolute ethyl alcohol

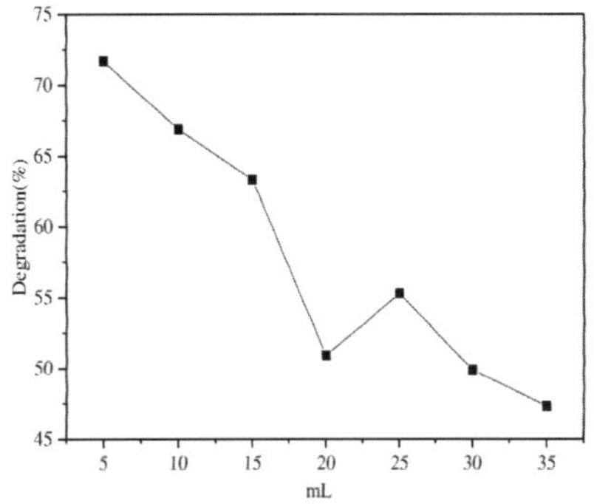

Figure 2. Effect of amount of distilled water

In the same condition, the amount of distilled water was 5mL, 10mL, 15mL, 20mL, 25mL, 30mL and 35ml, respectively. Obtained TiO_2 nanoparticle was employed to photodegrade the phenol solution. As seen from Figure 2, at the beginning, 5mL of distilled water was added, the degradation reached the top of 74.8%. With the increasing amount of distilled water, the degradation of phenol declined. This is mainly due to the excess distilled water accelerates the hydrolysis reaction of tetrabutyl titanate. Excess distilled water generates agglomerate and precipitate, leading to the increasing of the diameter size of TiO_2 nanoparticle. Thus, it makes the photodegradation efficiency worse.

As-prepared TiO_2 were calcined at 300°C, 400°C, 500°C, 525°C, 550°C, 575°C and 600°C, after grinding, obtained nanoparticle was employed to photodegradate the phenol solution. The XRD patterns of the TiO_2 nanoparticle was shown in Figure 3. When the calcination temperature was 300°C, the as-prepared TiO_2 did not generate crystal. With the rising of temperature, the crystal was generated little by little and when the temperature reached at 500°C, anatase (A) was formed well. Whereafter, a little rutile (R) appeared at 575°C and became more than antase at 600°C. The effect of calcination temperature on photodegradation was shown in Figure 4. With the rising of temperature, the degradation increased, reaching the top of 62.4% at 575°C, and then declined at 600°C. The reason of increasing degradation is that antanse, which was essential substance for photodegradation, formed gradually with the rising of temperature. The decline degradation at 600°C is mainly due to the rutile, which has low photocatalytic activity, became more than antase. Thus, the activity sites decreased and degradation declined.

The structure and diameter size of TiO_2 nanoparticle was evaluated by SEM and the results were shown in Figure 5, TiO_2 nanoparticle was alveolate and compact, and the diameter size was about 20~50nm.

3.2 Photocatalytic activity test

10mg, 25mg, 50mg, 75mg, 100mg and 125mg of TiO_2 nanoparticle prepared were employed to photodegrade the phenol solution and the results were shown in Figure 6. With the increasing amount of TiO_2 nanoparticle, the degradation was improved until 75mg of TiO_2 added, the degradation reached 51.9%. However, it was declined with the continuing increasing amount of TiO_2 nanoparticle. The increasing degradation is due to the increasing amount of TiO_2 nanoparticle increases the number of photoactive sites and also more phenol molecules absorbed. Moreover, the decline degradation is due to the excess TiO_2 nanoparticle hinders the penetration of light and reduces the specific activity of the catalyst.

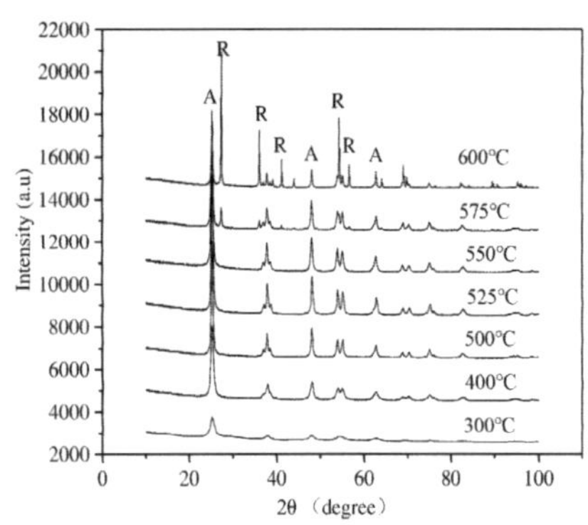

Figure 3. The XRD patterns of TiO_2 nanoparticle

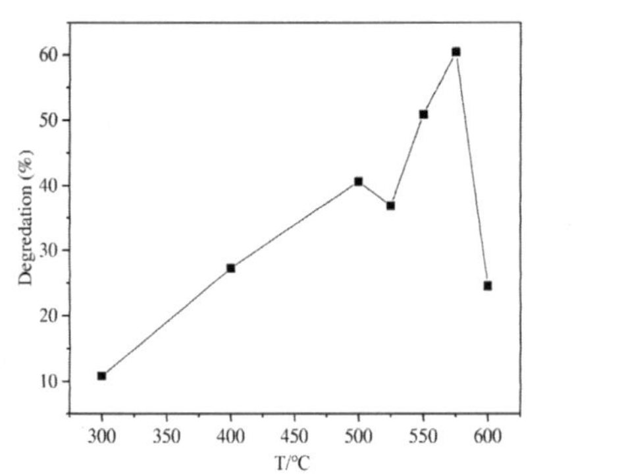

Figure 4. Effect of calcination temperature

Figure 5. SEM of TiO_2 nanoparticle

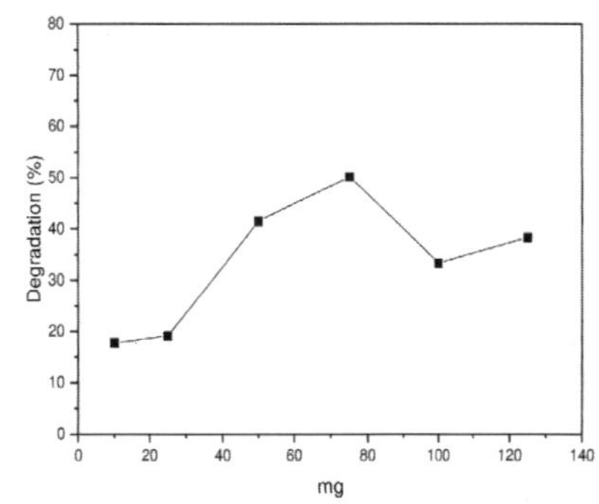

Figure 6. Effect of amount of TiO_2 nanoparticle

Hydrochloric acid and sodium hydroxide were employed to adjust the phenol solution pH value in 1, 2, 3, 4, 5, 6, 7, 8, and 9, respectively. Effect of phenol solution pH value was investigated and the results were shown in Figure 7. At the beginning, the degradation reached the top when pH value was 1 and then declined with the increasing of pH value, the highest degradation rate was 92.4%. According to previous reports [12, 13], the point of zero charge (pH_{pzc}) of TiO_2 is about 6.8. Thus, the TiO_2 surface is positively charged below this value and negatively charged above it. The results in Figure 7 indicate that the degradation is favored in an acidic solution, on the contrary, it is inhibited in a basic solution as compared with that of in neutral solution. In the acidic solution, phenol is primarily in its nonionic form, water solubility is minimized and the adsorption onto the catalyst is maximized. It is supposed that the density of •OH radicals is highest near the surface of TiO_2 nanoparticle and decreases rapidly with distance from the surface. At higher pH, phenol tends to exist as negatively charged phenolate species, which have extremely strong solubility in solution and do not be adsorbed on TiO_2 nanoparticle surface significantly. Also, more OH^- groups avoid the formation of •OH radicals, which have strong oxidizing property and play a significant role in the photodegradation of phenol. Moreover, high pH favors the formation of carbonate ions which are effective scavengers of OH^- ions. Through above results, the degradation efficiency is favorable in the acidic solution.

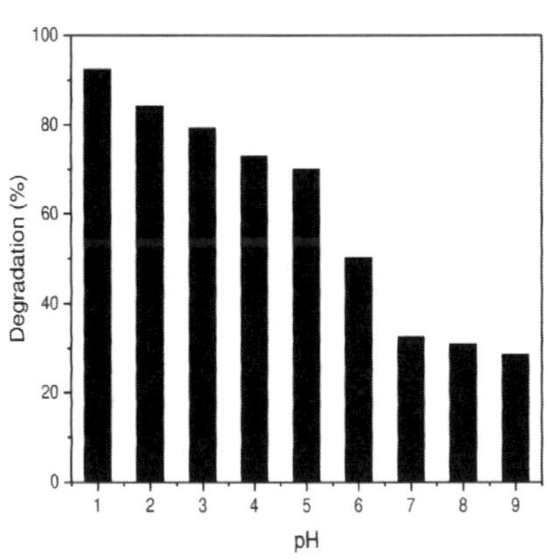

Figure 7. Effect of phenol solution pH value

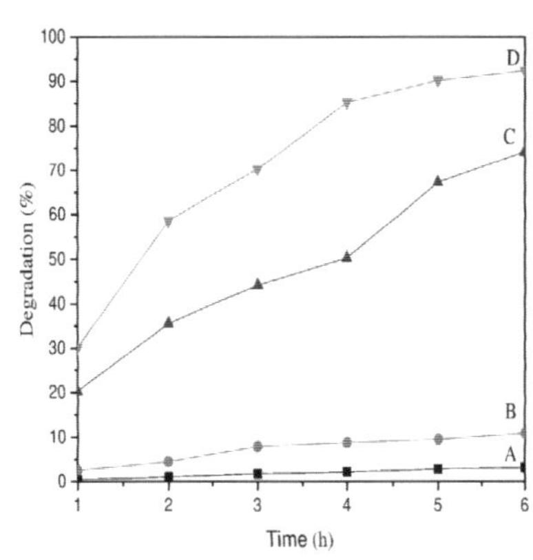

Figure 8. Comparison of degradation of phenol under different conditions

A: TiO_2 nanoparticle with dark, B: TiO_2 nanoparticle with visible light
C: TiO_2 nanoparticle with UV D: TiO_2 nanoparticle with UV when pH value was 1

The photocatalytic activity of TiO_2 nanoparticle was evaluated by monitoring the degradation of phenol in aqueous solution. As shown in Figure 8, the degradation phenol was low in the dark and reached 5.2% after 6 hours, the low degradation rate is mainly ascribed to the adsorption of phenol on TiO_2 nanoparticle and TiO_2 was not excited. Under UV irradiation, the degradation of phenol was 10.6% at a given time, higher than that with dark, it is mainly due to the TiO_2 nanoparticle are excited by a little UV-light, which exists in visible light, and photodegrades the phenol. The degradation of phenol with UV-light and at the pH value 1 were 72.4% and 92.4%. The reason was explained above.

4. Conclusion

Factors of TiO_2 preparation were investigated and the optimal amount of distilled water and absolute ethyl alcohol and calcination temperature were 5mL, 90mL and 575 ℃, respectively. In the photocatalytic activity test, 75mg of TiO_2 nanoparticle was best for photodegradation in 150mL of phenol (100mg/L) and degradation efficiency is favorable in the acidic solution. TiO_2 had a good performing of photodegradation of phenol solution and also had a potential application to the treatment of coal-chemical wastewater.

References

[1] Gupta S and Tripathi M 2011 Chem. Sci. Bull. 56 1639-57
[2] Matsuoka M and Anpo M 2003 J. Photochem. Photobio. 3 225-52
[3] Jiang W J, Zang J, Wang J, Liu Y F and Zhu Y F 2014 Applied Catalysis 160 44-50
[4] Wang Y S, Shen J H and Horng J J 2014 J. Hazardous Materials 274 420-27
[5] Xu L, Zang H and Zhang Q 2013 Chem. Eng. 232 174-82
[6] Alvarez P M, Jaramillo J and Plucinski P K 2010 App. Catal. 100 338-345
[7] Hankare P P, Patil R P, Jadhav A V, Garadkar K M and Sasikala R 2011 App. Catal. 107 333-39
[8] Gu H N, Liu S Y and Tian Y J 2012 Rock and Mineral Analysis 32 312-17
[9] Antonopoulou M, Karagianni P, Giannakas A, Makrigianni V, Mouzourakis E, Deligiannakis Y and Konstantinou I 2016 Catal. Today 280 114-21
[10] Zhang Y, Lu J, Wang X P, Xin Q, Cong Y Q, Wang Y and Li C J 2013 J. Colloid and Interface Sci. 409 104-11
[11] Adak A, Pal A and Bandyopadhyay M 2006 Physicochem. Eng. Asp. 277 63-8
[12] Kashif N and Ouyang F 2009 J. Environ. Sci. 21 527-33
[13] Wong C H, Tan Y N and Mohamed A R 2011 Environ 92 1-2

IOP Conf. Series: Earth and Environmental Science **59** (2017) 012047 doi:10.1088/1755-1315/59/1/012047

Analyzing the causes of urban waterlogging and sponge city technology in China

Yun-Fang Ning[1,2,*], Wen-Yi Dong[1], Lu-Sheng Lin[2], Qian Zhang[2]

[1]Harbin Inst Technol, Shenzhen Grad Sch, Shenzhen 518055, Peoples R China
[2]Bur Publ Works & Longgang Shenzhen Municipal, Shenzhen 518172, Peoples R China

Email: yue.886@163.com

Abstract. With the rapid development of social economy in China, increased urban population, and rapid urbanization cause serious problems, for example, a heavy rain in the city inevitably leads to waterlogging, which poses a great threat to the livelihood and property security. Disaster due to urban flood is a key problem that restricts the development of urban ecology in China. The reason is the sharp increase of impermeable surface ratio in urban areas, leading to a decrease in rainfall infiltration and increase in surface runoff. To effectively solve the urban waterlogging, China proposed the construction of sponge city. This paper analyzes and summarizes the reasons for the formation of urban waterlogging, and introduces the concept of the sponge city technology to prevent waterlogging.

1. Introduction

China experiences several disasters including flooding, storm, and drought due to its geographical location and monsoon climate. Climate change in China has modified the frequency of storm and flood, resulting in an annual summer waterlogging multiple times. Meanwhile, due to a high volume during peak flooding, majority of the flood are unused and infiltrated, leading to increased risk of intermittent and alternate river flooding.*[1]* In 2010, it was found in a survey of 351 cities that urban waterlogging in China has been increasing nationally*[2, 3]*.

It is important to note that building a "gray" infrastructure through a single-objective engineering approach has been widely used to solve complex and systemic water problems until today. However, the problem has not been solved, instead it goes into a vicious circle. Combined with the domestic and international theoretical achievements and practical experience in urban rainwater management, the "sponge city" theory encompassing the natural infiltration, and natural purification has been proposed and developed. *[4, 5]*

2. Causes of urban waterlogging

2.1 Geological and meteorological factors

With global warming, extreme weather events are more frequent than before. Urbanization has led to an increase in precipitation and heavy rainfall events in large cities. Urban heat island effect makes the atmosphere over the city unstable. In certain weather conditions, the development of convection is conducive to the formation of convective clouds and convective precipitation, and heavy rainfall events. Urban waterlogging caused by short duration of heavy rainfalls or long time precipitation of high rainfalls is influenced by direct meteorological factors.*[6, 7]*

2.2 Urbanization factor

2.2.1 Change in underlying surface condition. Urban development and construction led to the expansion of impervious area, reduced surface infiltration and reduced recharged groundwater, increased runoff, and increased peak flow, and increased peak flow in advance. With urbanization, the impervious surface area of the city increases continuously, while the storage facilities such as large paddy fields and fishponds decrease, leading to increased runoff coefficient. Permeability of the ground is conducive to guide the rainfall infiltration, and to form underground runoff, thereby resulting in the reduction of urban rainfall caused by the water accumulation. However, urbanization decreases the natural vegetation, changes the surface material, and hardens the bedding surface, and reduces the amount of surface runoff. The rainfall could not be well drained, leading to an increased runoff and peak discharge.*[8, 9]*

2.2.2 Artificial occupancy of the river section, and reduction in water storage capacity. River network as an important component of regional water cycle is formed by the interaction of natural factors and human activities. Since the reform and opening, the process of urbanization by leaps and bounds, a highly-urbanized region of the river network structure is simple, non-trunk river to reduce. The development and construction has changed the natural geographical environment. Large number of illegal buildings along the part of the river occupy the river flood section, aggravating the river water atrophy, and restricting the river tributaries.

2.3 Urban heat and rain island effect
Urban population, factories, and automobiles consume significant amount of energy. In addition to causing air pollution, they also release waste heat into the atmosphere. This man-made release of heat increases the average annual temperature of the city above 0.5 to 3 °C in comparison to the temperature in the suburbs. In terms of the spatial distribution of temperature, the city becomes similar to a warm island, known as the "heat island effect".

"Heat island effect" exacerbates the convection of air movement, leading to the formation of the easy showers over the city. Moreover, the dust in the city is far more than that in the suburbs. Thus, the atmosphere of the city can condense into the rain condensate nuclei compared to that of the suburbs. Therefore, it is more likely to rain in the city, which is known as the "rain island effect".*[10, 11]*

2.4 Drainage engineering construction lag
According to the information survey, the total length of urban drainage pipes in Germany in 2002 was 446,000 km, with an average density of more than 10 km/km^2. The length of urban drainage pipes in Japan was 350,000 km in 2004, and had the average density of 20 ~ 30 km/km^2, with the maximum of 50 km/km^2. The US urban sewer length in 2002 was about 1.5 million km, while the average density of urban drainage pipe network was more than 15km/km^2. However, the total length of urban drainage pipes in China reached 370,000 km by the end of 2010, and the density of urban drainage pipes was 9.0 km/km^2. The urban per capita drainage pipe was found to be only 0.57 m in 2010, for an urbanization rate of 46.6%. Before the release of the 2014 edition of the "Specification for Design of Outdoor Drainage Systems", the domestic urban waterlogging control could not be in accordance with the situation, suggesting that the city does not provide its due importance to the planning and construction of drainage works.*[12]*

2.5 Low terrain
Urban construction in low-lying areas often become waterlogging areas after rainfall, since the rain water cannot be discharged in time due to the terrain.

2.6 Lag-management
Drainage facilities are unevenly managed. Most urban villages and old villages have problems such as inadequate drainage facilities, lack of pipeline clogging and rain water collection facilities. More than

half of the waterlogging points are caused by this problem. During a rain event, these facilities cannot play its due role in the drainage.

Most communities, and villages in the city do not have the information about the pipe network. The disorder section of the pipeline in these places is widespread, such that some areas do not have adequate rainwater pipe network or collection facilities leading to inadequate drainage network capacity.

In some drainage facilities such as small village drainage pumping stations, and tidal gates, improper management leads to aging equipment and other problems such as rain cannot open the pump, shut the gate used to retain water, resulting in waterlogging.

Development of the city results in increased constructions, leading to difficulties during water law enforcement. There are some problems in the soil and water conservation and construction drainage in some subways, important transport facilities construction and area development and construction. The drainage problems are not given importance during the actual construction. Mud and silt from construction site block drainage culvert, resulting in waterlogging. In particular, stealing emission of mud can have a direct adverse impact on the municipal drainage system.

3. Sponge city technology

"Sponge City" is the integration of a series of specific rainwater management technology, and is the result of extensive practical experience. It covers concepts and technologies such as Low Impact Development (LID) in the United States, Sustainable Urban Drainage Systems (SUDS) in France, and Water-sensitive Urban Design (WSUD) in Australia. *[13]*

LID refers to a rainwater management method based on the principle of simulating natural hydrological conditions, and using source control concept to realize rainwater control and utilization. The main objective is to maintain or restore the previous hydrological characteristics of a development area, after the construction. Such measures include protection of water quality, replenishment of groundwater, reduction of flood peak, conservation of land, reduction of runoff accumulation time, prevention of river pollution, and reduction of pressure on municipal pipe network facilities. SUDS has evolved from a simple "emission" based drainage system to a sustainable and benign water cycle drainage system. It focuses on water quality, landscape value, entertainment and rest value of the urban environment.*[14, 15]*

4. Conclusions

This paper analyzes and summarizes the reasons for the formation of urban waterlogging in detail and introduces the concept of the sponge city technology for the prevention of waterlogging. Conventional urban drainage to prevent waterlogging can only deal with extreme storm waterlogging problems. The project focuses on the construction of large-scale storage tank, additional pumping stations, improving the drainage pipe network standards and other measures, but does not investigate the corresponding conditions conducive to waterlogging. In the construction of the sponge city, the generalized LID and green rainwater infrastructures can reduce the pressure on the traditional pipe network system to a certain extent in the control target, facility type and application scale. Further, it can replace the traditional gray drainage facilities under certain conditions.

References

[1] Q. Wu, X. X. Zheng, H. Xu, Y. F. Ying, Y. S. Hou, X. C. Xie, S. X. Wang, 2003 *SCIENCE IN CHINA SERIES D-EARTH SCIENCES* **46**, 74.
[2] F. Xue, M. Huang, W. Wang, L. Zou, 2016 *ADV METEOROL.*
[3] Y. Chen, H. Zhou, H. Zhang, Du Guoming, J. Chen, 2015 *Engineering Journal of Wuhan University* **48**, 608.
[4] N. Zhang, H. Chen, J. Chen, X. Chen, 2016 *COMPUTATIONAL INTELLIGENCE AND NEUROSCIENCE.*
[5] R. Quan, M. Liu, M. Lu, L. Zhang, J. Wang, S. Xu, 2010 *ENVIRON EARTH SCI* **61**, 1113.
[6] H. Peng, Y. Liu, H. Wang, L. Ma, 2015 *ENVIRON SCI POLLUT R* **22**, 15712.
[7] H. Long, C. Li, 2014 *Journal of Shenzhen University Science & Engineering* **31**, 593.

IOP Conf. Series: Earth and Environmental Science **59** (2017) 012047 doi:10.1088/1755-1315/59/1/012047

[8] Y. Shi, D. Wan, L. Chen, J. Zheng, 2014 *Water Resources and Power* **32**, 57, 12.
[9] X. Yang, X. You, M. Ji, C. Nima, 2013 *WATER SCI TECHNOL* **67**, 869.
[10] L. Xiaolei, W. Enjie, Z. Jie, C. Mingang, F. Zhongqi, 2014 *Proceedings of the 33rd Chinese Control Conference*, 6375.
[11] J. Holden, A. J. Howard, L. J. West, E. Maxfield, I. Panter, J. Oxley, 2009 *J ENVIRON MANAGE* **90**, 3197.
[12] Z. L. Liao, Y. He, F. Huang, S. Wang, H. Z. Li, 2013 *WATER SCI TECHNOL* **68**, 2559.
[13] Q. Peng, B. K. Greenfield, F. Dang, H. Zhong, 2016 *ENVIRON GEOCHEM HLTH* **38**, 169.
[14] X. Wang, 2016 *ADVANCES OF TRANSPORTATION: INFRASTRUCTURE AND MATERIALS, VOL 1*, 606.
[15] J. Zhang, L. Gao, 2015 *Building Energy Efficiency* **43**, 89.

Current research trend on urban sewerage system in China

Yun-Fang Ning[1,2,*], Wen-Yi Dong[1], Lu-Sheng Lin[2], Qian Zhang[2]

[1]Harbin Inst Technol, Shenzhen Grad Sch, Shenzhen 518055, Peoples R China
[2]Bur Publ Works & Longgang Shenzhen Municipal, Shenzhen 518172, Peoples R China
[*]Corresponding Author
Yun-Fang Ning

Email: yue.886@163.com

Abstract. The research emphasis has always been on sewerage treatment technology in China, though urban drainage system has gained little attention. In the context of urban drainage system and the problem associated with rain, the focus is still mainly toward the simple "emissions". While the relationship between conservation and utilization of rainwater resources and urban ecology are popular, the relationship between rainwater discharge and non-point source pollution are often neglected. The reasonable choice of sewerage system is dependent on the collection and discharge of urban sewerage, the applicability and economic benefits, along with the ability to meet the water quality requirements and environmental protection. This paper analyzes and summarizes the development of urban drainage system in china, and introduces different drainage forms. The choice of drainage system should be based on the overall planning of the city, environmental protection requirements, the local natural conditions and water conditions, urban sewerage and water quality, the original drainage facilities, and local climatic conditions. It must be comprehensive to meet the environmental protection requirements, through technical and economic comparison.

1. Introduction

With the rapid economic development and urbanization in China, there is a significant increase in the number of high-rise buildings in cities. However, the construction of underground sewerage system is generally ignored. Due to historical reasons, lack of scientific research, and shortage of funds, many fast-growing cities have largely focused on relatively rapid construction of sewage treatment plants, while the construction of underground sewerage system that collects and channels the sewerage to the treatment plant is slow.*[1]* This would result in many years of inadequate collection of sewage or low concentration of sewage. While there is a serious shortage of funds, and water pollution problems in these cities, heavily invested sewage treatment plants would not be utilized fully to their design capacity. *[2]*

Several old combined sewerage system in these cities have problems such as aging, small sewerage capacity, and rainy season overflow pollution. Thus, many cities are gradually phasing out the old system and designing a complete separate system. Some cities effectively transform and completely convert these old combined sewerage systems into complete separate systems, while other cities retain, modify and utilize these systems to maximize the return on investment. In this context, this paper analyzes and summarizes the development of urban drainage system in China, and discusses various types of drainage systems. Finally, the paper discusses an optimal sewage system for China.*[3]*

2. Urban sewerage system

The urban sewerage system, which is known as the urban conscience, is an engineering facility for the treatment, removal and discharge of municipal sewerage and rainwater. It is an integral part of urban public facilities. The purpose of the general urban drainage system is to timely discharge the rainwater, residential sewerage, industrial wastewater, and flood water from the city through the urban underground drainage system to protect urban residents and to ensure normal life in urban areas. The treatment and reuse of urban rainwater, sewerage, wastewater and flood can also save the water resources and protect the environment.*[4]*

2.1 Development of urban drainage system

2.1.1 First stage - "Creation". The initial stage of sewerage construction in China began in 1950s. The main goal of the construction of urban sewerage was to eliminate the hazards of rain waterlogging. After the founding of the New China, during three years of economic recovery, the government introduced the "work-for-relief" approach, where the residents involved in cleaning up the accumulation of waste and water through the patriotic health campaign. This generally solved the problem of drainage in some areas, and improved the environment of the city.

2.1.2 Second stage - "point source pollution treatment". Urban drainage facilities were constructed rapidly in the 1980s due to the acceleration of urbanization and increased awareness of urban water pollution. Government invested significant amount in laying sewerage pipes and constructing treatment plants to improve the sewerage collection and treatment rate. Furthermore, tail water emissions were strictly controlled in the industrial wastewater and sewerage treatment plant. It is also called "point source pollution" treatment.

"Provisions on technology to prevent water pollution" and other documents stipulate that a few large enterprises or enterprises away from the town can reuse or discharge their sewerage water after reaching the threshold values prescribed by the standards, while the others need to discharge their sewerage water into the urban sewerage pipe network. Enterprises only need to pre-treat the sewerage for heavy metals and toxic biodegradable substances, before discharging into the sewerage pipe network after meeting the city sewerage pipe discharge standards. The rest can directly discharge the sewerage into the city sewerage network, which is a centralized integrated sewage treatment plant, if they meet the city sewerage pipe discharge standards for industrial wastewater.*[5]*

2.1.3 Third stage - "Ecological concept of sponge city". Storm water and sewerage mixed water should to be treated and disposed before overflow to ensure that the water quality is within the control objectives after the overflow. Development of new regulation and treatment facilities such as detention pond, and sedimentation pond at appropriate locations in the existing secondary sewage treatment plants is a key supplementary measure to further reduce the pollution of urban water bodies. It can intercept the first "scouring" of pollutants during the storm at the wastewater treatment plant, reducing the number of mixed effluent discharges, and thereby improving the water quality, and balancing the amount of mixed sewerage entering the sewerage treatment plant. It can also undertake the initial treatment of rainwater. "Sponge City" is the integration of a series of specific rainwater management technology, and is the summary of vast practical experience. It covers concepts and technologies such as Low Impact Development (LID) in the United States, Sustainable Urban Sewerage Systems (SUDS) in France, and Water-sensitive Urban Design (WSUD) in Australia. *[6]*

3. Drainage form

Domestic sewerage, industrial wastewater and rainwater can be collected through one set of pipe network system or two or more, or can be a separate pipe network system to exclude the different form of sewerage system, known as the drainage form.

3.1 Straight emission combined sewerage system
This system discharges the mixed sewerage directly into the nearby water body without any treatment. As the sewerage is directly discharged without treatment, the receiving water bodies are highly polluted.

3.2 Intercepting combined sewerage system
On the basis of the straight emission combined sewerage system, the intercepting pipe is buried, and the lower than the intercept multiples sewage is drained to intercepting pipe before drained into the water body by the overflow well. Collected sewerage is transported to the sewerage treatment plant.*[7]*
The advantage of intercepting combined sewerage system is that the sewerage on sunny day can be collected through the intercepting well, and transported to the sewage treatment plant for the treatment of water pollutants. The drawbacks of the combined drainage system are as follows: (1) deposited sediment at the bottom of the pipe is washed away when the water velocity in the confluence pipe is higher; (2) mixed sewerage exceeding the capacity of the intercepting pipe can be discharged into the water bodies, resulting in their periodic pollution; (3) water quality during sunny and rainy days, and water volume in cutoff pipe significantly vary, resulting in changes in sewerage water quality, and water volume. It makes operation and management of sewerage treatment plant difficult.

3.3 Completely divided sewerage system
Completely divided sewerage system is divided into two separate pipe networks for sewerage and stormwater. The former collects and delivers domestic sewerage and industrial wastewater to the treatment plant, while the latter collects and discharges the stormwater and clean industrial wastewater into the nearest water bodies through a variety of drainage facilities. However, stormwater runoff, especially the first flush runoff, is highly polluted water.*[8]*

3.4 Intercepting divided sewerage system
In recent years, stormwater pollution, especially due to the initial rainwater runoff or first flush runoff, has attracted significant research attention. Therefore, it is necessary to strictly control the initial rainwater runoff. Intercepting divided sewerage system has both sewerage and rainwater pipe system, and it is different from the divided sewerage system, since it discharges the initial rainwater into the sewerage pipeline using special facilities, such as the rainwater interceptor wells. During light rain, stormwater and sewerage are drained together through the initial rainwater interceptor trunk into the sewage treatment plant. During heavy rain events, the stormwater is drained into the water body through the intercepting main pipe. The key feature of divided sewerage system is the initial rainwater intercepting well. It is necessary to ensure that the initial rainwater is drained into the interceptor, and the mid-rain directly into the water body, while the interception of the sewerage cannot spill into the water body. Because only the sewerage and the initial rainwater are accepted, the cut-off section of the interceptor is smaller than that of the intercepting combined sewerage system. The flow and water quality in the interceptor pipe are relatively stable, and the operation and management costs are also reduced in the sewerage treatment plant.

3.5 Choice of drainage system
The choice of drainage system should be based on the overall planning of the city, environmental protection requirements, local natural conditions and water conditions, urban sewerage and water quality, original drainage facilities, and local climatic conditions. It must be comprehensive to meet the environmental protection requirements through technical and economic comparison.

A city usually uses a mixed system of drainage, that is, both the divided sewerage system and the combined system, which are associated with different periods of urban development. In the early period of urban construction, the surrounding water body is good, and the project is restricted by the construction fund. Hence, the combined sewerage system is adopted. After the development of the city, the merging system is gradually transformed into the divided sewerage system.

Drainage network planning and design are based on overall urban planning. Therefore, the basic data on drainage planning and design should be based on the regional planning and design of urban and industrial enterprises. Overall, the divided sewerage system is more flexible than the combined system, and its construction can meet the needs of social development. The divided sewerage system is easier to adjust if there is an increase in the pollution load or changes in environmental requirements. However, it is feasible to use a combined system, where there are large water bodies in the vicinity, small towns with restricted development, or areas where the rainfall is scarce.

4. Conclusions

This paper analyzes and summarizes the development of urban drainage system in china, and introduces different drainage forms. The choice of drainage system should be based on the overall planning of the city, environmental protection requirements, local natural conditions and water conditions, urban sewerage and water quality, original drainage facilities, and local climatic conditions. It must be comprehensive to meet the environmental protection requirements through technical and economic comparison.

References

[1] W. Zhao, T. H. Beach, Y. Rezgui, 2016 *IEEE T SYST MAN CY-S* **46**, 659.
[2] N. Zhang, H. Chen, J. Chen, X. Chen, 2016 *COMPUTATIONAL INTELLIGENCE AND NEUROSCIENCE.*
[3] Y. Chen, H. Zhou, H. Zhang, Du Guoming, J. Chen, 2015 *Engineering Journal of Wuhan University* **48**, 608.
[4] H. Long, C. Li, 2014 *Journal of Shenzhen University Science & Engineering* **31**, 593.
[5] L. Xiaolei, W. Enjie, Z. Jie, C. Mingang, F. Zhongqi, 2014 *Proceedings of the 33rd Chinese Control Conference*, 6375.
[6] Y. Jian-min, G. Xiao-qing, C. Zhe, Z. Hui-qing, 2006 *Proceedings of the SPIE - The International Society for Optical Engineering* **6412**, 64121D.
[7] T. Li, W. Zhang, C. Feng, J. Shen, 2014 *WATER SCI TECHNOL* **69**, 422.
[8] C. C. Sun, V. Puig, G. Cembrano, 2016 *ENG APPL ARTIF INTEL* **49**, 126.

Spatial-temporal variation of grain production after the Grain for Green Project in the Loess Plateau, China

Zhiyuan Xu[1, 2], **Changhe Lu**[1, 2]

[1] Key Laboratory of Land Surface Pattern and Simulation, Institute of Geographical Sciences and Natural Resources Research, Chinese Academy of Sciences, 11A, Datun Road, Chaoyang District, Beijing, 100101, China
[2] University of Chinese Academy of Sciences, Beijing, 100049, China

xuzy.12b@igsnrr.ac.cn

Abstract. The Grain for Green Project (GGP) has been implemented for 17 years, promoted the regional environment significantly, but the impact on local food production is still an issue worth studying. This paper analysed the temporal and spatial variations of grain production in the Loess Plateau of China, using county-level statistics from 2000 to 2012. The results showed that the total grain yield and the unit area grain yield was continuously improved while the grain planting area was declined, which showed that ecological construction does not affect grain production. Besides, the regional differentiation of grain production efficiency in the Loess Plateau was decreasing.

1. Introduction

GGP began in August 1999, aiming at the conversion of cropland to forest, grassland and lake in a planned way, changing the unreasonable land use, such as steep land reclamation and deforested-land reclamation, suppressing the ecosystem degeneration. The soil erosion area of Loess Plateau is one of the key regions to carry out GGP. Since the implementation of the project, a lot of issues had been widely concerned by the society and scholars, mainly including eco-environmental effects [1, 2] , Land use and land cover change [3, 4], impacts on farmers' behaviour and livelihood [5, 6]. The effectiveness of GGP has not been well documented [7] and there was few studies paid attention to the spatial-temporal variation of grain production after the GGP.

The Loess Plateau is located in the area with high population density, fragile ecological environment and poor quality of cultivated land, the implementation of GGP increased the pressure of regional food security (Figure 1). The project ended in 2011 and its impact on local grain production is a problem worth of study.

2. Data and methods

The statistical data (yield, planting area and unit area yield of grain, wheat, corn, soybean and potato) used in this study were obtained from statistical yearbooks and statistical communiques of each city from 2000 to 2012, taking municipal districts, county-level cities and counties as the basic statistical units (hereinafter referred to as the county). In order to ensure the continuity of the data, this paper adopts the 2006 national county-level administrative zoning map as the base map and merged the municipal districts that belong to the same city, finally formed 283 counties.

Based on the county-level data from 2000 to 2012, and spatial analysis module in ArcGIS software, we can draw out the spatial distribution maps and Triangulated Irregular Network (TIN) images of the grain production in the Loess Plateau, wherein the method of drawing the TIN images of unit area grain yield including following 4 steps: 1) points arrangement. Points were set at counties' seat of government and were joined to the data of unit area grain yield; 2) Grid interpolation. Use of the inverse distance weighting (IDW) interpolation method to generate the grid surfaces; 3) Contour maps. Forming the contour maps using raster surface module in ArcGIS software; 4) Create TIN. Using TIN module of GIS 3D Analyst tools to draw the TIN images and adding the hill shades.

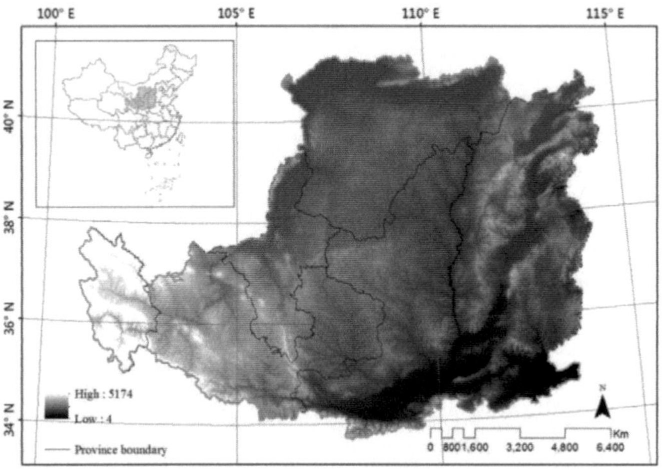

Figure 1. Location and DEM of the Loess Plateau

3. Spatial-temporal change of grain production in the Loess Plateau

3.1 Grain planting area

The fluctuation trend of grain planting area in the Loess Plateau was decreased first and then increased and gradually stabilized in the following year. From 2000 to 2004, with the fast development of GGP, grain planting area decreased sharply, reached the lowest value in 2003. Then with the structural adjustment of the project in 2004, the grain planting area began to increase gradually and back to the basic level in 2008. Since then, regional grain planting condition seemed to be stabilized from 2009-2012.

The standard difference coefficient of variation was 4.61% and the range variation coefficient was 14.15%. We divided the grain planting area in the Loess Plateau into 5 types in according to the above coefficients and analysed the spatial variations (Figure 2). From 2000 to 2004, 35.17% of the counties had a decreasing rate over 14.15%, located chiefly in the loess hilly region. 5.17% of the counties had an increasing rate of grain planting area over 14.15% from 2004 to 2008, while there were 24.83% of the counties have no significant changes in 2008-2012.

3.2 Grain yield

The total grain yield in the Loess Plateau generally increased by 47.83% (1379.53 million tons) from 2000 to 2012 and its proportion of the national total grain yield increased from 6.51% to 7.43%. The annual increasing rate is 3.31%, higher than national average level of 2.17%. According to grain yield, 84 counties' annual growth rate of grain yield is beyond 5%, mostly located in Shaanxi province and Shanxi province. But we can also find that 46 counties' annual growth rate is negative, mainly located in the key areas of GGP. In addition to the above two types, the annual growth rate of most counties is among 1% to 4%. The special distribution showed that regions used to have a low original grain yield led a higher growth rate, such as arid area and hilly and gully region, correspondingly, the Guanzhong

Plain and irrigation areas, which had a higher grain yield at the beginning, leading to a lower growth rate.

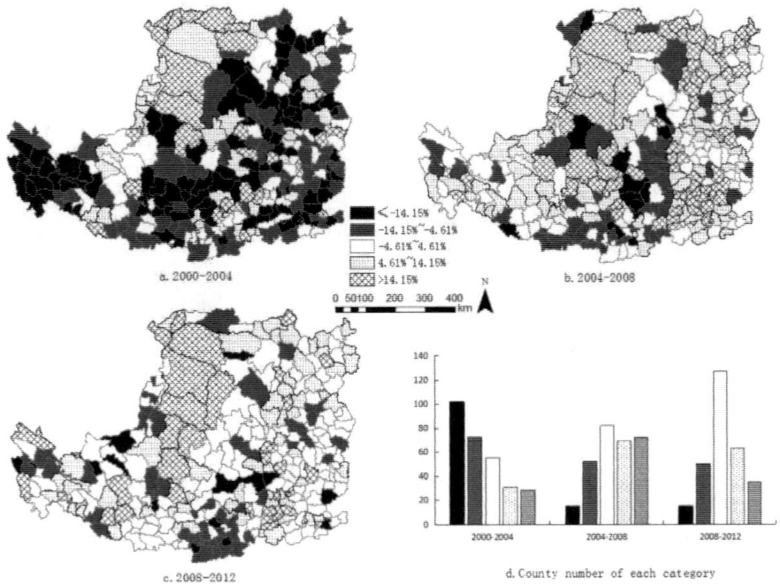

Figure 2. Spatial-temporal change of grain planting area in the Loess Plateau

Figure 3. Spatial-temporal change of the unit area grain yield in the Loess Plateau (kg/ha)

3.3 Unit area grain yield

The unit area grain yield of the Loess Plateau in general is at a low level, but grows rapidly (Figure 3). From 2000 to 2012, the unit area grain yield increased from 2858.6 kg/ha to 4165.50 kg/ha, while at the same period, China's average unit area grain yield increased from 4087.12 kg/ha to 5157.59 kg/ha.

According to the TIN images (Figure 4), Northwest and southeast region tended to have relatively higher levels of unit area grain yield and the middle of the region made a rapid growth ,it is clearly that the gap of unit area grain yield is narrowing, showing that regional food production is becoming more balanced.

The productivity of farmland can be divided into five levels, high level (≥6000 kg/ha), up-middle level (4500-6000kg/ha), middle level (3000-4500 kg/ha), below middle level (2000-3000 kg/ha) and low level (<2000 kg/ha). Seeing from the variation degree, the unit area grain yield is increasing, along with the obvious decrease in low level type. High level productivity farmland increased from 5.65% to 15.55% in 2000-2012. Up-middle level and middle level productivity farmland also increased, it jumped from 39.57% to 63.25% during the period。

4. Conclusions

GGP is an important policy to combat the degraded ecological environment in China, with the primary goal to replace farmland in fragile areas with trees, grass and lake, which covered 25 provinces, municipalities and autonomous regions, Loess Plateau is a typical region implementing the project. This paper took this area as an example to analyse the impact of grain production after GGP and the result showed that grain yield made a substantial increase while the grain planting area fluctuated under the influence of GGP. Due to space limitations, this paper mainly discussed the spatial and temporal variation of grain production after the GGP, an influencing factors analysis should be discussed at a micro scale further.

Acknowledgments

This paper was supported by the National Basic Research Program of China [grant number 2012CB955304]; the Junior Fellowship for CAST Advanced Innovation Think-tank Program (No.DXB-ZKQN-2016-033), funded by China Association for Science and Technology.

References

[1] Wang F, Li R and Wen Z 2002 *Bulletin of Soil and Water Conservation* Survey of Factors Affecting Eco-environmental Benefits of Cropland Conversion- A Case-based Study on Cropland Conversion Experimental Station in Ansai County **03** 1

[2] Yue Y, Yan W, Wang X, Shen Y, Qiu M, Zhou L and LI J 2014 *Journal of Arid Land Resources and Environment* Impacts of the ecological restoration program on regional ecosystem services --A case of Yanchi county, Ningxia Autonomous Region **02** 60

[3] Pan W 2001 *China Geology & Mining Economics* Problem and countermeasure on ecological deterioration and cultivated land protection in the west region of China **11** 23

[4] Ran S, Lu C and Wang Q 2010 *China Population, Resources and Environment* Impacts of Ecological Cropland -conversion on Land Use and Ecosystem Service Functions in Ansai County **03** 111

[5] XU Y, Ma D and Guo T 2006 *Research of Soil and Water Conservation* Effects of Actualized Ecological De - farming Policy and Its Influences on Farmer ' s Livelihood in the Loess Plateau **05** 255

[6] Ye H 2009 *Technology Economics* Study on Objectives Coupling of Grain Subsidy Policy and Policy of Returning Cultivated Land to Forest: Analysis Based on Behaviors of Farmer Households **07** 32

[7] Zhou D, Zhao S and Zhu C 2012 *Ecological Indicators* The Grain for Green Project induced land cover change in the Loess Plateau: A case study with Ansai County, Shanxi Province, China **23** 88

Risk management of PPP project in the preparation stage based on Fault Tree Analysis

Yuanzhi Xing[1], Qiuling Guan[2]

[1] College of Management, Shanghai University, Shanghai 200444, China.
[2] College of Management, Shanghai University, Shanghai 200444, China.

qiuling_guan@126.com

Abstract. The risk management of PPP(Public Private Partnership) project can improve the level of risk control between government departments and private investors, so as to make more beneficial decisions, reduce investment losses and achieve mutual benefit as well. Therefore, this paper takes the PPP project preparation stage venture as the research object to identify and confirm four types of risks. At the same time, fault tree analysis(FTA) is used to evaluate the risk factors that belong to different parts, and quantify the influencing degree of risk impact on the basis of risk identification. In addition, it determines the importance order of risk factors by calculating unit structure importance on PPP project preparation stage. The result shows that accuracy of government decision-making, rationality of private investors funds allocation and instability of market returns are the main factors to generate the shared risk on the project.

1. Introduction

The rational assessment of shared risk about the public-private partnership project (hereinafter referred to as "PPP project") is the basis of risk-response and risk-allocation mechanism[1]. The PPP project has a significant impact on regional economic and social development. At the same time, it is also the primary means of government to guide private investment. Due to the internal and external factors, PPP project involves many risks, which mainly involves risk losses, considerable uncertainties and increased risk events by the main investment part[2]. Therefore, it is necessary to identify the risk factors, improve the effectiveness of risk management, grasp the key points of risk management, and allocate resources rationally to reduce risk losses in the pre-project preparation stage. Besides, fault tree analysis (hereinafter referred to as "FTA") is a tree describing directed logic of accident and a logical deductive analysis tool, which is used to analyze the phenomenon, cause and result about accident, so as to find out the measures to prevent the accidents[3]. Consequently, in this paper, FTA method is used to analyze and assess the risk factors of PPP project in the preparation stage.

2. Literature review

At present, there are two main kinds of risk-sharing methods for PPP projects: one is to confirm the risk-sharing ratio of each project through questionnaire and expert interview statistics. The tendency of risk is reflected by comparing the mean. The other is to establish mathematical models to seek the optimal solution of risk-sharing so as to determine the number of key risk-sharing between government and private sector. Consequently, there are both advantages and disadvantages of two methods. The former is easy to operate, but it can not define the specific commitment number of shared risk in the government sector and private sector. While the latter can determine the amount of shared risk quantitatively. However, the model building process is complicated. What's more, most cases need to discard some conditions, and it is different from the real PPP project. Thus, the conclusions of reliability should be verified[4]. In addition, Thomas et al. proposed a fuzzy fault framework tree and Delphi method on the basis of risk probability and impact assessment, which included the process of key risk modeling and expert judgment system in the general case[5]. Besides, Zhang and Zou developed fuzzy analytic hierarchy process (AHP) model to evaluate the risk environment of joint venture.

However, as an emerging model, it appears to have many problems about PPP mode in China, such as delaying approvals, financing problems, lacking of market returns, project uniqueness, legal change, government credit and other risks[7]. Due to lacking of operational experience in the government and the private sector in PPP project, lots of projects failed to be identified and evaluated effectively[8]. Consequently, it is essential for government sectors and private investors to use scientific tools to assess the risk of PPP project and to provide a scientific basis for decision-making in the pre-preparation phase. Therefore, the purpose of this paper is to identify and assess the risks in PPP projects from the four perspectives of government sectors, private investors, macroeconomics and external environment. Moreover, this paper focus on finding out the main causes that lead to the occurrence of PPP projects so as to reduce the risk probability of different subjects and make risk assessment more closer to reality on the basis of uncertainty conditions.

3. The risk analysis process of PPP project

From a practical point of view, PPP project risk includes the probability of events and the seriousness of consequences. Thus, the analysis process can be divided into risk identification, risk estimation and risk assessment.

3.1. Risk identification

Risk identification, namely, identify all factors that may affect the progress of project to classify statistically and propose risk list. Hence, risk factors can be defined by checklist method and expert survey methods[9]. According to *PPP research report in 2016,* it pointed out that China has implemented more than 8,000 PPP projects, which involve water plants, power plants, bridges, tunnels, roads transportation, hospitals, sewage disposal and other fields[10]. As shown in Table 1, it summarizes the representative PPP projects and major risk factors for failure.

Table 1. PPP project problems and major risk factors for failure

Project / Risk	Government decision-making mistakes	Public object	Government credit	Force majeure	Insufficient market returns	Project uniqueness	Changes in market demand	Charges change
Jiangsu sewage treatment plant			√	√				
Shanghai Dachang Water Plant								
Beijing tenth water plant	√	√						
Hunan power plant			√	√				
Qingdao Veolia Sewage Treatment Project	√		√					
Hangzhou Bay Bridge					√	√	√	
Lianjiang Sino-French Water Supply Plant	√		√					
Fujian's Quanzhou Erythrina Bridge					√	√	√	
Tangxun Lake Sewage Treatment Plant								
Yan'an East Road Tunnel								√
Shenyang ninth water plant			√					

Beijing-Beijing Highway					√	√	√	
Beijing Metro Line 4		√	√		√			
National Stadium (Bird's Nest)		√	√		√			
Chongqing Water Group Project	√	√	√					
Beijing Subway Olympic Line BT Project	√	√						
Nanjing Metro Line 2 BT Project								
Guangzhou Guanghe hospital		√	√		√			
Shantou Chaonan People 's Hospital								
Beijing New Milestone Cancer Hospital		√	√		√			
Guangxi Laibin B Power Plant		√			√	√		
Beijing Yizhuang gas BOT project								
Hubei Shiyan City bus company project	√	√	√		√		√	

Source: PPP Study 2016

From the statistics of the reasons for failure, a half of the PPP project failures are related to the lack of government credit, insufficient market returns, public opposition, market demand changes and project uniqueness. They account for a high proportion, so special attention should be paid to such issues in PPP cooperation projects.

3.2. Risk Estimation

Identified risk factors may determine the number and probability of risk losses and it is necessary to find out the relation among them. the probability and the magnitude of loss to determine the degree of correlation is essential in the risk assessment. There are a lot of methods of risk estimation, such as planning evaluation technology(PERT), analytic hierarchy process(AHP) and fuzzy comprehensive evaluation, etc., which are analyzed the risk of PPP project from possibility of occurrence, severity of impact and controllable degree on the preparation phase[11].

3.3. Risk assessment

According to the estimation results, the appropriate methods are used to evaluate the risk factors of PPP projects and to determine the importance of factors and the level of risk, which can summarize the results of qualitative analysis so as to propose risk management recommendations for government departments and private investors to develop PPP risk management strategy in the project preparatory stage.

4. The construction of fault tree for PPP project

Bell Labs applied the method of Fault Tree Analysis (FTA) to militia missile launch control system and the research made great success in 1961. It is a method for reliability, security analysis and risk evaluation of large complex systems. At present, FTA has been widely used in aerospace, nuclear, electronics, machinery, chemical, mining and other fields[13-15]. On the whole, the FTA method not only can link the relevant factors of system failure analysis, but also can find the system fault spectrum and weak links easily. What's more, the system failure probability and other reliability parameters can be determined quantitatively in order to provide quantitative data to assess and improve the reliability of system. The basic steps of FTA method are shown in Figure1.

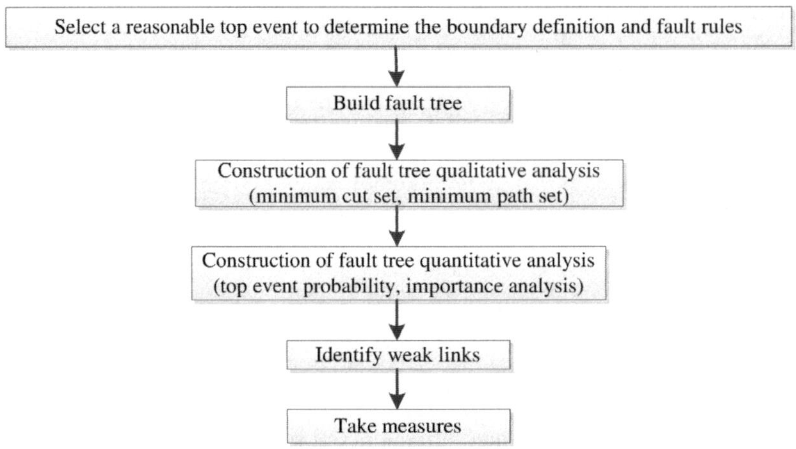

Figure 1. The basic steps of FTA method

5. The case study of PPP project

5.1. Project Overview

Yinxiang River Cultural Tourism Zone is located in the south of Shigu Town, Weibin District, Baoji City, Shaanxi Province. It has a total planned area of 67 square kilometers. Additionally, the planning area about the first centralized control area of Yinxiang River (hereinafter referred to as "Yinxiang river treatment area") is 9 square kilometers, of which construction area is 758 acres. Yinxiang river control area of 9 square kilometers is the scope of project. On the whole, the construction of project is divided into three parts, including infrastructure and public facilities construction (about RMB 1.272 billion), franchise facilities construction (about RMB 116 million), land acquisition and resettlement house construction (about RMB 1.141 billion). In addition, the initial investment is estimated about RMB 2.529 billion and its total construction period is 5 years. Besides, the project is started in 2016 and should be put into use in 2020.

5.2. The identification and analysis of project risk factors

Based on the long-term stability of cooperative relation between government and private investors, taking the risk management ability, project return mechanism and market risk management ability into account, it should be allocated the risk reasonably between government and social investors.

Firstly, due to land acquisition and demolition, the residents of resettlement, and archaeological heritage protection risk in the development process, the government departments is responsible for pre-formalities and administrative approval risks for the construction projects.

Secondly, the private investors are mainly responsible for operation risk and financial risks of the project, such as rising interest rates or financing costs, broken capital chain, construction timing less than expected, insufficient operating income and other risks.

Finally, the risk of force majeure should be shared by the two parties. At the same time, policy and legal risks, which is beyond the scope of government controllable legal changes and policy changes risk, should be shared by the two parties as well.

5.3. The construction of fault tree

There are many risks in PPP project. In the process of risk identification, the risk factors of whole system can be identified and decomposed one by one on the basis of project cycle, so that the network analysis and calculation can be carried out smoothly. And then, the risk factors that can be eliminated by general measures should be removed. According to the basic idea and principle of fault tree rendering, a part of the fault tree is shown in Figure 2. Especially, the fault tree consists of 13 intermediate events, 23 basic events, 8 logical OR gates and 6 logical AND gates in table 2. Based on the above risk estimation results, the risk of Yinxiang River PPP project was analyzed with Free FTA software in the project preparation phase.

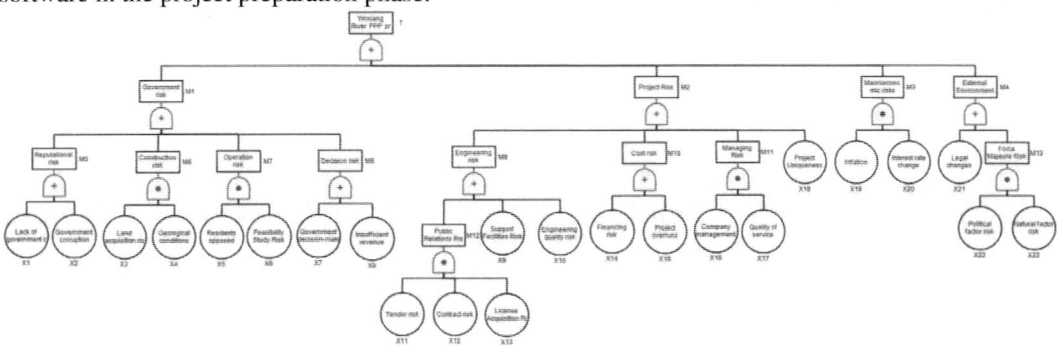

Figure 2. The fault tree analysis of PPP project for Yinxiang River

Table 2. The event table of PPP project for Yinxiang River

Notation	The event name	Notation	The event name
T	Yinxiang River PPP project failure	X_6	Feasibility study risk
M_1	Government risks	X_7	Decision-making mistakes
M_2	Project risks	X_8	Insufficient revenue
M_3	Macroeconomic risks	X_9	Supporting facilities
M_4	External environmental risks	X_{10}	Project quality problem
M_5	Reputational risk	X_{11}	Tender risk
M_6	Construction risk	X_{12}	Contract risk
M_7	Operational risk	X_{13}	License acquisition risk
M_8	Decision risk	X_{14}	Financing risk
M_9	Engineering risk	X_{15}	Project overruns
M_{10}	Financial risk	X_{16}	Company management risk
M_{11}	Managing risk	X_{17}	Quality of service
M_{12}	Public relations risk	X_{18}	Project uniqueness risk
M_{13}	Force majeure risk	X_{19}	Inflation
X_1	Lack of government credit	X_{20}	Interest rate change
X_2	Government corruption	X_{21}	Legal changes
X_3	Land acquisition risk	X_{22}	Political factor risk
X_4	Geological risk	X_{23}	Natural factor risk
X_5	Residents opposed		

5.4. Qualitative analysis of fault tree

5.4.1. Minimum cut sets. A cut set is a subset of the set that satisfies the following conditions in the bottom events of fault tree:

The subset is assumed to be $\{X_{i1}, X_{i2}, X_{i3},\ldots, X_{ik}\}$, i = 1,2,…,K (K is the number of cut sets), and the top event T must occur when all the bottom events contained in the subset are occurred. However, if the cut set that contains the bottom event is arbitrarily removed so that it does not become cut sets, namely, such cut set is a minimum cut set.

According to the fault tree, the minimum cut set can be obtained by Boolean algebra method.

$$
\begin{aligned}
T &= M_1 + M_2 + M_3 + M_4 \\
&= M_5 + M_6 + M_7 + M_8 + M_9 + M_{10} + M_{11} + X_{18} + X_{19} + X_{20} + M_{13} + X_{21} \\
&= X_1 + X_2 + X_3 + X_4 + X_5 + X_6 + X_7 + X_8 + X_9 + X_{10} + X_9 X_{10} M_{12} + X_{14} + X_{15} + X_{16} \\
&\quad + X_{17} + X_{18} + X_{19} + X_{20} + X_{21} + X_{21} M_{13} \\
&= X_1 + X_2 + X_3 + X_4 + X_5 + X_6 + X_7 + X_8 + X_9 + X_{10} + X_9 X_{10} X_{11} + X_9 X_{10} X_{12} \\
&\quad + X_9 X_{10} X_{13} + \ldots + X_{14} + X_{15} + X_{16} + X_{17} + X_{18} + X_{19} + X_{20} + X_{21} + X_{21} X_{22} + X_{21} X_{23}
\end{aligned}
$$

From the above results, we can get the conclusion that the top event is the union of n = 94 intersection, which is the minimum cut set of the fault tree, corresponding to the 23 fault occurrence patterns of the top event.

5.4.2. Minimum path set. The minimum path set of the fault tree is solved by the method to find the minimum cut set of the dual tree. Moreover, the minimum path set is solved by using the properties of dual system. There are three minimum path sets in the fault tree, and there are three paths in which the top event does not occur. Therefore, three kinds of processing schemes can be developed to make ensure that the top event does not occur.

5.4.3. The structure importance of the bottom event. The magnitude of importance about element is called the structural importance of element in the fault tree structure. Beside, the structural importance is independent of the probability about element. The formula is

$$
I_{\phi(i)} = 1 - \prod_{x_i \in k_j} \left[1 - \frac{1}{2^{N_j} - 1} \right] \tag{1}
$$

Where, k_j is j the minimum cut set; $N_j (j \in k_j)$ is the bottom event i that at the bottom event of the minimum cutset k_j; for the bottom event i , $x_i \in k_j$ is the smallest cutset j ;and then , $I_{\phi(i)}$ is the structural importance coefficient for the bottom event i .

The minimum cut set of the fault tree about PPP project failure has been determined, then the structural importance coefficient of the bottom event can be calculated according to Equation. (1). Besides, the importance of the structure is also calculated. And then, it can be sorted according to the size of coefficient about the structure importance. The coefficient is greater, the top event has a greater impact on the bottom event .

The structural importance of calculation results is:

$$
\begin{aligned}
&I(X_{21}) = I(X_{18}) = I(X_{15}) = I(X_{14}) = I(X_{10}) = I(X_9) = I(X_8) = I(X_7) = I(X_2) = I(X_1) > \\
&I(X_{23}) = I(X_{22}) = I(X_{20}) = I(X_{19}) = I(X_{17}) = I(X_{16}) = I(X_6) = I(X_5) = I(X_4) = I(X_3) > \\
&I(X_{13}) = I(X_{12}) = I(X_{11})
\end{aligned}
$$

The fault tree is composed of 23 bottom events. By analyzing the importance degree of the bottom event structure, we can find the bottom events clearly, which are more important in the system. Additionally, we can conclude that the structural importance values of X_1, X_2, X_7, X_8, X_9, X_{10}, X_{14}, X_{15}, X_{18} and X_{21} are equal by analyzing its distribution results. Thus, these bottom events have the greatest impact on the top event. The structural importance of X_3, X_4, X_5, X_6, X_{16}, X_{17}, X_{19}, X_{20}, X_{22}, X_{23} is greater, so these events have a greater impact on the top event. Therefore, it's necessary to strengthen the early warning and management of risk in M_1, M_2 and M_4 so as to avoid the maximum risk in the preparatory stage of PPP project.

5.5. Quantitative analysis of fault tree

5.5.1. The probability of the bottom event. Based on the collected cases to estimate the occurrence probability of the bottom events, the risks of PPP projects are decomposed into the event levels, and then the probability of the bottom events is calculated[16]. According to the calculation results, it can be concluded that the occurrence probability of the largest is $X_8 = 0.1600$, $X_3 = X_5 = X_9 = 0.0500$, and the smallest is $X_1 = X_2 = X_4 = X_{21} = X_{22} = X_{23} = 0.0010$ in the 23 base events.

5.5.2 The probability of the top event. The occurrence probability of fault tree, whose the top event is P (T) = 0.2378, is calculated by using the "down method". According to the calculation of intermediate event and top event probability, we can get the conclusion that the first level intermediate event, which has the greatest impact on the top event, is M_1 (the sum of occurrence probability about "government risk" is 0.2800). In addition, the second-floor event with the greatest impact on M_1 is X_2 ("government decision error" occurs with a probability of 0.0314). Therefore, it is necessary to be controlled to "OR" connection the events with high occurrence probability. Furthermore, reducing the occurrence probability of the top event is possible, which can be controlled the bottom event about X_3 (lack of government credit), X_6 (project non-uniqueness), X_{10} (insufficient market return) and X_{11} (market demand change).

5.5.3 The probabilistic importance of the bottom event. Based on the probability importance analysis of the bottom events, we can see that the probability of X_8 (income insufficiency) and X_9 (supporting facility risk) is the largest. When the above 23 bottom events occur, they have the greatest impact on the top event. At the same time, these bottom events are connected with the top event by "OR gate". Consequently, the change of probability about the bottom event has great influence on the top event. Therefore, the risk warning should be taken into account seriously to ensure that the occurrence probability of the top event is the least during the project implementation. However, the bottom event for "AND" connection is exactly opposite. The higher the probability of the bottom event, the lower the importance of probability. Therefore, it should focus on the occurrence probability of the smallest "AND" connection bottom event so as to avoid the top event effectively.

As the occurrence probability of the top event about Yinxiang River PPP project is large, the probability of adjustment event is P (T) = 0.1347. Table 3 shows the probability of improved bottom event and its importance ranking.

Table 3. The Ranking of Probabilistic Importance for the bottom Event

Bottom event notation	Probability importance	Bottom event notation	Probability importance
X_8	0.9615000	X_4	0.0087000
X_9	0.8741000	X_{19}	0.0026000
X_{14}	0.8723000	X_{20}	0.0024000
X_{15}	0.8723000	X_{16}	0.0009000
X_{10}	0.8697000	X_{17}	0.0009000
X_{18}	0.8697000	X_{22}	0.0009000
X_7	0.8662000	X_{23}	0.0009000
X_{21}	0.8662000	X_3	0.0004000
X_1	0.8658000	X_{11}	0.0000009
X_2	0.8657000	X_{12}	0.0000009
X_5	0.0104000	X_{13}	0.0000009
X_6	0.0087000		

5.5.4. The ratio of income to investment. To reduce the occurrence probability of risk events, namely, is to reduce the incidence of the top events in the Yinxiang River PPP project, then the relative reduction in the probability of the top event is income. Hence, the ratio of income to investment is:

$$r = \frac{|\Delta Q / Q|}{investment} \quad (2)$$

In the equation, Q is probability of the top event, which is the difference about occurrence probability between the improved top event and the original top event [15].

According to the analysis, it can be seen that the probability importance and structure importance of X_8 (income insufficiency) is large. After the fault tree risk controlling, the probability decreases from 0.1600 to 0.1200. In brief, the probability of the improved top event is 0.1347, and the probability of the initial event is 0.2378. Besides, the initial estimate of investment is about 2.529 RMB billion, which is calculated by equation (2):

$$r = \frac{(0.2378 - 0.1347)/0.1347}{25.29} = 0.0172$$

That is, compared to the original probability of X_8, the risk is reduced.

In short, on the basis of the existing cases, the mean value method was used to analyze the two important degrees of the bottom events comprehensively. To sort the importance of event, the conclusions can be drawn that the bottom events that have the greatest impact on the top events are X_8, X_9, X_{10}, X_{14}, X_{15}, X_{18}, which have higher probability and structure importance, that is, insufficient government revenue, supporting facilities, engineering quality problems and project unique risks have a great impact on Yinxiang River PPP project. Therefore, the necessary measures should be taken to reduce the probability of occurrence.

6. Control measures for risk management of PPP project

The basic principle of risk control measures about Yinxiang River PPP project is taken as the project risk contingency plan, which can reduce the adverse impact of the risk effectively on the project, so as to decrease the basic risk events fundamentally. Table 4 shows that probability of the bottom event and risk control measures for risk reduction of Yinxiang River PPP project.

Table 4. The risk-controlling measures of PPP Project for Yinxiang River

Bottom event notation	Risk control measures
X_1、X_2、X_4	Establish supervisal mechanism, sign the government default agreement, and strengthen geological survey in the early stage of project
X_3、X_5、X_9	Soothe emotions and give a certain amount of subsidy policy, and do a good job overall scheduling equipment
X_6	Increase expert assessment efforts
X_8	Increase the fee items
X_{10}	Control the quality of critical project review strictly
X_7、X_{11}、X_{12}、X_{13}	Grasp the macroeconomic factors, study the government documents in detail and maintain good public relations
X_{14}、X_{15}	Plan the project funds rationally, and control the expenditure strictly
X_{16}、X_{17}	Strengthen the management level of private investors and establish detailed rules and regulations
X_{18}	Sign liquidated damages agreement
X_{19}、X_{20}	Reserve profit margins

7. Conclusion

In this paper, the risk of PPP project is taken as the research object, and then the risk identification is carried out though building the fault tree in the preparation stage. Moreover, the four types of risk sharing are also confirmed. At the same time, the risk factors of different subjects are evaluated on the basis of risk identification, and the degree of risk influence is quantified to determine the importance order of risk factors, which have an influence on PPP project in the preparation stage. These results point out that the accuracy of government decision-making, the rationality of private investors' fund allocation and the instability of market return are the main factors that generate the shared risk about project. Secondly, the risk of project incompletion, insufficient facilities, financing and projects non-uniqueness, to some extent, will hinder the smooth implementation of PPP projects and there is a great deal of uncertainty. Finally, external environmental factors, such as legal change and force majeure risk, are less likely to occur in the preparation phase of PPP project.

References

[1] Wibowoa, Permana A, Kochendorfer B, et al.2012. Modeling contingent liabilities arising from government guarantees in indonesian BOT/PPP toll roads.*Journal of Construction Engineering and Management*, vol 138, no 12, pp 1403-1410.

[2] Xiaohua Jin and Guomin Zhang.2011. Modeling optimal risk allocation in PPP projects using artificial neural networks. *International Journal of Project Management*, vol 29, no 5, pp 591-603.

[3] Baidu Encyclopedia [EB/OL]. URL: www.baike.baidu.com.

[4] Liwei Huo 2010 The Risk Analysis and Countermeasure Research of PPP Project.Chongqing: Chongqing University, vol 4.

[5] Thomas A V, Kalidindi S N and Ganesh L S.2006. Modeling and assessment of critical risks in BOT road projects.*Construction Management and Economics*, vol 24,no 4, pp 407–424.

[6] Zhang, G M and Zou.2007. Fuzzy analytical hierarchy process risk assessment approach for joint venture construction projects in China. *Journal of Construction Engineering and Management*, vol 133, no 10, pp 771–779.

[7] Dongjun Wan, Yaowu Wang and Bin Yao.2006. Infrastructure PPP Financing Mode and Its Application in Small Towns. *Journal of China Civil Engineering*, vol 39, no 6.

[8] Xia Ji, Yongjian Ke and Shouqing Wang.2009. Analysis on the Main Risk Factors of China 's PPP Project Based on Cases. *Journal of China Soft Science*, vol 5.

[9] Wei Wu, Xiaobin Hui and Jianming Shen.2003. Research on schedule risk of major model project based on Monte Carlo simulation. *Journal of Air Force Engineering University(Natural Science Edition)*, no 5, pp 7-9.

[10] PPP Research Report 2016 (full version). 2016.

[11] Liwen Zhang, Yunbo Zhang, et al.2013. Application of FTA in Project Schedule Risk Analysis.*Journal of Wuhan University of Technology(Information & Management Engineering)*, vol 35, no 1, pp 94-97.

[12] Jiafeng Li.2011. Research on PSC Evaluation System Based on China 's National Conditions. Beijing: Tsinghua University, vol 5.

[13] Mingxuan Zhang, Yuejiao Zhu and Yujie Zhai.2008. Fault Tree Analysis Method of Fall Accidents in Building. *Journal of Coal Engineering*, no 2, pp 112-114.

[14] Junjie Zheng, Chifeng Lin and Dongan Zhao.2011. Cost Risk Assessment of Shield Tunneling Based on Fuzzy Fault Tree, *Chinese Journal of Geotechnical Engineering*, no 4, pp 501-508.

[15] Chunning Yan.2007. Risk Management. Shanghai: Shanghai University Press, vol 3.

[16] Hong Yan and Jing Xue.2010. Risk analysis and control of foreign BOT hydropower project by fault tree technique, *Journal of Northwest Hydropower*, pp 95-102.

IOP Conf. Series: Earth and Environmental Science **59** (2017) 012051 doi:10.1088/1755-1315/59/1/012051

China's fast growing CO_2 emissions driven by increasing consumption in 1992-2012: A structural decomposition analysis

Junping JI [1,2], **Guangxiao HU** [1], **Yiwei WU** [1,2], **Xiaoming MA** [1,2,a]

[1]Key Laboratory for Urban Habitat Environmental Science and Technology, School of Environment and Energy, Peking University Shenzhen Graduate School, Shenzhen, Guangdong, China

[2]College of Environmental Sciences and Engineering, Peking University, Beijing, China

[a]Corresponding author: xmma@pku.edu.cn

Abstract. China is the largest carbon emitter worldwide and subsequently the greatest contributor to increments in global carbon emissions. Uncovering the driving forces behind China's rapid CO_2 emission growth can provide a theoretical basis for formulating scientific emission reduction policies, which are of great significance for mitigating global climate change. Based on an updated noncompetitive input-output model and structural decomposition analysis, the driving factors and their mechanisms of action for the rapid growth of China's carbon emissions from 1992 to 2012 were explored in this study, with greater attention to the period 2007-2012. The main conclusions are as follows. (1) China's carbon emissions increased by 7424.2 $MtCO_2$ from 1992 to 2012. Particularly rapid growth of carbon emissions occurred during the periods 2002-2007 and 2007-2012, which respectively accounted for 43.33 % and 36.94% of the accumulated emission increments. (2) From 2007 to 2012, the growth of China's carbon emissions was almost entirely driven by the expansion of the final demand scale, whereas the decline in carbon intensity dominantly mitigated carbon emissions. The effects of input-output structure and final demand structure also contributed to abating carbon emissions, with the relatively small contributions.

1. Introduction

The Intergovernmental Panel on Climate Change (IPCC) declared that the rapid growth of anthropogenic greenhouse gas (GHG) emissions was quite likely the main cause of the current change in climate [1]. Reducing anthropogenic GHG emissions, therefore, is crucial for mitigating climate change. China is one of the most important carbon emitters worldwide. During the period 1992-2012, the share of China's carbon emissions in total global emissions increased from 12.0% to 27.5% [2], making China the world's largest carbon emitter. Reduction in China's carbon emissions is vital for mitigating global climate change.

To fight actively against climate change, the Chinese government has proposed several carbon emission reduction targets. In 2009, China announced its plan to cut CO_2 emissions per unit of GDP (carbon intensity) by 40%-45% by 2020 compared to the 2005 level. In 2014, China introduced a plan to reach its national carbon emission peak around 2030 in the "Sino-US Joint Statement on Climate Change." In 2015, China further committed to cut its carbon intensity by 60%-65% compared to the 2005 level in the "Sino-US Heads Joint Statement on Climate Change." In this context, it is of critical importance to uncover the main driving forces of China's carbon emission growth and accordingly formulate a targeted carbon emission reduction program.

With China's steadfast determination to mitigate carbon emissions, a growing number of researchers have applied structural decomposition analysis (SDA) to explore the driving factors of China's energy consumption and carbon emissions in recent years. According to research focus, the current literature generally can be classified into three categories. The first category encompasses the driving factors of China's national energy intensity or carbon intensity changes [3-7]. The second category focuses on the influential factors driving up energy consumption and carbon emissions of China's industrial sectors [8-17]. The third category of research varies slightly in that it regards the rapid growth of carbon emissions from residents' consumption and their driving factors in China [18-19]. However, it should be noted that the current literature is subject to certain defects, including: (1) most of the studies used the default emission factors published by the IPCC to conduct carbon emission accounting, which might lead to significant overestimation of China's carbon emissions [20]; (2) the overall impacts of driving factors were carefully examined whereas their impacts at the sectoral level rarely received attention; and (3) most of the studies depended on the Input-Output Tables of China 2007, the time lag of which cannot reflect the latest dynamics of the Chinese economic and production systems.

In this context, SDA was applied in this study to thoroughly and comprehensively investigate the drivers of and their impacts on China's carbon emissions growth. Firstly, to estimate carbon emissions more accurately, the latest carbon emission coefficients of fossil fuels at the subsectoral level of China and the revised data of energy consumption were used. Secondly, SDA secondary decomposition was introduced to reveal further the sectoral-level impacts of related driving factors on carbon emissions. Finally, based on the newly released Input-Output Tables of China 2012, the driving forces of and their mechanisms of action on China's carbon emission growth from 2007 to 2012 were explored.

2. Method and data

2.1. Environmentally extended input-output model

The environmentally extended input-output model (EEIOM) was first proposed by Leontief [21]. By incorporating the coefficients of pollutant emissions into the traditional input-output model, the EEIOM can capture the impacts of final demand and input-output structure on regional or national pollutant emissions. The structure of the EEIOM used in this study is:

$$c = \boldsymbol{Fx} = \boldsymbol{F}(\boldsymbol{I} - \boldsymbol{A})^{-1}\boldsymbol{y}, \quad (1)$$

where c refers to the amount of carbon emissions from all economic sectors; and $\boldsymbol{F}$ is a $(1 \times n)$ row vector of sectoral carbon intensity; $\boldsymbol{x}$ is a $(n \times 1)$ column vector of sectoral gross output; and $\boldsymbol{y}$ is a $(n \times 1)$ column vector of sectoral final demand; in all of which n denotes the number of economic sectors in China. Both $\boldsymbol{I}$ and $\boldsymbol{A}$ are $(n \times n)$ matrices; the former represents a unit matrix whereas the latter indicates a matrix of total production coefficients. $(\boldsymbol{I} - \boldsymbol{A})^{-1}$ is the total Leontief inverse matrix, the elements of which represent the complete consumption of the other sectors' products caused by increasing unit final demand of one specific sector.

2.2. Structural decomposition analysis

SDA has the potential to detect both direct and indirect impacts of related driving factors on the changes in the investigated indicators [22]. This method has two kinds of decomposition form. Additive decomposition is usually applied to analyze absolute indicators; the change in an absolute indicator is expressed as the sum of the effects of related factors. Multiplicative decomposition is widely used on relative indicators; the change in a relative indicator is denoted as the product of the effects of related factors [3]. To uncover the driving forces of China's carbon emissions, the form of additive decomposition was adopted in this study. We let $\boldsymbol{L}$ represent $(\boldsymbol{I} - \boldsymbol{A})^{-1}$ in equation (1) and expressed $\boldsymbol{y}$ as the product of the final demand structure (y_s) and final demand volume (y_v), thus the change in carbon emissions (Δc) within a specific interval was expressed as [12-17]:

$$\Delta c = \Delta \boldsymbol{F}\boldsymbol{L}\boldsymbol{y}_s y_v + \boldsymbol{F}\Delta \boldsymbol{L}\boldsymbol{y}_s y_v + \boldsymbol{F}\boldsymbol{L}\Delta \boldsymbol{y}_s y_v + \boldsymbol{F}\boldsymbol{L}\boldsymbol{y}_s \Delta y_v \, , \tag{2}$$

where $\boldsymbol{y}_s$ is the shares of sectoral final demand in the national aggregate final demand; the first term on the right side of the equation indicates carbon emission change caused by aggregated changes in the emission intensities; the second term represents the changes in the input-output structure; the third term denotes the changes in the final demand structure; and the fourth term indicates the changes in the final demand volume. In general, there are $n!$ forms of equivalent decompositions with n independent variables in the model [23]. In this study, all $n!$ forms were averaged using the method proposed by Li [24].

2.3. Data

We calculated CO_2 emissions from energy activities and industrial processes as Peters et al. [25] did. The data of energy consumption and industrial products produced are derived from various statistical yearbooks, including the *China Energy Statistical Yearbook* and the *China Industrial Economy Statistical Yearbook*. As the National Bureau of Statistics of China has revised China's energy data for 1996 to 2012 [26-27], the latest revision of data was used.

We followed the procedures in Liu and Peng [28] to construct China's input-output tables with 35 sectors at 2000 prices in 1992, 1997, 2002, 2007, and 2012. Many studies assume the emission intensities of imported products are the same as those of domestic products, which may lead to overestimation of the carbon emissions induced by final demand [29]. To avoid this problem, we converted the tables to noncompetitive tables using the method in Weber et al. [30], which subtracts imported products from intermediate demand and final demand proportionally. The 35 sectors of input-output tables are shown in Table 1.

Table 1. Sector classification used in the study

No.	Sector	No.	Sector
1	Agriculture	18	Transportation machinery and equipment manufacturing
2	Coal mining and dressing	19	Electric machinery and equipment manufacturing
3	Petroleum and natural gas extraction	20	Electronic and communication equipment manufacturing
4	Metals mining and dressing	21	Instruments, meters, and other measuring equipment manufacturing
5	Nonmetal minerals mining and dressing	22	Other manufacturing products
6	Other ores mining and support activities for mining	23	Utilization of wastes
7	Food, beverage, and tobacco processing	24	Repair service for metal products, machinery and equipment
8	Textile industry	25	Electricity and steam production and supply
9	Garments, leather, down, and other fiber products	26	Gas production and supply
10	Timber processing and furniture manufacturing	27	Water production and supply
11	Papermaking, printing, and cultural, educational articles	28	Construction

No.	Sector	No.	Sector
12	Petroleum processing, coking, and nuclear fuel processing	29	Transportation and warehousing
13	Chemical industry	30	Post and Telecommunications
14	Nonmetal mineral products	31	Wholesale and retail trade
15	Smelting and pressing of metals	32	Hotel and restaurants
16	Metal products	33	Finance and insurance
17	Industrial machinery and equipment manufacturing	34	Real estate
		35	Other services

3. Results and discussion

3.1. Primary decomposition of factor contributions to national carbon emissions

As seen in figure 1, the factor of carbon intensity made the largest contribution to mitigating the growth of China's carbon emissions, whereas the factor of final demand dominantly drove emissions up from 1992 to 2012. However, the contributions of driving factors exhibited different characteristics across the periods.

From 1992 to 2002, the amount of China's carbon emissions increased by 1465.1 $MtCO_2$, accounting for 19.7% of the 1992-2012 accumulated emission increments. The growth of China's carbon emissions was a race between booming consumption and improving efficiency within this period [17], in which the positive impacts of final demand were even stronger than the negative impacts of carbon intensity. During the period 2002-2007, the growth of China's carbon emissions accelerated with an absolute increment of 3217.0 $MtCO_2$. The input-output structure was another key factor driving up China's carbon emissions [12]. Emission increments caused by the input-output structure factor were almost 75% of those caused by the final demand factor, which implied that China had experienced extensive development, as featured by the heavy energy consumption and carbon emissions during the period. From 2007 to 2012, China's carbon emissions increased by 2742.1 $MtCO_2$, which were almost entirely contributed by the expansion of final demand. All of the carbon intensity, input-output structure, and final demand structure factors contributed to offsetting carbon emissions, the aggregate effects of which reduced emissions by 817.9 $MtCO_2$. The factor of carbon intensity still had the greatest effects on mitigating carbon emissions, but its contributions were just 22.0% of those during the period 2002-2007.

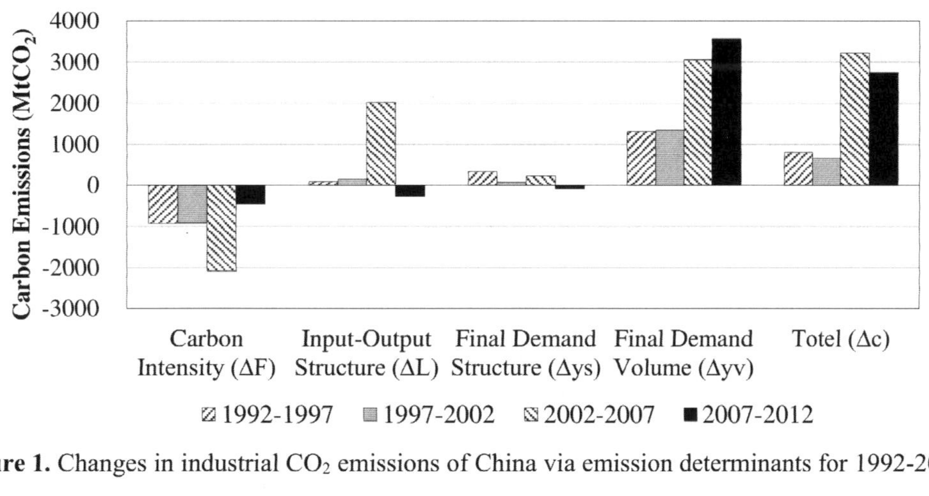

Figure 1. Changes in industrial CO_2 emissions of China via emission determinants for 1992-2012

3.2. Carbon intensity effects

Although still the largest contributor to slowing down the growth of carbon emissions, the effect of carbon intensity factor significantly declined during the 2007-2012 period compared with that during the three other periods. As seen in figure 2, carbon intensities declined in the majority of sectors and hence contributed to offsetting national carbon emissions. The sectors of nonmetal mineral products (sector 14); chemical industry (sector 13); transportation and warehousing (sector 29); as well as food, beverage, and tobacco processing (sector 7) were the four biggest contributors to the effects of carbon intensity on offsetting emissions, with individual contributions of 65.42%, 37.82%, 17.77%, and 11.87%, respectively, to the total carbon intensity effects, resulting in an integrated contribution of 132.87%.

However, it should be noted that carbon intensity of the electricity and steam production and supply sector (sector 25) showed an upward trend in the same period, which was the main reason for the weakened aggregate-inhibiting effects of the carbon intensity factor. More specifically, the carbon intensity of this sector increased by 0.66 $tCO_2/10^4$RMB, which caused the 425.5 $MtCO_2$ of emission increments and negatively contributed (-92.59%) to the total carbon intensity effects. The main reason for this was that China was at a critical period of rapid urbanization and industrialization during 2007 to 2012, with booming demand for electricity. The "rich coal, lack of oil, and less gas" energy resource endowment of China determined that the increased supply of electricity were mainly generated from coal, which eventually led to the rise of carbon intensity in the electric and heat power sector. Judging from this, optimization of the electricity structure and enhancement of the low-carbon power generation mode would have great potential for reducing carbon intensity and carbon emissions in this sector.

The dynamics of China's carbon intensity can be further analyzed by the changes in energy intensity and energy structure. From 2007 to 2012, although China's energy consumption was still coal based, the proportion of coal consumption in the total energy consumption decreased to 68.5% in 2012. In addition, for the Eleventh Five-Year Plan period, the Chinese government invested heavily in clean energy, renewable energy, and other non-fossil energy technologies, causing rapid development of clean energy (including hydropower installed capacity with an average annual growth rate of 12.9% to the Eleventh Five Planning End), which reached 220 million kilowatts, ranking first in the world. Meanwhile, the 2010 nuclear power installed capacity reached 10.82 million kilowatts, and wind power was growing annually by an average of 89.8%. The reduction in China's carbon emission intensity was mainly due to an increase in the proportion of clean energy use and decreased energy consumption per unit GDP.

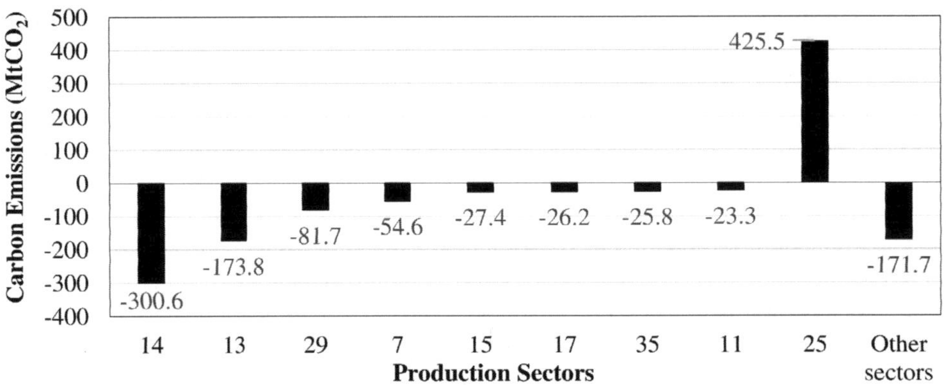

Figure 2. CO_2 emission changes caused by sectoral emission intensity changes for 2007-2012

3.3. Final demand volume effects

As seen in figure 3, fixed capital formation leads to the largest increment of total carbon emissions, followed by exports and urban residents' consumption. These accounted for approximately 90% of the emissions from the final demand volume effects. The construction sector (sector 28) accounted for about 25% of the emission increase. This conclusion is consistent with those of Peters et al. [17] and Minx et al. [12]. The main reason for this is the vigorous promotion of China's urbanization from 1992 to 2012, with construction products existing in the form of fixed capital formation. The construction sector required large amounts of carbon-intensive products, such as electric power, steel, and cement, thus indirectly stimulating the rapid growth of carbon emissions.

The emission increment induced by exports accounted for 26.14% of the total effect of final demand volume change in the period 2007-2012. Export product distribution is more dispersed, wherein the electronic and communication equipment manufacturing industry (sector 20) had the largest share of 2.37%, followed by the industrial machinery and equipment manufacturing industry (sector 17), accounting for 1.92%. China, as the "world's factory", produced abundant carbon-intensive goods for the world, such as electronics, communication equipment, steels and others. These products greatly increased China's carbon emissions.

The increase in carbon emissions stimulated by urban residents' consumption accounted for 16.94% of the effect of final demand volume change for 2007-2012. This is mainly because the State Council vigorously promote urbanization, including encouraging the rural labor to work in cities and towns. The Eleventh Five-Year Plan included a proposal for the urbanization rate increase to 47% and a target to increase the urban labor force to 45 million. This initiative promoted an increase in urban population and an increase in urban incomes. Lifestyles became more energy consumptive with subsequent increased carbon emissions, and therefore the changes in consumption behavior of urban residents sped up the rate of carbon emissions. The increase in urban consumption was mainly in sectors closely related to daily living, such as other services (sector 35); food, beverage and tobacco processing (sector 7); and real estate (sector 34).

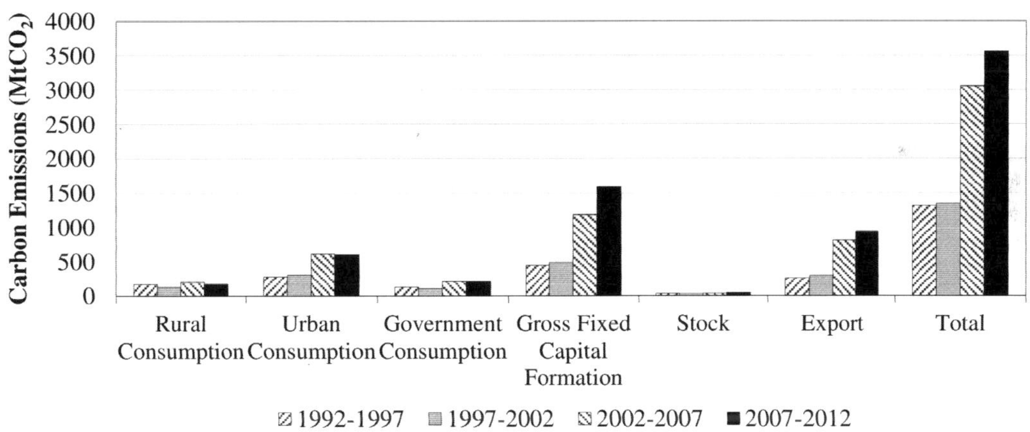

Figure 3. CO_2 emission changes caused by final demand volume change for 1992-2012

3.4. Input-output structure effects and final demand structure effects

The factor of input-output structure represents the generalized technological advances, including science and technology, management system, and industrial structure change. The effect of the input-output structure factor on carbon emission growth was positive from 1992 to 2012, but negative (-271.8 $MtCO_2$) during the subperiod 2007-2012. This implies that the economic development of China changed from "extensive" mode to "intensive" mode, which placed more emphasis on low carbon production. From 2007 to 2012, the nonmetal mineral products industry (sector 14) caused the most carbon emissions (59.8 $MtCO_2$), followed by the smelting and pressing of metals industry (sector 15, 52.1 $MtCO_2$); the transportation and warehousing industry (sector 29, 48.3 $MtCO_2$); and agriculture sector (sector 1, 39.6 $MtCO_2$). These sectors are mainly in the primary and secondary industries, which need to invest a lot of energy and resources for production.

The effects of the final demand factor have shifted from driving up emissions during the period 1992-2007 to offsetting emissions for 2007-2012. It was demonstrated that the structure of final demand in China has been gradually turning from high-carbon products to low-carbon products in recent years. The effects in various sectors of final demand structure on carbon emissions are quite different from 2007 to 2012. Among them, the smelting and pressing of metals industry (sector 15) had the largest reduction (-190.7 $MtCO_2$), accounting for 220.43% of the total final demand structure factor reductions, followed by the sectors of electricity and steam production and supply (sector 25, -145.0 $MtCO_2$) and textiles (sector 8, -79.62 $MtCO_2$). These sectors accounted for the gradual decreasing proportion of final demand, reduced carbon emissions. Meanwhile, sectors causing increased carbon emissions were mainly construction (sector 28, 298.2 $MtCO_2$), transportation machinery and equipment manufacturing (sector 18, 147.7 $MtCO_2$), and industrial machinery and equipment manufacturing (sector 17, 101.0 $MtCO_2$). These sectors made increasing contributions to the final demand effects and hence led to the growth of carbon emissions.

4. Conclusions

The carbon emissions of China increased by 7424.2 $MtCO_2$ during the period 1992-2012, to which carbon intensity and final demand volume were the largest negative and positive contributors, respectively. The rapid growth of carbon emissions was witnessed during the subperiods 2002-2007 and 2007-2012, which accounted for 43.33% and 36.94%, respectively, of the 1992-2012 emission increment. More importantly, it should be noted that the growth of China's carbon emissions was almost entirely driven by the expansion of the final demand scale for 2007-2012, with an absolute contribution of 3560.0 $MtCO_2$. All of the carbon intensity, input-output structure, and final demand structure factors contributed to mitigating carbon emissions, but their combined inhibitory effects were just 22.97% of the simulative effects of final demand volume. Based on these facts, final demand volume is proposed as the key factor in abating China's carbon emissions at the current stage.

The factor of carbon intensity played the dominant role in mitigating the growth of China's carbon emissions during the period 1992-2012. The inhibitory effects of carbon intensity factor within 2007-2012 were only half of those within 1992-2002 and 22.04% of those within 2002-2007, which presented a significant downward trend across the subperiods. The growth of carbon intensity of the electricity and steam production and supply sector was the main reason for the weakened 2007-2012 inhibitory effects of the carbon intensity factor. In the same period, the sectors of nonmetal mineral products; chemical industry; transportation and warehousing; as well as food, beverage, and tobacco processing ranked as the four biggest sectoral contributors to the inhibitory effects of carbon intensity, with a combined contribution of 132.87% to the total carbon intensity effects.

The factor of final demand volume played an increasingly important role in driving up carbon emissions in China. The emission increments for 2007-2012 caused by the final demand volume factor were 16.56% larger than they were for 2002-2007, and 264% larger than they were for 1997-2002. Gross fixed capital formation, exports, and residents' consumption dominantly affected and thus drove up emissions, the combined effects of which accounted for 87.68% of the total effects of final demand volume. Furthermore, the emission increments driven by gross fixed capital formation mainly stemmed from the construction sector, whereas that driven by exports were mainly caused from the communications equipment, computers and other electronic equipment manufacturing sector.

References

[1] IPCC 2014 *Climate change 2014: Synthesis report. Contribution of working groups I, II and III to the fifth assessment report of the intergovernmental panel on climate change* (Geneva, Switzerland: Intergovernmental Panel on Climate Change)

[2] WRI 2015 *CAIT climate data explorer* (Washington, DC: World Resources Institute) http://cait.wri.org accessed on 2016-10-20

[3] Su B, Ang B W 2015 Multiplicative decomposition of aggregate carbon intensity change using input-output analysis *Appl. Energy* **154** 13-20

[4] Zeng L, Xu M, Liang S, Zeng S, Zhang T 2014 Revisiting drivers of energy intensity in China

during 1997-2007: A structural decomposition analysis *Energy Policy* **67** 640-7
[5] Nie H, Kemp R 2013 Why did energy intensity fluctuate during 2000-2009? A combination of index decomposition analysis and structural decomposition analysis *Energy Sustain. Dev.* **17** 482-8
[6] Xia Y, Yang C, Chen X 2012 Structural decomposition analysis on China's energy intensity change for 1987-2005 *J. Syst. Sci. Complex.* **25** 156-66
[7] Zhang Y 2009 Structural decomposition analysis of sources of decarbonizing economic development in China: 1992-2006 *Ecol. Econ.* **68** 2399-405
[8] Li H, Wei Y 2015 Is it possible for China to reduce its total CO_2 emissions? *Energy* **83** 438-46
[9] Xie S 2014 The driving forces of China's energy use from 1992 to 2010: An empirical study of input-output and structural decomposition analysis *Energy Policy* **73** 401-15
[10] Zhang G, Liu M 2014 The Changes of Carbon Emission in China's Industrial Sectors from 2002 to 2010: A Structural Decomposition Analysis and Input-Output Subsystem *Discrete Dyn. Nat. Soc.* 798576-9
[11] Fan Y, Xia Y 2012 Exploring energy consumption and demand in China *Energy* **40** 23-30
[12] Minx J C, Baiocchi G, Peters G P, Weber C L, Guan D, Hubacek K 2011 A "carbonizing dragon": China's fast growing CO_2 emissions revisited *Environ. Sci. Technol.* **45** 9144-53
[13] Zhang H, Qi Y 2011 A structure decomposition analysis of China's production-source CO_2 emissions: 1992-2002 *Environ. Resour. Econ.* **49** 65-77
[14] Zhang Y 2010 Supply-side structural effect on carbon emissions in China *Energy Econ.* **32** 186-93
[15] Guan D, Peters G P, Weber C L, Hubacek K 2009 Journey to world top emitter: An analysis of the driving forces of China's recent CO_2 emissions surge *Geophys. Res. Lett.* **36**
[16] Guan D, Hubacek K, Weber C L, Peters G P, Reiner D M 2008 The drivers of Chinese CO_2 emissions from 1980 to 2030 *Glob. Environ. Chang.* **18** 626-34
[17] Peters G P, Weber C L, Guan D, Hubacek K 2007 China's growing CO_2 emissions - a race between increasing consumption and efficiency gains *Environ. Sci. Technol.* **41** 5939-44
[18] Zhu Q, Peng X, Wu K 2012 Calculation and decomposition of indirect carbon emissions from residential consumption in China based on the input－output model *Energy Policy* **48** 618-26
[19] Wang Z, Liu W, Yin J 2015 Driving forces of indirect carbon emissions from household consumption in China: An input-output decomposition analysis *Nat. Hazards* **752** S257-72
[20] Liu Z, Guan D, Wei W, Davis S J, Ciais P, Bai J, et al. 2015 Reduced carbon emission estimates from fossil fuel combustion and cement production in China *Nat.* **524** 335-8
[21] Leontief W 1970 Environmental repercussions and the economic structure: An input-output approach *Rev. Econ. Stat.* **52** 262-71
[22] Rose A, Casler S 1996 Input-output structural decomposition analysis: A critical appraisal *Econ. Syst. Res.* **8** 33-62
[23] Dietzenbacher E, Los B 1998 Structural decomposition techniques: Sense and sensitivity *Econ. Syst. Res.* **10** 307-24
[24] Li J 2005 A decomposition method of structural decomposition analysis *J. Syst. Sci. Complex.* **18** 210-8
[25] Peters G Weber C Liu J 2006 *Construction of Chinese energy and emissions inventory* (Trondheim, Norway: Norwegian University of Science and Technology)
[26] Department of Energy Statistics of National Bureau of Statistics 2015 *China energy statistical yearbook 2014* (Beijing, China: China Statistics Press)
[27] Department of Energy Statistics, Department of Comprehensive Statistics of National Bureau of Statistics 2009 *China energy statistical yearbook 2008* (Beijing, China: China Statistics Press)
[28] Liu Q, Peng Z 2010 *China's input-output tables in constant prices 1992-2005* (Beijing, China: China Statistics Press)

IOP Conf. Series: Earth and Environmental Science **59** (2017) 012051 doi:10.1088/1755-1315/59/1/012051

[29] Su B, Ang B W 2013 Input-output analysis of CO_2 emissions embodied in trade: Competitive versus non-competitive imports *Energy Policy* **56** 83-7
[30] Weber C L, Peters G P, Guan D, Hubacek K 2008 The contribution of Chinese exports to climate change *Energy Policy* **36** 3572-7

IOP Conf. Series: Earth and Environmental Science **59** (2017) 012052 doi:10.1088/1755-1315/59/1/012052

Comparative analysis of driving forces behind carbon emissions among regions in China during 2002-2014

Siqin Xiong[1,2], Yushen Tian[1,2], Xiaoming Ma[1,2,3]

[1]School of Environment and Energy, Peking University Shenzhen Graduate School, Shenzhen 518055, China;
[2]Key Laboratory of Urban Environment, School of Environment and Energy, Shenzhen Graduate School, Peking University, Shenzhen 518055, China;
[3]College of Environmental Sciences and Engineering, Peking University, Beijing 100871, China.

Abstract. By using the logarithmic mean Divisia index (LMDI) method, this paper decomposed the factors that affect carbon emissions both at a national and multi-regional level, and comparatively analysed the difference of driving factors between 2002-2008 and 2008-2014. It is found that economic growth and the energy intensity are two major factors that drive up carbon emissions in the two periods. The economic structure effect and the energy structure effect had little influence on national carbon emissions, and the inhibitory impact was more obvious in 2008-2014 than the first period in most regions. Both the change of economic structure and energy structure can result in the change of national carbon emissions. Also, the variety of different regions can be attributed to the effect of economic growth and the energy intensity.

1. Introduction

Since China has become the largest carbon emitter, the pressure to reduce emissions is greater than ever. Therefore, China has issued a series of emission reduction policies and promised to peak its carbon emissions by 2030. China is a vast territory with big differences in economic scale and resource endowment among different regions. Thus, it is necessary to elucidate the characteristics of regional carbon emission and provide guideline for governments to formulate corresponding policies.

Some studies have examined the driving factors related to carbon emissions. Fan et al. (2007) studied the factors that influence carbon emissions in China, and concluded that the economic output effect is a key factor that promotes carbon emissions, whereas the energy intensity has an adverse effect. Lei (2011) adopted the LMDI method to analyse the increase of China's carbon emissions and suggested that the decrease of energy intensity is the main factor to restrain the growth of carbon emissions, while the changes of industrial structure and energy structure promote the increase of carbon emissions. Meng et al. (2011) used panel model to analyse the characteristics of China's regional carbon emissions. Deng et al. (2014) decomposed the carbon emission into population scale, economic development level, energy intensity and energy structure to elucidate the carbon emission characteristics of eight regions in China from 1995 to 2010, and explored the causes and patterns of regional differences.

In general, most researchers suggest that the amount of carbon emission is influenced by economic development, economic structure, energy structure and technological progress. But more researches are still needed in this field. These studies mentioned have investigated the driving forces in a long-time period, but a few conducted comparative analysis of the difference in driving forces during different periods and take the economic status into account. And due to the access of data, the analysis of the driving factors at a regional level is still rare and worth of in-depth research.

Decomposition model, input-output model and metric methods are the main methods to study what may drive the carbon emissions up. LMDI, one of the decomposition model, is wildly used for the lack of an unexplainable residual and its availability even though the data is incomplete.

In this study, from the multi-regional perspective, we adopted the LMDI multiple method to decompose the factors that influence carbon emissions with respect to seven regions of China from 2002-2008 and 2008-2014, and analysed the distinctions between different periods and different regions. The paper is organized as follows: Section Two describes the methodology and data. Section 3 presents the empirical analysis. Section 4 gives our conclusions and policy implications.

2. Methodology and data

2.1. Methodology

From Ang (2005), the LMDI decomposition model is as follows:

$$C=\sum_{i=1}^{n} C_i=\sum_{i=1}^{n} GDP\times\frac{\Delta GDP}{GDP}\times\frac{\Delta E_i}{\Delta GDP}\times\frac{C_i}{\Delta E_i}=\sum_{i=1}^{n} Q\times V_i\times I_i\times F_i$$

where C represents the total carbon emissions in China; i represents each region; Q represents the regional gross domestic product; Vi represents the proportion of the gross domestic product of industry i accounted for GDP; Ii represents the energy intensity of industry i; Fi represents carbon emissions caused by energy consumption in industry i.

Using the LMDI method, this paper decomposed carbon emissions into four driving factors, including economic growth, energy intensity effect, economic structure effect and energy structure effect. The decomposition model is as follows and X represents Q, V, I, F, respectively.

$$\Delta C_x=\sum_i \frac{C_i^T-C_i^0}{ln\,C_i^T-ln\,C_i^0}ln\left(\frac{X^T}{X^0}\right)$$

2.2. Data

The paper regrouped China's 30 provinces and municipalities into seven regions, taking the geographic position into account combined with the economic development level(Table.1). Due to the access to the data, Hong Kong, Macao, Tibet, and Chinese Taiwan were not included in our analysis.

Table 1. The grouping of 30 provinces and municipalities

Region	Provinces or Municipalities
East China	Shanghai, Jiangsu, Zhejiang, Anhui, Jiangxi, Shandong, Fujian
North China	Beijing, Tianjin, Shanxi, Hebei, Inner Mongolia
Central China	Henan, Hubei, Hunan
South China	Guangdong, Guangxi, Hainan
Southwest	Chongqing, Sichuan, Guizhou, Yunnan
Northwest	Shanxi, Gansu, Qinghai, Ningxia, Xinjiang
Northeast	Heilongjiang, Jilin, Liaoning

The time series data was from 2002 to 2014 and divided into 2002-2008 and 2008-2014. The energy data used was from The Energy Statistical Yearbook and different energy types were converted into standard coal. GDP data was obtained from the China Statistical Yearbook and was converted into standard prices using a price index (2002,100). Since we could not obtain the energy-balance table for Hainan and Hunan provinces in 2002, we used the table in 2003 to represent the table in 2002. Carbon emissions were calculated based on the carbon-emission coefficients provided by IPCC (2006).

3. Empirical results

3.1. Driving factors of China's carbon emissions

3.1.1. The total carbon emissions in China between 2002-2014. From 2002 to 2014, the national total carbon emissions increased from 3580.83MTe to 8933.06MTe.Due to the long-term birth control policy, China's population growth rate was relatively stable, and the growth of per capita carbon emissions had the same trend as the growth of total carbon emissions . In 2002, the energy intensity was 2.81 tce/10^4RMB and continuously increased to 4.10 tce/10^4RMB until 2007.But from 2008 to 2012, the energy intensity showed a downward trend in 2014.

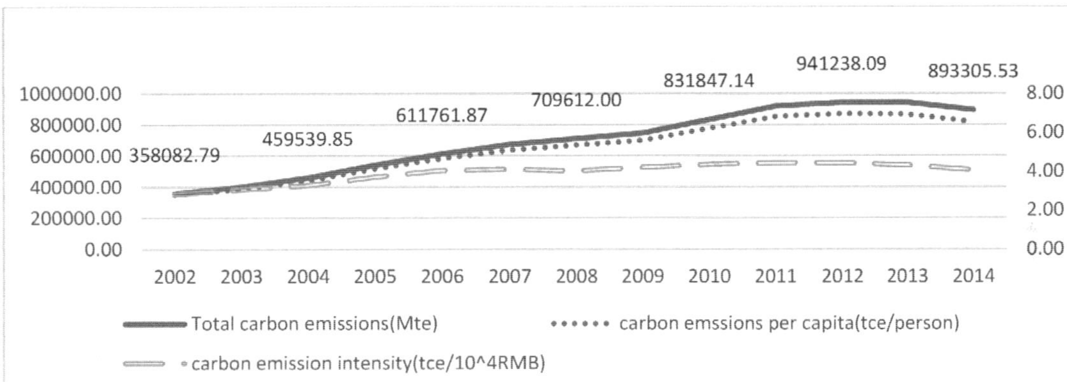

Figure 1. The carbon emissions in China between 2002-2014

3.1.2. Impact of driving factors on China's carbon emissions during 2002-2014. The effect of GDP and energy intensity played a stimulating role in promoting China's carbon emissions whereas the effect of economic structure and energy structure reduced carbon emissions during 2002-2014. Before 2008, China's economy grew rapidly, which was mainly attributed to the secondary industry growth, and led to the year by year increase of carbon emissions. But in 2008 due to the financial crisis, the accumulated carbon emission was 49.70MTe. After that, China shift the economic growth mode, and subsequently entered a new normal stage. Therefore, the contribution rate of the GDP effect dropped since 2012. The energy intensity effect also led to an increase in China's carbon emissions by 2013, but had inhibitory impact in 2014, which suggests a technological progress in emission reductions.The economic structure effect increased carbon emissions before 2006 but was transformed from stimulative to inhibitive impact since then, and played a major inhibitory effect after 2012. The effect of energy structure played a different role in different periods, and the contribution rate was relatively small.

Table 2. Impact of driving factors on China's carbon emissions during 2002-2014

	2002-2003	2003-2004	2004-2005	2005-2006	2006-2007	2007-2008	2008-2009	2009-2010	2010-2011	2011-2012	2012-2013	2013-2014
Economic growth effect	103.6	292.9	192.3	224.1	516.1	502.7	49.7	574.3	715.0	248.1	251.9	67.8
Economic structure effect	6.2	3.4	64.7	18.2	-121.7	-47.2	-241.3	-54.6	-104.2	-271.1	-267.4	-211.2
Energy intensity effect	277.1	336.0	537.5	435.8	202.3	9.8	546.3	219.5	262.5	362.5	43.0	-184.1
Energy structure effect	49.2	-53.9	15.4	34.2	4.9	-88.3	7.7	120.7	-8.7	-110.1	-24.8	-154.5

3.1.3. Impact of driving factors on China's carbon emissions during 2002-2008 and 2008-2014. In 2008, the economy slowdown due to the financial crisis, after that, the government began to optimize economic structure and develop innovation-oriented industries. Therefore, comparative analysis of the driving factors between the period of 2002-2008 and 2008-2014 is useful for the country to implement corresponding emission reduction policies in accordance with different economic status .

From 2002to 2008,the energy intensity effect played a major role in stimulating the growth of carbon emissions in China. From 2008-2014, the stimulus effect of the energy intensity effect slightly weakened, and the cumulative increase in carbon emissions was 1235.86 MTe. Following the energy intensity effect, the GDP effect also played a stimulating effect from 2002-2008,and the contribution in 2008-2014 increased slightly,up to 1773.08 MTe. Adjustment of industrial structure played an inhibitory role in China's carbon emissions. During 2008-2014, the inhibitory effect of this factor was more obvious, with the emission reductions of 1045.55 MTe. Before 2008, the effect of energy structure had inhibition of carbon emissions, with carbon emissions decreased 12.65 Mte. During 2008-2014, the contribution rate increased to -6.9%, with the cumulative reductions increased by nearly 10 times, up to 126.46MTe.

3.2. *Driving factors of carbon emissions in seven regions*

We decomposed China's regional carbon emissions over the period 2002-2008 and 2008-2014 based on the LMDI method for a better understanding of the changes in driving forces between different periods.

3.2.1. Impact of driving factors on seven regions during 2002-2008. From 2002 to 2008, the results demonstrate that the growth of GDP had stimulating effects on carbon emissions in all regions, with the contribution rate ranged from 33.5%-54.1%. The energy intensity effect was also the key factor that increased emissions. The energy intensity in East China and North China had relatively greater impacts on national carbon emissions than the remaining regions. The economic structure effect in North China, Northwest and Central China increased the emissions whereas decreased the emissions in remaining regions. The change of energy structure had a positive pulling effect on the carbon emissions of East China, Southwest China and Northwest China, on the contrary, exerted a negative effect for the North China, South China, Central China and Northeast regions.

3.2.2. Impact of driving factors on seven regions during 2008-2014. The results in 2008-2014 show that economic growth effect and energy intensity effect still had a great impact on the growth of carbon emissions in each region. Compared to 2002-2008, the adjustment of economic structure had a negative effect on the carbon emissions in all regions, and the contribution rate was about -6.2% - 22.6%. In addition to South China and Central China, the remaining five regions by the energy structure adjustment led to their carbon emission reduction.

Table 3. Impact of driving factors in different regions between 2002-2008 and 2008-2014

	Economic growth effect		Economic structure effect		Energy intensity effect		Energy structure effect	
	2002-2008	2008-2014	2002-2008	2008-2014	2002-2008	2008-2014	2002-2008	2008-2014
China	1675.60	1773.08	-38.12	-1045.55	1890.46	1235.86	-12.65	-126.46
Northeast	102.30	104.85	-19.83	-126.94	198.59	298.07	-11.76	6.03
North China	334.73	176.28	22.69	-276.87	435.78	650.32	-10.95	-5.59
East China	399.83	284.53	-38.68	-338.08	829.34	214.71	1.66	-71.44
South China	97.99	71.40	-31.09	-71.21	230.51	163.39	-17.27	3.30
Central China	208.62	157.62	2.42	-135.05	274.06	136.57	-10.15	12.82
Northwest	134.82	140.37	29.59	-114.27	71.19	394.41	13.50	-4.22
Southeast	135.44	137.50	-3.68	-145.23	191.51	241.00	21.55	-67.36

4. Conclusions and Policy implications

This paper identified the driving factors in carbon emissions at a national and multi-regional level and comparatively analysed the differences between 2002-2008 and 2008-2014. Here are main conclusions and policy implications:

(1) China's economy expanded fast during 2002-2014, resulting in the increase of carbon emissions. But the growth rate declined during 2008-2014, indicating that the economic structure became less carbon-intensive and the efforts to optimize the energy structure achieved success to some degree.

(2) Differences in national carbon emissions between different periods mainly came from the economic structure and energy structure. During 2008-2014, the government implemented economic optimization and energy restructuring policies, which significantly inhibited carbon emission growth.

(3) Differences in carbon emissions between different regions mainly came from the economic effect and energy intensity. Economic growth effect on the East, North and Central China are more obvious than that in other regions. These regions are the major regions for booming China's economic growth, therefore, the government should pay more attention to decouple the economic growth and energy consumption. Energy intensity effect had relatively small difference between regions, indicating that some actions to improve energy efficiency should be made in all regions. The economic structure played an inhibitory effect in all regions during 2008-2014, which shows the optimization of economic structure has achieved an initial success. Compared with other factors, the energy structure effect was small, but it still shows that the adjustment of energy structure is conducive to reducing China's carbon emissions. Thus, China should still support the development of renewable energies in the future.

References

[1] Ang, B.W., 2005. The LMDI approach to decomposition analysis: a practical guide. Energy Policy 33, 867-871.

[2] Ang B.W., 2015. LMDI decomposition approach: a guide for implementation. Energy Policy 86, 233–238.

[3] Deng, J., Liu, X., Wang, Z.,2014. Characteristics Analysis and Factor Decomposition Based on the Regional Difference Changes in China's CO2 Emission. Journal of Natural Resources 29,189-200.

[4] Fan, Y., Liu, L.C., Wu, G., Tsai, H.T., Wei, Y.M., 2007. Changes in carbon intensity in China: empirical findings from 1980-2003. Ecol. Econ. 62, 683-691.

[5] Meng, L., Guo, J., Chai, J., Zhang, Z., 2011. China's regional CO2 emissions: characteristics, interregional transfer and emission reduction policies. Energy Policy39, 6136-6144.

[6] Paul, S., Bhattacharya, R.N., 2004. CO2 emission from energy use in India: a decomposition analysis. Energy Policy 32, 585-593.

[7] Zhang, Y., Zhang, J., Yang, Z., Li, S., 2011. Regional differences in the factors that influence China's energy-related carbon emissions and potential mitigation strategies. Energy Policy 39, 7712-7718.

Riparian soil physicochemical properties and correlation with soil organic carbon of an inflowing river of Taihu Lake

Jin Qian[a,b,*], Jingjing Liu[a,b], Peifang Wang[a,b], Chao Wang[a,b], Kun Li[a,b], Mengmeng Shen[a,b]

[a] Key Laboratory of Integrated Regulation and Resources Development on Shallow Lakes, Ministry of Education, Hohai University, Nanjing, People's Republic of China, 210098
[b] College of Environment, Hohai University, Nanjing, People's Republic of China, 210098

The corresponding author's e-mail address: hhuqj@hhu.edu.cn

Abstract. Profile method and the classic measurement method were used to analyse the vertical distribution of riparian soil physicochemical properties of an inflowing river (Yincungang River) of Taihu Lake. The results showed that the soil physicochemical properties distribution of the riparian zones with three different vegetation communities were showed a similar regular change tendency with the increase of soil depth. But the maximum of soil organic carbon (SOC), total nitrogen (TN) and total phosphorus (TP) were occurred in the riparian zone with *Arundo donax* community, and the maximum of soil bulk density (BD) and pH value were occurred in the riparian zone with *phragmites australis* community. Besides, there was significant positive correlation between SOC and TN, TP, BD, while the opposite result was observed for pH.

1 Introduction

Soil organic carbon (SOC) pool, as one of the most important carbon stocks on the earth, plays a key role in the global carbon cycle. Globally, about 1500 ~2000 Pg SOC is stored in the upper 100 cm of the soil[9][15]. Riparian zone is the interface between land and a river or stream[12]. Riparian SOC is strongly influenced by soil structures, soil properties, aquatic plants root depth, effective water retention capacity and soil biological diversity [3][6][21]. Soil nitrogen (N), phosphorus (P), pH value, bulk density（BD） and other soil physicochemical characteristics indexes in accurately reflect the functional variability of riparian ecosystem, and it is important to determine the change of soil element in riparian ecosystem, which has a significant effect on the response of soil C cycle [18]. Analyzing the relationship between SOC and N, P, pH values and BD in riparian zone, can let us understand the condition of soil fertility of riparian zone, the composition of soil organic matter and the condition of soil quality as well as the supply ability of nutrients [18][23].

The interactions between the cycle of soil N and P with the soil carbon cycle affect the distribution of SOC and then affect the environmental effects of riparian zone. Wang et al. [23] studied the distribution characteristics of soil carbon and nitrogen in different Riparian zone of Yangze River estuary and Zhang et al. [28] studied the pattern and causes of soil carbon distribution in riparian zone in arid desert area. However, the relationship between soil carbon characteristics and soil physicochemical properties of riparian zone in Taihu Lake Basin has not been reported. Yingcungang

River is one of the main inflowing rivers of Taihu Lake, China. Yingcungang riparian zone is an important part of the ecological system of riparian zone in Taihu Area [25]. *Phragmites australis*, *Arundo donax*, *Canna indica* is the zonal typical aquatic plants in the region. This study was designed to examine the distribution characteristics of SOC in Yingcungang riparian zone with different vegetation types, analysis the correlation between SOC and soil physicochemical properties. The objectives of this study were to explore the distribution pattern of riparian soil physicochemical properties and the response of SOC to soil physicochemical properties of an inflowing river of Taihu Lake.

2 Materials and methods

2.1 Study sites

The study area, located in Yingcungang riparian zone, is characterized as the subtropical monsoon climate with a mean annual temperature of 15.7℃, the mean annual rainfall is approximately 1177 mm. The area was mainly used for agricultural production and the main crops were rice, wheat,, soybean and so on. Near the river bank, emergent plants was dominated by *Phragmites australis*, *Arundo donax*, *Canna indica* and other aquatic plants [14][25].

2.2 Soil sampling design and field work

The sample sites were located at latitude 31° 27′ N and longitude 119° 59′ E. Soil samples were collected from three regions growing *phragmites australis*, *Arundo donax* and *Canna indica* in the riparian zone, respectively in June 2016,. We choose two blocks (10 m×10 m) in every region and recorded the slope, coverage and average height aquatic of the plants (Table 1). By the S-shaped sampling strategy, each group has three parallels.

Three soil samples (0–15, 15-30, 30-45cm) were collected in each sample site. To avoid compaction and degeneration, the soil samples were put into hard aluminum boxes and transported to laboratory for the following analyses.

Table 1 Information of plots in Riparian zone

Plant type	Slope	Average Height (cm)	Taproot Length (cm)	Diameter (cm)	Coverage (%)
Phragmites australis	South	258.7	43.7	2.6	95
	North	102.8	39.2	1.7	75
Arundo donax	South	285.3	37.5	3.1	97
	North	258.8	35.8	2.7	80
Canna indica	South	102.1	21.3	2.1	95
	North	65.9	14.5	1.4	75

2.3 Laboratory procedures

Each soil sample was weighed for bulk density determination, then passed through an 8-mm mesh sieve by gently breaking soils along natural fracture planes [4]. Soil was digested using H_2SO_4–$HClO_4$ and TP was measured by the spectrophotometric method with a continuous flow automated analyzer at 700 nm [1]. SOC was measured using acid dichromate wet oxidation as described by Nelson and Sommers [13], TN was tested with potassium persulfate oxidation [2]. A soil sub-sample was air-dried and pH was measured on a 1:2.5 soil–water suspension [11].

2.4 Statistical analysis

All statistical calculations were performed with Office excel 2013 and SPSS 19.0, indicating the relationship between SOC and soil physicochemical properties in Riparian Yingcungang. To analyze the significant relationship between SOC and physicochemical properties (BD, pH value, TN and TP), one-way analysis of variance (ANOVA) was used with post hoc test (Duncan's test, with a critical significance level of 0.05). Levene's test was also used for determining equality of variances.

3 Results and Discussion

3.1 Distribution pattern of soil physicochemical properties

The soil physicochemical properties of the riparian zones with three plant communities were changed regularly with the increase of soil depth (Table 2). SOC, TP and TN decreased with the increase of soil depth, but BD and pH value decreased after an increase along with the increasing of soil depth. Overall, with the increase of soil depth, soil acidity enhanced, soil nutrient contents reduced, the soil became tight. The results were in accordance with the findings of Kong et al. [8].

As illustrated in Table 2, the maximum of SOC, TN and TP was occurred in *Arundo donax* community, and the maximum of BD and pH value was occurred in *phragmites australis* community. The results showed that the differences of physicochemical properties of soil in riparian zone are closely related to vegetation community composition. The characteristic factors of the physicochemical properties of soils are mainly affected by root exudates [8][23]. The differences of decomposition speed, decomposition quantity, and chemical composition of the decomposition products caused by the exudates composition, led to the differences of soil physicochemical properties of the riparian zone with three different vegetation communities.

Table 2 Soil properties in different soil layers in Riparian zone

Vegetation type	Depth (cm)	SOC (g kg^{-1})	TN (mg g^{-1})	TP (mg kg^{-1})	BD (g cm^{-3})	pH value
Phragmites australis	0-15	11.24±0.42	343.86±17.82	420.09±4.61	1.35±0.10	7.22±0.20
	15-30	8.76±0.24	341.19±11.27	402.81±5.79	1.66±0.06	7.81±0.16
	30-45	9.44±0.27	513.23±14.64	398.73±7.82	1.45±0.11	7.44±0.06
	Mean	9.81±0.31	399.43±14.58	407.21±9.11	1.48±0.09	7.79±0.14
Arundo donax	0-15	17.37±0.62	1324.82±31.87	504.75±5.64	1.36±0.04	5.14±0.03
	15-30	15.84±0.73	1048.41±26.84	507.64±3.64	1.39±0.01	5.60±0.02
	30-45	15.32±0.45	899.11±18.36	498.76±6.84	1.30±0.06	4.76±0.02
	Mean	16.17±0.60	1090.78±25.69	503.72±5.37	1.35±0.04	5.16±0.03
Canna indica	0-15	11.80±0.48	803.63±22.67	433.11±3.72	1.32±0.06	6.17±0.14
	15-30	8.52±0.32	884.82±9.74	427.43±7.65	1.46±0.02	5.95±0.07
	30-45	7.65±0.36	1009.55±13.26	403.83±4.86	0.98±0.06	5.87±0.04
	Mean	9.32±0.58	899.33±15.22	421.46±5.41	1.23±0.05	6.01±0.08

3.2 Relationship between soil nutrients and SOC

There was a significant positive logarithmic correlation between soil TN and SOC (R^2=0.8608, P<0.01, n=24) in the riparian zone. The result was similar to the finding obtained by Kong et al. [8] and Knoepp et al. [7] (Fig.1). SOC had a great influence on soil TN. The appropriate ratio of C/N was beneficial to the nutrient release and the increase of available nitrogen during the decomposition of microorganism on organic matter. To some extent, the amount of SOC was determined by the soil TN

[17][26].

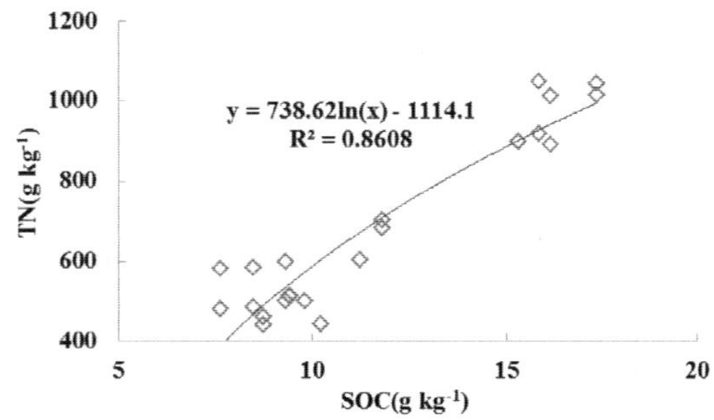

Fig.1 Relationship between TN and SOC in riparian soil

As an important nutrient element in soil, phosphorus was closely related to SOC. The decomposition of SOC can promote the release of phosphorus and the increase of phosphorus can contribute to the accumulation of organic carbon in soil. There was a significant positive correlation in logarithmic function (R^2=0.7101, P<0.01, n=24) between TP and SOC (Fig.2). This was in accordance with the results of Qiu K L et al. [16].

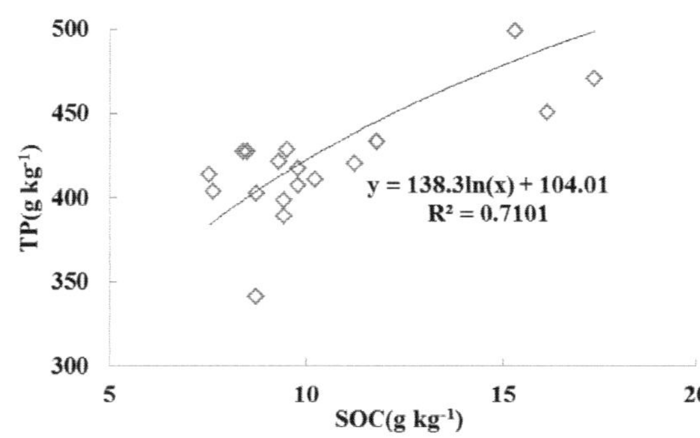

Fig.2 Relationship between TP and SOC in riparian soil

3.3 Correlation between BD, pH value and SOC

Fig.3 showed a significant positive power function correlation between BD and SOC in the study area (R^2=0.6374, P<0.01, n=24). The results showed that the higher BD was, the higher the content of SOC was. The main reason of this phenomenon was that the increase of SOC content resulted in the increase of soil carbon density, soil bulk density [16][27].

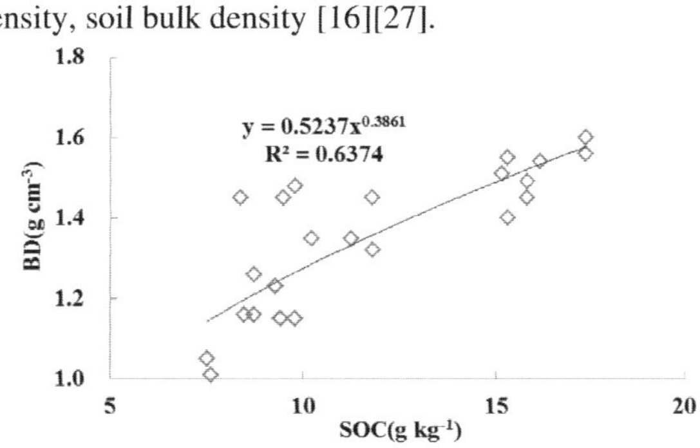

Fig.3 Relationship between BD and SOC in riparian soil

It was generally showed a negative correlation between soil pH value and SOC. Some research suggested that it was not meaning for the relationship between pH and SOC if pH value was not defined in certainty range [22]. In the study area, the soil is weak acidity (Table 2), and it showed linear negative correlation between pH and SOC (R^2=0.4182, P<0.01, n=24) (Fig.3). This was in accordance with the findings of Li S J et al. [10] and it showed that pH can affect the type and activity of soil microorganisms, further to mineralization of SOC. The lower the soil pH value was, the faster the mineralization rate of soil and the lower the SOC content would be.

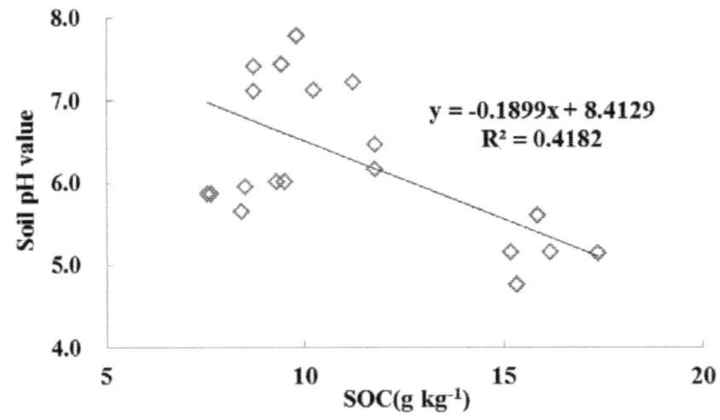

Fig.4 Relationship between pH value and SOC in riparian soil

4 Conclusion

Firstly, the soil physicochemical properties distribution of the riparian zones with three different vegetation communities in the Yincungang River were showed a similar regular change tendency with the increase of soil depth. With the increasing of soil depth, soil nutrient content reduced, the soil becomes tight, and soil acidity increased gradually. Secondly, the obvious differences of soil physicochemical properties existed in the riparian zone with three different plant communities. The maximum of SOC, TN and TP was occurred in the riparian zone with *Arundo donax* community, and the maximum of BD and pH value was occurred in the riparian zone with *phragmites australis* community. Finally, it was showed that there was significant positive correlation between SOC and TN, TP, and BD, while the opposite result was observed for pH.

Acknowledgements

We acknowledge the National Natural Science Foundation of China (No. 51379062), the National Key Plan for Research and Development of China (No. 2016YFC0401703), the Key Program of National Natural Science Foundation of China (No. 41430751), the National Natural Science Foundation of China (No. 51579073), the Priority Academic Program Development of Jiangsu Higher Education Institutions (PAPD) for funding this research.

References

[1] Bao, S.D.,. Soil and agricultural chemistry analysis. third ed. Agriculture Press, Beijing,2008 .(In Chinese)

[2] Bremner, J.M., 1996. Nitrogen-total. In: Sparks, D.L. (Ed.), Methods of Soil Analysis.Part 3. Chemical Methods, No. 5. ASA and SSSA, Madison, WI, pp. 1085–1121.

[3] Capon, S.J., Chambers, L.E., Mac Nally, R., Naiman, R.J., Davies, P., Marshall, N., Pittock, J., Reid, M., Capon, T., Douglas, M., Catford, J., Baldwin, D.S., Stewardson, M., Roberts, J., Parsons, M., Williams, S.E., 2013. Riparian ecosystems in the 21st century: hotspots for climate change adaptation. Ecosystems 16, 359–381.

[4] Cheng, X., Luo, Y., Xu, X., Sherry, R., Zhang, Q., 2011. Soil organic matter dynamics in a North

America tallgrass prairie after 9 yr of experimental warming. Biogeosciences 8, 1487–1498.

[5] Chimel D S. Terrestrial ecosystems and the carbon cycle.Global Change Biology, 1995, 1(1): 77-91. (In Chinese)

[6] Feller, C., Beare, M.H., 1997. Physical control of soil organic matter dynamics in the tropics. Geoderma, 79, 69–116.

[7] Knoepp JD,Clinton BD. Riparian zones in southern Appalachian headwater catchments: Carbon and nitrogen responses to forest cutting. Forest Ecology and Management, 2009, 258(10): 2 282–2 293.

[8] Kong Tao, Zhang De-sheng, Kou Yong-ping ,et al. Soil Organic Carbon, Total Nitrogen and Total Phosphorus Distribution of Typical Vegetation Riparian Zones in Upper Reaches of Hun River. Soils, 2014, 46(5): 793 - 798. (In Chinese)

[9] Lal, R., 2004. Soil carbon sequestration impacts on global climate change and food security. Science 304, 1623–1627.

[10] Li S J, Qiu L P, Zhang X C. Mineralization of soil organic carbon and its relations with soil physical and chemical properties on the Loess Plateau.Acta Ecologica, 2010, 30(5): 1217-1226. (In Chinese)

[11] Lu, R., 2000. Analysis of Soil Agrochemistry. Chinese Agricultural Science and Technology Press, Beijing.

[12] Mander, U., Kuusemets, V., Krista, L., Lõhmus, K., Mauring, T., 1997. Efficiency and dimensioning of riparian buffer zones in agricultural catchments. Ecol. Eng. 8: 299-324.

[13] Nelson, D.W., Sommers, L.E., 1996. Total carbon, organic carbon and organic matter.In: Sparks, D.L. (Ed.), Methods of Soil Analysis. Part 3. Chemical Methods, No. 5.ASA and SSSA, Madison, WI, pp. 961–1010.

[14] Niu Yong,Yu Hui,Niu Yuan,et al.Pollution of Nutrients and Heavy Metals in Sediments from Yin Cun Gang River of Lake Taihu Basin,China. Agricultural journal of environmental science, 2015， 34(8):1557-1562. (In Chinese)

[15] Post W M, Emanuel W R, Zinke P J et al. Soil carbon pools and life zones. Nature, 1982, 298: 156-159.

[16] Qiu K L, Chen W D, Peng P H, et al. Analysis on correlation between soil organic carbon and other factors in Neijiang Section of Tuojiang River Basin.Research of Soil and Water Conservation, 2013,20(3): 28-31. (In Chinese)

[17] Raven J A, Handley L L, Andrews M. Global aspects of C/N interactions determining plant-environment interactions[J]. Journal of Experimental Botany, 2004, 55(394):11-25.

[18] Robin Hale , Paul Reich , Tom Daniel ,et al. Scales that matter: guiding effective monitoring of soil properties in restored riparian zones. Geoderma 228–229 (2014) 173–181.

[19] Schlesinger W H. Carbon storage in the calishe of arid soils: a case study from Arizona.Soil Science, 1982,133: 247-255.

[20] Susan, S., 2007. Cambridge University Press, Cambridge.

[21] Thomson, J.R., Bond, N.R., Cunningham, S.C., Metzeling, L., Reich, P., Thompson, R.M., Mac Nally, R., 2012. The influences of climatic variation and vegetation on stream biota: lessons from the Big Dry in southeastern Australia. Glob. Chang. Biol. 18, 1582–1596.

[22] Wang Q K, Wang S L, Feng Z W, et al. Active soil organic matter and its relationship with soil quality. Acta Ecologica, 2005, 25(3): 513-519. (In Chinese)

[23] Wang Zhi et al. Distribution Characteristics of Soil Carbon and Nitrogen in Different Riparian Zone in Chongming Island. Journal of Anhui Agri. Sci. 2013, 41(22):9266-9269. (In Chinese)

[24] Watson, R.T., Noble, I.R., Bolin, B., Ravindranath, N.H., Verardo, D.J., Dokken, D.J., 2000. Land use, land-use change and forestry. IPCC Special Report. Cambridge University Press, Cambridge, UK.

[25] Yang Xiao-ying, Luo Xing-zhang, Zheng Zheng,et al. Explore the spatial and temporal patterns of water pollution in the Yincungang canal of the Lake Taihu Basin, China. Environmental

Science, 2012, 33(9): 3051-3056. (In Chinese)
[26] Yu S,Chen W, He X Y, et al. Soil physical and chemical properties in four types riparian forests around Dahuofang Reservoir. Journal of Northeast Forestry University, 2015, 43(3): 87-89. (In Chinese)
[27] Zhang Xue-ni, Lv Guang-hui, Gong lu, et al. Pattern and Causes of Soil Carbon Distribution in Riparian Zone in Arid Desert Area. Chinese Journal of Soil Science, 2013, Vol.44, No.6.(In Chinese)
[28] Zhang Y, Qin J H, Zhao Y C, et al. Change of soil physical and chemical properties, organic carbon and nutrients of different soil profile on different forest in Binggou of Heihe Basin. Journal of Soil and Wster Conservation, 2013, 17(2): 126-130.

IOP Conf. Series: Earth and Environmental Science **59** (2017) 012054 doi:10.1088/1755-1315/59/1/012054

Comparison of several ethanol productions using xylanase, inorganic salts, surfactant

Yan Wu [1], Jie Lu [1,*], Rui-feng Yang [1], Wen-jing Song [1,*], Hai-ming Li [1,2], Hai-song Wang [1], Jing-hui Zhou [1]

[1] Dalian Polytechnic University, Dalian, China
[2] State Key Laboratory of Pulp and Paper Engineering, South China University of Technology, Guangzhou , China

[*]Corresponding author: lujie@dlpu.edu.cn, songwj@dlpu.edu.cn

Abstract. Liquid hot water (LHW) pretreatment is an effective and environmentally friendly method to produce bioethanol with lignocellulosic materials. Corn stover was pretreated with liquid hot water (LHW) and then subjected to semi-simultaneous saccharification and fermentation (S-SSF) to obtain high ethanol concentration and yield. The present study aimed to confirm the effect of several additives on the fermentation digestibility of unwashed WIS of corn stover pretreated with LHW. So we also investigated the process, such as enzyme addition, inorganic salts, surfactant and different loading Triton. Results show that high ethanol concentration is necessary to add xylanase in the stage of saccharification. The ethanol concentration increased mainly with magnesium ion on fermentation. Comparing with Tween 80, Span 80 and Polyethylene glycol, Triton is the best surfactant. In contrast to using xylanase and Triton respectively, optimization can make up the lack of stamina and improve effect of single inorganic salts.

1. Introduction

Bioethanol has been widely used as a substitute for fossil fuels. The use of bioethanol produced from lignocellulosic material can reduce our dependence on fossil fuels [1-2]. Pretreatment is crucial to determine conversion efficiency. Studies have investigated and proposed many pretreatment materials/ methods, such as steam explosion [3], alkaline [4], ammonia fiber expansion [5], organic solvent [6], and diluted acid [7]. One of the most promising pretreatment processes of lignocelluloses material is liquid hot water (LHW) pretreatment. LHW pretreatment has been considered as an environmentally friendly technology. Studies have indicated that additives can improve enzymatic hydrolysis and bioethanol fermentation of lignocellulosic biomass [8-10].

The present study aimed to confirm the effect of several additives on the fermentation digestibility of unwashed WIS of corn stover pretreated with LHW. So we also investigated the process, such as enzyme addition, inorganic salts, surfactant and different loading Triton. The paper presented the results.

2. Materials and methods

2.1 Materials

Corn stover was collected from a field near Jinzhou New District (Dalian, China). Corn stover was manually cut into pieces of 4 cm to 7 cm in our laboratory. Corn stover was milled to particle sizes ranging from 20 mesh to 80 mesh by using a laboratory ball mill (Taijihuan Nanometer Limited Company, Qinhuangdao, China) Samples were then homogenized and stored in a plastic bag for subsequent experiments.

2.2 LHW pretreatment

LHW pretreatment was conducted in a 200 mL steel tank about 10 g of corn stover and 60 mL of deionized water were loaded in the small tanks. Then it is putted it in oil bath kettle. The pretreatment temperature was controlled at 195 °C and the pretreatment time was set at 20 min. After pretreatment, the water-insoluble solids (WISs) and the pre-hydrolyzates were separated by filtration by using a Bŭchner funnel. The WISs were used for subsequent ethanol fermentation.

2.3 S-SSF

The WISs from LHW pretreatment experiments were used as substrates. About 1 g of WISs was added into 100 mL Erlenmeyer flasks. Each flask contained 10 mL of pH 4.8 buffer. Enzyme loading was 1 g, and the medium temperature was maintained at 50 °C during the pre-hydrolysis phase. After pre-hydrolysis, the medium temperature was adjusted to a constant fermentation temperature and then maintained all throughout the following SSF phase. Afterward, about 1 mL of the activated yeast was added into the medium. The experiments were performed in a constant-temperature incubator for 72 h. The flasks were sealed with rubber stoppers and equipped with syringe needles to remove the generated carbon dioxide. The samples were collected at 0, 12, 24, 36, 48, 60, and 72 h for glucose and ethanol concentration determination.

2.4 Analysis methods

The contents of xylan, Klason lignin, ash, and benzene–alcohol (2:1) extractives were determined by using the Chinese National Standard methods, namely, GB/T2677.9-1994,GB/T2677.8-1994, GB/T2677.3-1993, and GB/T2677.6-1994, respectively. Acid-soluble lignin content was determined by using the method described in GB/T10337-1989. Glucan content was then determined according to National Renewable Energy Laboratory. Glucose content and ethanol concentration were identified by using the SBA-40D Biological Sensing Analyzer (Biology Institute of the Shandong Academy of Sciences, Jinan, China).

3. Results and discussion

3.1 Effect of enzyme addition on ethanol concentration

The effect of enzyme addition on bioethanol concentration was investigated. The xylanase loading was 0.5g. It was added in the saccharification and fermentation stage. The concentration of bioethanol was measured. The result is shown in Figure 1.

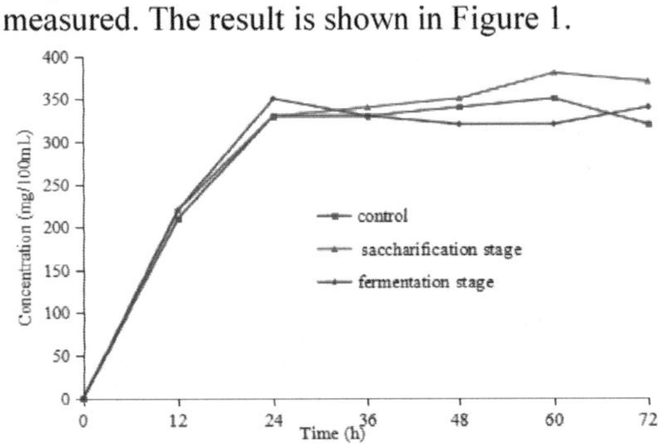

Figure 1. Effect of enzyme addition on bioethanol concentration.

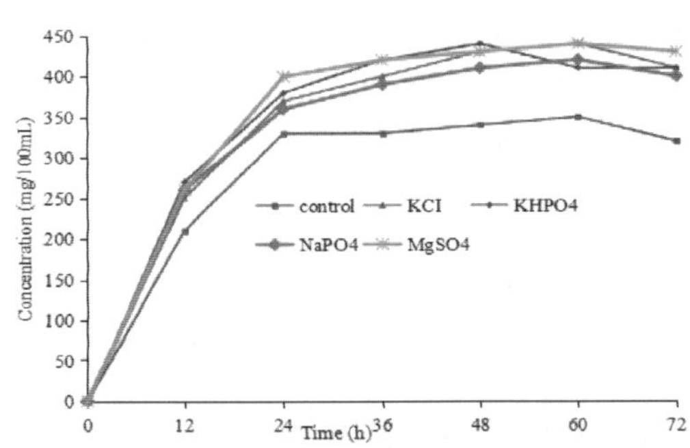

Figure 2. Effect of inorganic salts on bioethanol concentration.

Figure 1 shows that adding xylanase in the stage of saccharification produced higher ethanol concentration compared with fermentation process. Adding in saccharification can promote the decomposition of hemicellulose into glucose for fermentation.

3.2 Effect of inorganic salts on ethanol concentration

The changes in ethanol concentration with different inorganic salts were investigated comparatively. The concentration of ethanol was measured. The result is shown in Figure 2. Figure 2 shows that potassium, phosphorus and magnesium ions have obvious promoting effect on the fermentation process. For comparison, inorganic salts were added under the same stage. Results show that ethanol concentration increased mainly with magnesium ion on fermentation.

3.3 Effect of surfactant on ethanol concentration

The effect of surfactant on ethanol concentration was investigated. Tween 80, Span 80, Polyethylene glycol and Triton were added. The concentration of ethanol was measured. The result is shown in Figure 3.

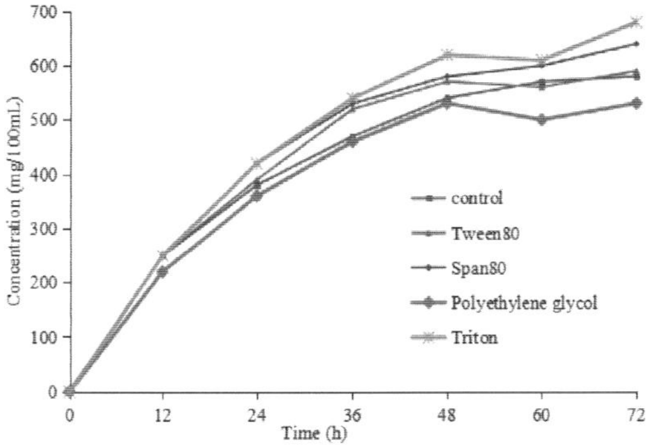

Figure 3. Effect of surfactant on ethanol concentration.

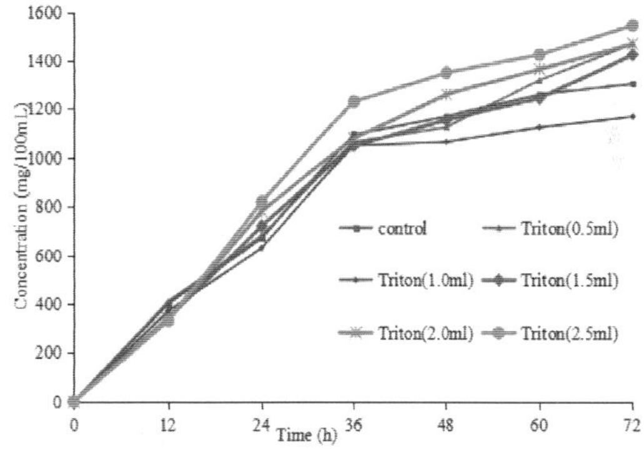

Figure 4. Effect of different loading Triton on bioethanol concentration.

Figure 3 shows that most of the surfactants showed a significant promoting effect. Results show that Triton has maximal effect on ethanol concentration, whereas ethanol produced less with Polyethylene glycol.

3.4 Effect of different Triton loading on ethanol concentration

The effect of different loading Triton on ethanol concentration was investigated. Triton loading was 0.5 mL, 1.0 mL, 1.5 mL, 2.0 mL, 2.5 mL. The concentration of ethanol was measured. The result is shown in Figure 4. Figure 4 shows that the ethanol concentration dropped significantly with Triton 1.0 mL, lower than the other groups. And the effect increased with the increase of the amount of adding. Therefore, Triton 2.5 mL was adopted in subsequent experiments.

3.5 Optimization

Results show that magnesium ion showed good performance compared with others. Xylanase and Triton have certain promoting effect on the saccharification process. Combining these beneficial substances, and extracting an optimized programme. The concentration of ethanol was measured. The result is shown in Figure 5. Figure 5 shows that optimization has a significant effect on the preparation of cellulose ethanol. A higher concentration of ethanol was achieved when the optimization was implemented. In contrast to using xylanase and Triton respectively, optimization can make up the lack of stamina and improve effect of single inorganic salts.

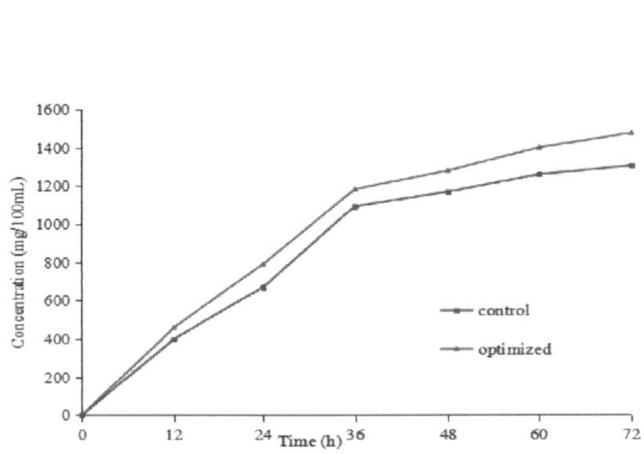

Figure 5. Optimization of ethanol process.

4. Conclusions

The positive effect of xylanase, inorganic salts, surfactant was observed. Results show that high ethanol concentration is necessary to add xylanase in the stage of saccharification. The bioethanol concentration increased mainly with magnesium ion on fermentation. Triton is the best surfactant. In contrast to using xylanase and Triton respectively, optimization can make up the lack of stamina and improve effect of single inorganic salts.

Acknowledgments

This study was financially supported by the National Science Foundation of China (No. 31370584, 31370582), the Basic Research Projects of Liaoning Education Department (No.2016J001), the Natural Science Foundation of Liaoning (No. 2015020640), the Key Laboratory of Liaoning (No.LZ2015005) and the State Key Laboratory of Pulp and Paper Engineering Open-end Foundation (No.201113).

References

[1] Lu J, Li XZ, Yang RF, Yang L, Zhao J, Liu YJ, and Qu YB. 2013. Fed-batch semi-simultaneous saccharification and fermentation of reed pretreated with liquid hot water for bio-ethanol production using Saccharomyces cerevisiae. *Bioresour. Technol.* 144, 539-547.

[2] Guerriero G, Hausman JF, Strauss J, Ertan H, and Siddiqui K S. 2016. Lignocellulosic biomass: Biosynthesis, degradation, and industrial utilization. *Eng. Life Sci.* 16, 1-16.

[3] Yu ZD, Zhang BL, Yu FQ, Xu GZ, and Song AD. 2012 . A real explosion: The requirement of steam explosion pretreatment. *Bioresour Technol.* 121:335-341.

[4] Kim I, and Han JI. 2012. Optimization of alkaline pretreatment conditions for enhancing glucose yield of rice straw by response surface methodology. *Biomass Bioenerg.* 46, 210-217.

[5] Li BZ, Balan V, Yuan YJ, and Dale BE. 2010. Process optimization to convert forage and sweet sorghum bagasse to ethanol based on ammonia fiber expansion (AFEX) pretreatment. *Bioresour. Technol.* 101,1285-1292.

[6] Sathitsuksanoh N, Zhu ZG, and Zhang YHP. 2012. Cellulose solvent- and organic solvent-based lignocellulose fractionation enabled efficient sugar release from a variety of lignocellulosic feedstocks. *Bioresour. Technol.* 117, 228-233.

[7] Hsu TC, Guo GL, Chen WH, and Hwang WS. 2010.Effect of dilute acid pretreatment of rice straw on structural properties and enzymatic hydrolysis. *Bioresour. Technol.* 101, 4907-4913.

[8] Lu J, Li XZ, Yang RF, Zhao J, and Qu YB. 2013. Tween 40 pretreatment of unwashed water-insoluble solids of read straw and corn stover pretreated with liquid hot water to obtain high concentrations of bioethanol. *Biotechnol. Biofuels* 6, 159-170.

[9] Shen JC, and Agblevor FA. 2010. Modeling semi-simultaneous saccharification and fermentation of ethanol production from cellulose. *Biomass Bioenerg.* 34, 1098-1107.

[10] Lu XQ, Zheng, XJ, Li XZ, and Zhao J. 2016. Adsorption and mechanism of cellulase enzymes onto lignin isolated from corn stover pretreated with liquid hot water. *Biotechnol. Biofuels* 9, 118-129.

Effect of xylanase, urea, Tween and Triton additives on bioethanol production of corn stover

Xiu Xin[1], Jie Lu [1,*], Rui-Feng Yang [1], Wen-jing Song [1,*], Hai-ming Li [1,2], Hai-song Wang [1], Jing-hui Zhou [1]

[1]Dalian Polytechnic University, Dalian, China
[2]State Key Laboratory of Pulp and Paper Engineering, South China University of Technology, Guangzhou ,China

[*]Corresponding author: lujie@dlpu.edu.cn, songwenjing@dlpu.edu.cn

Abstract. Corn stover is a potential source of renewable biomass for conversion to bioethanol. Fed-batch semi-simultaneous saccharifcation and fermentation (S-SSF) of corn stover pretreated by liquid hot water (LHW) was investigated. The present study aimed to confirm the influence of xylanase, urea, Tween and Triton additives on bioethanol. Results show that the positive effect of xylanase, urea, Tween was observed. High ethanol concentration requires the addition of xylanase in the stage of saccharification. The optimal amount of xylanase was 0.2 g/g biomass and addition of Triton (Triton X-100) increases the effect of xylanase. Urea has a promotion effect on the whole fermentation process.When adding 0.1% urea in the fermentation stage,the best promoting rate is 24.2%. In the longitudinal comparison of the Tween series, under the same experimental conditions, the promoting effect of Tween series: Tween 40 > Tween 80 > Tween 20 > Tween 60.

1. Introduction

Currently, bioethanol as a clean alternative source of fuel has raised more attention due to the diminishing fossil fuel storage [1-2]. However, the limited quantity of food stuff in China and their comparatively high prices greatly restrict large scale production of bioethanol [3]. Thus, the conversion of lignocellulosic material to bioethanol has been a research focus in China [4-5]. There are two production methods for bio-ethanol production: separate hydrolysis and fermentation (SHF), and semi-simultaneous saccharification and fermentation (SSF) [6]. And the ethanol conversion process of lignocellulosic materials generally includes four steps, namely, pretreatment, enzymatic hydrolysis, fermentation, and distillation [7-8]. One of the most promising pretreatment processes of lignocelluloses material is liquid hot water (LHW) pretreatment. Studies have indicated that additives can improve enzymatic hydrolysis and bioethanol fermentation of lignocellulosic biomass [9-10].

The present study aimed to confirm the influence of several additives on the cellulosic ethanol: such as enzyme, Triton, urea and Tween. The paper presented the results.

2. Materials and methods

2.1 Materials

Corn stover was collected from a field near Jinzhou New District (Dalian, China). Corn stover was manually cut into pieces of 4 cm to 7 cm in our laboratory. Corn stover was milled to particle sizes ranging from 20 mesh to 80 mesh by using a laboratory ball mill (Taijihuan Nanometer Limited

IOP Conf. Series: Earth and Environmental Science **59** (2017) 012055 doi:10.1088/1755-1315/59/1/012055

Company, Qinhuangdao, China). Samples were then homogenized and stored in a plastic bag for subsequent experiments. The urea, Tween and Triton (Triton X-100) were obtained from Chinese medicine group chemical reagent Co., Ltd.

2.2 LHW pretreatment

LHW pretreatment was conducted in a 200 mL steel tank, and 10 g of corn stover and 60 mL of deionized water were added in the tanks. And Put it in oil bath kettle. The pretreatment temperature was controlled at 195 ℃ and keep 20 min. After pretreatment, the water-insoluble solids (WISs) were separated by filtration using the B üchner funnel. The WISs were used for subsequent ethanol fermentation.

2.3 Semi-simultaneous saccharification and fermentation (S-SSF)

The WISs were used as substrates of the semi-simultaneous saccharification and fermentation (S-SSF). About 1 g of WISs and 1 g of enzyme were put in 100 mL Erlenmeyer flasks. Each flask contained 10 mL of pH 4.8 buffer. And the medium temperature was kept at 50 ℃ during the pre-hydrolysis phase. After pre-hydrolysis, the medium temperature was adjusted to a constant fermentation temperature and then maintained all throughout the following SSF phase. Afterward, about 1 mL of the activated yeast was added into the medium. The experiments were performed in a constant-temperature incubator for 72 h. The flasks were sealed with rubber stoppers and equipped with syringe needles to remove the generated carbon dioxide. The samples were collected at 0, 12, 24, 36, 48, 60, and 72 h for ethanol concentration determination.

2.4 Analysis methods

Ethanol content and ethanol concentration were measured by using the SBA-40D Biological Sensing Analyzer (Biology Institute of the Shandong Academy of Sciences, Jinan, China).

3. Results and discussion

3.1 Effect of enzyme addition on ethanol concentration

In the first group of assays, we only change the amount of enzyme added. We choose a set of enzyme content gradient: 0 g, 0.2 g, 0.4 g, 0.6 g, 0.8 g were added to the reaction system, to explore the effect of the amount of enzyme added to the fermentation process. The concentration of ethanol was measured. And the promotion rate (p) are calculated to equation (1):

$$p = \frac{C_0 - C_e}{C_0} * 100\% \quad (1)$$

Where p is the promotion rate compared with the control. C_0 and C_e are the initial and equilibrium concentration of ethanol (mg / 100 mL). The result is shown in Figure 1.

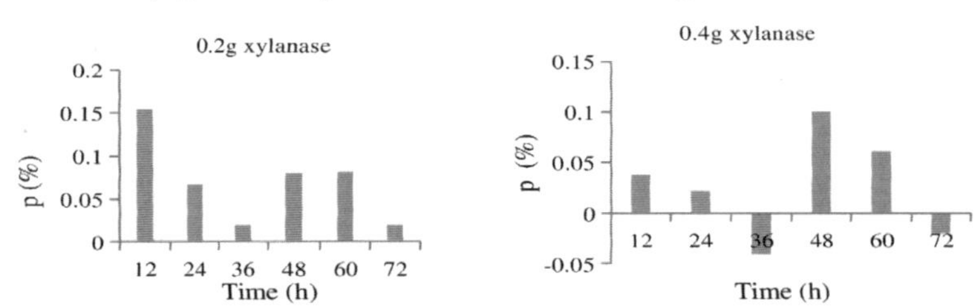

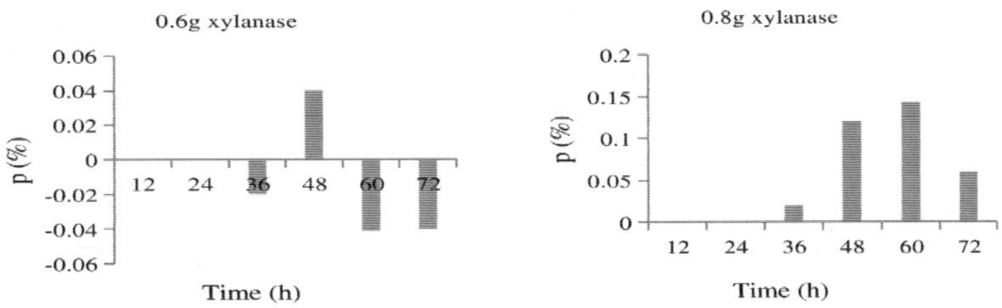

Figure 1. Effect of enzyme addition on ethanol concentration.

Figure 1 shows that 0.4 g and 0.6 g content of xylanase in the fermentation process can inhibit the phenomenon, while 0.2 g and 0.8 g can promote. And the promoting effect of xylanase on 0.2 g was earlier than that of other enzymes. Owing to the causes of economic benefit, recommended dosage by 0.2 g xylanase.

3.2 Effect of Triton combined with xylanase on ethanol concentration

Based on adding different contents of xylanase, 1mL 0.1% Triton was added. The concentration of ethanol was measured to explore the effect of Triton. The result is shown in Figure 2. Figure 2 shows that based on adding different contents of xylanase, the addition of Triton make the promoting effect improve a lot. But when the enzyme content is too high, then adding Triton but not conductive to fermentation.

3.3 Effect of urea on ethanol concentration

The effect of urea on ethanol concentration was investigated. 0.1% concentration of urea was added. The concentration of ethanol was measured. The result is shown in Figure 3.

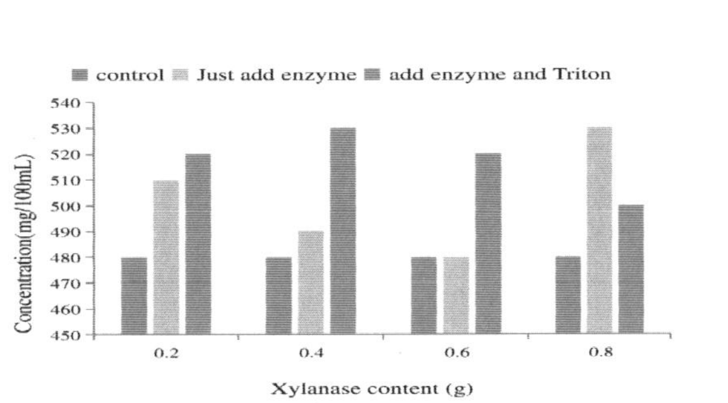

Figure 2. Effect of Triton combined with different content Xylanase on ethanol concentration.

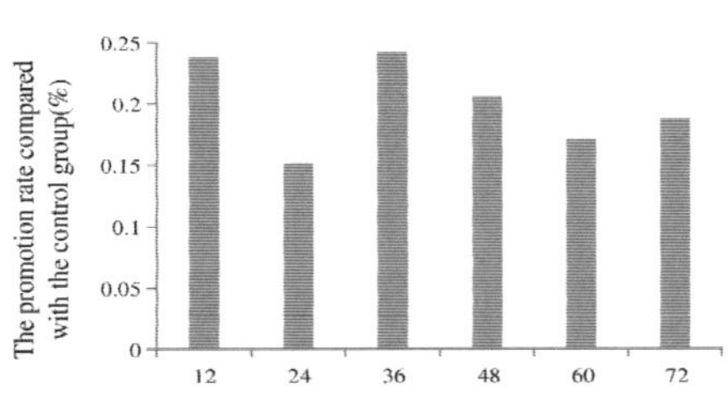

Figure 3. Effect of urea on ethanol concentration.

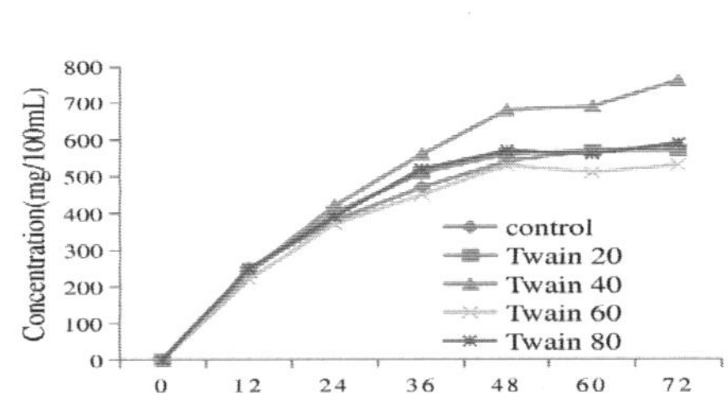

Figure 4. Effect of Tween series on ethanol concentration.

Figure 3 shows that urea showed a significant promoting effect. Ethanol concentration increased with the increase of time. Results show that urea in the fermentation of 36 h to achieve the best promotion rate of 24.2%.

3.4 Effect of Tween 20, 40, 60, 80 on ethanol concentration

Tween 20, Tween 40, Tween 60, and Tween 80 were investigated. The concentration of bioethanol was measured. The results are shown in Figure 4. Figure 4 shows that most of Tween has shown a clear promoting effect. Under the same experimental conditions, the effect of Tween series: Tween 40 > Tween 80 > Tween 20 > Tween 60. The effect of Tween 40 is very significant. It was started from 24 h. The ethanol concentration is significantly greater than the blank sample. With the fermentation time increasing, the difference increased gradually.

4. Conclusion

The positive effect of xylanase, urea, Tween series was observed. When the enzyme content is 0.2 g and 0.8 g, the promoting effect is the best, and especially when 0.2 g enzyme combines with Triton, its promoting effect is improved significantly. With the reasons of economic benefits, adding 0.2 g xylanase is the best choice. When adding urea as surfactant, ethanol concentration increased with the increase of time, and urea has achieved the best promotion rate of 24.2% in the fermentation of 36 h. For the Tween series, Tween 40 has the best effect starting from 24h, the ethanol concentration is significantly greater than the blank sample, and with the time increasing, the difference is also gradually increasing. And the effect of Tween 80 and Tween 20 is slight; Tween 60 does not show a promotion effect.

Acknowledgements

This study was financially supported by the National Science Foundation of China (No. 31370584, 31370582), the Basic Research Projects of Liaoning Education Department (No.2016J001), the Natural Science Foundation of Liaoning (No. 2015020640), the Key Laboratory of Liaoning (No.LZ2015005) and the State Key Laboratory of Pulp and Paper Engineering Open-end Foundation (No.201113).

References

[1] S.Ye, Y.C.Jia, Hydrolysis of lignocellulosic materials for ethanol production :a review, Bioresou. Technol.83 (2002) 1-11.

[2] Buruiana CT, Viziranu C, Garrote G, Parajó J C. 2014. Bioethanol production from hydrothermally pretreated and delignified corn stover by fed-batch simultaneous saccharification and fermentation. Energy & Fuels 28 (2): 1158-1165.

[3] Jing Zhao, Liming Xia. Simultaneous saccharification and fermentation of alkaline-pretreated corn stover to ethanol using a recombinant strain.Fuel Processing Technology. 90(2009)1193-1197.

[4] Fang X, Shen Y, Zhao J, Bao XM, Qu YB: Status and prospect of lignocellulosic bioethanol production in China. Bioresour Technol 2010, 101:4814-4819.

[5] Arvaniti E, Bjerre AB, Schmidt JE. 2012. Wet oxidation pretreatment of rape straw for ethanol production. Biomass and Bioenergy 39: 94-105.

[6] Dionisi D, Anderson JA, Aulenta F, McCue A, Paton G. 2015. The potential of microbial processes for lignocellulosic biomass conversion to ethanol: a review. Journal of Chemical Technology and Biotechnology 90(3): 366-383.

[7] LimayemA, Ricke SC: Lignocellulosic biomass for bioethanol production: current perspectives, potential issues and future prospects. Progress in Energy and Combustion Science 2012, 38:449-467.

[8] Kristensen JB, Börjesson J, Bruun MH, Tjerneld F, Jørgensen H: Use of surface active additives in enzymatic hydrolysis of wheat straw lignocelluloses. Enzyme Microb Technol 2007,

40:888-895.

[9] Lu J, Li XZ., Yang RF, Zhao J, and Qu YB: Tween 40 pretreatment of unwashed water-insoluble solids of read straw and corn stover pretreated with liquid hot water to obtain high concentrations of bioethanol. Biotechnol Biofuels 2013, 6:159-170.

[10] Yu, Q., Zhuang, X. S., Yuan, Z. H., Qi, W., Wang, W., Wang, Q., and Tan, X. S. (2013). " Pretreatment of sugarcane bagasse with liquid hot water and aqueous ammonia, " Bioresource Technol. 144, 210-215. DOI: 10.1016/j.biortech. 2013.06.078.

IOP Conf. Series: Earth and Environmental Science **59** (2017) 012056 doi:10.1088/1755-1315/59/1/012056

Research on analysis method for lateral displacement of buried pipeline

Liu Yuqing Zhang Zhenyong Liu Shaoxing

China Petroleum Pipeline Engineering Co.,Ltd., Langfang, Hebei, China 065000
liu_yuqing@cnpc.com.cn

Abstract. Oil and gas pipeline will inevitably pass through the area such as active fault, mine goaf and soft soil where ground movement is likely to happen. In such area, the pipeline is easy to move laterally along the direction vertical to axial direction with the movement of ground. Displacement will occur at the interface of external constraint section and section with large displacement due to continuity of deformation of pipeline, i. e. there is a deflection at the end of external constraint section. In this paper, deflection equation for lateral movement of buried pipeline is derived depending on fundamental principle in structural mechanics for elastic foundation beam. It is verified by means of finite element method that the equation has high accuracy and can reflect the deformation of buried pipeline. The equation presented in this paper lays a foundation for establishing analysis method for lateral displacement of pipeline and its engineering application.

1 Introduction

The pipeline needs to pass inevitably pass through the area such as active fault, mine goaf and soft soil where ground movement is likely to happen[1][2][3]. As the ground moves, the pipeline laid in the ground will move along the direction vertical to the axial direction of pipeline, resulting to increased stress on pipeline that may threaten to the safety of pipeline. Currently, deformation calculation and stress analysis of pipeline can only by means of finite element method[4]. This method is complex and not convenient for the application by personnel of pipeline design, construction and operation. A simple and easy-to-use deflection curve equation for lateral displacement of buried pipeline is derived in this paper on the basis of mechanical theory to facilitate engineering application.

2 Simplified mechanical model

Displacement will occur at the interface of external constraint section and section with large displacement due to continuity of pipeline deformation, i.e. there is a deflection at the end of external constraint section. When the section with large displacement of buried pipeline deflects, an inverse force proportional to the deflection is imposed on pipeline by soil. This force can be represented with elastic foundational beam [5] in structural mechanics. The mechanical model of buried section and lateral displacement for lateral movement of pipeline is shown in figure 1.

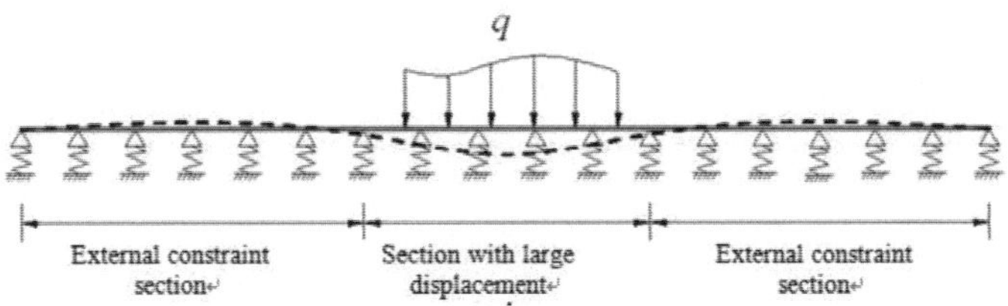

Figure 1 mechanical model for lateral movement of pipeline

The equivalent model can be established by breaking a point in mechanical model and adding moment, M0, and support force, R0, equal in size and opposite in direction at both sides. It is known from the condition of continuous displacement [5,6] that both buried constraint section and laterally deformed section will produce the same rotational angle, θ, and displacement, Δ at the broken location. A deflection curve is derived in this paper on this basis of simplified model.

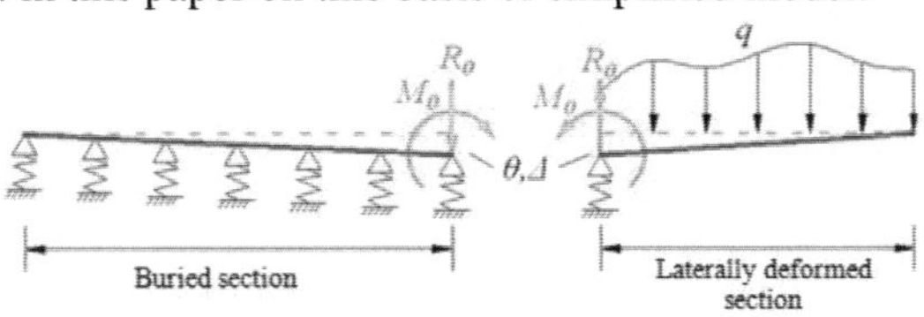

Figure 2 Equivalent mechanical model for lateral displacement of pipeline

3 Analysis on deflection curve for buried section of pipeline

3.1 Basic assumptions

Following basic assumptions are made for deriving the deflection curve for buried constraint section of pipeline: (1) The buried section of pipeline appears very small deformation and the deformation follows geometric linear relationship;(2) The pipe material does not yield, conforming to Hook's law; (3) The soil does not yield. Displacement of soil and inverse force follows linear relationship;(4) soil-pipeline interaction is assumed as the linear zone of horizontal lateral soil spring in GB50470-2008[3].

3.2 Mechanical model and coordinate system

The coordinate system defining buried constraint section of pipeline is shown in figure 3.

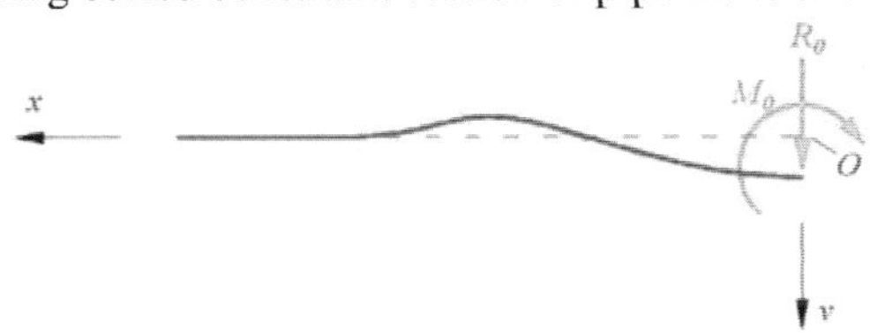

Figure 3 Mechanical model and coordinate system for buried constraint section of pipeline

3.3 Deflection curve for buried pipeline

Equation 1[3,4] is a differential equation for bend beam:

$$EI\frac{d^4v}{dx^4} = q_v \qquad \text{(equation 1)}$$

where,E = elasticity modulus, in Pa;I = inertial moment, in m^4;v = deflection, in m;x = the distance along axial direction of pipeline, in m;q_v = distributed payload in unit length vertical to axis of pipeline, in N/m.

For buried pipeline, the distributed payload $q_v = -kv$ satisfies when the pipeline deflects. Substitute $q_v = -kv$ into equation 1, then

$$EI\frac{d^4v}{dx^4}+kv=0 \quad \text{(equation 2)}$$

$$k=p_u/X_u \quad \text{(equation 3)}$$

Where,k = stiffness/ stiffness coefficient of soil surrounding the pipe, in N/m^2;p_u = lateral action force between pipeline and soil (unit length), in N/m;X_u = yield displacement of horizontal lateral soil spring, in m.p_u and X_u can be calculated as per GB50470-2008[3].
The general solution of equation 2 is

$$v=e^{\alpha x}(B_1\cos\alpha x+B_2\sin\alpha x)+e^{-\alpha x}(B_3\cos\alpha x+B_4\sin\alpha x) \quad \text{(equation 4)}$$

Where, B_1, B_2, B_3 and B_4 are undetermined coefficients; α is characteristic value of pipeline, in 1/m. the expression for α is as below

$$\alpha=\sqrt[4]{\frac{k}{4EI}} \quad \text{(equation 5)}$$

The relationships respectively between moment, M, and shield force, N, at the cross section of any point in pipeline is as follows:

$$M=EI\frac{d^2v}{dx^2} \quad \text{(equation 6)}$$

$$N=EI\frac{d^3v}{dx^3} \quad \text{(equation 7)}$$

The buried pipeline is supposed as a structure with infinite length. If x tends to infinite, the moment and shield force of the cross section, M and N, created by the moment and support force, M_0 and R_0, at point O tends to 0. Combining equation 6 and 7, the undetermined coefficients in equation 3 and 4, B1 and B2, are all equal to zero, i.e. $B_1=B_2=0$. Then, equation can be transformed into equation 8:

$$v=e^{-\alpha x}(B_3\cos\alpha x+B_4\sin\alpha x) \quad \text{(equation 8)}$$

Suppose that the moment and support force at cross section be M_0 and R_0, then:

$$M_{x=0}=EI\frac{d^2v}{dx^2}\Big|_{x=0}=-2EI\alpha^2B_4=M_0 \quad \text{(equation 9)}$$

$$N_{x=0}=EI\frac{d^3v}{dx^3}\Big|_{x=0}=2EI\alpha^3(B_3+B_4)=R_0 \quad \text{(equation 10)}$$

Combining equation 9 and 10, following equations can be derived:

$$B_3=\frac{R_0+\alpha M_0}{2EI\alpha^3} \quad \text{(equation 11)}$$

$$B_4=-\frac{M_0}{2EI\alpha^2} \quad \text{(equation 12)}$$

Now, the expression of deflection curve for buried pipeline is derived.

4 Precision verification of deflection curve

Table 1 Verification parameters for deflection curve

parameter	symbol	unit	value
diameter	D	m	1.219
wall thickness	t	m	0.0184
elasticity modulus	E	Pa	2.10×10^{11}
Horizontal lateral action force between soil and pipeline	p_u	kN/m	185.103

Yield displacement of horizontal lateral soil spring	X_u	m	0.096
pipe tip lateral force	H0	N	400000
pipe tip moment	M0	N.m	260000

Table 2 summarizes the displacements of all points along axis of pipeline calculated with the analytical equations derived in this paper and the results calculated with finite element method. The displacement curve for all points along the axis of pipeline is plotted in figure 4.

Table 2 comparison of calculation results between analytical method and finite method

x (m)	Displacement (m)		x (m)	Displacement (m)		x (m)	Displacement (m)		x (m)	Displacement (m)	
	analytical	Finite element		analytical	Finite element		analytical	Finite element		analytical	Finite element
0	-0.0519	-0.0508	16	0.0029	0.0028	32	0.0010	0.0010	48	-0.0002	-0.0001
2	-0.0394	-0.0383	18	0.0036	0.0035	34	0.0006	0.0006	50	-0.0001	-0.0001
4	-0.0281	-0.0272	20	0.0037	0.0036	36	0.0003	0.0003	52	-0.0001	-0.0001
6	-0.0186	-0.0180	22	0.0035	0.0034	38	0.0001	0.0001	54	-0.0001	-0.0001
8	-0.0111	-0.0106	24	0.0031	0.0029	40	0.0000	0.0000	56	-0.0001	-0.0001
10	-0.0054	-0.0051	26	0.0025	0.0024	42	-0.0001	-0.0001	58	-0.0001	0.0000
12	-0.0013	-0.0012	28	0.0020	0.0019	44	-0.0001	-0.0001	60	0.0000	0.0000
14	0.0013	0.0013	30	0.0014	0.0014	46	-0.0002	-0.0001			

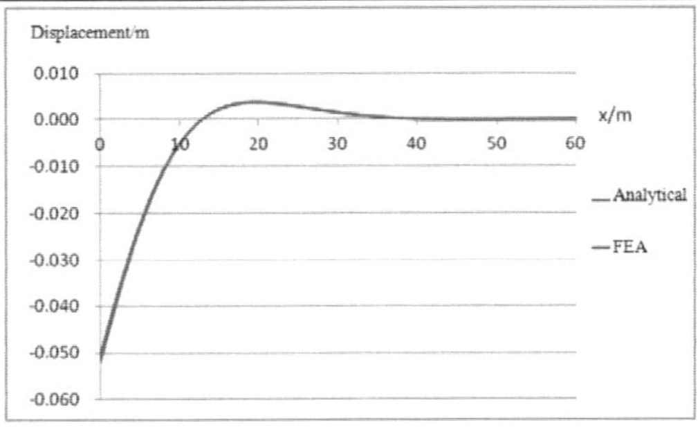

Figure 4 Displacement curve for all points along the axis of pipeline

It is indicated from the calculated results that the tendency of displacement and rotation angle calculated from analytical method derived in this paper is consistent with those by finite element method. The maximum error ratio is 2.17%, which is less than 5%, indicating better precision.

5 Conclusion

Equations of deflection curve for displacement of buried pipeline are derived in this paper on the basis of fundamental principle of elastic foundation beam in structural mechanics. The precision of these equations is verified by means of finite element method. The results show that the equations deflection curve derived in this paper have high precision. The deformation tendency is consistent with the results from finite element method, which can reflect the deformation of buried pipeline.

References

[1] GB50251-2015.Code for Design of Gas Transmission Pipeline Engineering[S]. Beijing: China plan press, 2015.
[2] GB50253-2014.Code for Design of Oil Transmission Pipeline Engineering[S]. Beijing: China plan press, 2015.
[3] GB 50470-2008.Seismic Technical Code for Oil and Gas Transmission Pipeline Engineering [S]. Beijing: China plan press, 2009.
[4] Shuai Jian. Pipeline Mechanics [M]. Beijing：Science Press，2010.

[5] Chen Tieyun,Chen Bozhen. Structural Mechanics of Ships [M]. Shanghai：Shanghai Jiao Tong University Press，1991.
[6] Long Yuqiu,Bao Shihua,Yuan Si. structural mechanics [M]. Beijing：Higher Education Press，2012.
[7] Liu Yuqing,Zhong Guixiang,Zhao Zifeng. Improvement of elastic laying bending stress formula in oil and gaspipeline standard [J]. Oil & Gas Storage and Transportation,2014,33(4):429-432.

Risk assessment of drought-induced water scarcity in upper and middle reaches of Xiu River

L N Liu and W L Liu

Jiangxi Provincial Key Laboratory of Hydrology-Water Resources and Water Environment, Nanchang Institute of Technology, Nanchang 330099, China

Abstract. In this paper, an index-based approach was presented to assess water scarcity risk induced by drought based on relationship between water use and water supply. Firstly, the index system of water scarcity risk was established from the viewpoint of hazard, exposure and vulnerability, which included drought hazard index (frequency, duration, and severity of droughts), exposure index (water use) and vulnerability index (relative water shortage). Then, an integrated risk index, namely water scarcity risk index (*WSRI*), was proposed using a multiplicative formula of drought hazard index, exposure index and vulnerability index to quantitatively assess drought-induced water scarcity risk. Finally, the *WSRI*-based approach was applied to a case study of assessing water scarcity risk in the upper and middle reaches of Xiu River, China. The results show that the drought-induced water scarcity risk of upper and middle reaches of Xiu River was at a low level, which was in line with the situation about the level of water resources exploitation and utilization in the study area. This study can serve as a reference for further water resources planning in the Xiu River Basin, and other relevant studies.

1. Introduction

Water resources are important strategic resources and play a critical role in human society and healthy ecosystems. However, with the rapid socio-economic development, water resources are facing a more and more serious situation of shortage, and water scarcity has already been the outstanding problem restricting the socio-economic sustainable development. Therefore, how to resolve the problem has attracted extensive attention from scholars at home and abroad [1-3].

China is one of the countries most threatened by water scarcity. Especially in recent years, along with high-speed economic development, population growth, rapid urbanization and industrialization, water scarcity is becoming more and more severe [4, 5]. Many researches have been done to evaluate and cope with water scarcity in Northern China. For example, Ruan et al. [3] assessed the water scarcity risk of capital area in China including Beijing and Tianjin with the traditional fuzzy comprehensive assessment method. Wang et al. [6] analyzed the water scarcity risk of Beijing from 1979 to 2005 based on fuzzy probability. Zhang et al. [7] studied water balance in the Haihe River Basin and evaluated water scarcity risk during 1994-2007. However, there is not enough attention paid to temporary water scarcities in Southern China. In such areas, the water scarcity occurs in drought conditions due to prolonged absence of precipitation and soil moisture deficits. Although not all water resources systems suffer water scarcity under a given drought situation, drought conditions can impose additional stress on water supply systems, which have to face the risk of water scarcity induced by drought. Therefore, it is still a necessary consideration for sustainable development of water resources to develop methods to analyze and control the risk of water scarcity during drought periods, with the ultimate goal of decreasing the loss before the occurrence of risk.

Many scholars present their understanding about the risk of water scarcity and also develop several indices to evaluate the risk of water scarcity. For instance, Hashimoto et al. [1] proposed three evaluation indices (reliability, resiliency and vulnerability) to evaluate the possible performance of water resource systems. Ruan et al. [3] put forward the indexes including the risk rate, weakness, possibility of recovery, period for reappear and risk level to assess water scarcity risk. Wang et al. [6] defined the risk as the product of the probability and the impact degree due to water scarcity risk. These indices prove to be useful and helpful in giving us an evaluation of water scarcity risk. Nevertheless, the indices do not address water scarcity risk induced by drought. In fact, measuring potential water-supply problems caused by droughts needs to jointly consider severity and spatial extension of droughts, water use relative to reliable water supply, and drought-resistant measures, which few studies are devoted to [8-11]. Furthermore, most of the risk indices established in such studies are based on the evaluation indices proposed by Hashimoto, but these indices reflect the performance of the water resources system, and the nature of risk is not taken into account [12,13]. Therefore, a suitable performance index is necessary to quantify the characteristics of water scarcity induced by drought. In this paper, the integrative risk index, namely water scarcity risk index (*WSRI*), was built and applied to assess the risk of water scarcity induced by drought in the upper and middle reaches of Xiu River.

2. Risk assessment of drought-induced water security

Water scarcity risk is affected by meteorological factors and non-meteorological factors. So it is difficult to combine all these factors to analyze the risk of water scarcity. In this paper, the drought as the main factor was taken into account on the research of water scarcity risk. Based on the natural disaster risk index established by Mileti [14] and the performance evaluation system for water resources system proposed by Hashimoto, here, water scarcity risk induced by drought is regarded as the consequence of the interaction of drought hazard, exposure and vulnerability of hazard-affected body. In terms of indices of hazard, exposure, and vulnerability, a multiplicative formula is proposed to link these three indices to define water scarcity risk during drought periods.

2.1. Drought hazard index

In this paper, the hazard of water scarcity risk during drought periods is defined as the probability of failure in different drought intensity, including drought severity and the drought probability. The standardized precipitation index (SPI), developed by McKee et al. [8, 15], is employed to detect occurrences of droughts.

The SPI is computed firstly by fitting precipitation data summed over the time scale to a suitable statistical distribution from which probabilities are transformed to the standard normal distribution, which can be expressed as follows:

$$SPI = \phi^{-1}(F_Y(y)) \tag{1}$$

where $\phi^{-1}(\cdot)$ is the inverse of a standard normal function; $F_Y(\cdot)$ is the fitted cumulative distribution function (CDF) of precipitation; y is the aggregated amount of consecutive monthly precipitation for a fixed time scale of interest (e.g. 1, 3, 6, 9, 12 ... months).

Based on SPI series for multi time scales, cumulative SPI value of a drought event is computed as follows:

$$S = -\sum_{i=1}^{D} \mathrm{SPI}_i \tag{2}$$

where D and S are the drought duration and severity of a drought event, respectively.

To compare the impacts of droughts at different regions, the probability of all drought events in each region during the analysis period is summed to represent overall impacts of droughts, which is given by

$$P_k = -\sum_{i=1}^{N_k} P(d_i, s_i) \tag{3}$$

where P_k is the summed probability of all drought events in the kth region; $P(d_i, s_i)$ is the probability of a drought event with drought severity s_i and drought duration d_i which can be estimated by a copula-based bivariate distribution function; N_k is the number of drought events of the kth region.

For easy comparison, the summed probability of droughts at different regions is rescaled between 0 and 1, which is defined as the hazard index of water scarcity risk, that is

$$HI_k = 1 - \left(\frac{\max(P) - P_k}{\max(P) - \min(P)}\right)^\alpha \tag{4}$$

where HI_k is the hazard index of water scarcity risk in the kth region; α is a rescale parameter.

2.2. Exposure index

The exposure factor of water scarcity risk measures level of assets, resources and populations exposed to threats of water scarcity induced by droughts. Since this study aims to evaluate water scarcity risk induced by drought based on the relationship between water use and water supply, the exposure index of water scarcity risk is defined as the total water use in a region which is a reasonable indicator to represent level of resources exposed to droughts. In this study, water uses mainly involve with agricultural water use, industrial water use and domestic water use. Compared with the agricultural water use, the domestic and industrial water uses have a very limited tolerance to water scarcity. Thus, a total water use is defined as the weighted total water use in which the weight on the agricultural water use is less than one.

$$U = U_d + U_i + \beta U_a \tag{5}$$

where U is the weighted total water use; U_a, U_i and U_d is agricultural, industrial and domestic water use, respectively; β is the weight on agricultural water use.

Also, the same rescale scheme is adopted to rescale the value of the weighted total water use ranged between 0 and 1, which is defined as the exposure index (EI) of water scarcity risk in this study. That is

$$EI_k = 1 - \left(\frac{\max(U) - U_k}{\max(U) - \min(U)}\right)^\alpha \tag{6}$$

where EI_k is the exposure index of the water scarcity risk of the kth region.

2.3. Vulnerability index

The vulnerability of water scarcity risk measures the potential loss of water resources under drought stress. In a certain period of water supply, the more serious the water scarcity of drought is, the more vulnerable the water supply system is, which means the more severe the loss of drought is. The ratio of the average water scarcity and the average water demand is an important criterion to evaluate the water scarcity vulnerability of that region. That is

$$V = \frac{\sum_{i=1}^{m}(D_i - S_i)}{\sum_{i=1}^{m} D_i} \tag{7}$$

where V denotes the ratio of the average water shortage and the average water demand; D and S are the water demand and the water supply, respectively.

A high value of this index V means that the system is not reliable to satisfy its demands and is prone to water scarcity. Also, the value of V is rescaled between 0 and 1, which is called as the vulnerability index of water scarcity risk (VI) in this study.

$$VI_k = 1 - \left(\frac{\max(V) - V_k}{\max(V) - \min(V)}\right)^{\alpha} \quad (8)$$

where VI_k is the vulnerability index of the water scarcity risk of the kth region.

2.4. Water scarcity risk index

The water scarcity risk consists of three components: hazard, exposure, and vulnerability. Thus, water scarcity risk index ($WSRI$) is defined as the product of drought hazard index, exposure index and vulnerability index:

$$WSRI_k = HI_k \times UI_k \times VI_k \quad (9)$$

where $WSRI_k$ denotes the water scarcity risk of the kth region.

Obviously, the value of $WSRI$ ranges between 0 and 1. A risk-prone region is that water use is significantly affected by droughts, and/or severity of droughts is greater, and/or reliable water supply is lack. Null water scarcity risk occurs only for the case of no water use ($EI = 0$) and/or fully reliable water supply ($VI = 0$).

3. Case study

3.1. Study area

The developed $WSRI$ was applied to assess water scarcity risk during drought periods in the upper and middle reaches of Xiu River, China. The Xiu River basin is located in the west of Jiangxi Province, China. It has a drainage area of 14,797 km^2, and situated between latitudes 28°40′-29°30′N and between longitudes113°55′-116°01′E. The length of the main stream is 357 km and the mean annual runoff is approximately 135×10^8 m^3. In view of the representation of sites and accessibility of data, the upper and middle reaches of Xiu River were selected as the study area. The region covers a total area of 9,243.8 km^2, and includes three counties, namely, Xiushui, Wuning and Tonggu in Jiangxi province. This area is dominated by a subtropical humid monsoon climate. The mean annual temperature is about 17°C, and the mean annual precipitation is approximately 1629 mm, about 40%of which is received from April to July. Because of the impact of high solar radiation and the circulation of monsoon weather patterns, the streamflow in the area varies significantly by season, which makes more precipitation occur during the monsoon period each year, while less precipitation in late autumn and winter. In general, the study area is characterized by abundant precipitation. However, due to uneven spatial and temporal distribution of rainfall, occasional droughts may lead to limited water availability during such periods. With the rapid socio-economic development, demand for water is also increasing in the region. Therefore, effective evaluation of water scarcity risk during drought periods in the upper and middle reaches of Xiu River is of great importance for a sustainable social and economic development in the region.

3.2. Water scarcity risk assessment

According to the available water resources and water consumption in the upper and middle reaches of Xiu River, a model of water resources optimum allocation was built, and the results for 2010 was presented in figure 1. From figure 1, it can be seen that, because of agriculture-based region, the ratio of agricultural water use accounted for a great proportion of total water consumption, which was about 80%, while the ratio for the domesticity and industry sector was low. It also shows that the water supply basically satisfied the needs of water use for all sectors except for agriculture sector, in which there was slightly water shortage with 3.45 million m^3 in the study area, 1.85 million m^3 in Xiushui and 1.6 million m^3 in Wuning county, respectively. Since the agricultural responses to the drought strongly which is the main factor of agricultural water shortage, here, we employed the *WSRI* to assess water scarcity risk deduced by droughts in the upper and middle reaches of Xiu River.

Three representative rainfall gauge stations with 52-year (1957–2010) daily precipitation were selected to calculate drought condition of the upper and middle reaches of the Xiu River. Based on the long-term precipitation series of three gauge stations, corresponding SPI series for multi time scales such as 1-, 3-, 6-, 12-month, were calculated for each station. Here, we defined drought events as a period during where SPI values remained below a threshold of -0.5. Thus, the drought duration (*D*)was the number of consecutive days for which SPI was less than -0.5, and the drought severity (*S*) was the cumulative deficit of the absolute value of SPI for that duration. Then, the drought duration and severity were abstracted from the SPI series of each rainfall gauge station's record. And all paired drought duration and severity from all regions were pooled to construct the joint probability distribution by empirical copula. Figure 2 shows the cumulative probability curve of the joint drought duration and severity based on the Copulas Frank function for Xiushui county. Thus, the probability of each drought event for all regions was estimated. According to equation (3) and equation (4), the hazard index of each region was obtained to evaluate the relative threats to water scarcity (Table 1).

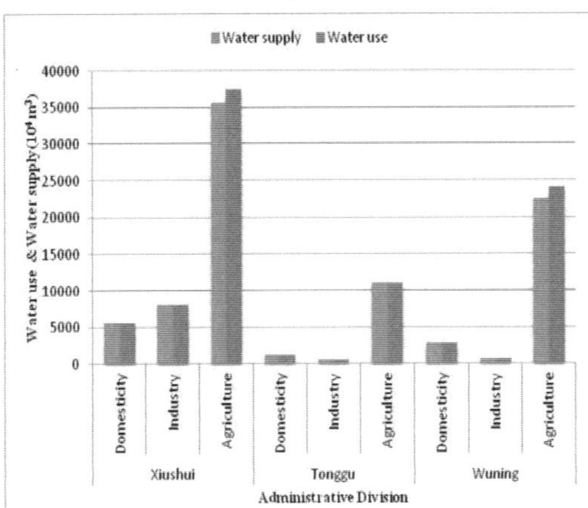

Figure 1. The allocation results of water resources in the upper and middle reaches of Xiu River.

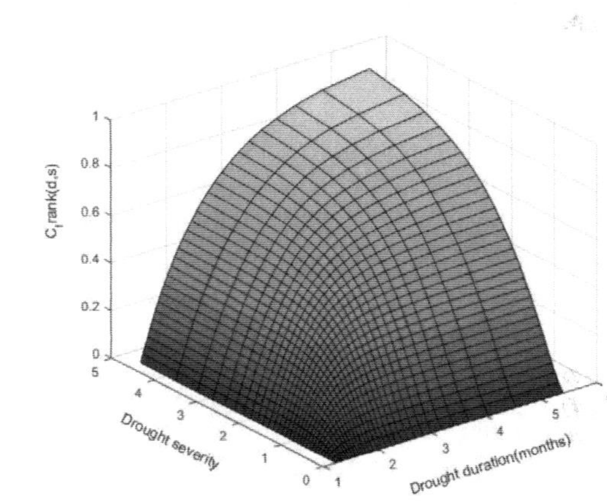

Figure 2.. The copula-based joint probability curve for Xiushui county.

According to the results of water resources optimum allocation, the exposure index of each region was calculated in the upper and middle reaches of Xiu River based on equation (5) and equation (6), and also the vulnerability index of each region was calculated based on equation (7) and equation (8). Finally, according to equation (9), the value of the *WSRI* for each region was obtained (Table 1). Table 1 shows all risk indices during drought periods in the upper and middle reaches of Xiu River.

The *WSRI* value of each region demonstrates the status of water scarcity and depicts the real risk conditions (Table 1). Table 1 shows that *EI* value of each region was relatively close. Compared with *HI* values in Xiushui county and Tonggu county, *HI* values in Wuning county was relatively weakest. For the *VI* values and EI values, there was a large difference among three counties. The calculated *EI* ranged between 0.03 and 0.93, which the highest was 0.93 in Xiushui county, the minimum was 0.03 in Tonggu county. And the value of *VI* was ranged between 0 (Tonggu county) and 0.5 (Xiushui county). The *WSRI*, revealing the risk degree of water scarcity, had a range between 0 (Tonggu county) and 0.17(Xiushui county), which, in general, implied a low level of water shortage problem in the upper and middle reaches of Xiu River during drought periods. Although Xiushui county and Tonggu county had similar drought risk, water scarcity risk in Xiushui county was higher than in Tonggu county due to the less vulnerability of water scarcity risk in Tonggu county. Meanwhile, in terms of the *WSRI* values, there was a certain degree of the water scarcity risk in Xiushui county and Wuning county, but it was at a low level, as long as we take the corresponding measures, the water scarcity risk can be reduced. And the results were in line with the situation about the level of exploitation and utilization and scarcity of water in the study area.

Table 1. Risk indices during drought periods in the upper and middle reaches of Xiu River.

Zone	*P*	*HI*	*U*	*EI*	V	*VI*	*WSRI*
Xiushui	1.345	0.57	51199.09	0.93	0.04	0.33	0.17
Tonggu	1.342	0.57	12899.66	0.03	0.00	0.00	0.00
Wuning	1.290	0.50	27529.86	0.50	0.06	0.50	0.12

4. Conclusion

Although not all water resources systems suffer water scarcity under a given drought situation, drought conditions can impose additional stress on water supply systems, which have to face the risk of water scarcity induced by drought. Therefore, it is still a necessary consideration for sustainable development of water resources to develop methods to analyze and control the risk of water scarcity during drought periods. In this paper, an integrated risk index (*WSRI*), was proposed using a multiplicative formula of drought hazard index, exposure index, and vulnerability index to quantitatively assess drought-induced water scarcity risk based on relationship between water use and water supply. The proposed approach was applied to assess water scarcity risk during drought periods in the upper and middle reaches of Xiu River, China. The results show that the water scarcity risk of the upper and middle reaches of Xiu River during drought periods was at a low level, which was consistent with the situation about the level of exploitation and utilization and scarcity of water in the study area. Therefore, the proposed approach can provide a tool for drought-induced water scarcity risk analysis, future water-resources planning and economic developments in the Xiu River Basin, and other relevant or similar studies.

Acknowledgements

This work was financially supported by CRSRI Open Research Program(Program SN CKWV20132 18/KY), National Natural Science Foundation of China(51309130) and the Natural Science Foundatio n of Jiangxi Province(20132BAB213025).

References

[1] Hashimoto, T, Stedinger J R and Loucks, D P 1982 *Water Resour. Res.* **18** 14-20
[2] Rijsberman F R 2006 *Agr Water Manage* **80** 5–22
[3] RUAN, B Q, HAN, Y P, Wang, H and JIANG, R F 2005 *J Hydraul Eng* **36** 906-12
[4] Wang H and Wang J H 2012 *Bull Chin Acad Sci* **3** 351-58
[5] Gleick, P H, Allen, L, Cohen, M J, Cooley, H, Christian-Smith, J, Heberger, M and Schulte, P 2011 *China and Water* **7** 79-100
[6] Wang, H R, Qian, L X, Xu, X Y and Wang, Y 2009 *J Hydraul Eng* **40** 813-21
[7] Zhang, S, Meng, X, Hua, D, Chen, J, Li, J, Zhang, Y and Xia, J 2012 J. *Resour. Ecol.* **2** 362-69
[8] Shiau, J T and Hsiao, Y Y 2012 *Natural hazards* **64** 237-57
[9] Jinno, K, Xu, Z X, Kawamura, A, and Tajiri, K 1995 *International Journal of Water Resources Development* **11** 185-204
[10] Merabtene, T, Kawamura, A, Jinno, K and Olsson, J 2002 *Hydrological Processes* **16** 2189-208
[11] Feng P 1998 *Journal of Natural Resources* **13** 139-43
[12] Qian, L, Wang, H and Zhang, K 2014 *Water Resour. Res.* **28** 4433-47
[13] Renfroe, N A and Smith, J L 2002 *Threat/vulnerability assessments and risk analysis* (Cambridge: Applied Research Associates Inc)

[14] Mileti D S 1999 *Disasters by Design: A Reassessment of Natural Hazards in the United States* (Washington D C: Joseph Henry Press)
[15] Łabędzki, L 2007 *Irrigation and Drainage* **56** 67-77

Study on the desalination process and improvement effect of FGD-gypsum improving coastal saline-soil

Sanyan Du[1,2], Xiaohua Chen[2] and Meifang Hou[1,*]

[1]Institute of Ecological Technology and Engineering, Shanghai Institute of Technology, Shanghai 201418, China
[2]Shanghai Academy of Environmental Sciences, Shanghai 200233, China

Email: 562578846@qq.com

Abstract. To provide theoretical basis for the improvement of Shanghai coastal saline-soil, a soil column leaching experiment was carried out to study the process and improvement effect of soil desalination after soil applying 0.5%, 1%, 2.5%, and 5% mass ratio of FGD-gypsum, aimed at the characteristics of Shanghai coastal saline-soil. The result showed that, compared with the control, applying FGD-gypsum could increase soil saturated hydraulic conductivity by 5.1-6.5 times, increase leaching sodium quality by 11.0%-22.2%, decrease soil pH 0.41-0.11 units, decrease exchangeable sodium percentage (ESP) by 25.1%-76.7%. When FGD-gypsum improved the coastal saline-soil, the change rules of soil's physical and chemical properties were that, soil saturated hydraulic conductivity increased before leveling off, leaching sodium quality gradually increased, soil pH declined before leveling off, ESP reduced with increasing ratios of FGD-gypsum; soil leaching sodium ion concentration decreased sharply and then tended to equilibrium over time, and soil leaching out sodium was active in the first 2 days of leaching. In the process of FGD-gypsum speeding up soil desalination, soil leaching calcium quality and leaching sodium quality had relationship with soil hydraulic conductivity and the ratio of FGD-gypsum, and the leaching out sodium of FGD-gypsum-soil system was more accelerated when calcium ion concentration in soil leacheate was higher.

1. Introduction

The coastal saline-soil is an important reserve resource for the development of coastal regional economies. In China, there are around 66,700 km^2 coastal saline-soil, which is distributed in coastal areas, such as Shandong, Hebei, Liaoning, Zhejiang, Fujian, Guangdong [1]. The coastal areas are at the junction of the sea and the land, which has high groundwater level, tidal erosion, frequent supply of seawater and evaporation-concentration, thereby a large expanse of saline-soil is formed [2, 3]. The physical and chemical properties of coastal saline-soil are poor due to the high content of salt, low natural desalination rate, heavy salinization. It is important to improve coastal saline-soil for improving ecological environment, development and utilization of land resources and relaxing man-land contradiction in coastal areas.

In recent years, FGD-gypsum as a saline-soil amendment is popular for its low cost, large output and fast repair rate [4]. FGD-gypsum improving saline-soil is native to mineral gypsum improving saline-soil, whose principle is that calcium ions could promote the aggregation of clay particles and displace excess sodium on soil colloids [4, 5]. Compared with mineral gypsum, FGD-gypsum has better physical and chemical properties and economic effects, such as finer particles, higher solubility, higher purity, greater production, lower cost and so on. The United States has applied FGD-gypsum in agriculture. In 2008, the United States EPA issued a paper of Agricultural Uses for Flue Gas Desulfurization (FGD) Gypsum. In the United States, the agricultural utilization rate of FGD-gypsum increased from 2% to 5% in the total output of FGD gypsum from 2006 to 2013. In 2015, the USDA officially put the FGD-gypsum as a soil conditioner as the best national agricultural practice [5, 6].

Shanghai is on the tip of the Yangtze Delta, which has distinctive physicochemical properties, has around 960 km^2 coastal saline-soil which accounts for more than 15% of the city's land area. Shanghai as a globalizing metropolis is developing at a very high speed, thereby it is important to improve coastal saline-soil for Shanghai's ecological environment and development and utilization of land resources. Meanwhile, with the promotion of FGD technology, there are millions of tons of FGD-gypsum needed to be treated in Shanghai every year. Studies on FGD-gypsum improving saline-soil have many at home and abroad, but saline-soil in different areas has its unique physical and chemical properties and we can not blindly copy the results of previous studies, thereby it is needed to be treated separately according to concrete conditions of different types of soil for FGD-gypsum improving saline-soil. In this paper, FGD-gypsum improving Shanghai coastal saline-soil is explored through soil column leaching experiment. In soil column leaching experiment, soil saturated hydraulic conductivity, leaching concentration and quality of calcium and sodium during the soil column leaching process were measured and analyzed for exploring the desalination process. And soil pH, exchangeable sodium, ESP before and after FGD-gypsum improving soil were measured and analyzed for exploring the improvement effect.

2. Materials and Methods

2.1. Soil samples and FGD gypsum

Soil samples were taken from Chongming Island (E 121° 57'20", N 30° 31'25") in Shanghai, China, which is located in the Yangtze River. The region has a typical subtropical monsoon climate, the annual average temperature is 15.3℃, the hottest months are July and August (monthly average temperature 26.8-26.9℃), and the annual average evaporation capacity (1346 mm) is higher than the annual average precipitation (1078 mm) [6]. The soil samples were collected at plough layer, then air-dried, disaggregated, sieved (2 mm), and mixed well for soil column leaching experiment. The soil physicochemical properties were 2.64 $g{\bullet}cm^{-3}$ for soil density, 1.2 $g{\bullet}cm^{-3}$ for bulk density, 1.092 $mS{\bullet}cm^{-2}$ for conductivity, 38% for moisture content, 1.29×10^3 $mg{\bullet}kg^{-1}$ for soil total salt, 9.07 for pH, 30.61% for ESP, which belonged to typical severe saline-soil, and the soil main soluble salt was sodic salt [6].

FGD-gypsum was chosen from a steelworks with 94% $CaSO_4{\bullet}2H_2O$. The heavy metals contents in FGD-gypsum are shown in Table 1, which were determined by SGS Standard Technical Service Co. Ltd. Hg and As of FGD-gypsum were determined by atomic fluorescence spectrometry (AFS) while other heavy metals were determined by inductively coupled plasma atomic emission spectroscopy (ICP-AES) after FGD-gypsum being pretreated according to US EPA Method 3052. When examining the heavy metals contents of FGD-gypsum, it was apparent that all heavy metals contents were far lower than the Class II of Environmental Quality Standard for Soils of China (GB15618-1995). Therefore, using FGD gypsum for improving coastal saline-soil has a low ecological risk.

Table 1. Heavy metals contents in FGD-gypsum.

Sample	Heavy metals contents ($mg{\bullet}kg^{-1}$)							
	As	Cd	Cr	Cu	Ni	Pb	Zn	Hg
FGD-gypsum	2.2	0.37	5.5	8.7	13.2	19.8	11.3	0.36

Chinese environmental quality secondary standard for soils (GB15618–1995)	20	0.60	350	100	60	350	300	1.0

2.2. *Soil column leaching experiment*
Soil column leaching device is shown in Figure 1. Soil column leaching experiment was conducted with fine treatments, which were the natural soil, FGD-gypsum mixeded with natural soil under 0.5%, 1%, 2.5%, and 5% mass ratio of FGD-gypsum, each treatment was repeated thrice. After soil column filled, soil column was leached by deionized water, and deionized water leaching volume was designed to 1100 mL for simulating the annual average rainfall of Shanghai.

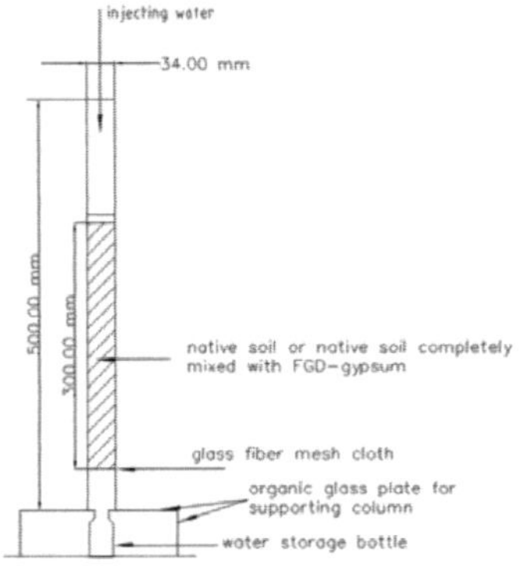

Figure 1. Soil column leaching experiment device.

2.3. *Analysis and method*
Selected soil characteristics were determined in accordance with the NY/T 1121-2006 Agricultural Standard of People's Republic of China-Soil Testing. The calcium ion and sodium ion concentration in the leachate samples were determined by a flame furnace atomic absorption spectrophotometer. Soil pH was determined on a 1:5 (soil:water, W/W) mixture using a glass electrode. The exchangeable, sodium ions, cations, were extracted using 1 mol/L NH_4OAc, whose concentrations were determined using an flamephotometer. The ESP was calculated by the percentage of exchangeable sodium in soil exchangeable cations where soil exchangeable sodium ions and exchangeable cations were expressed in $cmol\bullet kg^{-1}$. The soil saturated hydraulic conductivity was determined through the collected volume of leachate from soil column per hour [7].

2.4. *Analysis and method*
ORIGIN 9.0 and SPSS 17.0 were used to deal with the experimental data. Unless special instructions, the chart data are the average values of repeated treatment.

3. Results and Discussion

3.1. *Soil saturated hydraulic conductivity*
When examining figure 2, it is apparent that FGD-gypsum accelerate soil saturated hydraulic conductivity, which increases before leveling off with increasing ratios of FGD-gypsum. Compared with the control, applying 0.5%, 1%, 2.5%, and 5% of FGD-gypsum could increase soil saturated hydraulic conductivity by 5.1, 6.5, 6.5, 6.2 times respectively, soil saturated hydraulic conductivity had no significant difference among 1%, 2.5%, and 5% of FGD-gypsum ($p>0.05$), and the maximum soil saturated hydraulic conductivity was reached to 15.2 $mL\bullet h^{-1}$.

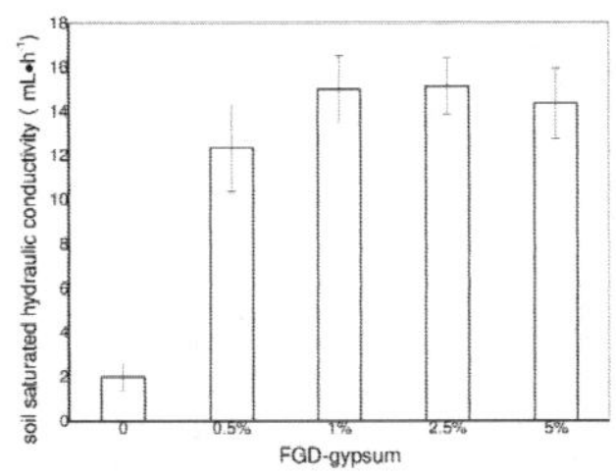

Figure 2. Soil saturated hydraulic conductivity under different ratios of FGD-gypsum application.

Study of Cheng et al. [7] has shown that FGD-gypsum could increase soil saturated hydraulic conductivity, whose change role is similar to that of the study. This is due to that, FGD-gypsum increases calcium ions in soil and positively charged calcium ions could reduce the dispersion of soil particles by promoting the aggregation of negatively charged clay particles, which improves the stability of soil structure and prevents soil crusting. Better soil particle aggregation and less crusting allow for greater soil hydraulic conductivity [4, 5].

3.2. Changes of calcium and sodium in desalination process

Sodium ions accumulation in soil solid phase or liquid phase is the main reason that causes the bad physical and chemical properties of saline-soil. Removal of sodium ions from soil is the key to the improvement of saline-soil. FGD-gypsum can improve saline-soil, whose main principle is that calcium ions act on saline-soil to remove sodium ions in the leaching process. Therefore, in this paper, only soil leaching calcium and sodium are involved in soil desalination process.

3.2.1. Calcium ion. When examining figure 3, it is apparent that, calcium ion concentration in soil leacheate decreases sharply and then tends to dynamic equilibrium (or continues to fall) over time, and the decreasing slope of calcium ion concentration increases with increasing ratios of FGD-gypsum, calcium ion concentration tending to dynamic equilibrium or continuing to fall is emerged in the first 2 days of soil column leaching. Meanwhile, figure 3 also shows that leaching calcium quality increases with increasing ratios of FGD-gypsum, whose increasing extent gradually decreases with increasing ratios of FGD-gypsum. Compared with the control, applying 0.5%, 1%, 2.5%, and 5% of FGD-gypsum could increase leaching calcium quality by 17.4, 41.8, 48.8, 49.7 times respectively.

Study of Cheng et al. [7] is consistent with our study for the change rule of calcium ion concentration in soil leacheate. Calcium ion concentration decreasing sharply is due to the decrease of FGD-gypsum with the leaching and reaction with soil and the inhibition function of more and more sulfate ion from FGD-gypsum for dissolution of FGD gypsum [8]. Then calcium ion concentration tending to dynamic equilibrium is due to calcium ion could get relatively adequate supplement from FGD-gypsum, or calcium ion concentration will fall. Leaching calcium ion is the kind of calcium ion that has not enough time to react with soil with the leaching.

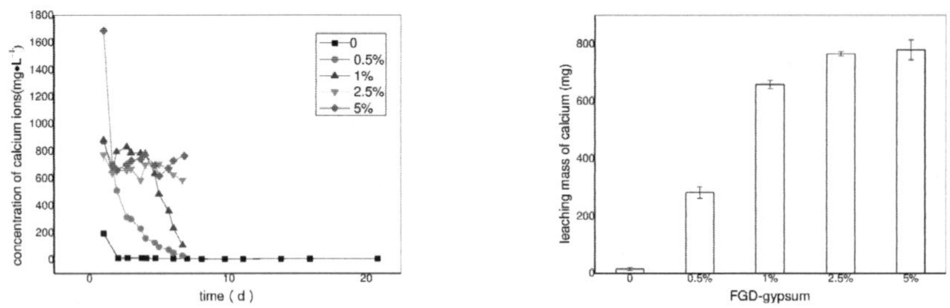

Figure 3. Leaching calcium quality and the relationship of leaching calcium ion concentration and time under different ratios of FGD-gypsum.

Salt moving with water moving, within a certain range, the faster soil hydraulic conductivity, the more leaching calcium quality of soils is, thereby the leaching calcium quality is affected by soil hydraulic conductivity. In figure 3, the loss of calcium quality slightly increases when the soil saturated hydraulic conductivity tends to equilibrium, which indicates that the loss of calcium quality is also affected by the ratio of FGD-gypsum and soil hydraulic conductivity is dominant for accelerating the loss of calcium in this study.

3.2.2. Sodium ion. When examining figure 4, it is apparent that, leaching sodium ion concentration decreases sharply and then tends to equilibrium over time, and the decreasing slope of sodium ion concentration increases with increasing ratios of FGD-gypsum, sodium ion concentration tending to dynamic equilibrium is emerged in the first 2 days of soil column leaching. Meanwhile, figure 4 also shows that leaching sodium quality increases with increasing ratios of FGD-gypsum. Compared with the control, applying 0.5%, 1%, 2.5%, and 5% of FGD-gypsum could increase leaching sodium quality by 11.0%, 16.8%, 18.0%, 22.0% respectively.

Combined figure 3 and Figure 4, the change rule of sodium ion concentration decreasing sharply is similar to that of calcium ion concentration, which is in good agreement with studies of Cheng et al. and Cui et al. [7, 9] that the leaching sodium of FGD-gypsum-soil system is more accelerated when the calcium ion concentration in soil leacheate is higher. In the stage of sodium ion concentration decreasing sharply, the soil leaching sodium includes water-extractable sodium and exchangeable sodium replaced by calcium ions from FGD-gypsum [9]. However, the soil leaching sodium only includes exchangeable sodium replaced by calcium ion in the stage of sodium ion concentration tending to equilibrium [7].

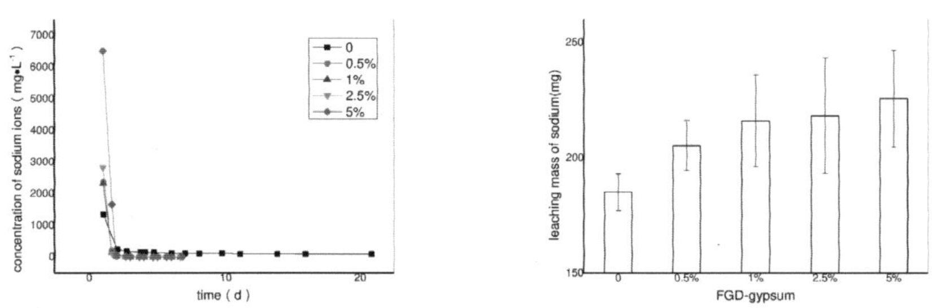

Figure 4. Leaching sodium quality and the relationship of leaching sodium ion concentration and time under different ratios of FGD-gypsum.

Salt moving with water moving, within a certain range, the faster soil hydraulic conductivity, the more leaching sodium quality of soils is, so the leaching sodium quality is affected by soil hydraulic conductivity. In figure 4, leaching sodium quality continues to increase when the soil saturated hydraulic conductivity tends to equilibrium, which indicates that the leaching sodium quality is also affected by the ratio of FGD-gypsum. Therefore, the leaching sodium quality of soil is affected by soil hydraulic conductivity and the ratio of FGD-gypsum [10].

3.3. Soil pH

When examining figure 5, it is apparent that, FGD-gypsum reduces obviously soil pH which decreases and then tends to equilibrium with increasing ratios of FGD-gypsum. Compared with the control, applying 0.5%, 1%, 2.5%, and 5% of FGD-gypsum could decrease soil pH by 0.41, 0.51, 1.11, 1.06 units respectively. Applying 0.5%, 1% of FGD-gypsum cannot decrease soil pH to below 8.0 which is suitable for plant growth. But 2.5%, 5% of FGD-gypsum could decrease soil pH to about 7.6, and soil pH has no significant difference between 2.5% and 5% of FGD-gypsum ($p>0.05$).

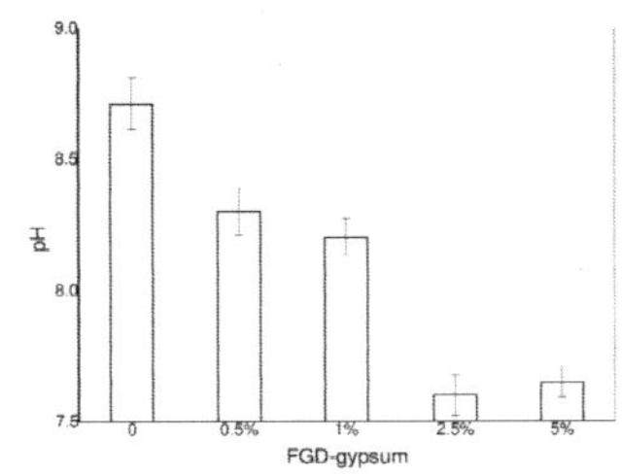

Figure 5. Soil pH under different ratios of FGD-gypsum application.

FGD-gypsum could decrease soil pH, but the decreasing extent of soil pH is not increased with increasing ratios of FGD-gypsum, which is consist with study of Sakai et al. [10]. Saline-soil's high pH is controlled by exchangeable sodium and water-soluble CO_3^{2-}+HCO_3^-. FGD-gypsum decreasing soil pH is due to that calcium ion from FGD-gypsum reacts with CO_3^{2-}+HCO_3^- to generate $CaCO_3$ deposition and replaces exchangeable sodium which is leached out [10].

3.4. Soil alkalinity

When examining figure 6, it is apparent that, the cation exchange capacity (CEC) of all treatments have no significant difference ($p < 0.05$), which means FGD-gypsum will not reduce the water retain and fertilizer of soil. Meanwhile, figure 6 also shows that, FGD-gypsum could reduce soil ESP and exchangeable sodium, the change rule of ESP is similar to that of exchangeable sodium, which reduces with increasing ratios of FGD-gypsum. Compared with the control, applying 0.5%, 1%, 2.5%, and 5% of FGD-gypsum could reduce soil ESP by 25.1%, 64.7%, 67.2%, 76.7% respectively. When the ratio of FGD-gypsum was more than 1%, soil ESP would not continue to decrease sharply, and soil exchangeable sodium decreased by up to 76.8%.

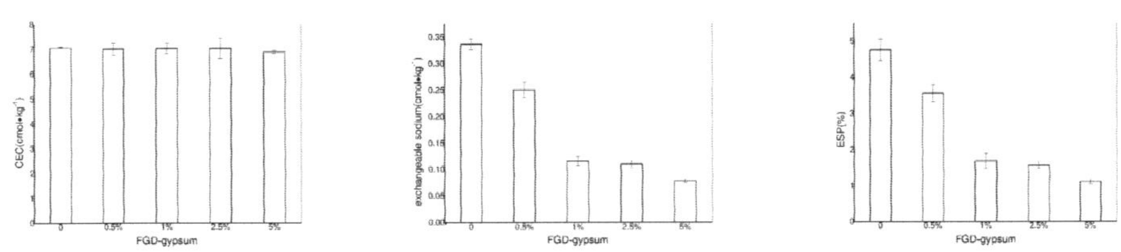

Figure 6. Soil CEC, exchangeable sodium, ESP under different ratios of FGD-gypsum application.

Our result showing FGD-gypsum would not reduce the water retain and fertilizer of soil is consistent with the study of Mao et al. [6]. Meanwhile, study of Mao et al. [6] also has shown that FGD-gypsum could reduce soil ESP and exchangeable sodium and the change rule of ESP is similar to that of exchangeable sodium. The key to the improvement of saline-soil is the reduction of soil ESP. FGD-gypsum reducing soil ESP is due to that, calcium ion's flocculation ability for soil clay particles is superior to that of sodium ion [4], calcium ions from FGD-gypsum could replace exchangeable sodium on soil colloids, then soil ESP is reduced.

4. Conclusions

FGD-gypsum could improve the physical and chemical properties of Shanghai coastal saline-soil in soil column leaching experiment. When FGD-gypsum improved the coastal saline-soil, the change rules of soil's physical and chemical properties were that, soil saturated hydraulic conductivity increased before leveling off, leaching sodium quality gradually increased, soil pH declined before leveling off, ESP reduced with increasing ratios of FGD-gypsum, soil leaching sodium ion concentration decreased sharply and then tended to equilibrium over time, and soil leaching out sodium was active in the first 2 days of leaching. In the process of FGD-gypsum speeding up soil desalination, soil leaching calcium quality and leaching sodium quality had relationship with soil hydraulic conductivity and the ratio of FGD-gypsum, and the leaching out sodium of FGD-gypsum-soil system was more accelerated when calcium ion concentration in soil leacheate was higher.

References

[1] Xiao X L et al 2014 Construction and application of evaluation index system for green vegetation in saline land of coastal areas *Guangdong Agricultural Science* **41** (23) 145-9

[2] Rengasamy P 2006 World salinization with emphasis on Australia *Journal of experimental botany* **57** (5), 1017-23

[3] Li X P et al 2014 Effects of Flue Gas Desulfurization (FGD) Gypsum on Desalination of Reclaimed Tidal Flat Soil *Journal of Environmental Engineering Technology* **4** (6) 502-7

[4] Walworth J 2012 Using gypsum and other calcium amendments in Southwestern Soils *College of Agriculture & Life Sciences* AZ1413

[5] US EPA 2008 Agricultural uses for flue gas desufurization (FGD) gypsum (EPA 530-F-08-009)

[6] Mao Y M, Liu X P and Dick W A 2016 Remediation of saline-sodic soil with flue gas desulfurization gypsum in a reclaimed tidal flat of southeast China *Journal of environmental sciences* **45** 224-232

[7] Cheng J R, Chen X H and Liu Z H 2014 The experimental study on the process and effect to the FGD-gypsum as an improvement in coastal saline-alkali soil *China Environmental Science* **34** (6) 1505-13

[8] Liu J and Yuan J J 2010 Calcium sulfate scaling and its advances in application of control techniques *Tianjin Chemical Industry* **24** (2) 10-4

[9] Cui Y et al 2016 Effect of desulfurized gypsum on leaching of major ions in heavy soda saline soil *Journal of Soil and Water Conservation* **30** (1) 310-14

[10] Sakai Y, Matsumoto S and Sadakata M 2010 Alkali Soil Reclamation with Flue Gas Desulfurization Gypsum in China and Assessment of Metal Content in Corn Grains *Soil and Sediment Contamination: An International Journal* **13** (1) 65-80

IOP Conf. Series: Earth and Environmental Science **59** (2017) 012059 doi:10.1088/1755-1315/59/1/012059

Distribution and Variation Law of Drought and Flood in Multi-scale in Poyang Lake Basin Based on SPI

Li-ying XIAO[1], Wen-wen ZHANG[1]

[1] School of Hydraulic and Ecological Engineering, Nanchang Institute of Technology, Nanchang, 330099, Jiangxi, China

Correspondence to: Li-ying XIAO (Liying-xiao@163.com)

Abstract. The Poyang lake basin has been affected seriously by drought and flood. The monthly precipitation data from 1951 to 2013 were used to analyze the variation law of drought and flood in Poyang lake basin based on standardized precipitation index in different time scales. Results indicated that the Poyang lake basin is suffered with frequent droughts and floods and the standardized precipitation index is suitable to analyze the law of drought and flood for the Poyang lake basin.

1 Introduction

Droughts and waterlog have strong social, economic and environmental impacts. This study focuses on the drought and waterlog over the Poyang lake basin by using the Standardized Precipitation Index (SPI) technique. The SPI (McKee et al. 1993[1]; Hayes et al. 1999[2]) is widely used to reveal droughts. It has been proven to be a useful and convenient tool just because the computation of SPI value is based on precipitation data only. Wu et al. (2001[3]) indicated that the SPI was developed to detect droughts and flood periods at different time scales. Morid et al. (2006[4]) showed by a case study that SPI was able to detect the onset of a drought, its spatial and temporal variation consistently. In this case, the SPI method is accepted in this study to analyze drought and waterlog variations.

2 Study region and data

Poyang Lake Basin is situated between 28° 24′ – 29° 46′ N longitude and 115° 49 ′ – 116° 46 ′ E latitude. The total drainage area of the water systems is 16.22×104 km^2. The basin is located on the south shore of the Yangtze River, in the northern part of Jiangxi Province. Poyang Lake basin has a subtropical flood climate characterized with a mean annual precipitation of 1,680 mm for the period of 1960–2007. Daily precipitation data from 14 rain gauging stations in Poyang lake basin were analyzed, and the data were obtained from the National Climate Center (NCC) of the China Meteorological Administration (CMA). Locations of the meteorological stations are shown in Fig. 1.

Table.1 The details of 14 meteorological stations in Poyang Lake Basin

	Meteorological station	Location		Annual Mean Rainfall (*mm*)	Length (year)	Period
		Latitude (E)	Longitude (N)			
1	Boyang	116.41°	29.00°	1613	59	1955～2013
2	Ganxian	115.00°	25.52°	1437	63	1951～2013
3	Guangchang	116.20°	26.51°	1728	60	1954～2013

IOP Conf. Series: Earth and Environmental Science **59** (2017) 012059 doi:10.1088/1755-1315/59/1/012059

4	Guixi	117.13°	28.18°	1858	61	1953～2013
5	Jian	114.55°	27.03°	1487	62	1952～2013
6	Jingdezhen	117.12°	29.18°	1786	61	1953～2013
7	Lushan	115.59°	29.35°	1938	59	1954～2013
8	Nanchang	115.55°	28.36°	1604	62	1951～2013
9	Nancheng	116.39°	27.35°	1679	61	1953～2013
10	Shuichuan	114.30°	26.20°	1448	63	1951～2013
11	Xiushui	114.35°	29.02°	1586	61	1953～2013
12	Yichun	114.23°	27.48°	1602	61	1953～2013
13	Yushan	118.15°	28.41°	1828	63	1951～2013
14	Zhangshu	115.33°	28.04°	1642	57	1957～2013

3 Methodology

SPI allows an analyst to determine the rarity of a drought at a given time scale of interest for any rainfall station with historic data. The gamma distribution is defined by its frequency or probability density function[5]:

$$g(x) = \frac{1}{\beta^{\alpha}\Gamma(\alpha)} x^{\alpha-1} e^{-x/\beta} \quad (1)$$

$$\Gamma(x) = \int_0^{\infty} x^{\alpha-1} e^{-x} dx \quad (2)$$

where: $\alpha > 0$ α is a shape parameter , $\beta > 0$ β is a scale parameter;$x > 0$ x is the precipitation amount , $\Gamma(\alpha)$ is the gamma function.The maximum likelihood solutions are used to optimally estimate α and β :

$$\hat{\alpha} = \frac{1+\sqrt{1+4A/3}}{4A} \qquad \hat{\beta} = \frac{\bar{x}}{\alpha} \qquad A = \ln\bar{x} - \frac{1}{n}\sum_{i=1}^{n}\ln\bar{x}_i$$

where:n = number of precipitation observations. The cumulative probability is given by:

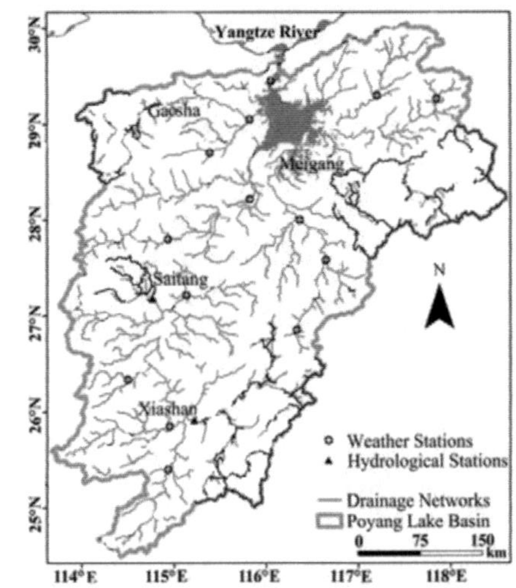

Fig. 1 Location of the study area

$$G(x) = \int_0^{x} g(x)dx \quad (3)$$

$$G(x) = \frac{1}{\beta^{\alpha}\Gamma(\alpha)} \int_0^{x} x^{\alpha-1} e^{-x/\beta} dx \quad (4)$$

Letting $t = x/\beta$, the above equation becomes the incomplete gamma function:

$$G(x) = \frac{1}{\Gamma(\alpha)} \int_0^{x} t^{\alpha-1} e^{-t} dt \quad (5)$$

Since a precipitation distribution may contain zeros, the cumulative probability becomes:

$$H(x) = q + (1-q)G(x) \quad (6)$$

where q is the probability of no rainfall on specified time scale. If m is the number of zeros in a precipitation time series, q can be estimated by m/n.The cumulative probability, $H(x)$, is then transformed to the standard normal random variable Z with mean zero, which is the value of the SPI.

$$For: 0 < H(x) \le 0.5, Where: t = \sqrt{\ln\left\{\frac{1}{[H(x)]^2}\right\}}, then: Z = SPI = -\left(t - \frac{c_0 + c_1 + c_2 t^2}{1 + d_1 t + d_2 t^2 + d_3 t^3}\right) \quad (7)$$

$$For: 0.5 < H(x) < 1, Where: t = \sqrt{\ln\left\{\frac{1}{[1+H(x)]^2}\right\}}, then: Z = SPI = \left(t - \frac{c_0 + c_1 t + c_2 t^2}{1 + d_1 t + d_2 t^2 + d_3 t^3}\right) \quad (8)$$

Where: $c_0 = 2.515517$, $c_1 = 0.802853, c_2 = 0.010328$,

Where: $d_1 = 1.432788$, $d_2 = 0.189269$, $d_3 = 0.001308$.

The drought classification[6] adopted in this study is defined in Table. 2.

Table. 2 Classification of droughts and floods by SPI

Classification	SPI	Classification	SPI
Extremely drought	≤-2.0	Slightly wet	[0.5, 1.0)
Severe drought	(-2.0, -1.5]	Moderately wet	[1.0, 1.5)
Moderately drought	(-1.5, -1.0]	Very wet	[1.5, 2.0)
Mild drought	(-1.0, -0.5]	Extremely wet	≥2.0
Near normal	(-0.5, -0.5)		

4 Results and discussion

4.1 Applicability of SPI method in the study of droughts in the Poyang Lake River Basin

The precipitation data from 14 stations were used to calculate the SPI value of each station. And then the mean SPI value of the basin can be calculated. Therefore, the distribution and variation law of drought and flood can be analyzed by SPI in different time scales (1,3,6,12 months). The SPI evolution in different time scales in Poyang Lake Basin during 1951-2013 is shown in Fig. 2.

The SPI in short time scales has been affected more by short time precipitation. With the growth of the time scale, SPI value is less affected by rainfall. Waterlog occurred four times, respectively in 1953, 1954, 1998; while heavy flood happened three times, in 1973, 1975 and 2010, respectively. The SPI value which is highest in 1954 reached by 2.12. The most serious drought is in 1963. Overall, the drought and flood appear alternately in the Poyang lake basin with a rising trend.

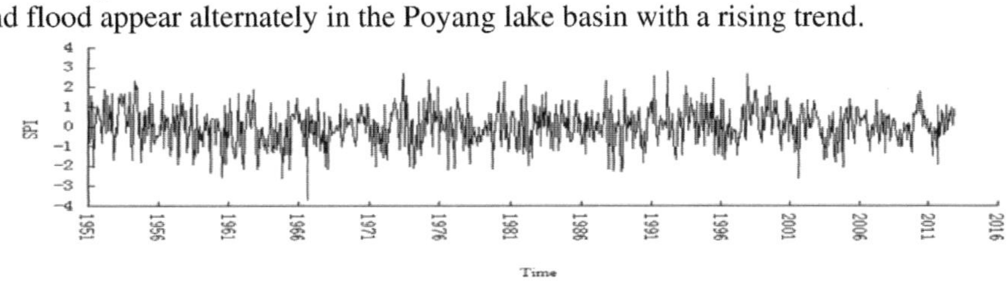

(a) SPI evolution in 1 month scales

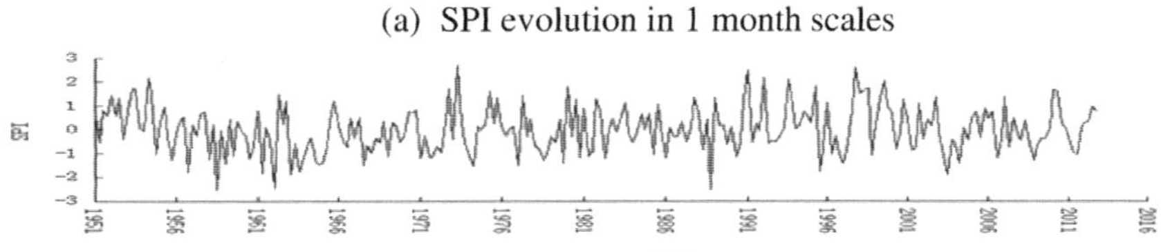

(b) SPI evolution in 3 months scales

(b) SPI evolution in 6 months scales

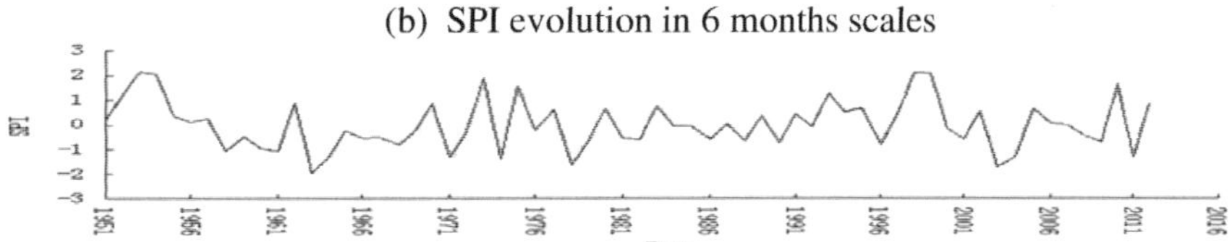

(c) SPI evolution in 12 months scales

Fig.2 SPI evolution in different time scales in Poyang Lake Basin during 1951-2013

4.2 Change law of drought and flood in multi-scales in Poyang lake basin

Studies have shown that [6]: In season drought analysis, the SPI index in 3 months scale can better represent the agricultural drought conditions. One year can be divided into: January to March for the first quarter, April to June for the second quarter,and so on. The SPI along the temporal variation characteristics of each period can explore the change trend of drought and flood situation along the time sequence.

In Fig. 3, the SPI value within the first season and fourth season is a rising trend, and it is not a significant decline in the second quarter. The trend is not obvious in the third season, but there has an overall rise during 1951-2013. Related studies have shown that [7] the SPI is on a rise trend during nearly 50 years in the Poyang lake basin. The result is consistent with Fig. 3.

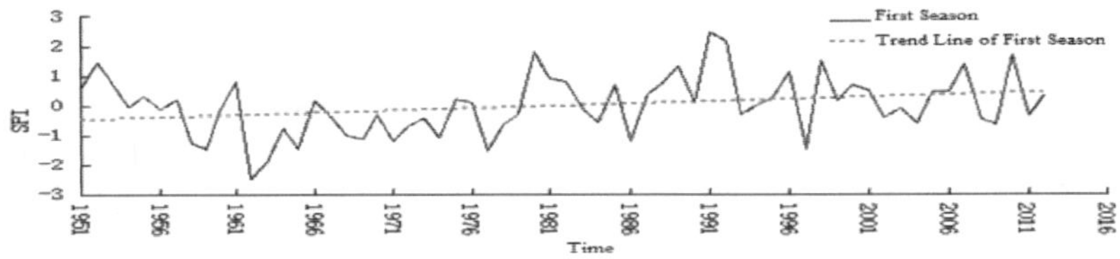

(a) The First Season

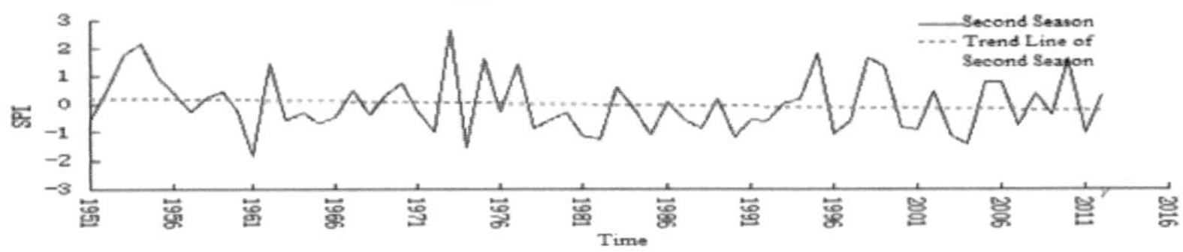

(b) The Second Season

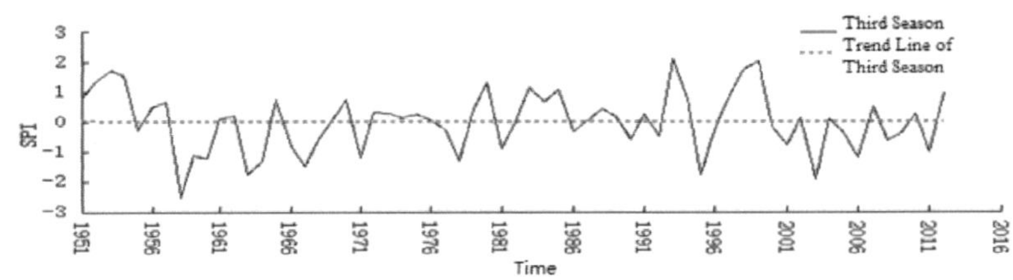

(c) The Third Season

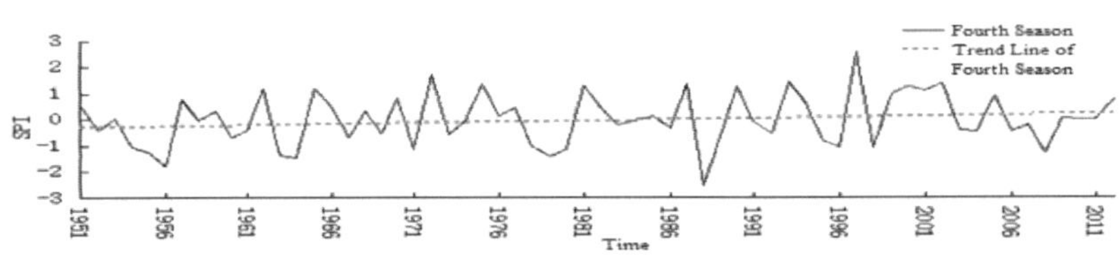

(d) The Fourth Season

Fig.3 SPI changes in time series in Poyang Lake Basin during 1951-2013

5 Conclusions

The change law of drought and flood in the Poyang lake basin is analyzed based on SPI. Results show that the conversion between drought and flood, with regional change trend. Drought increased from south to north, from east to west within the basin. Flood is intensified from south to north and from the east to west. The results are consistent with the actual basic, and are similar to some research achievements[7].All these show that the method is practical and effective.

The major objectives of this study are to investigate changing drought and flood properties by SPI. Results indicated that the SPI is suitable to analyze the law of drought and wet for this basin. The generation of drought and flood are inseparable with hydrological factors, therefore, it is the research direction to analysis the relationship between the drought and flood and the hydrological elements for the Poyang lake basin in the future.

Acknowledgement

This research was financially supported by the CRSRI Open Research Program (Program SN CKWV2013218/KY) , and supported by the Project of science and Technology Department of Jiangxi Province (No. 20161BAB216109) , and supported by the Project of science and Technology Department of Jiangxi Province (No. 20151BBG70008) ,and by Training Program of Innovation and Entrepreneurship for Undergraduates of Jiangxi Province (No.2014-16).

References

[1] McKee T.B., Doesken N.J. and Kleist J. (1993) The relationship of drought frequency and duration to time scale. In proceedings of the eighth conference on applied climatology, American meteorological society. Anaheim (CA), 17–22 January 1993, AMS, pp. 179–184

[2] Hayes MJ, Svoboda M, Wilhite DA, et al. Monitoring the 1996 drought using the SPI[J]. Bull.Amer.Meteor.Soc.,1999,80: 429-438.

[3] Wu H, Hayes MJ, Wilhite D A, et al. The effect of the length of record on the standardized precipitation index calculation[J]. International Journal of Climatology, 2005, 25:505-520.

[4] Morid S, Smakhtin VU, Moghaddasi M (2006) Comparison of seven meteorological indices for drought monitoring in Iran. Int J Climatol 26:971–985

[5] HONG Xingjun, GUO Shenglian, MA Hongxu and LIU Dedi. Spatial and temporal analysis of drought and its relationship with water level of Poyang Lake. Journal of China Hydrology, 2014, 34(2): 25-31. (in Chinese)

[6] YUAN Wenping, ZHOU Guangsheng. Comparison between Standardized Precipitation Index and Z-index in China. Chinese Journal of Plant Ecology, 2004, 28(4): 523-529. (in Chinese)

[7] CAI Zhe,ZHANG Yizhi,HE Yongfeng,et al.Analysis of spatiotemporal trend characteristics of droughts and floods in Jiangxi Province[J].,2013,22(02):144-149. (in Chinese)

Electrodeposition of CuSe Nanosheets films and its growth mechanism

Lingzhi Li, Jiangfeng Gong,* Weihua Zhu

College of science, Hohai University, Nanjing 210098, People's Republic of China

E-mail: jfgong@hhu.edu.cn (J.F. Gong)

Abstract. Copper selenide nanosheets films were synthesized successfully by simple electrodeposition method with controlling various experimental parameters. CV was used to study its growth process. It demonstrated that copper selenium is formed by an induced co-deposition mechanism where selenium is deposited first and then the reduction of Cu^+ is induced, finally the formation of copper selenide nanostructures is finished.

1. Introduction

As a class of multifunctional semiconductors with layered structure, CuSe is identified as p-type semiconductor due to the existence of copper vacancies within crystal lattice, and widely used in solar cells, [1] gas sensors, [2] and photo-detectors. [3] CuSe nanomaterials have been prepared by diverse methods including hydrothermal method, solvothermal method, microwave-assisted polyol process, and mechanical alloying route. [4-6] However, to date, synthesis of copper selenide nanosheets has rarely been reported.

In this paper, novel two-dimensional (2D) CuSe nanosheets films were successfully synthesized via electrodeposition method. The growth mechanism was study carefully using Cyclic voltammograms(CV) technique. It demonstrated that copper selenium is formed by an induced co-deposition mechanism, where selenium is deposited first and then the reduction of Cu^+ is induced, finally the formation of copper selenide nanostructures is finished.

2. Experimental details

For the deposition of cupper selenium nanosheets film, the electrolyte was prepared by dissolving $CuCl_2 \cdot 2H_2O$ (2 mM/L), SeO2 (4.5 mM/L) and KCl(99%)(0.1 M/L) in distilled water. The pH of the solution was adjusted to 1.5 by using 0.2M HCl solution. The electrolyte was not stirred during all the electrochemical measurements and depositions. The electrodeposition was carried out in a conventional three-electrode electrochemical cell with a platinum counter electrode and a saturated calomel electrode (SCE) as reference electrode, and gold coated polyethylene terephthalate (PET) was used as working electrode. To obtain high quality nanosheets film, a series of experiments on the basis of orthogonal experimental design were performed to optimize the deposition parameters. The deposition temperatures were varied from 20 ℃ to 80 ℃, and the deposited potentials were varied from -0.02 V to -0.20 V.

3. Results and Discussion

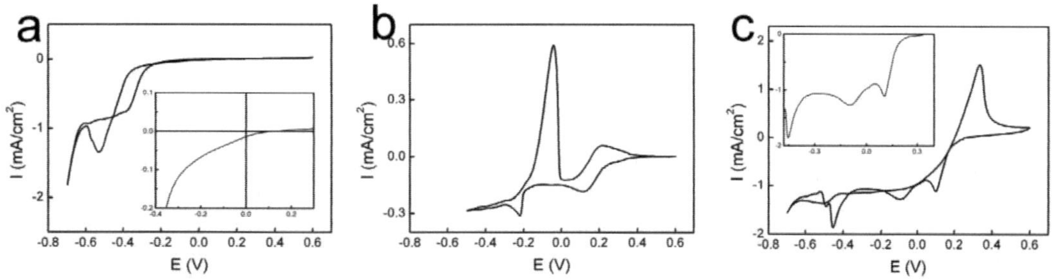

Figure 1. CV of the relative electrolytes. a) 4.5 mM SeO_2+0.1M KCl. b) 2.5mM $CuCl_2$+0.1M KCl. c) 4.5 mM SeO_2+2.5mM $CuCl_2$+0.1M KCl. The PH values of the electrolytes were adjusted to 1.5. The scan rates were kept at 10mV/S.

To study electrochemical features of Cu and Se induced co-deposition, a series of CV were done on Au coated PET substrates. Figure 1a shows the typical CV containing 4.5mM SeO_2 in 0.1M KCl solution. A weak cathodic current begins to flow at +0.12 V (inset in Figure 1a), while the current changes slowly from +0.12 to -0.36 V, after which it increases rapidly and a very strong reduction peak appears at -0.52 V. Combined with the previous results, the assignment of the cathode polarization process can be described as follow. The weak reduction current in the potential ranging from +0.12 V to -0.2 V indicating slow deposition kinetics, corresponds to bulk selenium deposition through four-electron reduction of Se^{4+} to Se^0, [7-9] and the process is described by Eq. 1. This characterization qualitatively is similar to the under-potential deposition (UPD) of metals, involving the formation of the first monolayer selenium before the bulk deposition. The followed deposition of Se is, however, prevented by slow deposition kinetics due to the passivating effect of the selenium film. It is verified by the fact that in the second scan, the cathodic current density is much smaller than that observed at the Au surface. For the electrodeposition Se at -0.52 V, it goes through a two-step reaction, the first step of which corresponds to the six electron reduction from $HSeO_3^-$ to H_2Se (Eq. 2), and the second step is a chemical reaction between $HSeO_3^-$ and H_2Se according to Eq. 3. The electrodeposition mechanism of Se in the potential ranging from -0.20 V to -0.52 V is attributed to the competition between four-electron reaction and six-electron reaction. Additionally, the deposition process is influenced by the substrate surface and additional ions, which can change its deposition potential.

$$HSeO_3^- + 5H^+ + 4e^- = Se + 3H_2O \qquad \text{Eq. 1}$$

$$HSeO_3^- + 7H^+ + 6e^- = H_2Se + 3H_2O \qquad \text{Eq. 2}$$

$$HSeO_3^- + 2H_2Se + H^+ + 6e^- = 3Se + 3H_2O \qquad \text{Eq. 3}$$

Figure 1b depicts the CV obtained in 2.5 mM $CuCl_2$+0.1 M KCl solution. The reduction current starts to flow at +0.40V and formed a reduction peak at +0.13V, which corresponds to the transformation of Cu^{2+} to Cu^+ as described by Eq. 4. According to Eq.5, the subsequent peak at -0.22V corresponds to the reduction of Cu^+ to metallic Cu, and the later process is verified the metallic copper deposition with an unsymmetrical oxidation peak at -0.05 V. [10]

$$Cu^{2+} + e^- = Cu^+ \qquad \text{Eq. 4}$$

$$Cu^+ + e^- = Cu \qquad \text{Eq. 5}$$

The electrodepositon of CuSe was conducted on Au-PET substrate using 4.5 mM SeO_2, 2.5mM $CuCl_2$, and 0.1M KCl as the electrolyte. The CuSe formation mechanism was studied first by CV as shown in Figure 1C. When both Cu^{2+} and Se^{4+} ions are present in the solution, significant change occurs during the initial scan. The first cathodic peak is seen at about +0.1V, which has a small potential shift compared with the reduction of Cu^{2+} to Cu^+ measured containing only $CuCl_2$. The phenomena were very common if additional ions were added. The second cathodic peak is at about -

0.1V, which corresponds to the reduction of Cu^+ induced by $Se^{2-}$and thus the CuSe formed before the deposition of metallic Cu. The formation of CuSe is processed by the Eq. 6. As mentioned above, in the first monolayer, Se formed at +0.12 V through UPD-like process, then the Cu^+ in the electrolyte will be combined by as-deposited Se. Meanwhile, Se is being deposited on CuSe at the same deposition potential, which runs the reaction continually. Since the reduction of Se and the remaining of the Cu^+ in the electrolyte can take place at the same time during a large deposit potential about from +0.1 V to -0.2V, the electrodeposition of CuSe may undergo in the same deposit potential range. Thus, it concludes that copper selenium is formed by an induced co-deposition mechanism that selenium is deposited first then the Cu^+ reduction is induced and finally the formation of copper selenide nanostructures is finished. The $HSeO_3^-$ ions shift the reduction potential of Cu^+ significantly to the negative direction, and the cathodic current increases rapidly at – 0.3V forming a cathodic peak at – 0.45V, which corresponds to the formation metallic Cu and is not suitable for formation CuSe , which do not discussed here.

$$Se + Cu^+ + e^- = CuSe \qquad \text{Eq. 6}$$

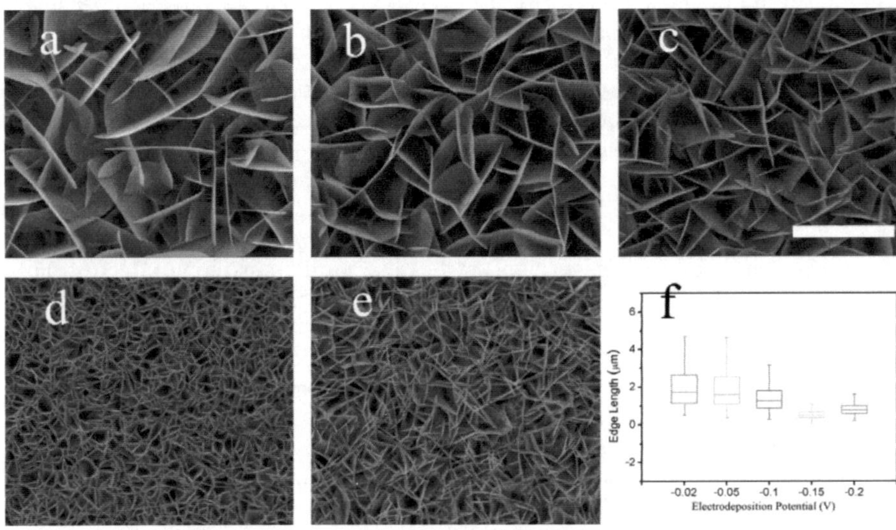

Figure 2. SEM images of the CuSe nanosheets obtained at different deposition potential. a) -0.02 V; b) -0.05 V; c) -0.10 V; d) -0.15 V; e) -0.20 V. f) The statistical distribution of the nanosheets. The deposition time was 30 min and deposition temperature was 60 ℃. The scale bar is 5 μm.

Based on the CV results, a series of experiments were carried out to observe the effect of deposition potential on the growth of CuSe nanosheets. Figure 2a-e show the SEM images of the deposits obtained at different deposition potential. The as-prepared nanosheets cross each other and almost perpendicular to the substrate. It is found that the deposition potential has a remarkable influence on the as-prepared products. When the deposition potential is -0.02 V, the CuSe nanosheets are unevenly sized. Some of them can reach 5 um while others only several hundred nanometers. With increasing the deposition potential, the deposited products trend uniformed gradually. For the deposition potential of -0.15 V, the nanosheets are uniformed with the average size about 700 nm. However, when changing the deposition potential to -0.20 V, the size of the nanosheets increases. The specific statistical distribution is shown in Figure 2f. At more negative cathodic deposition potentials, some bubbles will come out from the cathodic substrate, which corresponds to the formation of H_2Se. Since the bubbles would block the growth of the nanosheets, only a few nanoparticles were found on the substrate.

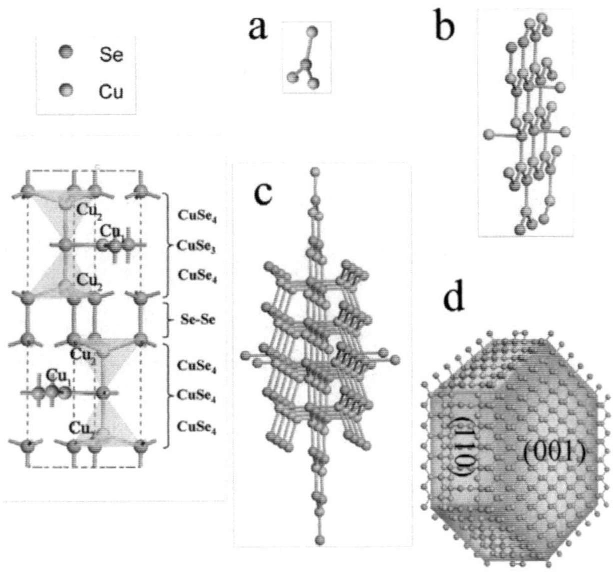

Figure 3. Schematic illustration of the growth of CuSe nanosheet. Left hand is the unit cell of the klockmannite CuSe. (a-d) Atomic models of CuSe nanosheet at different growth stage.

CuSe nanosheets were synthesized during the electrodeposition under a wide range experiment condition, which indicated that the morphology of the product is affect by its special crystal structure. To understand the growth mechanism of CuSe nanosheets, a series atomic model were proposed as illustrated in Figure 3. The left-hand side is the unit cell of the klockmannite CuSe, where the green atoms are copper and brown atoms are selenium. The Cu atoms are in two different environment, $CuSe_3$ (triangular planes) and $CuSe_4$ (tetrahedron). The unit cell can be seen as slabs connected by Se−Se bonds. The tetrahedral units are linked by vortex through triangular layers. The Cu_1-Se bond at triangular planes is 2.275 Å, while the Cu_2-Se bonds at tetrahedron are 2.408 Å and 2.375 Å respectively. It is worth noting that the Cu_1-Se bonds are striking shorter than Cu_2-Se bonds, which suggests Cu_1-Se bonds strength is much larger than Cu_2-Se bonds. The Se-Se bond length at connect layer is 2.378 Å and combined with covalent chemical bond. [[11]]

When the deposition potential is lower than +0.13V, Se atoms are formed on the cathodic substrate by the redox of Se^{4+} ion. At the same time, Cu^{2+} ion turns into Cu^+ ions at the same potential. The deposited Se atoms combine with Cu^+ by sacrifice electrons supplied by the cathode, and form CuSe dipoles through ionic bonding as illustrated in Figure 3a. The main reason is that the Cu_1-Se bond length is much smaller than the other bond, which makes them combined more easily. The feature makes them grow fast in the Cu_1-Se triangular plane. When the slab is extend to a certain degree, scattered Cu^+ will be combined with Se out of the plane to reduce its surface energy (Figure 3b), meanwhile, the dangling bonds of Cu will be the site adsorbing Se to form the Cu_4Se tetrahedron layer. In this layer, Cu_2-Se bonds are no longer in-plane, but zigzag form as shown in Figure 3c. Once the Cu_2-Se bonds form, the Cu_4Se tetrahedron layer will grow fast duo to the polarization of the positive and negative ionic charges on the Cu^{2+} and Se^{2+} terminated surfaces. According to its atomic model, Se atoms will combine with Se atoms to form Se-Se in the following growth process. Because the binding ability of Se-Se is much lower than Cu-Se, the growth rate of Se-Se layer is much slower than Cu-Se layer, which leads to the formation anisotropic 2D CuSe nanosheets. When the electrolyte temperature is low, Cu^+ is difficult to combine with Se^{2+} to enter its lattices on the viewpoint of thermodynamic.

IOP Conf. Series: Earth and Environmental Science **59** (2017) 012060 doi:10.1088/1755-1315/59/1/012060

4. Conclusions

In summary, 2D cupper selenide nanosheets films were synthesized on Au coated PET substrate by the simple electrodeposition. CV was used to study its growth process. It demonstrated that copper selenium is formed by an induced co-deposition mechanism, where selenium is deposited first and then the reduction of Cu^+ is induced, and finally the formation of copper selenide nanostructures is finished.

Acknowledgment
This work is supported the Fundamental Research Funds for the Central Universities (Grant No. 2015B22313).

References

[1] Seo, Y.-H.; Lee, B.-S.; Jo, Y.; Kim, H.-G.; Choi, Y.; Ahn, S.; Yoon, K.; Woo, K.; Moon, J.; Ryu, B.-H.; Jeong, S 2013 *J. Phys. Chem. C* **117** 9529-9536.
[2] Xu, J.; Zhang, W.; Yang, Z.; Ding, S.; Zeng, C.; Chen, L.; Wang, Q.; Yang, 2009 S *Adv. Funct. Mater.* ***19*** 1759-1766.
[3] Seo, Y.-H.; Lee, B.-S.; Jo, Y.; Kim, H.-G.; Woo, K.; Moon, J.; Choi, Y.; Ryu, B.-H.; Jeong, S 2013 *Acs App. Mater. Inter.* **5** 6930-6936.
[4] Wu, YM; Korolkov, I; Qiao, X; Zhang, XH; Wan, J; Fan, XP 2016 *J. Solid State Chem.* **238** 279-283
[5] Wu, XJ; Huang, X; Liu, JQ; Li, H; Yang, J; Li, B ; Huang, W; Zhang, H 2014 *Angew. Chem. Int. Edit* **53** 5083-5087
[6] Gu, YJ; Su, YJ; Chen, D; Geng, HJ; Li, ZL; Zhang, LY; Zhang, YF 2014 *Crystengcomm* **16** 9185-9190
[7] Dhanam, M.; Manj, P. K.; Prabhu, R. R. 2005 *J. Cryst. Growth* **280**, 425-435.
[8] Kemell, M.; Ritala, M.; Saloniemi, H.; Leskela, M.; Sajavaara, T.; Rauhala, E. 2000 *J. Electrochem. Soc.* **147**, 1080-1087.
[9] Kemell, M.; Saloniemi, H.; Ritala, M.; Leskela, M 2000 *Electrochim. Acta*, **45**, 3737-3748.
[10] Sebastian, P.; Valles, E.; Gomez, E 2014 *Electrochim. Acta*, **123**, 285-295.
[11] Morales-Garcia, A.; Soares, A. L., Jr.; Dos Santos, E. C.; de Abreu, H. A.; Duarte, H. 2014 A *J.Phys. Chem. A* **118**, 5823-5831.

Urban scenic spot interest space research based on transportation junction buffer

ZHOU Xiao, WANG Li, CAO Sida, LI Shijie, LI Xu, XUE Chongchong
Information Engineering University

zhou_laoge@163.com

Abstract. According to city tourism features, this paper studies the urban scenic spot interest space based on transportation junction buffer. City tourism features and the transportation junction radiation impact on scenic spots are studied. Take Zhengzhou city as an example, urban scenic spots are sampled and classified. Research range and objects are confirmed. By setting up buffer model and interest field data model, tourists' interests on scenic spots within buffers are studied quantitatively. Scenic spot interest space is studied and analyzed. Meanwhile, tourism decision support projects relying on scenic spot interest space are provided for tourists to refer to.

1. Introduction

Urban scenic spot spatial distribution plays an important role on attracting tourists to visit the city. Not all the scenic spots have the same appeal on different tourist groups. Usually, the first destinations are bus stations, railway stations and airport, etc[1]. There is a close relationship between transportation junction and scenic spots. Tourists visit scenic spots relying on their interests. Thus, scenic spot spatial distribution within transportation junction buffer has a great impact on the tourist motive[2]. In this paper, interest field of three tourist groups is set up. The geographic spatial analysis model set up on the basis of transportation junction buffer is brought forward and used to study scenic spot interest space, and then decision support projects are provided for tourists.

2. Research objects sampling

In Zhengzhou, there are Zhengzhou railway station and Zhengzhou east railway station, Central bus station, Long-distance central bus station, East bus station, North bus station and West bus station. They are set as research objects Ⅰ. Relying on attribute features, scenic spots can be divided into Park and green field, Amusement, Venue and Shopping center, labelled by A, B, C and D and set as research objects Ⅱ.

A={A_1: Zhengzhou botanic garden; A_2: Bishagang park; A_3: Renmin park; A_4: Zijingshan park; A_5: Lvcheng square; A_6: Forest park; A_7: Zhengzhou zoo; A_8: Lvyin park; A_9: Wenhua park; A_{10}: Yueji park; A_{11}: Xiliuhu park};

B={ B_1: Century play yard; B_2: Water park; B_3: Fount world; B_4: Bar street};

C={ C_1: Henan museum; C_2: Zhengzhou museum; C_3: Zhengzhou science and technology museum; C_4: Erqi memorial; C_5: Zhongyuan tower; C_6: Zhengzhou aquarium};

D={ D_1: Wangfujing mall; D_2: Erqi Uanda; D_3: Zhongyuan Uanda; D_4: International trade 360; D_5: CC mall; D_6: Erqi Dehua; D_7: Big Shanghai city; D_8: Renminlu Dennis}。

Arcgis9.0 software is used to draw the spatial distribution of transportation junctions and scenic spots as Figure.1 shows. Class A is rectangle, Class B is triangle, Class C is diamond and Class D is roundness. Railway station is texture roundness while bus station is texture rectangle.

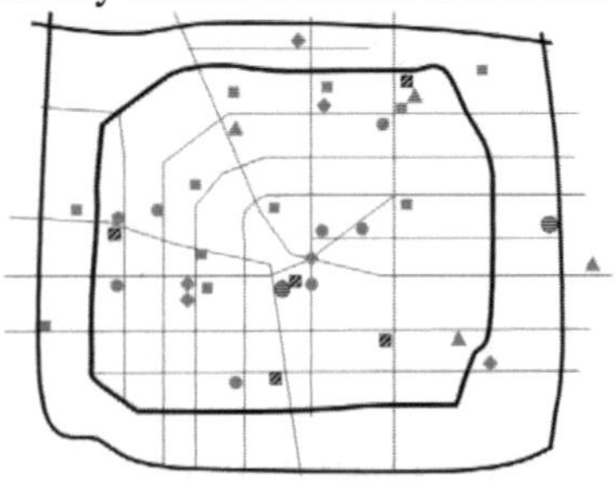

Figure.1 Transportation junctions and scenic spots spatial distribution

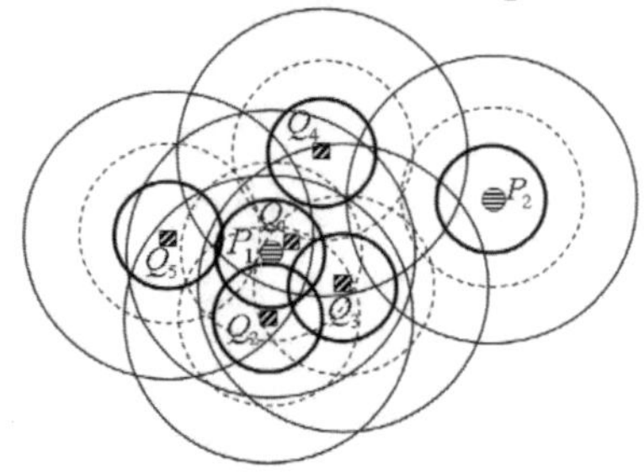

Figure.2 Transportation junction buffer topology net and buffer radiation impact model

3. Buffer and interest field modeling

Buffer model is used to research spatial relationship. Three-level buffers are founded. Questionnaire survey is used to obtain data to set up three groups' interest field and study the relationship between tourist interests and transportation junction radiation impact as well as scenic spots distribution.

3.1 Buffer modeling

Point buffer research method is used. As to a certain object W_i , its buffer area is defined as follows.

$$P = \{x \mid d(x, W_i) \le R\} \quad (1)$$

Object W_i buffer area with neighborhood radius R is the collection of points whose distance d to object W_i are less than R [3]. Objects I are set as center, R_1 3 *km*、 R_2 5 *km* and R_3 8 *km* are taken as radius to found buffer model. Two railway stations are set as P_1 and P_2 . Bus stations as Q_1 、 Q_2 、 Q_3 、 Q_4 and Q_5 . Transportation junction buffer topology net and buffer radiation impact model are set as Figure.2 shows. The thickest inner ring is R_1 , imaginary line is R_2 and the outer ring is R_3 .

3.2 Interest field data modeling

Enough samples are used to do questionnaire survey. Each group contains 100 samples and the ages are averaged. Each sample has certain tourism interest and all samples take part in[4,5]. Data is shown in Table 1. Interest field is shown in Figure.3. Each scenic spot data is arranged from left to right. Red is the middle and old group. Blue is young adults group. Brown is children group.

Table 1. Scenic spot interest data of different groups

Scenic spots	A_1	A_2	A_3	A_4	A_5	A_6	A_7	A_8	A_9	A_{10}	A_{11}	B_1	B_2	B_3	B_4
The middle and old	86	75	91	79	56	23	78	43	33	36	47	23	25	19	11
Young adult	45	52	58	66	34	10	70	26	19	29	33	44	65	50	46
Children	22	46	32	31	20	11	80	12	10	19	27	96	93	86	—
Scenic spots	C_1	C_2	C_3	C_4	C_5	C_6	D_1	D_2	D_3	D_4	D_5	D_6	D_7	D_8	
The middle and old	66	58	24	76	26	37	34	45	39	50	29	61	50	34	
Young adult	54	49	32	74	34	55	87	86	93	81	70	95	88	79	
Children	30	38	88	34	31	65	45	50	44	39	31	45	25	33	

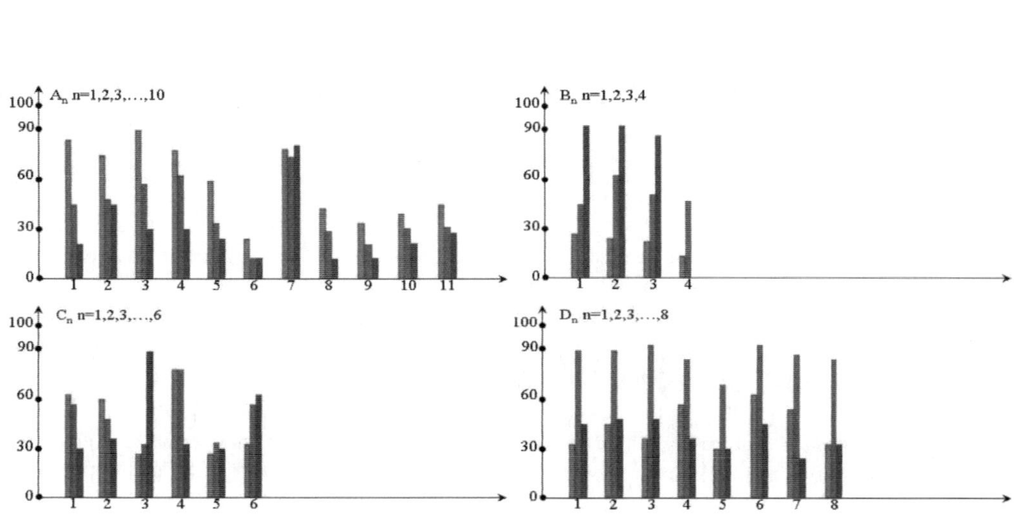

Figure.3 Scenic spot interest data of different groups

Relying on the data, park and green field attracts the middle and old most, and then attracts the young adults; it attracts children least. Playground yard attracts children most except Bar Street, and then attracts young adults; it attracts the middle and old least. Each venue attracts three groups fairly. Zhengzhou science and technology museum and Zhengzhou aquarium attract children most. Erqi memorial attracts the middle and old and young adults most. Transportation junction buffer's impact on each scenic spot is greatly related to tourist group's interest field.

4. Decision support of scenic spot interest space

4.1 Scenic spot interest space analysis

Scenic spots spatial structure within buffer in the range of transportation junction radiation impact is analyzed and Table.2 scenic spot spatial distribution data within three-level buffers is obtained.

Table 2. Scenic spot spatial distribution data within three-level buffers

	P_1	P_2	Q_1	Q_2	Q_3	Q_4	Q_5
$0<r\le 3km$	$A_3C_4D_6$ D_7D_8	— —	A_3C_4 $D_6D_7D_8$	$C_4D_2D_6$	C_4D_6	$A_7A_8A_9$ $B_4C_1D_4$	$A_2A_{11}D_1$ D_3D_5
$3<r\le 5km$	$A_2A_4A_5$ $B_2C_1C_2C_3$	— —	$A_2A_4A_5$ $B_2C_1C_2C_3$	$A_2A_5C_2C_3$	$A_2A_3A_4A_5$ $C_2C_3D_2D_7D_8$	A_3A_6 B_2C_6	$A_5A_{10}C_2C_3$
$5<r\le 8km$	$A_7A_8A_9$ $A_{10}B_4D_1$ $D_2D_3D_4D_5$	A_6、 B_1、 C_5	$A_7A_8A_9A_{10}$ $B_4D_1D_2$ $D_3D_4D_5$	$A_3A_4A_{10}$ $B_1C_5D_1D_7D_8$	$A_7A_{10}B_1B_2B_4$ $C_1C_5D_1D_3D_4D_5$	$A_4A_{10}C_4$ $D_6D_7D_8$	$A_1A_3B_2C_4$ $D_2D_6DD_7$

Main scenic spots and transportation junctions' spatial relationship in urban Zhengzhou as well as impact on scenic spots are shown in table.2. The following conclusion is obtained.

- Railway transportation.

Zhengzhou railway station P_1 is located in the center downtown and has great radiation impact. Within 3 *km* buffer, park and green field, venue and shopping center attract the middle and old and young adults. Within 3~5 *km* buffer, playground yard turns out to attract children. Within 5~8 *km* buffer, park and green field and shopping center take up a great quantity.

Zhengzhou east railway station P_2 is far away from downtown. Only within 5~8 *km* buffer, a few scenic spots meet the interest needs. Its radiation impact is inferior to Zhengzhou railway station.

- Bus transportation.

Central bus station Q_1 is close to Zhengzhou railway station and has the fair radiation impact.

Long-distance central bus station Q_2 is located in south downtown and impacts south scenic spots. Within 3 *km* buffer, shopping center and venue meet the needs of the middle and old and young adults.

Within 3~5 *km* buffer, there exist park and green field and venue. Zhengzhou science and technology museum meets the needs of most children.

East bus station Q_3 is located in the east downtown and impacts east scenic spots. Within 3 *km* buffer, only one venue and one shopping center meet interest needs. Within 3~5 *km* buffer, park and green field, venue and shopping center meet the needs of the middle and old and young adults. Within 5~8 *km* buffer, plenty of playground yard meet the needs of children while other scenic spots are also plentiful. It illustrates that East bus station has better and greater radiation impact.

North bus station Q_4 is located in the north downtown and impacts north scenic spots. Within 3 *km* buffer, all kinds of scenic spots meet all groups' needs. Within 3~5 *km* buffer and 5~8 *km* buffer, fewer scenic spots meet three groups' interest needs.

West bus station Q_5 is located in the west downtown and impacts west scenic spots. Within 3 *km* buffer, park and green field and shopping center meet the middle and old and young adults. Within 3~5 *km* buffer, venue turns out. Within 5~8 *km* buffer, plentiful scenic spots meet all groups' needs.

4.2 Decision support analysis

Transportation junction and scenic spot geographic spatial distribution are studied. Relying on interest space model and analysis, decision support projects are provided as follows.

As to long-distance tourists, choosing Zhengzhou railway station or Central bus station can reach scenic spot in short time, which is convenient. If choosing Zhengzhou east railway station, tourists should spend more time and fee on taxi, subway or public bus to reach scenic spot.

Besides, short-distance tourists can choose other bus stations in downtown. These bus stations are far away from city center, which can avoid traffic jam as much as possible. Meanwhile, as to each station, within three-level buffer, there exist different scenic spots to meet tourists' interest needs.

In order to get maximum motive benefit, tourists should rely on their own interests, transportation junction radiation impact and scenic spot spatial distribution to choose proper station and scenic spots.

5. Conclusion

This paper studies the urban scenic spot interest space distribution set up by transportation junction buffer. According to different tourist group interests, scenic spots are divided into four categories. It researches the transportation junction's radiation impact on scenic spots. By studying scenic spot interest spatial distribution, decision support projects are provided for tourists. In order to get maximum tourist motive benefit, tourists should rely on their own interests, transportation junction radiation impact and scenic spot spatial distribution to choose proper station and scenic spots.

References

[1] WANG C Y, QU H L. Tourist motive, destination image and tourism expectation. Tourism Tribune, 2013, 28(6) , p.26-34.

[2] LI X J, CUI S S, ZHANG K. Neural net method of best tourism route choice. Transportation and Computer, 2006, 24(5) , p.103-106.

[3] Manuel V, Robertico R C(2000) . Evaluation of demand US tourists to Aruba. Annals of Tourism Research'27(4) , p.946-963.

[4] Vansteenwegen P, Van Oudheusden D(2007) . The mobile tourist guide: an OR opportunity. OR Insight, 20(3) , p.21-27.

[5] Bosque I R, Martin H S. Tourist satisfaction: A cognitive affective model [J]. Annals of Tourism Research, 2008, 35(2) , p.551-573.

How to monitor and adjust in real time the total water consumption and water use efficiency: Earned value method

Zhong Du[1,2,a], Zengchuan Dong [1,b], Huixiu Wu [2,c], Lin Yang [2,d]

1 College of Hydrology and Water Resources, Hohai University, Nanjing 210098, P. R. China
2Hydrology and Water Resources Survey Bureau of Liaoning Province, Shenyang 110003, P. R. China

[a]duzhongmail@126.com, [b]dongzengchuan@163.com, [c]wuhuixiumail@126.com,[d]604697442@qq.com

Abstract. The evaluation indexes of total water consumption and water use efficiency have the characteristics of post feedback. In this paper we introduce the basic concept and specific theory of Earned value method (EVM) from project management, and reconstruct parameters in the method to adapt to water resources monitoring. The case of Dandong was studied, by analyzing the industry and irrigation water utilization. Although the total water consumption of two aspects reaches standards, the industrial added value and water use efficiency of irrigation are not up to standard. The results show that PV can be used as a baseline for real-time monitoring and adjustment, and the advantage of the EVM is that it can be an organic unity of water consumption and efficiency, so we can analyze comprehensively water utilization process.

1 Introduction

Water is essential for human being, economic activities, ecology and environment. In recent years, because of the over-exploitation and low efficiency use of water resources, water resources shortage has become a major social, environmental and ecological problem [1].

In order to avoid further deterioration of water resources, the government of China introduced the most stringent water resources management system, and implemented “three red line” assessment indexes. But the shortage of controlling indexes, such as water consumption and water use efficiency, are only post feedback at the end of the year.

How to monitor and adjust in real time the total water consumption and water use efficiency? By foreign experience, this paper introduces the basic concept and specific theory of Earned value management (EVM) from project management.

The goals of this study are: (1) to design suitable earned value method of water resources; and (2) to monitor and adjust the index of water consumption and water use efficiency, and to evaluate the performance.

2 Materials and methods

2.1 Earned value method

Earned value method (EVM) as a systematic technical measure, plays an important role in project investment and schedule management. It integrates project scope, cost, and schedule measures to help the project management team to assess in real time and measure project performance and progress [4-6].

EVM develops and monitors three key dimensions for each work package and control account as follows:

Plan value (PV) is the authorized budget for the work to be accomplished, and the total planned value for the project is also known as budget at completion.

PV=BC*WS (1)

Earned value is the authorized work that has been completed, plus the approved budget for such completed work. Managers monitor EV, both incrementally determine current status and cumulatively determine the trends of long-term performance.

EV=BC*WP (2)

Actual cost is the total cost actually incurred and recorded in accomplishing work which the EV measured.

AC=RC*WP (3)

Where, there are four indexes in the Formulas: BC is the value of budgeted cost; WS is the value of work scheduled; RC is the value of real cost; WP is the value of work performed.

Variances from the approved baseline will also be monitored. Scheduled variance (SV) is a measure of schedule performance of a project, as an useful metric which can indicate whether a project falls behind its baseline schedule or not. If the value of SV is less than 0, it's indicated that the work we completed was less than we planted. On the contrary, if the value of SV is greater than 0, it's indicated that the work we completed was more than we planted.

SV=EV-PV (4)

Cost variance is also a measure of cost performance of a project. If the value of CV is less than 0, the cost overrun for work completed is indicated. On the contrary, if the value of CV is greater than 0, the cost under-run of performance to date is indicated.

CV=EV-AC (5)

2.2 Water consumption earned value

The principles of EVM can be applied to all projects, in any industry. In this paper, the purpose of introducing earned value method is to use the functions of dynamic monitoring and pertinent adjustment.

Water consumption earned value is a water resources management technique. The three key dimensions and four indexes were endowed with new meaning. PV is the water consumption which is planned monthly or annually; EV is the work, such as industry, irrigation, service and domestic, which has been completed, plus the authorized water quota or water use efficiency for completed work; AC is the actual monthly or annually water consumption. Moreover, in the next section, the work of WS, WP is per ten thousand RMB Yuan industrial added value and mu of irrigation farmland. The quota of the BC, RC is water consumption of per ten thousand RMB Yuan industrial added value and irrigation consumption of water per mu of farmland.

2.3 Study region and data

Dandong is located in the southeast of Liaoning Province, where the average annual precipitation is 1000mm, and the water resources quantity is 8.9 billion. The GDP is 98.5 billion RMB yuan (2015), rising by -3%, and the industrial added value is 40.2 billion RMB yuan, rising by -8.8%, (current price) [7].

The basic data for PV was referred to through "Dandong city water resources bulletin" and "Dandong national economic and social development statistical bulletin", and the real-time data for EV and AC is statistic in bureau of statistic and water resources of Dandong.

3 Results

In this paper we focus on two aspects of industrial added value and irrigation farmland with the Earned value model. Industry and irrigation water use efficiency are important constraint indexes of the most stringent water resources management in china.

WS of industry is based on the average monthly data in the national economic bulletin of 2013-2014, plus the increase by 7%, which is listed in the Outline of 12th Five-Year Plan. BC (water consumption of per ten thousand RMB Yuan industrial added value) of 2014 is 24.7, decreased by 7%, so BC of 2015 is 22.97 as shown in Table 1. The actually data of 2015 is taken from statistics of bureau, but the data of RC is obtained by indirect calculation with AC and WP, and the price should be converted into the comparable price of the year of 2010.

Table 1. Earn value of water consumption of per ten thousand RMB Yuan industrial added value

	Jan.	Feb.	Mar.	Apr.	May.	June.	July.	Aug.	Sept.	Oct.	Nov.	Dec.
WS	40	37	40	43	46	50	43	43	45	42	35	29
BC	22.97	22.97	22.97	22.97	22.97	22.97	22.97	22.97	22.97	22.97	22.97	22.97
PV	919	850	919	988	1057	1149	988	988	1034	965	804	666
WP	46	38	37	37	38	39	38	36	36	32	25	18
RC	22.83	22.11	22.43	22.43	22.11	22.18	22.11	22.22	21.67	22.19	22.00	22.22
EV	1057	873	850	850	873	896	873	827	827	735	574	413
AC	1050	840	830	830	840	865	840	800	780	710	550	400

Unit: WS, hundred million RMB yuan; BC,m^3/ten thousand RMB yuan; PV,ten thousand m^3; WP, hundred million RMB yuan; RC,m^3/ten thousand RMB yuan; EV, ten thousand m^3; AC, ten thousand m^3.

Table 2. Earn value of irrigation consumption of water per mu of farmland

	Jan.	Feb.	Mar.	Apr.	May.	June.	July.	Aug.	Sept.	Oct.	Nov.	Dec.
WS				107	107	107	107	107	107			
BC				11	182	179	148	138	31			
PV				1177	19474	19153	15836	14766	3317			
WP				100	100	100	91	100	91			
RC				10	199	185	160	145	32			
EV				1100	18200	17900	13394	13800	2806			
AC				1000	19850	18500	14500	14500	2900			

Unit: WS, ten thousand mu; BC,m^3/mu; PV, ten thousand m^3; WP, ten thousand mu; RC,m^3/mu; EV, ten thousand m^3; AC, ten thousand m^3.

Table 2 shows the irrigation data from April to September during the irrigation period. The effective irrigation area is still 107 ten thousand mu. The BC of irrigation is the same as the industry, by the decrease of 5%, but the distribution of monthly value is average monthly data from the water resources bulletin of 2013-2014.

Monthly cumulative values of PV, EV, AC are shown as the thick solid line, dotted line and thin solid line in Figure 1 and Figure 2.

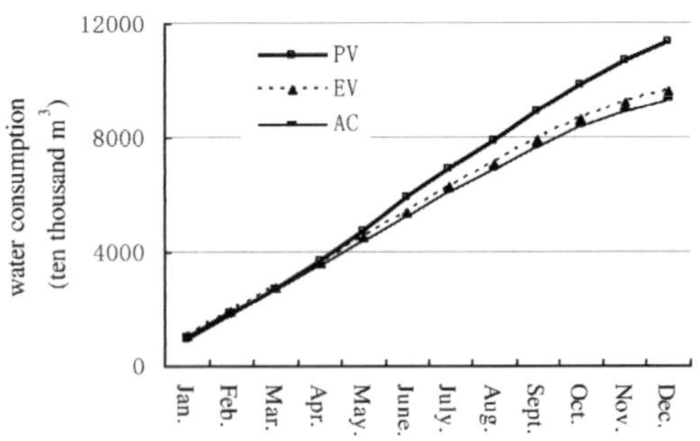

Figure 1. Earned value, planned value, and actual costs of water consumption of per ten thousand RMB Yuan industrial added value.

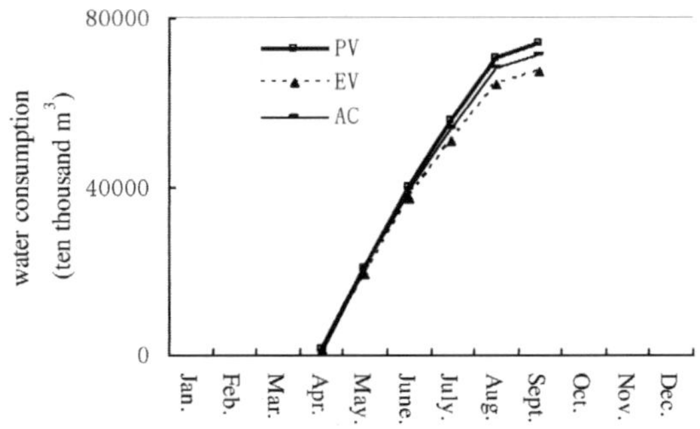

Figure 2. Earned value, planned value, and actual costs of irrigation consumption of water per mu of farmland.

4 Discussion

The expected performance of water managers in theory is the relationship of PV≈EV> AC, which indicates that both the total water consumption and the water use efficiency are better than the planned values. It shows the critical time of adjustment when the relationships were broken.

In this section, we focus on the application of the Earned value method to the industry. Table 1 shows that the unexpected result occurs in the first month (January), and lasts until February. The relation of EV>AC>PV indicates that not only the ten thousand RMB Yuan industrial added value is greater than the baseline, but also its efficiency is better than the planned value, but the total water consumption is beyond the critical value.

However, we must balance economic growth and water utilization, in order to promote economic and water resources sustainable development. In this paper, we suggest that the adjustment time may be in the midyear or the time the variance is greater than 10%-30% of total water consumption, at the same time, it is difficult to adjust the water utilization structure. So we don't adjust in February.

The accumulation of EV is greater than PV from January to March as shown in Figure 1, but from April to December, the relation of cumulative value is PV>EV>AC. The reason is that the industrial added value is less than the growth rate of 7%, but the efficiency is average 0.65m^3/ten thousand RMB yuan less than the planned value.

As shown in Table 2 and Figure 2, except for April, the rest months from May to September have the same monthly and cumulative trends, PV>AC>EV, and although the total water consumption reaches standards, the water use efficiency is not up to standard. By the supervision of irrigation process, it is found that drought weather leads to the reduction of available water supply and the increment of irrigation water requirement, so it's a feasible way to reduce irrigation area in order to control the total water consumption.

5 Conclusions

The shortage of water resources controlling index of the most stringent water resources management system is only post feedback. Water resources earned value management can monitor in real time and adjust in the procedure of water utilization.

Furthermore, PV can be used as a baseline or alert value. And according to the range of its variance, it can be a real-time warning and we can make adjustment for reaching standard of PV. The results of the case show that the water consumption of both industry and irrigation meet the requirement, but the efficiency of irrigation does not reach the standard during the whole process except April. Fortunately, we have not made a prediction, forewarning and adjustment scheme by the actual water resources and economic environment.

Finally, the advantage of the earned value method is that it can be an organic unity of water consumption and water use efficiency, so we can do comprehensive analysis of water resources utilization.

References

[1] Dao zhi Gong,Xu rong Mei, Wei ping Hao, 2016. Comparison of multi-level water use efficiency between plastic film partially mulched and non-mulched croplands at eastern Loess Plateau of China, Agricultural water management,17(9):215-226.

[2] Qian ying Mao, 2010.Research status and Prospect of earned value method, project management technology, 8(2):19-23.

[3] David A. Wood, 2017. high-level integrated deterministic, stochastic and fuzzy cost-duration analysis aids project planning and monitoring, focusing on uncertainties and earned value metrics, journal of natural gas science and engineering,37(11):303-326.

[4] Project management institute, 2012. A guide to the project management body of knowledge-fourth edition, create space independent publishing plat,:257-262.

[5] Roger D.H.,Warburton, Denis F. Cioffi, 2016. Estimating a project's earned and final duration, international journal of project management,34(11):1943-1504.

[6] Qing Chang, Ge di Ji, Chang qing Li,2006 .The research to second-level earned value method for improve performance measurement of engineering project, Chinese journal of management science,14(2):65-71.

[7] Dandong water resources bureau, 2015. Dandong city water resources bulletin,:5-25.

Quantitative calculation model of dilution ratio based on reaching standard of water function zone

Zhong Du[1,2,a], Zengchuan Dong [1,b], Huixiu Wu [2,c], Lin Yang [2,d]

1 College of Hydrology and Water Resources, Hohai University, Nanjing 210098, P. R. China

2Hydrology and Water Resources Survey Bureau of Liaoning Province, Shenyang 110003, P. R. China

[a]duzhongmail@126.com, [b]dongzengchuan@163.com, [c]wuhuixiumail@126.com, [d]604697442@qq.com

Abstract. Dilution ratio is an important indicator of water quality assessment, and it's difficult to calculate quantitatively. This paper proposed quantitative calculation model of dilution ratio based on the permissible pollution bearing capacity model of water function zone. The model contains three parameters of concentration. Particularly, the 1-D model has three river characteristics parameters in addition. Applications of the model are based on the national standard of wastewater discharge concentration and reaching standard concentration. The results show the inverse correlation between the dilution ratio and the C_P and C_0, and the positive correlation with C_s. The quantitative maximum control standard of dilution ratio is 12.50% by 0-D model and 22.96% by 1-D model. Moreover, we propose to choose the minimum parameter and find the invalid pollution bearing capacity.

1 Introduction

Dilution ratio (ratio of waste water to runoff) can be used for evaluating water quality, calculating the local water resources sustainable capacity, evaluating the sustainable development of the region, etc [1] [2].

However, determination on the reasonable value of the ratio is difficult, particularly, integrated quantity-quality assessment of available water resources in water function zone. The reason is that different water quantity, quality, pollutant capacity, concentration, and reaching standard, etc comprehensively influenced the dilution ratio [3]. Qiang Huang [1] improves the dilution-ratio calculation method and its way for successful application. Their results could be better convinced if they constructed unified model based on water function zone.

Water function zone is the basis for water resources development and protection. The "Code of practice for computation on permissible pollution bearing capacity of water bodies" was based on reaching standard of water function zone [4].

This paper we preliminarily focus on the quantitative calculation model of dilution ratio based on reaching standard of permissible pollution bearing capacity of water function zone.

2 Models

In this paper, the dilution ratio model is generally designed by the identity of following equation at the pollution discharge section.

IOP Conf. Series: Earth and Environmental Science **59** (2017) 012063 doi:10.1088/1755-1315/59/1/012063

$$M = C_P Q_P \tag{1}$$

Where, M is the capacity on reaching standard of water function zone (g/s), whose models are from the "Code of practice for computation on permissible pollution bearing capacity of water bodies".

The types of the dilution ratio model are divided by the dimension and the section. Zero dimensional model is simpler than one dimensional model. Three types of the 1-D model are divided by different discharge section and standard section.

2.1 0-D model

Zero dimensional model is suitable for the rivers whose stream has uniform mixing contamination. The formula of pollutant concentration is given below:

$$C = (C_P Q_P + C_0 Q)(Q + Q_P) \tag{2}$$

Where, C: pollutant concentration in discharge section (mg/l); C_p: pollutant concentration in sewage outfall section (mg/l); C_0: initial concentration in upper section (mg/l); Q : initial flow of upstream (m^3/s); Q_P: sewage discharge of outfall (m^3/s).

Focus on the reaching standard of water function zone, the identical equation is as follows:

$$C_s = C \tag{3}$$

Where, C_s is concentration of reaching standard in mixing section (mg/l).

By Formula 1 and Formula 2, the quantitative dilution rate model can be derived as below:

$$\frac{Q_P}{Q} = \frac{C_s - C_0}{C_P - C_s} \tag{4}$$

2.2 1-D model

One dimensional model is suitable for the rivers whose cross-section with uniform mixing contamination.

The 1-D model needs to consider the location combination of the reaching standard section and the discharge section. The model is divided into three categories according to the combination of different position of sections, namely, sections in the same position, discharge in middle section and standard section in lower, existed outfall and sections in the same position.

2.2.1 All sections in lower or medium. The capacity model refers to the same code in the last section of this paper.

$$M = C_s(Q + Q_P) - C_O \exp(-K\frac{L}{u})Q \tag{5}$$

Where, L is the length of river (m); K is the degradation coefficients(s^{-1}); u is velocity of river(m/s).

By Formula 1 and Formula 5, the model also can be derived as below:

$$\frac{Q_P}{Q} = \frac{C_s - C_0 \exp(-K\frac{L}{u})}{C_P - C_s} \tag{6}$$

If all sections in medium of the function zone, the model is same as Formula 6,

$$\frac{Q_P}{Q} = \frac{C_s - C_0 \exp(-K\frac{L}{2u})}{C_P - C_s} \tag{7}$$

2.2.2 Discharge in middle section and standard section in lower. The capacity model is as follows:

$$M = C_s \exp(K\frac{L}{2u})(Q + Q_P) - C_O \exp(-K\frac{L}{2u})Q \tag{8}$$

By Formula 1 and Formula 8, the model also can be derived as below:

$$\frac{Q_P}{Q}=\frac{C_s\exp(K\frac{L}{2u})-C_0\exp(-K\frac{L}{2u})}{C_P-C_s\exp(K\frac{L}{2u})} \quad (9)$$

Furthermore, in order to compare with the Formula 9, we propose a new way to the model. As the capacity of lower section, the value of the Formula 1 is 0, so we can get:

$$M=C_s(Q+Q_P)-C_O\exp(-K\frac{L}{u})Q-C_p\exp(-K\frac{L}{2u})Q_P=0 \quad (10)$$

And the follow model is essential as well as Formula 9:

$$\frac{Q_P}{Q}=\frac{C_s-C_0\exp(-K\frac{L}{u})}{C_P\exp(-K\frac{L}{2u})-C_s} \quad (11)$$

2.2.3 Existed outfall. In this section, we focus on the special conditions that a sewage outfall was in the middle section, so how much can the pollution bear permissiblely in the lower section to reaching standard.

This paper defined the basic assumptions to the pollutants, with two parameters of C_{Po} (mg/l), and Q_{Po} (m^3/s), so the capacity model modified is as follows:

$$M=C_s(Q+Q_{P0}+Q_P)-C_O\exp(-K\frac{L}{u})Q-C_{P0}\exp(-K\frac{L}{2u})Q_{P0} \quad (12)$$

Also with Formula 1, we derive the model as follows, too:

$$\frac{Q_P}{Q}=\frac{C_s-C_0\exp(-K\frac{L}{u})+\frac{(C_s-C_{P0}\exp(-K\frac{L}{2u}))Q_{P0}}{Q}}{C_P-C_s} \quad (13)$$

Essentially, the Formula 13 is the same as the Formula 6. The differences are the value of initial concentration and the effective degradation distance.

2.3 Summary

The basic structure of the quantitative model is like the Formula 4, with three parameters of concentration and characteristics of river, such as L, K, u in the 1-D model. However, all the models have the unreasonable assumptions: 0-D model has the mixing in the whole stream and 1-D model has the mixing in the whole section of the contamination.

The quantitative model indicates that the ratio is independent of stream flow (Q). But in fact, the other influencing factors of 1-D model, such as K and u, were affected indirectly by Q.

3 Result and discussion

This paper assumes that all the sewage is reaching the standard of discharge through the sewage treatment plant. Table 1 shows the different levels of emission values from the simplified code of "Discharge standard of pollutions for municipal wastewater treatment plant" (GB 18918—2002). The reaching standard of water function zone is decided by the standard of "Environmental quality standards for surface water" (GB3838-2002).

Table 1. The national standard for wastewater discharge guideline, highest emission concentration (unit: mg/L)

NO.	Parameter	First grade		Second grade	Third grade
		First A	First B		

1	COD	50	60	100	120
2	BOD_5	10	20	30	60
3	NH_4-N	5（8）	8（15）	25（30）[a]	-

[a]External value of the bracket is the control index of water temperature >12^{0}C.

The performance of the 0-D model was investigated at different COD standard concentration. Table 2 shows that the minimum of the ratio is 12.5% by the class II (water quality standard, 15mg/L), First grade- B (discharge standard, 60mg/L), and class III (reaching standard, 20mg/L). And the maximum of the ratio is 33.33%, by the class III (water quality standard, 20mg/L) , Second grad (discharge standard, 100mg/L), and class V (reaching standard, 40mg/L) .The type of the discharge combinations were met the requirements of the discharge standard. The result clearly confirm that the dilution ratio has inverse correlation relationship with C_P and C_0. Contrary, there is a positively correlated between the ratio and C_s.

Table 2. Percentage of dilution of 0-dimension model (unit: mg/L)

C_S		C_O		C_P		Q_P/Q
Grade	Value	Grade	Value	Grade	Value	Percentage
III	20	II	15	First A	50	16.67%
III	20	II	15	First B	60	12.50%
IV	30	II	15	Second	100	21.43%
IV	30	III	20	Second	100	14.29%
V	40	III	20	Second	100	33.33%
V	40	IV	30	Second	100	16.67%

Table 3 indicates that the three parameters of COD, BOD and NH_4-N has different dilution ratio in the same conditions. The reason for the different value is that the national standard value of each parameter is different and disproportionate. Thus, generally, we propose to choose the minimum as the emission standard, so it leads to some parameters which exist invalid pollution capacity [5] [6].

Table 3. Percentage of dilution of different water quality parameters of 0-dimension model (unit: mg/L)

	C_S	C_O	C_P	Q_P/Q
	III	II	First B	100%
COD	20	15	60	12.50%
BOD	4	3	20	6.25%
NH_4-N	1	0.5	8	7.14%

The data used in the 1-D model was taken from a local technical report in which the K value of COD, BOD and NH_4-N is 0.18d^{-1} 0.27d^{-1} and 0.28d^{-1}(unit conversion to S^{-1}); the length of case river is 50km(unit conversion to m), the velocity is 0.5m/s, the stream flow is 60m^3/s; the existing sewage flow is 3m^3/s and pollutant concentration is 60mg/L.

Compared with the dilution ratio, Table 4 shows that the effective degradation distance leads that the ratio of 1-D model is more than 0-D model. The dilution ratio is also significantly affected by the location of sewage outfall, and the ratio of medium section is 3.41% which is more than lower section. Table 4 also indicates that the value by Formula 9 is the same as Formula 11.Similarly, the ratio of lower section discharge (2) and lower section discharge (1) is 15.29% and 19.55%, respectively, and the ratio of the existed outfall in initial condition is 5%(3 m^3/s /60 m^3/s), so the ratio of lower section discharge (2) is more than lower section discharge (1) by 0.74%.

However, under the assumption that the pollutants are mixed instantaneously in section, the 1-D model is only feasible in theory, so the sewage outfall is only discharged at upper reaching standard section. Besides, 22.96% of middle section discharge in Table 4 is reasonable and should be used as the control standard.

Table 4. Comparison of dilution-ratio of different discharge section of 1-dimension model (unit: mg/L)

Zero-dimension	One-dimension, lower section standard			
	lower section discharge(1)	Middle section discharge(1)	Middle section discharge(2)	lower section discharge(2)
12.50%	19.55%	22.96%	22.96%	15.29%

4 Conclusions

1. The basic structure of the quantitative model is like Formula 4, the dilution ratio is dependent on three concentration parameters but independent of stream flow (Q). But 1-D model has other influencing factor of river characteristics, such as L, K and u, which was affected indirectly by Q. However, all the models in this paper have the assumptions of contamination mixing in the whole stream or section.

2. Dilution ration has inverse correlation relationship with C_P and C_0, and positively correlation relationship with C_s. The maximum of the control standard is 12.50% of 0-D model and 22.96% of 1-D model.

3. Dilution ration of COD, BOD and NH_4-N has different values with the same characteristics of river. The reason is that the national standard value of each parameter is different and disproportionate. Moreover, we propose to choose the minimum as the emission standard and give up invalid pollution bearing capacity.

4. Theoretical derivation of the quantitative model is under the assumption that the pollutants are mixed instantaneously in stream or section, so further studies will be on the 2-D model and the application to the case.

References

[1] Qiang Huang,Ze-zhong Zhang, Kuan Wang, 2008. Improvement to the dilution-ratio calculation method and its application, journal of safety and environment, 8(1):37-40.

[2] Shuhua Hou, Xiaoling Song,Liming Yao, 2015. A Bi-Level Waste Load Allocation Model Based on Water Function Zoning for Sichuan-Neijiang, Proceedings of the Ninth International Conference on Management Science and Engineering Management, (1): 209-220.

[3] Xiqin Wang, Changing Liu, Yuan Zhang, 2006.Water quantity-quality combined evaluation method for rivers' water requirements of the instream environment in dualistic water cycle: a case study of Liaohe river basin, Journal of geographical sciences, 61(11):1132-1140.

[4] Hock-Lye Koh, Poh-Eng Lim, Hooi-Ling Lee, 1997. Impact Modeling of Sewage Discharge From Georgetown of Penang, Malaysia on Coastal Water Quality, Environmental Monitoring and Assessment, 44(2): 199–209.

[5] Koh, H.L, 1986. Mathematical Modeling in a Sewage Outfall Design, In Proceedings of the Autumn Course Research Seminar in Mathematical Ecology,(11): 284–300.

[6] Alla Choban,Igor Winkler, 2011. Problems of Regulation of Anthropogenic Load on the Water Objects with Various Dilution Ratio in Ukraine, Advanced Water Supply and Wastewater Treatment: A Road to Safer Society and Environment, (1):313-318.

Analysis on the restriction factors of the green building scale promotion based on DEMATEL

Hong Wenxia, Jiang Zhenyao, Yang Zhao

Qingdao Technological University, No.2, Changjiang Road, Huangdao District, Qingdao, Shandong Province, China

1181148336@qq.com

Abstract. In order to promote the large-scale development of the green building in our country, DEMATEL method was used to classify influence factors of green building development into three parts, including green building market, green technology and macro economy. Through the DEMATEL model, the interaction mechanism of each part was analyzed. The mutual influence degree of each barrier factor that affects the green building promotion was quantitatively analysed and key factors for the development of green building in China were also finally determined. In addition, some implementation strategies of promoting green building scale development in our country were put forward. This research will show important reference value and practical value for making policies of the green building promotion.

1. Introduction

With the rapid development of China's construction industry, environmental problems have become increasingly prominent and paramount. In the process of construction, large amount of waste water, waste gas and waste residue are discharged into the environment, resulting in serious pollution of air, water and soil and destruction of the ecological environment. Therefore, green building as a new concept of architecture has gradually been recognized and respected by the public. Green building combines the sustainable concept with the field of architecture, which can significantly improve energy efficiency and reduce energy consumption. In our country,green building is defined as the building which can maximize the conservation of resources, protect the environment and reduce pollution in its whole life cycle, and thus providing healthy, suitable and efficient use of space for people and being in harmony with nature. With the rapid development of China's economy, the factors affecting the large-scale promotion of green building are changing all the time. With the aid of DEMATEL method, the mutual influence degree of each obstacle factor was revealed quantitatively, and the key factors restricting the development of green building were also determined, which could simultaneously provide scientific suggestions for urban sustainable development.

2. Analysis of green building scale promotion restricting factors

2.1. Green building market

The amount of green buildings is mainly determined by the market demand. The level of consumers' cognition is the main obstacle of the green building market.

2.1.1. The cognition level of green building (S_1). The government and relevant institutions haven't transformed the qualitative effects into quantitative effects that makes most of the public consciousness fuzzy, which lead to the green building being difficult to promote and affect the status of green building in the society to a certain extent.

2.1.2. Market requirement of green building (S_2). Consumer demand mainly refers to the purchasing intention and purchasing power for green building. Consumer demand is usually determined by the individual age, education and culture level, income level, personal awareness of consumers.

2.2. Green Technology
The construction of green building is determined by the support of green technology and technical bottlenecks lead to the low degree of green building scale directly.

2.2.1. Application and diffusion level of green technology (S_3). The application of green technology is the integration of development, improvement and promotion process. Also, the cost of green building technology is the common concern of developers, consumers and designers and its application was determined by the exploit investment, application and diffusion of green technology.

2.2.2. Applicability level of green building (S_4). The designer should pay attention to local conditions, instead of implementing blindly "piling products" and "cold spell" policies. It is necessary to consider the regional environment suitability, resource saving and the economic applicability of the technology.

2.2.3. Quality performance level of green building (S_5). The level of quality performance comes from the integration of design, construction and appropriate technology system, which directly relates to the recognition of green building by consumers.

2.3. Macro economy
The development of green building is inseparable from the development of economy. The overall scale of the local economy and the activity of green building market are main factors that developers should take into account.

2.3.1. City GDP (S_6). The total amount of GDP is the measurement standard of local economy. When an area's economy has the development superiority and the development potential, it can provide the foundation for the green building development.

2.3.2. Income level of residents (S_7). Green building consumers must have the willingness to purchase the green buildings. The income levels determine the purchase power of consumers. Digging out the potential demand of consumers for green buildings is an effective measure to promote the development of green building in a large scale.

2.3.3. Attraction of urban residence (S_8). Attraction of urban residence refers to the total amount and quality of residence which meet the needs of urban residents.

2.3.4. Investment amount of green building (S_9). The investment amount of green building can reflect the economic development level and the scale of green building to a certain extent, which should be one of the variables considered by the macroeconomic subsystem.

3. Summary of DEMATEL model
The full name of DEMATEL method is decision-making and trial evaluation laboratory. The DEMATEL method is a methodology used to screen complex systems and simplify the process of system structure analysis. This method is to make full use of the experts' experience and knowledge to deal with complex social problems, especially for those systems with uncertain elements. This method

can analyze the relationship among various factors quantitatively through the relationship between factors and their strength to simplify complex problems.

The steps of the DEMATEL method are as follows:

(1) Set the system influence factors to $S_1, S_2, S_3, \ldots\ldots, S_m$.

(2) Setting up the direct influence matrix: X (X=$[x_{ij}]_{m\times m}$). $x_{ij} = 0,1,2,3,5$.

(3) Translate the direct influence matrix into normalized influence matrix: Y ($Y = [y_{ij}]_{m\times m}$).

$$Y = \frac{X}{\max\limits_{1\le i\le m} \sum_{j=1}^{m} x_{ij}} \tag{1}$$

(4) Calculate the comprehensive influence matrix of system influence factors: T ($T = [t_{ij}]_{m\times m}$).

$$T = Y(I - Y)^{-1} \tag{2}$$

(5) Calculate the influence degree and influenced degree of each factor.

$$p(i) = \sum_{j=1}^{m} t_{ij} \tag{3}$$

$$q(i) = \sum_{i=1}^{m} t_{ij} \tag{4}$$

(6) Calculate the centre degree and reason degree.

$$e(i) = p(i) + q(i) \tag{5}$$

$$f(i) = p(i) - q(i) \tag{6}$$

4. Green building scale promotion influence factors DEMATEL analysis

4.1. Data acquisition and process

The relationship among factors was determined by Delphi method. The influence degree, influenced degree, centre degree and reason degree (Table 1.) are calculated on this basis.

4.2. Result interpretation

From Table 1. we can draw some conclusions as follows:

(1)The factors which have more important influence degree are city GDP(S_6), applicability level of green building(S_4), quality performance level of green building(S_5). This result represents that these factors have great influence on other factors. As a result, we should promote the economic development of the city and think about the local conditions and the economic applicability at the same time. Furthermore, improving the level of quality performance of green building and meeting consumer demand for environmental quality and residential suitability requirements are also important.

(2)The factors which are influenced easily include market requirement of green building(S_2), the investment amount of green building(S_9)and the cognition level of green building(S_1). These data represent that these factors are influenced by other factors easily. Improving the public cognitive level is very important to promote the development of green building in a large scale.

(3)The factors which have large centre degree are market requirement of green building(S_2), the investment amount of green building(S_9)and city GDP (S_6). These three factors are the most direct factors hindering the development of green building that should arouse our concern.

(4)The factors which get large reason degree are city GDP (S_6), applicability level of green building(S_4)and income level of residents(S_7). The level of urban economic development is the key factor affecting the promotion of green building, accelerating the development of urban economy is the urgent requirement for green building development.

Table 1. Green building scale promotion restriction factor comprehensive relationship

	S_1	S_2	S_3	S_4	S_5	S_6	S_7	S_8	S_9
Influence degree p(i)	0.2011	0.4077	0.5090	1.1627	0.8658	1.4772	0.6963	0.2107	0.3446
Influenced degree q(i)	1.0589	1.1865	0.8566	0.1066	0.4709	0.0000	0.2381	0.8169	1.1404
Centre degree e(i)	1.2600	1.5942	1.3656	1.2693	1.3367	1.4772	0.9344	1.0276	1.4850
Reason degree f(i)	-0.8578	-0.7788	-0.3476	1.0561	0.3949	1.4772	0.4582	-0.6062	-0.7958

5. Conclusion

According to the DEMATEL model, we can analyze the influence degree, influenced degree, centre degree and reason degree of every factor. Through analysis of the model, city economic development level and the application level of green building proved to be the main factors influencing the development of green building. At present, China is in a period that the urbanization is in the high speed development and the construction industry is about to certainly entering into a booming situation. However, the construction industry of our country is still in an extensive operation stage with high input, high consumption, high pollution but low benefit. This situation has become an important factor affecting the construction of low-carbon economy. Therefore, in order to achieve low-carbon target, the construction industry must transform into green building, reform the traditional production mode, promote green building vigorously and take the road of sustainable development practically. It's our future research direction to design technical path and implementation strategy of the green building promotion according to quantitative analysis method.

References

[1] Li Jihua. Research on restrictive factors and countermeasures of housing industrialization in China[D].Chongqing University.2012.
[2] Luan Shihong, Wang Zhongxiu, Song Daijun. Analysis of factors affecting sustainable

development of real estate enterprises based on DEMATEL[J]. Journal of Shenyang Jianzhu University,2011(4):422-425.
[3] Qin Xuan, Jing Lei. Assessment and analysis of green building life cycle risk factors based on questionnaire survey[J]. China civil engineering journal, 2013(08):123-135.
[4] Wang Jun, Wang Youwei, Lin Haiyan, etc. Research progress of low carbon building technology in China[J]. Building science,2013(10):2-9+33.
[5] Zhang Xin. Application of green strategy in high-rise civil building design[J].Jiangxibuilding materials, 2015(2):46.

Study on vacuum pyrolysis of coffee industrial residue for bio-oil production

Nanwei Chen[1,2], Jie Ren[1], Ziwei Ye[1], Qizhi Xu[1], Jingyong Liu[1], Shuiyu Sun[1,2,*]

[1] School of Environmental Science and Engineering, Guangdong University of Technology, Guangzhou, 510006, P. R. China.
[2] Guangdong Polytechnic of Environmental Protection Engineering, Foshan, 528216, P. R. China.

[*] Corresponding author. E-mail: sysun@gdut.edu.cn.

Abstract: Coffee industrial residue (CIR) is a biomass with high volatile content (64.94 wt.%) and heating value (21.3 MJ·kg^{-1}). This study was carried out to investigate the pyrolysis condition and products of CIR using thermogravimetric analyser (TGA) and vacuum tube furnace. The influence of pyrolysis temperature, time, pressure and heating rate on the yield of pyrolysis products were discussed. There was an optimal pyrolysis condition: CIR was heated from normal temperature to 400 °C for 60 min, with 10 °C·min^{-1} heating rate and a pressure of 30 kPa$_{abs}$. In this condition, the yields of bio-oil, char and non-condensable gas were 42.29, 33.14 and 24.57 wt.%, respectively. The bio-oil contained palmitic acid (47.48 wt.%), oleic acid (17.45 wt.%), linoleic acid (11.34 wt.%), octadecanoic acid (7.62 wt.%) and caffeine (5.18 wt.%).

1. Introduction

Biomass is a clean and renewable energy source leading to environmental, technical and economical benefits. With the concern of environmental protection, the energetic recovery of biomass via pyrolysis, gasification or combustion has attracted increasing worldwide interest. The pyrolysis is a thermal decomposition process producing char with ash content, volatiles and low molecular weight gases. Without char and volatiles oxidation or less, it's the better way to recycle energy from biomass (e.g. microalgae, wood, straw and coffee waste) (Bridgwater, 2012; Collard and Blin, 2014).

Coffee is considered as the second largest traded commodity in the world. Coffee plants are cultivated in over 70 countries with an average annual production of 8 million tons in the world (ICO, 2014). Coffee industrial residue (CIR) is the by-product from instant coffee manufacturing process, left behind at the factory. By the statistics, instant coffee production consumes around 20% of the global production of green coffee beans (González et al., 2013) and for each kilogram of instant coffee produced 0.91 kg of CIRs (Silva et al., 1998). In addition, the CIR contains great amounts of organic compounds (i.e. fatty acids, lignin, cellulose, hemi-cellulose and other polysaccharides) (Pujol et al., 2013), while its comprehensive utilization has become a new study in recent years. For example, CIR has been investigated for biodiesel production (González et al., 2013), as production of activated carbon (Tsai et al., 2012; Tehrani et al., 2015), as raw material of fatty acids (Cruz et al., 2014).

Pyrolysis can be achieved via heating under inert gas (e.g. nitrogen and argon) or vacuum. The former is frequently-used, but the latter (vacuum pyrolysis) has had a high interest for several years (Jongh et al., 2011; Fan et al., 2014). In this study, both of them were considered. The parameters of pyrolysis include feedstock, particle size, pyrolysis temperature, pressure, time and heating rate etc. In general, different combinations of parameters lead to different product yields or chemicals. About pyrolysis conditions of CIRs, there were some studies from other researchers (Romeiro et al., 2012; Li et al., 2014; Tsai and Liu, 2013). The pyrolysis temperature and heating rate of CIRs had been discussed, but the effect of pyrolysis pressure and time were not considered. All of the reactors used inert gas for pyrolysis, leaving vacuum pyrolysis out of account.

This paper presents a study about pyrolysis of CIR, experimenting in thermogravimetric analyser at nitrogen atmosphere and tube furnace under vacuum. The thermogravimetry was carried out to investigate the changes of weight loss in pyrolysis. And the experiments in vacuum tube furnace were carried out for counting yields of products and providing bio-oil for characterization. The effect of pyrolysis factors on product yields were researched, including pyrolysis temperature, time, pressure and heating rate. Those results would make a contribution on the utilization of CIR.

2. Materials and methods

2.1. Materials

The coffee industrial residue (CIR) used for this study was collected from a coffee processing plant in Dongguan, Guangdong province, China. The original sample was oven-dried at 105 °C for 24 h and then became the sample of experiments in vacuum tube furnace. Dried CIR was grounded with a micromill (Yili QE-300, China) to pass through a 150 μm sieve, taken as the sample for characterization and thermalgravimetric analyzer.

2.2. Characterization of Materials

2.2.1. Proximate analysis. The moisture analysis was conducted according to DD CEN/TS 14774-3: 2004. The ash content was determined according to DD CEN/TS 14775: 2004. The volatile matter content was analyzed according to DD CEN/TS 15148: 2005. The fixed carbon was calculated using Eq.(1).

$$FC = 100\% - V - A - M \quad (1)$$

where FC, V, A and M are the content (wt.%) of fixed carbon, volatile, ash and moisture, respectively.

2.2.2. Ultimate analysis. The C, H, N and S contents in the sample were measured using an elemental analyzer (Perkin Elmer Series II 2400, USA). The O content was calculated using Eq.(2).

$$O = 100\% - C - H - N - S - A - M \quad (2)$$

where O, C, H, N, S, A and M are the content (wt.%) of O, C, H, N, S, ash and moisture, respectively.

2.2.3. Heating value. The higher heating value (HHV) was determined by an isothermal-jecket bomb calorimeter (SHR-15, China).

2.3. Non-isothermal thermogravimetric analysis

The experiments were carried out on a thermalgravimetric analyzer (Henven HCT-3, China). Pyrolysis was performed under a carrier gas flow (nitrogen) of 30 mL·min^{-1}, with a sample mass of approximately 10 mg, which was heated up to 800 °C at a heating rate of 10 °C·min^{-1}.

2.4. Pyrolysis

2.4.1. The experiments in thermogravimetric analyser. The experiments in thermogravimetric analyser were implemented before the experiments in vacuum tube furnace, in order to study the changes of weight loss on different pyrolysis conditions. The experiments were performed using a thermalgravimetric analyzer (Henven HCT–3, China). The sample was approximately 10 mg in each experiment. Pyrolysis was conducted with variable temperatures (300–500 °C), heating rates (5–25 °C·min^{-1}) and pyrolysis times (60 or 120 min) under a nitrogen gas flow of 30 mL·min^{-1}.

2.4.2. The experiments in vacuum tube furnace. The experimental setup is shown in Fig. 1. The tube furnace (SK-G06123K, China) consists of a 1 m long quartz tube (diameter 60 mm), heated by well-insulated, controllable heating elements. The quartz tube is connected to condensation pipes, bottles and a circulating water vacuum pump. The circulating water vacuum pump removed the organic vapors and other gas products from the reactor through the condensation pipes. The condensable vapors were then condensed in the vacuum traps and recovered as liquid, but non-condensable gas was absorbed by alkaline (NaOH) liquor in the third bottle or escaped. A control program was used to control the heating rate, final pyrolysis temperature and pyrolysis time, and the valve before circulating water vacuum pump was used to control pyrolysis pressure.

The experiments were conducted in batch runs with variable temperatures (300–500 °C), heating rates (5–20 °C·min^{-1}), pyrolysis times (20–100 min) and pressures (10–50 kPa$_{abs}$). Once the quartz tube was loaded (quartz boats with 40 g of dried CIR), heating started after the quartz tube was sealed and evacuated to the desired operating pressure. The quartz tube was natural cooling after control program finished. The setup was stopped cooling when the control program was over, but under vacuum until the sample temperature reached low temperature (approximately 100 °C). Then the quartz boats with char and the bottle with bio-oil were removed and weighed. As the empty quartz boats and bottle had been weighed, the weights of char and bio-oil were obtained. At last, the yields of char, bio-oil and non-condensable gas were estimated using Eqs.(3)–(5).

$$\eta_S = \frac{m_S}{m_0} \times 100\% \tag{3}$$

$$\eta_L = \frac{m_L}{m_0} \times 100\% \tag{4}$$

$$\eta_G = \frac{m_G}{m_0} \times 100\% = \frac{m_0 - m_S - m_L}{m_0} \times 100\% \tag{5}$$

where η_S, η_L and η_G are the yield (wt.%) of char, bio-oil and non-condensable gas, respectively; m_0, m_S, m_L and m_G are the weight (g) of sample, char, bio-oil and non-condensable gas, respectively.

2.5. Characterization of bio-oil

In characterization of bio-oil, pH meter (Leici PHS-25, China) was used to determine the pH of bio-oil. Dynamic viscosity of bio-oil was determined by rotary viscometer (Changji NDJ-8S, China). Moisture of bio-oil was measured using infrared moisture meter (Guanya SFY-20A, China). Organic matters in the bio-oil were analyzed by gas chromatography/mass spectrometry (GC/MS: Agilent 7890A/5975C, USA). A HP-5MS capillary column (30 m×250 μm×0.25 μm) was employed to separate the constituents. Bio-oil was diluted in acetone and passed through the organic filter membrane (aperture size 0.22 μm) using an injector, then 0.2 μL sample solution was injected into column at 250 °C, with a split ratio of 10:1. Helium was used as carrier gas at flow rate of 20 mL·min^{-1}. Oven temperature was programmed from 100 to 240 °C at 10 °C·min^{-1}, held at initial and final temperatures for 2 min. The detector and ion source temperature were 150 °C and 230 °C, respectively.

3. Results and discussion

3.1. Characterization of Materials

The results of proximate analysis, ultimate analysis and heating value about CIR were presented in Table 1. The volatile content of CIR is very high (approximately 65 wt.%), implicating high yield of bio-oil in pyrolysis. The CIR contains numbers of fixed carbon (about 21 wt.%), which would make a contribution to the yield of char in pyrolysis. In ultimate analysis, the main elements are C (46.62 wt.%) and O (23.90 wt.%). The higher heating value of CIR is 21.3 MJ·kg^{-1}, closed to wood (about 20 MJ·kg^{-1}) (Skreiberg et al., 2011), between sewage sludge (3–16 MJ·kg^{-1}) and coal (approximately 30 MJ·kg^{-1}) (Otero et al., 2002). In short, the CIR is an energy source containing much organic matter, and it's quite significant to recycle energy from CIRs with pyrolysis technology.

Table 1. Proximate analysis, ultimate analysis and heating value of the CIR (dry basis).

Proximate analysis (wt.%)				Ultimate analysis (wt.%)					Higher heating value ($MJ \cdot kg^{-1}$)
Moisture	Volatile	Fixed carbon	Ash	C	H	O	N	S	
6.98	64.94	21.03	7.05	46.62	8.87	23.90	5.64	0.94	21.3

3.2. Non-isothermal thermogravimetric analysis

Fig. 2 shows the typical TG and DTG curves of CIRs pyrolysis. TG curve indicates the relation between temperature and weight, and its derivative with respect to time is DTG curve, which shows the change speed of weight over time. The pyrolysis process could be divided into three stages: stage I (0–150 °C), stage II (150–550 °C) and stage III (550–800 °C). In stage I, there was a small weight loss that represented the evaporation of water, corresponding to a peak in DTG curve. Stage II was characterized by a major mass loss, approximately 65% of total weight, which was caused by the release of organic matters. And only one syncretic peak was observed in DTG curve. During stage III, the loss of weight was very slow, while the carbonization of pyrolysis residues was underway. Deduced from DTG curve in non-isothermal thermogravimetric analysis, the optimum pyrolysis temperature for CIRs might be between 300 and 500 °C.

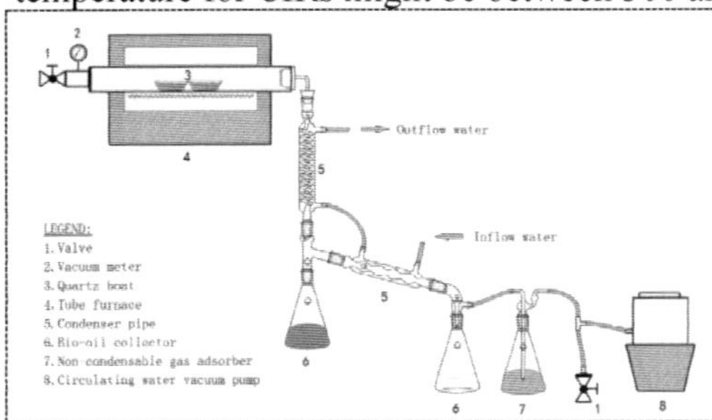

Fig. 1. Experimental setup.

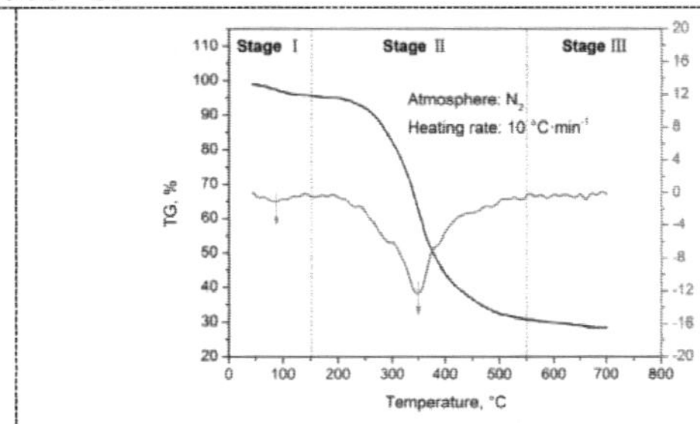

Fig. 2. Typical TG and DTG curves of CIRs pyrolysis (heated up to 800 °C with heating rate of 10 $°C \cdot min^{-1}$ and 30 $mL \cdot min^{-1}$ nitrogen flow).

3.3. Product yields on different pyrolysis conditions

3.3.1. Effect of pyrolysis temperature. Fig. 3a shows the changes of weight loss from the sample pyrolysis at different final temperatures. The weight of final solid products decreased as temperature increased, i.e. from 49.21 wt.% at 300 °C to 35.22 wt.% at 400 °C. The loss components were moisture and volatile, and different volatiles volatilized according to temperatures. The weight loss curves of different temperatures coincided at earlier stage as their heating processes were the same. And the curves of 450 °C and 500 °C completely coincided, which indicates that all volatiles had been lost. However, the final weight of 400 °C and 450 °C curves were much closed, which means that most volatiles had been lost at 400 °C.

Fig. 4a presents the yields of char, non-condensable gas and bio-oil from CIRs pyrolysis at different temperatures. As temperature increased, the yield of char decreased from 60.00 wt.% at 300 °C to 27.24 wt.% at 500 °C, and the yield of bio-oil increased to a maximum (approximately 43 wt.%). During the sample pyrolysis, it included the evaporation of water, the escape of volatiles and the charring of solid product. At 400 °C, the yield of bio-oil had been closed to 43 wt.%, that is to say, nearly all volatiles had gone out. Moreover, the charring of solid product had happened to a certain degree, while higher temperatures (450 or 500 °C) caused the loss of char. Therefore, the optimum pyrolysis temperature of CIRs was 400 °C, which was much closed to pyrolysis temperature (380 °C) of a study about soluble coffee ground (Romeiro et al., 2012).

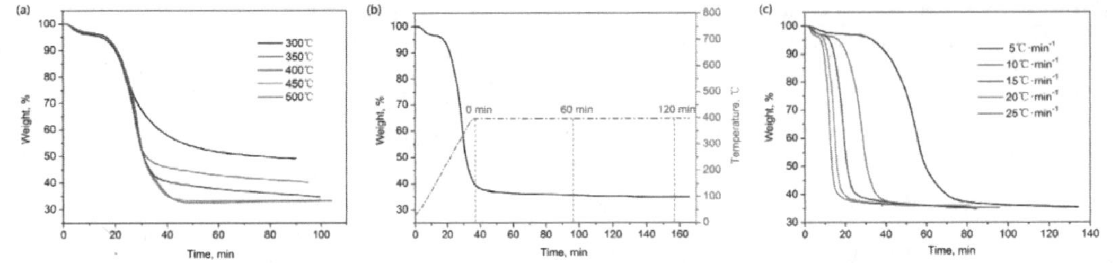

Fig. 3. Weight loss curves of CIRs pyrolysis: (a) at different final temperatures (with 10 °C·min^{-1} heating rate and 30 mL·min^{-1} nitrogen flow); (b) for different time (with 10 °C·min^{-1} heating rate to 400 °C with 30 mL·min^{-1} nitrogen flow); (c) with different heating rates (heated up to 400 °C for 60 min with 30 mL·min^{-1} nitrogen flow).

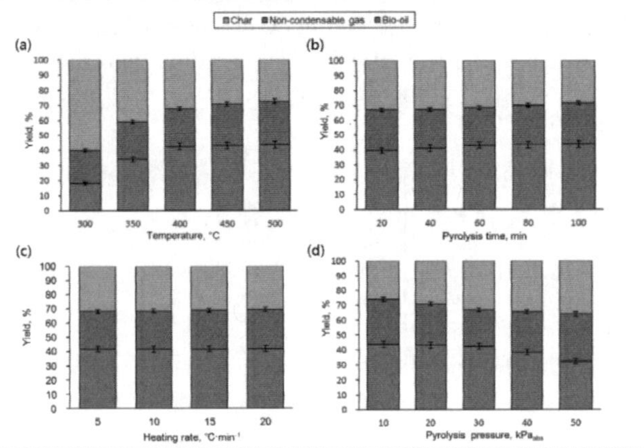

Fig. 4. Effect of pyrolysis factor on the yield of pyrolysis products: (a) pyrolysis temperature (with heating rate of 10 °C·min^{-1} for pyrolysis time of 60 min at pressure of 10 kPa$_{abs}$); (b) pyrolysis time (with 10 °C·min^{-1} heated up to 400 °C at 10 kPa$_{abs}$); (c) heating rate (heated up to 400 °C for 60 min at 10 kPa$_{abs}$); (d) pyrolysis pressure (with 10 °C·min^{-1} heated up to 400 °C for 60 min).

3.3.2. Effect of pyrolysis time. Fig. 3b presents the changes of weight loss from the sample pyrolysis related to pyrolysis time. The weights of solid product were 38.95 wt.% in 38 min (0 min at final temperature), 35.52 wt.% in 98 min (60 min at final temperature) and 34.84 wt.% in 158 min (120 min at final temperature), respectively. In other words, the pyrolysis of samples had been finished when the weight decreased to 35 wt.%.

Fig. 4b shows the yields of char, non-condensable gas and bio-oil from CIRs pyrolysis with different pyrolysis times. With pyrolysis time extended, the yield of char decreased slowly, and the yield of bio-oil increased to a maximum. More pyrolysis time, more heat and mass transfer happened, such as the evaporation of water, the escape of volatiles and the charring of solid product. Water and volatiles became bio-oil after cooling, and the solid turned to char, while charring too long caused the decrease of char. When pyrolysis time was 60 min, all volatiles had separated out, thus the yield of bio-oil reached a maximum value. As a result, the optimum pyrolysis time of CIRs was 60 min.

3.3.3. Effect of heating rate. Fig. 3c shows weight loss curves of the sample pyrolysis with different heating rates. The final weights of all curves were about 35 %, which means that the yield of solid product was not affected by heating rate. Except curve of 5 °C·min^{-1}, curves converged between 40 and 60 min, in other words, their pyrolysis processes began to synchronize. Obviously, the heating rate determined the speed to reach final pyrolysis temperature, but was not necessary to be too high as heat and mass transfer also needed some time.

As shown in Fig. 4c, the yields of char, non-condensable gas and bio-oil from CIRs pyrolysis with different heating rates were almost the same. That is to say, the heating rate did not influence product yields. However, the operating instruction manual of vacuum tube furnace suggests that the heating rate should be less than or equal to 10 °C·min^{-1}, in order to prolong furnace life. Based on the small-sized experiments, 10 °C·min^{-1} was suitable. Consequently, the optimum heating rate of tube furnace for CIRs pyrolysis was 10 °C·min^{-1}.

3.3.4. Effect of pyrolysis pressure. The experiments were just carried out in the vacuum tube furnace, because the thermogravimetric analyser in this work could not control pressure. Fig. 4d illustrates the yields from pyrolysis of CIRs as a function of pressure. As pyrolysis pressure reduced from 50 kPa_{abs} to 10 kPa_{abs}, the char yield decreased from 36.03 wt.% to 25.98 wt.%, and the bio-oil yield increased from 32.34 wt.% to 43.81 wt.%, while a maximum value was approximately 43 wt.%. This is an intuitive result, since it is commonly accepted that at higher pressures, there will be more recondensation reactions which would lead to more of the vapors becoming trapped on the char, thereby increasing the char yield. At the lower pressure, volatiles extract from tube faster, and there are fewer reactions of volatiles bringing about more bio-oils. A possible mechanism could be that higher gas density as a result of higher pressure improves the number of reactions between gases and solids, and causes char yield decreased (Jongh et al., 2011). Among 10 kPa_{abs}, 20 kPa_{abs} and 30 kPa_{abs}, the bio-oil yields were almost the same, but char yield at 30 kPa_{abs} was the greatest and 30 kPa_{abs} was the easiest to achieve. For this reason, the optimum pyrolysis pressure of CIRs was 30 kPa_{abs}.

3.3.5. Product yields on the optimal pyrolysis condition. Among pyrolysis factors, pyrolysis temperature and pressure are more important since they had most significant effect on yields of products. Pyrolysis time should be enough ensuring the sufficient transfer of mass and heat. And heating rate is allowed to adjust on the basis of machinery and equipment. According to results above, the optimal pyrolysis factors of experiments were 10 °C·min^{-1} (heating rate), 400 °C (pyrolysis temperature), 60 min (pyrolysis time) and 30 kPa_{abs} (pyrolysis pressure). On the optimal condition, the yields of char, non-condensable gas and bio-oil were 33.14, 24.57 and 42.29 wt.%, respectively.

3.4. Characterization of bio-oil

The bio-oil was from CIRs pyrolysis on optimum condition. It was an acidic liquid (pH=3.12), so its reservoir must be acid-proof. The dynamic viscosity of bio-oil was 3.6 mm^2·s^{-1}, which was much higher than water with a dynamic viscosity of 1.0 mm^2·s^{-1}. Simultaneously, the moisture of bio-oil was 17.86 wt.%, while it should be noted that the higher water content contributed to the lower viscosity. The problem of bio-oil translocation must be considered when comes to industrialized application. GC/MS was employed to identify the chemical compounds presenting in bio-oil. Table 2 provides the major chemical constituent of bio-oil. The chemical compounds were palmitic acid (47.48 wt.%), oleic acid (17.45 wt.%), linoleic acid (11.34 wt.%), octadecanoic acid (7.62 wt.%) and caffeine (5.18 wt.%). These organic matters have value to chemical industry.

Table 2. Major chemical constituent of bio-oil on optimum condition.

NO.	Retention time (min)	Chemical compound	Area (%)	Quality (%)
1	8.7	Pentadecane	3.42	96
2	12.8	Caffeine	5.18	97
3	13.2	Hexadecanenitrile	2.76	96
4	13.4	Methyl palmitate	2.53	98
5	13.8	Palmitic acid	47.48	95
6	15.4	Linoleic acid	11.34	94
7	15.5	Oleic acid	17.45	95
8	15.7	Octadecanoic acid	7.62	99
9	15.8	Octadecanamide	2.22	93

4. Conclusions

According to experimental results, the optimal pyrolysis condition was as follow: dried CIRs were heated from normal temperature to 400 °C for 60 min, with 10 °C·min^{-1} heating rate, under a pressure of 30 kPa_{abs}. In this condition, more bio-oil and char were prepared and less non-condensable gas was exhausted, in other word, yields of product were well-content. The bio-oil has value to chemical industry, containing a large number of palmitic acid and oleic acid.

Acknowledgments
This study was financially supported by the Science and Technology Planning Project of Guangdong Province (No. 2014B020216009), the Special Fund of University Discipline Construction of Department of Education of Guangdong Province (No. 2014KTSP022), the Major Achievements Cultivating Foundation of Guangdong University of Technology (No. 201206).

References

[1] Bridgwater, A.V., 2012. Review of fast pyrolysis of biomass and product upgrading. Biomass and Bioenergy 38, 68–94.
[2] Collard, F.X., Blin, J., 2014. A review on pyrolysis of biomass constituents: Mechanisms and composition of the products obtained from the conversion of cellulose, hemicelluloses and lignin. Renewable and Sustainable Energy Reviews 38, 594–608.
[3] Cruz, M.V., et al, 2014. Production of polyhydroxyalkanoates from coffee industrial residues oil obtained by supercritical fluid extraction technology. Bioresource Technology 157, 360–363.
[4] Fan, Y.S., et al, 2014. Rape straw as a source of bio-oil via vacuum pyrolysis: Optimization of bio-oil yield using orthogonal design method and characterization of bio-oil. Journal of Analytical and Applied Pyrolysis 106, 63–70.
[5] González, A.S., et al, 2013. Post-combustion CO2 capture adsorbents from spent coffee grounds. Energy Procedia 37, 134–141.
[6] Ico.org [Internet], London: International Coffee Organization; c1963– [updated 2014 Mar 3; cited 2015 Mar 19]. Available from: http://www.ico.org/news/annual-review-2012-13-e.pdf.
[7] Jongh, W.A.de, et al, 2011. Vacuum pyrolysis of intruder plant biomasses. Journal of Analytical and Applied Pyrolysis 92, 184–193.
[8] Li, X.F., et al, 2014. Energy recovery potential analysis of spent coffee grounds pyrolysis products. Journal of Analytical and Applied Pyrolysis 110, 79–87.
[9] Otero, M., et al, 2002. Analysis of the co-combustion of sewage sludge and coal by TG-MS. Biomass and Bioenergy 22, 319–329.
[10] Pujol, D., et al, 2013. The chemical composition of exhausted coffee waste. Industrial Crops and Products 50, 423–429.
[11] Reffas, A., et al, 2010. Carbons prepared from coffee grounds by H3PO4 activation: Characterization and adsorption of methylene blue and Nylosan Red N-2RBL. Journal of Hazardous Materials 175, 779–788.
[12] Romeiro, G.A., et al, 2012. A study of pyrolysis oil from soluble coffee ground using low temperature conversion (LTC) process. Journal of Analytical and Applied Pyrolysis 93, 47–51.
[13] Silva, M.A., et al, 1998. The use of biomass residues in the Brazilian soluble coffee industry. Biomass Bioenergy 14, 457–467.
[14] Skreiberg, A., et al, 2011. TGA and macro-TGA characterisation of biomass fuels and fuel mixtures. Fuel 90, 2182–2197.
[15] Tehrani, N.F., et al, 2015. Coffee extract residue for production of ethanol and activated carbons. Journal of Cleaner Production 91, 64–70.
[16] Tsai, W.T., et al, 2012. Preparation and fuel properties of biochars from the pyrolysis of exhausted coffee residue. Journal of Analytical and Applied Pyrolysis 93, 63–67.
[17] Tsai, W.T., Liu, S.C., 2013. Effect of temperature on thermochemical property and true density of torrefied coffee residue. Journal of Analytical and Applied Pyrolysis 102, 47–52.

IOP Conf. Series: Earth and Environmental Science **59** (2017) 012066 doi:10.1088/1755-1315/59/1/012066

The effect of substrate morphological structure on photoelectrical conversion performance of silicon solar cell

Yongtao Li[1,2,3], Xiaomeng Sun[2], Yang Xia[1,2]

1 Key Laboratory of Microelectronics Devices & Integrated Technology, Institute of Microelectronics, Chinese Academy of Sciences, Beijing 100029, China
2 Beijing Key Laboratory of IC Test Technology， 100088, Beijing, China

liyongtao@ime.ac.cn

Abstract. A novel method is proposed to evaluate electrical characteristics of silicon solar cell at real operating conditions. Silicon solar cells with different substrate morphological structures have various photoelectrical performances. The effect of substrate morphological structure on photoelectrical conversation performance of silicon solar cell has been investigated by illustration analysis, mathematical model and I-V test system. Results show that the solar cell with the porous-sponge like substrate has better electrical characteristics than those with conical like substrate morphological structure. The output power of porous substrate solar cell can exceed by 11% at 40 degree incident angle of sunlight compared to conical substrate solar cell.

1. Introduction

The study on photoelectrical conversion performance of solar cell is an importance research subject of photovoltaic industry. Many researches have been performed to assess electrical characteristics of solar cell in recent years [1-8]. But most of these evaluation methods focus on models of solar cells under the standard test conditions (STC), which means the incident simulated sunlight beam is directed at an angle of 90° relative to the solar cell plane. However these conditions only represent a very limited fraction of actual, realistic operating conditions for solar cells. The incidence angle of the sunlight indeed varies between different geographical locations and throughout the duration of a day and a year. Silicon solar cells, which boast more than 70% of the market share of the photovoltaic products, show a more cosine like angular dependence [9].

This paper presents an angle-resolved characterization approach of silicon solar cell and measured angle-resolved photovoltaic properties of porous substrate solar cells in comparison with conical substrate solar cells.

2. Experiments and model

The texturing of the solar cell surface has been shown to yield superior reflectance properties over a broad range of incident angles [10–12]. In our previous work, plasma immersion ion implantation (PIII) has been put forward to produce black silicon to texture the silicon solar cell surface. Hence, the silicon solar cells used for this work was commercially available boron-doped p-type multi-crystalline silicon wafers obtained from the ingot by wire sawing with thickness ~200 μm, area 156mm × 156 mm, and resistivity 1~3Ωcm. The black silicon was prepared by plasma immersion ion implantation process on domestic equipment and subsequently subjected to acid etching in HF/HNO3 solutions.

After that, the black silicon wafers were phosphorus doped using phosphorous oxychloride (POCl3) as the dopant source and then subjected to edge etching through reactive ion etching and removing phosphosilicate glass (PSG) layer with diluted HF. Silicon-nitride layer for passivation was grown by plasma enhanced chemical vapor deposition (PECVD) process. Finally, the front and back metallization of all the wafers were carried out by screen-printing technique and followed by baking and co-firing in a conveyer belt furnace.

The microstructure of the silicon solar cells were characterized by scanning electron microscope (SEM). Figure 1 and figure 2 shows different substrate surface of solar cells respectively. The microstructures seen in figure 1 are porous-sponge like concaves arbitrarily spread throughout the solar cell surface. The microstructures seen in figure 2 are conical-like hillocks randomly distributed across the entire solar cell surface.

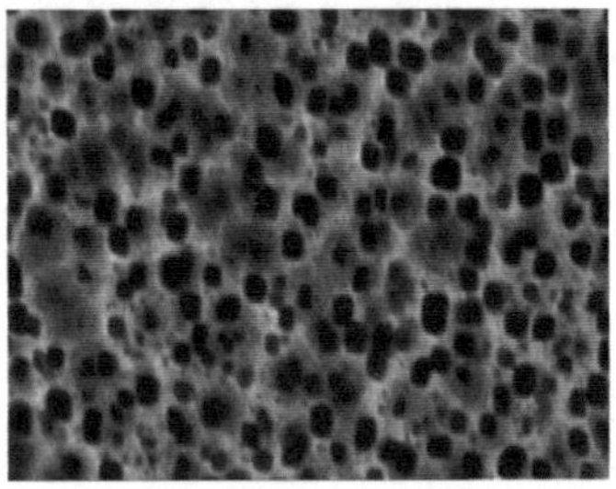

Figure 1. SEM-image of porous-sponge like substrate surface of solar cell.

Figure 2. SEM-image of conical like substrate surface of solar cell.

The electrical equivalent circuit module of solar cell is depicted in figure 3, which contains a photocurrent source, a p-n junction diode, a p-n junction capacitance, a series resistor and a shunt resistor. Due to the p-n junction capacitance is relatively small, and the time constant of the PV system is large enough, the influence of the p-n junction capacitance can be ignored. Applying Kirchhoff's current law (KCL) and Shockley equation to the junction point of these two resistors gives the characteristic equation of the solar cell, which is a nonlinear transcendental equation, as follows:

$$I = I_{ph} - I_0\left\{\exp[\frac{q(U+IR_s)}{AKT_c}]-1\right\} - \frac{U+IR_s}{R_{sh}} \tag{1}$$

Where U and I are respectively the output voltage and current of the solar cell, I_{ph} is the photocurrent which is proportional to the irradiance; I_0 is the diode saturation current, R_s is the series resistance, R_{sh} is the parallel resistance, A is the ideality factor of the diode which change with the output voltage, K is the Boltzmann constant (1.38×10^{-23} J/K) and T_c is the absolute temperature.

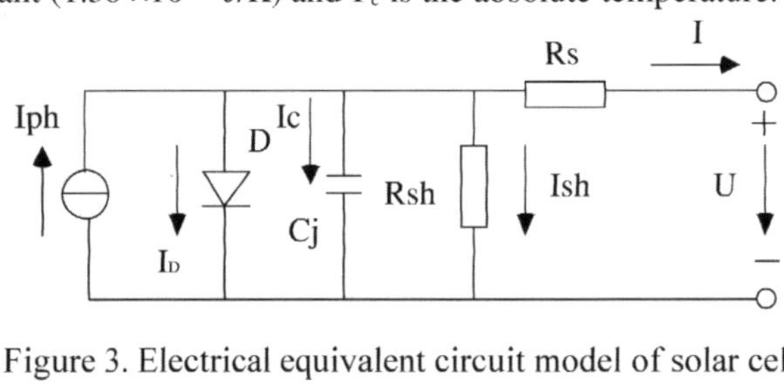

Figure 3. Electrical equivalent circuit model of solar cell

The presence of R_{sh} corresponds to the leakage current in the p-n junction. In practice, R_{sh} is much larger than $U+IRs$ .Therefore R_{sh} was ignored. The exponential term is much larger than 1. Thus, Equation (1) can be further simplified as equation (2):

$$I = I_{ph} - I_0 \exp[\frac{q(U + IR_s)}{AKT_c}] \tag{2}$$

For a given PV module, in order to evaluate A, R_s, I_{ph}, I_0 in a particular environmental condition, a system of non-linear equations is generated by forcing to fit the I-V curve expressed by equation (2) in four points. Two particular points are used to determined two equations: open circuit point ($V=V_{oc}$, $I=0$), short circuit point ($V=0$, $I = I_{sc}$). In these cases, equation (2) becomes:

$$0 = I_{ph} - I_0 \exp[\frac{qU_{oc}}{AKT_c}] \tag{3}$$

$$I_{sc} = I_{ph} - I_0 \exp[\frac{qI_{sc}R_s}{AKT_c}] \tag{4}$$

In equation (4), because the equivalent series resistance R_s is relative smaller than the p-n junction on-resistance, the exponential term approximates to 0. Then the equation becomes:

$$I_{sc} \approx I_{ph} \tag{5}$$

The diode saturation current can derived from equation (3).

$$I_0 = I_{ph} \exp[-\frac{qU_{oc}}{AKT_c})] \tag{6}$$

From equation (5) and (6), equation (2) becomes:

$$I = I_{ph}(1 - \exp[\frac{q(U + IR_s - U_{oc})}{AKT_c}]) \tag{7}$$

The photocurrent I_{ph} can be influenced by the irradiance as the equation (9) described:

$$I_{ph} = I_{phref} F(\theta) \tag{8}$$

Where I_{phref} is the photocurrent under one sun illumination (1000 W/m^2, AM1.5G) at 298K temperature, $F(\theta)$ is defined as a function of the sunlight incident angle θ on the surface of solar cell, θ is incident angle of the sun light due to solar cell normal axis, seen in figure 4 and figure 5.

In equation (7), because product item IR_s is much smaller than the difference of U and U_{oc}, combined with equation (8), equation (7) becomes:

$$I \approx I_{phref}(1 - \exp[\frac{q(U - U_{oc})}{AKT_c}])F(\theta) \tag{9}$$

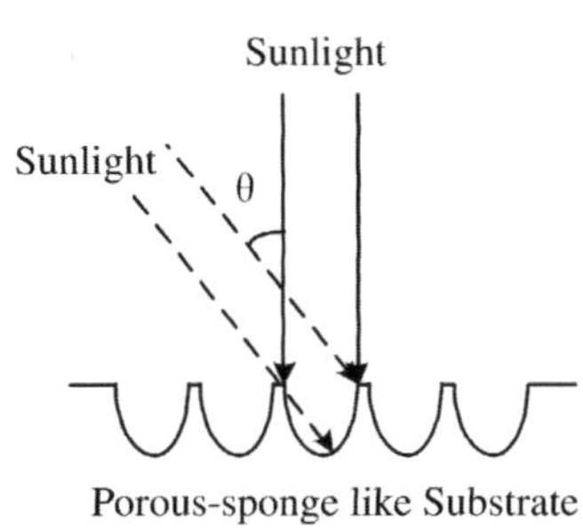

Figure 4. Schematic view of porous substrate solar cell radiated at various incident angle of sunlight

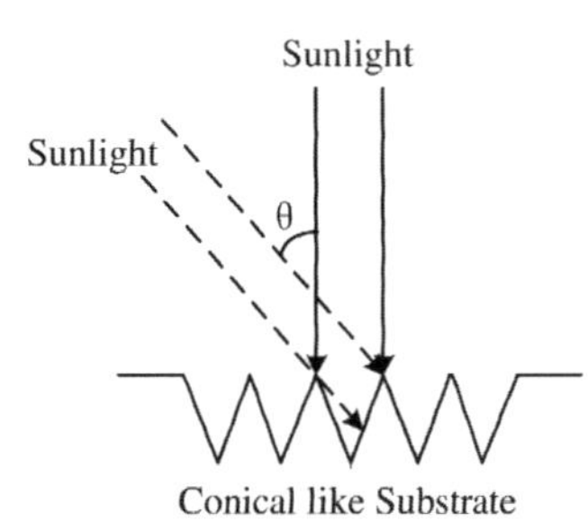

Figure 5. Schematic view of conical substrate solar cell radiated at various incident angle of sunlight

Figure 4 shows schematically the porous substrate solar cell radiated at various incident angle of sunlight, and figure 5 shows schematically the conical substrate solar cell radiated at various incident angle of sunlight. When sunlight is deflected a contain angle, i.e. incident angle θ, the irradiance area on the solar cell changes. The bigger the incident angle θ, the smaller the irradiance area. It results in photocurrent I_{ph} declines. Thus, $F(\theta)$ is a monotone decreasing function of θ.

It is seen in figure 4 and 5 that the irradiance area of the porous substrate solar cell is a bit larger than that of the conical substrate solar cell at the same of incident angle θ. Accordingly, it should be inferred that the photoelectrical conversion performance of porous substrate solar cell is superior to the conical substrate solar cell.

3. Results and discussion

The I–V curves and photoelectrical conversion properties including short circuit current I_{SC}, open circuit voltage V_{OC}, fill factor, and output power P, were measured on solar cells under one sun illumination (1000W/m2, AM1.5G) using a Newport Oriel solar simulator and a Keithley 2400 source meter. The solar simulator was fixed during all measurements. The solar cell test stage consists of an upper probe station and a bottom rotating table. The incidence angle was varied using two PMC100 stepping motors connected to the bottom rotating table.

Based on the experiment data, it is found that the open circuit voltage V_{OC} and fill factor do not change significantly for the incident angles below 80 degree. However, the short circuit current I_{SC} and consequently output power P decrease with increasing incident angle.

The measured angle-dependent short circuit currents for different substrate solar cells are plotted in figure 6. The I_{SC} of the porous substrate solar cell is always higher than that of the conical substrate solar cell below 60 degree incident angle of sunlight. The difference between the short circuit currents of two kinds of solar cells achieves the maximum at 40 degree, about 0.538 A.

Figure 7 shows the measured angle-dependent output power for different substrate solar cells. The P of the porous substrate solar cell is always higher than that of the conical substrate solar cell below 70 degree incident angle of sunlight. The difference between the short circuit currents of two kinds of solar cells achieves the maximum at 40 degree, about 0.38 W. The maximum 0.38W takes up 11% of the output power of the conical substrate solar cell.

Therefore, the solar cell with the porous-sponge like substrate has better electrical characteristics than those with conical like substrate morphological structure.

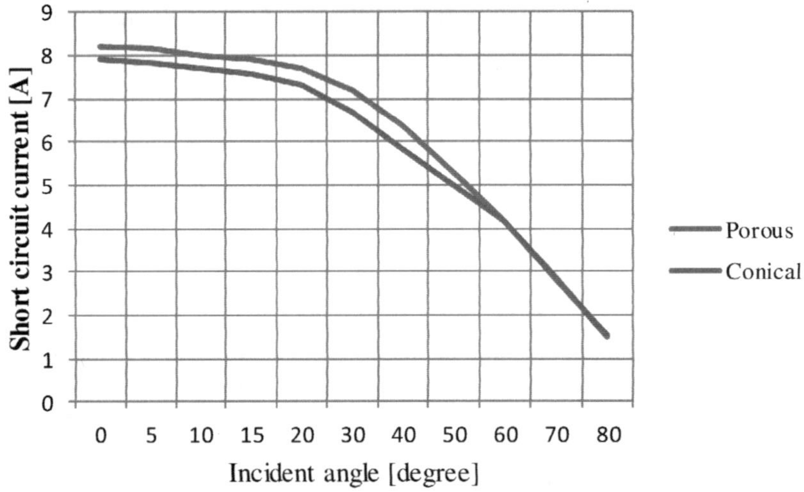

Figure 6. Short circuit current curves of different solar cells at various incident angle of sunlight

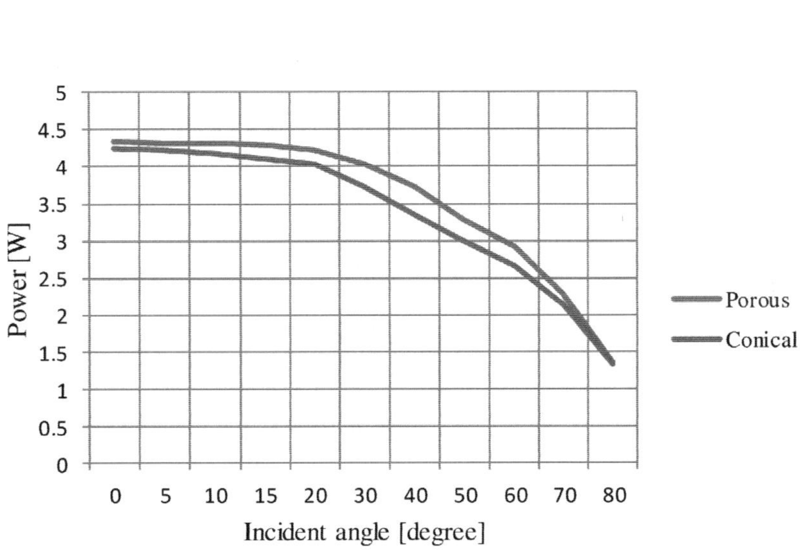

Figure 7. Power output curves of different solar cells at various incident angle of sunlight

4. Conclusions

A novel method for angle-resolved characterization of solar cells and estimated angle-dependent photoelectrical conversion performances of different substrate surface of solar cells has been presented. Silicon solar cells with different substrate morphological structures have various photoelectrical performances at real operating conditions. Compared to the solar cell with conical like substrate morphological structure, the solar cell with the porous-sponge like substrate has better photoelectrical characteristics over a broad range of incident angles due to larger irradiance area in general. The short circuit current and the output power of porous substrate solar cell can respectively exceed by 0.538A and 0.38W at 40 degree incident angle of sunlight.

Acknowledgements

This work was financially supported by the National Natural Science Foundation of China (Grant No. 61274059).

References

[1] Saloux Etienne, Teyssedou Alberto, Sorin Mikhaïl 2011 *Solar Energy* **85** 713.
[2] Nema RK, Nema Savita, Agnihotri Gayatri 2009 *International Journal of Recent Trends in Engineering* **1** 151.
[3] Khazaei J, Miao Z, Piyasinghe L, Fan L 2015 *Electric Power Systems Research* **123** 85.
[4] Lo Brano V, Orioli A, Ciulla G, Di Gangi A 2010 *Solar Energy Materials and Solar Cells* **94** 1358.
[5] Laudani A, Mancilla-David F, Riganti-Fulginei F, Salvini A 2013 *Solar Energy* **97** 122.
[6] Orioli A, Di Gangi A 2013 *Applied Energy* **102** 1160.
[7] Kashif Ishaque, Zainal Salam, Hamed Taheri, Amir Shamsudin 2011 *Solar Energy* **85** 1768.
[8] Xiaofang Yuan, Yongzhong Xiang, Yuqing He 2014 *Solar Energy* **108** 238.
[9] Balenzategui J L, Chenlo F, 2005 *Solar Energy Materials and Solar Cells* **86** 53.
[10] X. Liu, Coxon P R, Peters M, Hoex B, Cole J M, Fray D J 2014 *Energy & Environmental Science* **7** 3223.
[11] Rahman A, Ashraf A, Xin H, Tong X, Sutter P, Eisaman M D, Black C T 2015 *Nature Communications* **6** 5963.
[12] Jihun Oh, Hao-Chih Yuan, Howard M B 2012 *Nature Nanotechnology* **7** 743.

Investigation and analysis of medical waste generation in Enshi area of Hubei Province, China

Jin Dengchao[1,a], Teng Hongjun[2,b], Bao Zhenbo[1,c] and Li Yang [1,d]

[1]Engineering and Technology College, Tianjin Agricultural University, Tianjin, China, 300384

[2]Clinic, Tianjin Agricultural University, Tianjin, China, 300384

[a]jindengchao@163.com; [b]hongjun-teng@163.com; [c]baozhenbo@sohu.com; [d]liyang.tjac@gmail.com

Abstract. Based on medical waste collecting data of Enshi medical waste disposal center. The generation of medical waste and its change trend in Enshi area were both studied. The influencing factors and changing rules of medical waste generation were also analyzed. It can be found that the amount of medical waste in Enshi area is increasing year by year, the average annual growth rate of about 6.14% between 2011-2014. It was also found that the output of medical wastes varied regularity by seasons. February was the lowest month for medical waste, March and July were the peak months. By statistical analysis, average annual medical waste production per 10000 people was 4.5 ton and per bed average annual production was 133.58 kg.

1. Introduction

Medical waste is hazardous waste, more and more attention had been attracted to medical waste management and safely disposal in many countries. In December 2008, Asian countries held an international conference on health policy in Asia and medical waste management at the university of Tokyo[1], calling for treatment of medical waste in the world. The related literature about medical waste the management and treatment in many countries, such as Greece[2], Sudan[3], Iran[4], provided useful references for the management and disposal of medical wastes in local and other countries.

Since the outbreak of SARS in 2003, China began to focus on medical waste disposal and to build medical waste centralized treatment center in many area. Before each medical waste treatment center is constructed, it is necessary to estimate the amount and generation trend of local medical waste. This paper analyzes the data of medical waste collection in Enshi area of 2012-2014 years , the regularity of medical waste generation and its influencing factors were studied, in order to provide reference for similar project.

Enshi area is located in the west of Hubei province, local population is of 3.29 million, the geographical area is of 24,000 square kilometers. A centralized medical waste treatment center had been finished in 2008, which adopted steam sterilization technology to treat medical waste. Due to longer operating time and perfect management system, medical waste within the region can be fully collected and treated. Therefore, the project is representative.

The data used in this paper was the actual collection records of Enshi medical waste treatment center. Includes the total amount of medical waste collected between 2012-2014 year and the monthly medical waste generated by each hospital.

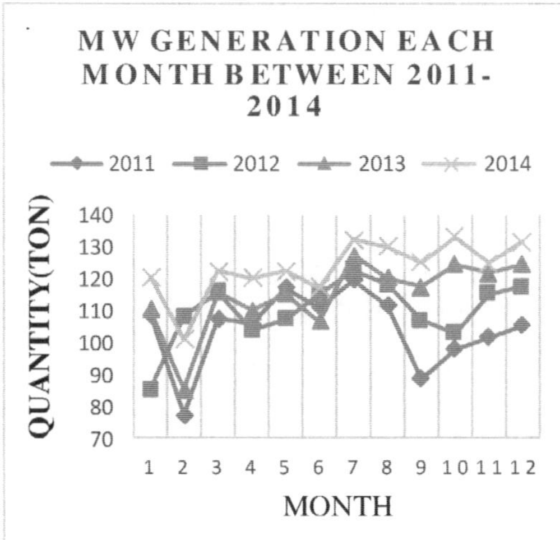

Figure 1. Medical waste generation each month between 2011-2014 in Enshi area

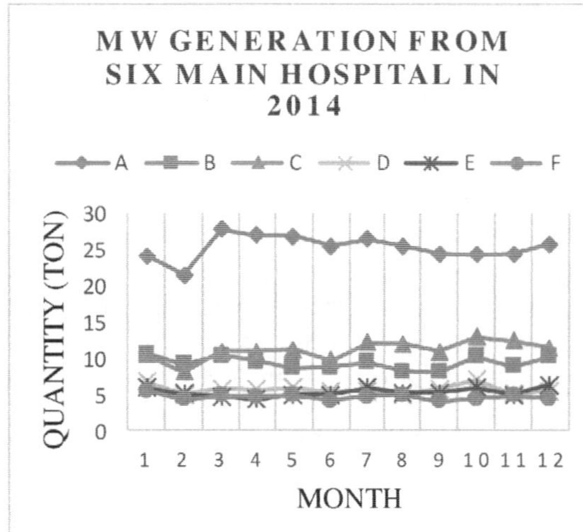

Figure2. Medical waste generation in six main hospital in 2014

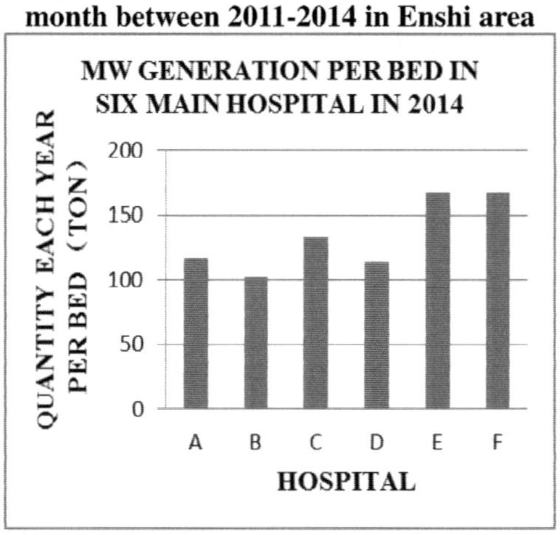

Figure 3. Medical waste generation per bed in six main hospital in 2014

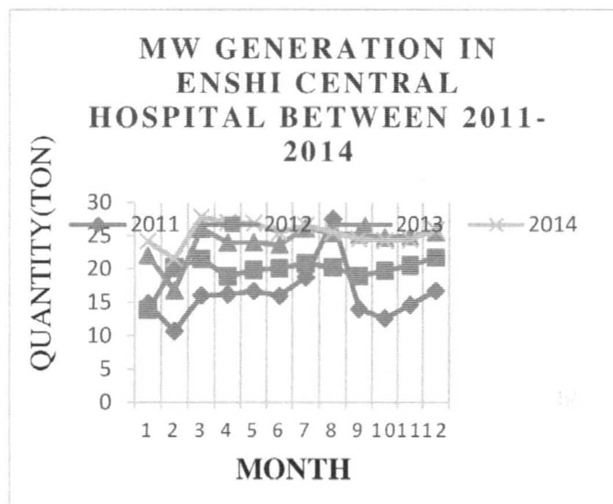

Figure4. Medical waste generation in Enshi central hospital between 2011-2014

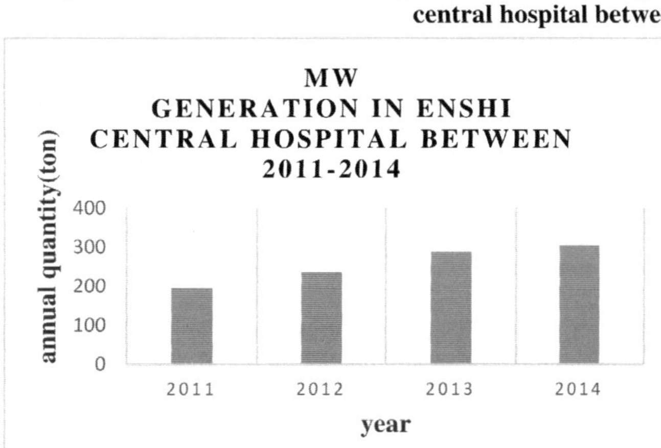

Figure 5. Medical waste generation annually in Enshi central hospital between 2011-2014

2. Method

The data used in this paper was the actual collection records of Enshi medical waste treatment center. Includes the total amount of medical waste collected between 2012-2014 year and the monthly medical waste generated by each hospital.

3. Results and discussion

Figure 1 describes the monthly medical waste production between 2011-2014, Figure 2 and Figure 3 are the results of medical waste production in the top six hospitals, Figure 4 and Figure 5 describe the generation of medical waste in the Enshi central hospital.

By analysis of medical waste production in Enshi area, it can be found:

3.1 Medical waste generation showed a gradual increase trend.

Based on the comprehensive analysis of data, it can be found that the medical wastes have been increasing year by year, this trend is reflected in the total collection of medical waste in Enshi between 2011-2014, from 2011 to 2014, annual amount of medical waste were: 1251.11 tons, 1317.64 tons, 1378.24 tons and 1481.54 tons, with an average annual growth rate of: 6.14%.

From the data of medical waste in Enshi central hospital, more obvious reflected the trend of increasing the amount of medical waste year by year, Enshi central hospital, in 2011-2014, the average annual growth rate of medical waste was 16.86%.

The trend of increasing medical waste production is very similar to that reported in Brazil[5]. The main reasons maybe as follows:

(1) With the continuous development of Chinese economy, people paid more and more attention to health, the ability to seek medical treatment was stronger, more and more people had been receiving hospital treatment when they were ill, resulting in the rise of medical waste.

(2) With the continuous promotion of government and environmental protection departments, people's awareness of medical waste collection continues to improve, the amount of medical waste into the collection has gradually increased.

3.2 Medical waste produced less in February of each year.

After the discussion with the hospital, the reason was found that February is the month of the Spring Festival in China, according to Chinese traditional customs, people generally avoid the Spring Festival hospitalization, therefore, the amount of medical waste generated was relatively less than other months.

3.3 For four consecutive years, the peak of medical waste production in March and July

After the discussion with several typical hospitals, the reasons maybe as follows:

(1) March is the first month after the Spring Festival, more patients visit hospital. In addition, March is the season changing month in this region, easily lead to disease, indirectly result in the increase in the amount of medical waste.

(2) Enshi is a famous tourist area in China, the tourism season is July and August, the influx of large numbers of tourists increased the number of local migrants. In 2015, for example, the annual tourists 37.5 million passengers, of which tourists received a total of 5.8 million passengers in July, Tourism population accounted for 174.7% of the local population. A large increase in population increased the number of hospital patient and medical waste production in the tourist season.

3.4 In 2014, the per bed annual production of medical waste in Enshi's main hospitals was between 102-170 kg, and the average per bed production capacity of medical waste was 133.58 kg. This value is close to the medical waste data of Gansu province[6].

3.5 In addition, by calculation, we can get each ten thousand people's medical waste production capacity of 4.5 tons in Enshi area in 2014.

4. Conclusion

With the economic growth and people's health requirements for the improvement of medical condition, medical waste in Enshi area gradually increased, the growth rate is about 6%.

The amount of medical waste in Enshi area changed significantly with the month, minimum amount of medical waste was generated during Spring Festival. Affected by seasonal changes and tourism, medical waste generated a larger amount in March and July.

Acknowledgements

The project has been carried out in Enshi medical waste treatment center of great help, thanks in this.

References

[1] Haruyo Nakamura, Moazzam Ali, Yoshihisa Shirayama. A Symposium on International Health Policy and Medical WasteManagement Research in Asian Region, *BioScience Trends,* 2008, 2(4), pp. 133-134.

[2] Komilis D, Katsafaros N, Vassilopoulos P. Hazardous medical waste generation in Greece: case studies from medical facilities in Attica and from a small insular hospital. *Waste Management & Research.* 2011, 29 (8), pp. 807-814.

[3] Suhair A. Gayoum Saad. Management of hospitals solid waste in Khartoum State. *Environ Monit Assess* .2013, 185, pp. 8567–8582.

[4] Majid Sartaj, Raheleh Arabgol. Assessment of healthcare waste management practices and associated problems in Isfahan Province (Iran). *J Mater Cycles Waste Manag.* 2015, 17, pp. 99–106.

[5] Moreira AM, Assessment of medical waste management at a primary health-care center in SãoPaulo, Brazil. *Waste Management* , 2013,33 (1), pp. 162-167.

[6] Zhang HJ, Zhang YH, Wang Y. Investigation of medical waste management in Gansu Province, China. *Waste Management & Research.* 2013, 31 (6), pp. 655-659.

Adsorption mechanisms and impact factors of oxytetracycline on activated sludge

Song Xiancai[1,2], Liu Dongfang[2*] , Zhao Lejun[1]

[1]Tianjin Municipal Engineering Design & Research Institute,Yingkou road,239,300051, Tianjin, China
[2]School of Nankai University,Weijin road 94,300071, Tianjin, China

421092143@qq.com

Abstract: The adsorption mechanisms and the effect of Oxytetracycline (OTC) onto activated sludge were studied. The results show that the adsorption of Oxytetracycline (OTC) onto activated sludge was coincident with the Pseudo-second-order kinetic model which suggested that chemical adsorption mechanism was dominant. The influences including pH and metal ions on the OTC were examined. It was demonstrated that the adsorption process was highly pH-dependant, which indicate that cationic exchange mechanisms may play an important role in the adsorption process. Na^+, K^+, Ca^{2+}, Mg^{2+} and Cd^{2+} ions more or less inhibited the adsorption of OTC on activated sludge while Cu^{2+} enhanced the adsorption ability. The phenomenon may reflect the result that a surface complexation mechanism could involved in the adsorption.

1. Introduction

Tetracycline (TC), has been used extensively world wide as a feed additive to treat diseases and thus improve the growth rate of livestock. As in the case with other antibiotics, tetracycline is poorly dsorbed in the digestive tract. As a result, 50 – 80% of the initial TCs dose was discharged in livestock waste [1]. TCs residues have been found everywhere in the environment, including soils [2], sediments [3], surface water[4], and wastewater[5-6]. The presence of low levels of antibiotics and their byproducts in the environment is favorable for the transfer and spread of antibiotic resistant determinants among microorganisms, and cause public health issues [7]. Therefore, it is of great importance to fully understand the occurrence, fate, transport and transformation of antibiotics in the environment [1].

The activated sludge process is the most common form of secondary treatment employed in China. Tetracycline originating from livestock is discharged into the sewer (with or without in situ pretreatment) and ends up in municipal wastewater treatment plants (WWTPs)[8]. The problem of Tetracycline pollution has drawn great attention of researchers all over the world in recent years. However, knowledge gaps still exist on the interaction between TCs and activated sludge. Some previous studies on pure clays, soils, and sediments had proven that the adsorption capacity of sorbent to TCs was influenced by pH due to different mechanisms[9-11], but little research has been focused on the effect of the adsorption of OTC on activated sludge[12]. So, it is of significant importance to understand the factors that affect tetracycline adsorption by activated sludge and the adsorption mechanism.

In this study, Oxytetracycline, was selected as a representative compound for the investigation of the adsorption behavior of TCs on activited sludge. The study investigates the effect of pH and metals on adsorption - in aqueous solution using batch experiments. The overall objectives of this study were

(1) To evaluate the kinetics of OTC adsorption onto activatedsludge, (2) To investigate effects of pH and heavy metals on the adsorption of OTC on activated sludge, and (3) to discuss the interaction mechanisms between OTC and activated sludge based on OTC molecular structure and the results of FTIR analyses.

2. Materials and methods

2.1 Sorbents

The activated sludge used in this study was obtained from the stage 1 aeration basin in the wastewater treatment plant for the Technical and Economic Development Area (TEDA), Tianjin, China. After collection, the aerobic activated sludge samples were aerated in a well mixed 12L sequencing batch reactor (SBR). Conditions in the reactor were maintained at an agitation rate of 150 rpm, temperature of 25℃, pH of 6.5-6.8, hydraulic retention time (HRT) of 24 h, sludge retention time (SRT) of 15 days and dissolved oxygen (DO) from 2 to 4 mg/L. Synthetic wastewater[13] was fed into the reactor to keep the organic loading rate of the system at about 500 mg COD/L·d. The mixed liquor suspended solids (MLSS) and mixed liquor volatile suspended solids (MLVSS) concentration were 1.2 ± 0.1 and 0.9 ± 0.1 g/L, respectively. After 30 days, the activated sludge in SBR reactor achieved stable properties and adsorption efficiency.

In sorption tests, mature activated sludge was used. The sludge was washed three times using deionized water to remove the surface soluble ions, and then sealed into flasks for storage at 4 ℃. The treated sludge was diluted to the required experimental concentrations in subsequent adsorption experiments [14].

2.2 Chemicals

Hydrochloride salt of OTC (98% purity), acetonitrile (HPLC grade), and oxalic acid (99% purity) were purchased from sangon biotech. (Shanghai, China). All other chemicals, including Cadmium, sodium, potassium, copper, calcium, and magnesium salts, were A.R. grade bought from Sinopharm Chemical Reagent Co. (Shanghai, China).

2.3 Analysis of OTC

The concentration of OTC in the supernatant filtrate was analyzed directly using a high performance liquid chromatography system HPLC (Waters 2695, USA) equipped with a Supercol C18 reverse-phase column (5um, 4.6 × 200 mm) with column temperature at 30℃. The mobile phase consisted of acetone trile (23%) and 10 MM oxalic acid (77%) at a flow rate of 1.0 mL/min. OTC was analyzed by a UV detector at 355nm. The detection limit of HPLC was 0.1 mg/L. The calibration curve was statistically significant and linear in the concentration range of 0.5-20 mg/L ($p<0.05$, R2=0.9996).

3. Results and discussion

3.1 Adsorption kinetics of OTC

The adsorption kinetic parameters of OTC onto activated sludge are illustrated in table 1. The high correlation coefficient ($R^2>0.999$) shows that pseudo-second-order kinetic model was more suitable for description of sorption kinetics for the adsorption of OTC than pseudo-first-order kinetic model, while the theoretical values of qe calculated from pseudo-second-order kinetic corresponed well with the experimental value of maximum sorption amount. The pseudo-second-order kinetics model is used to describe chemisorptions and has been widely applied to the adsorption of pollutants from aqueous solutions in recent years [6]. The pseudo-second-order rate equation is as follows:

$$\frac{dq_t}{dt} = k_2(q_e - q_t)^2 \quad (1)$$

Where k_2 (g mg^{-1} h^{-1}) is the rate constant of adsorption, qe (mg g^{-1}) is the amount of OTC adsorbed at equilibrium, and q_t (mg g-1) is the amount of OTC adsorbed on the surface of the adsorbed at any time.

Rearranging Eq. (1) into a linear form, we obtained

$$\frac{t}{q_t} = \frac{1}{k_2 q_e^2} + \frac{1}{q_e} t \quad (2)$$

Where $k_2 q_e^2$ (h) is the initial rate (mg g^{-1} h^{-1}). The pseudo-second-order rate constants k_2 and q_e, were calculated from the slope and intercept of the plots of t /q_t versus t. As shown in the Fig. 1, the plots showed good linearity under different conditions of MLSS and R^2s were greater than 0.999. Kinetics of OTC adsorption on activated sludge followed the pseudo-second-order model, suggesting that the biosorption of OTC onto activated sludge is a chemical-sorption process. The process involves exchange or sharing of electrons mainly between cation and functional groups (hydroxyl and carboxyl groups) of the biomass cells[8]. The equilibrium adsorption capacity (q_e, mg g^{-1}) of activated sludge is varies depending on the initial concentration of adsorbent. Table.1 and Fig. 1 show that, adsorption capacities decreaes with an increase in the initial concentration of adsorbent. A certain amount of activated sludge can adsorb a certain ratio of OTC. If the initial concentration of adsorbent increase while OTC amount kept constant, the specific removal efficiency of OTC by activated sludge is decreased.

Table1. Pseudo-first-order and pseudo-second-order adsorption rate constants of OTC on activated sludge in different MLSS.

MLSS	First-order constants				Second-order constants			
mg·L^{-1}	c_0/mg·L^{-1}	q_e/mg·g^{-1}	K_1/h^{-1}	R^2	K_2/g·(mg·h)$^{-1}$	q_e/mg·g^{-1}	R^2	h/mg·h^{-1}
400	10	13.9	0.216	0.962	0.107	25.00	0.999	66.7
800	10	4.98	0.169	0.885	0.390	12.66	0.999	62.5
1600	10	1.51	0.197	0.850	1.976	6.49	1	83.3
2000	10	0.91	0.154	0.692	3.725	5.18	1	100.0

The initial adsorption rates were 62.5, 66.7, 83.3 and 100 mg g^{-1} h^{-1}, when the initial MLSS concentration was 400 mg/L, 800 mg/L, 1600 mg/L, and 2000 mg/L, respectively. This is may be because the mass transfer driving increases with increasing MLSS concentration resulting in OTC molecule adsorbing to the sludge surface in a short period of time[7]. These initial rates are slow when compared to TC adsorption on palygorskite [9] and smectites[9] due to a lower TC adsorption capacity of 25 mg g^{-1} on activated sludge compared to 30mg g-1 on paly-gorskite[7] and 441 mg g^{-1} on smectite [9]. However, the fitted rate constant was 3.725 g·(mg·h)$^{-1}$, when the initial MLSS concentration was 2000 mg/L, which was much larger than 0.13 and 0.43 g·(mg·h)$^{-1}$ for TC adsorption on palygorskite[8] and smectite[7]. Such a large rate constant indicates that activited sludge is a good adsorbent for OTC removal.

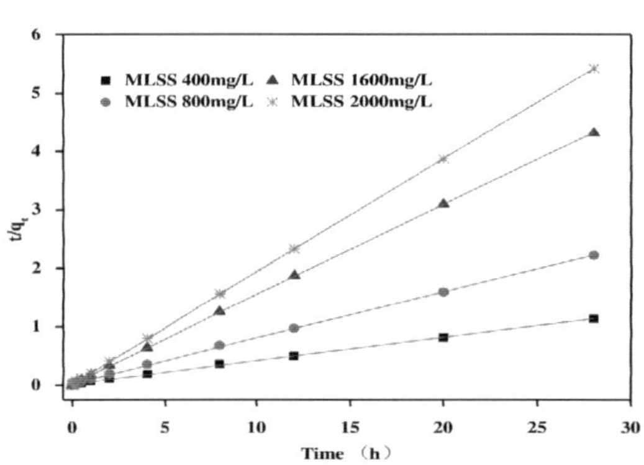

Fig.1. Curve of the Sorption kinetic of oxytetracyline (OTC) fitted into the pseudo-second-order kinetics model.

3.2 The effect of metal ions on Adsorption of OTC on activated sludge

In Figure 2, compared with the control group, tests containing Na^+, K^+, Ca^{2+}, Mg^{2+} and Cd^{2+} suppressed the adsorption of OTC on activated sludge to varying degrees. Cations initially present on the activated sludge surface affect OTC adsorption by competing for surface sites. These may also to interfere with cation exchange mechanism between OTC and the activated sludge surface in low pHs. Generally, the replacing power of a cation increases with its charge, which may explain why Ca^{2+}, Mg^{2+} and Cd^{2+} suppressed the adsorption of OTC on activated sludge more than Na^+ and K^+. This phenomenon was consistent with previous study examining OTC adsorbing to clay [15-16].

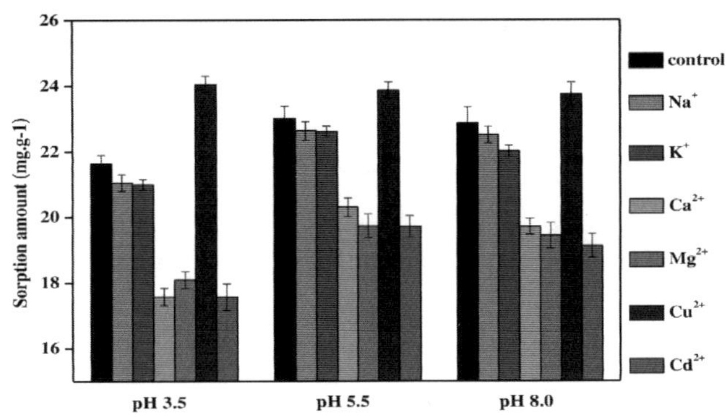

Fig .2. Amount of oxytetracycline (OTC) adsorbed on activated sludge at three pH values as affected by metal ions.

As seen in Fig 2, Cu^{2+} enhanced the adsorption of OTC to activated sludge at all pH values. Previous studies showed that Cu^{2+}could form strong inner-sphere complexes (binding with amide I group) with activated sludge [17] and that OTC can easily complex with metal cations [18]. It is known that Cu^{2+} chelates with OTC at a ratio of 1:1 and the stability constant of the chelate complex is 1012.4 [15].This value is much higher than those of theCa^{2+} and Mg^{2+} ions with OTC (106.4, and 105.8, respectively) [19]. Therefore,Cu^{2+} might act as a bridge between activated sludge surfaces and OTC by forming surface species. Ca^{2+}, Mg^{2+} and Cd^{2+} cannot act as abridge ion to increase the adsorption of OTC[20]. So, different effects of metalions on the adsorption of OTC derive from the interaction mechanisms of metalions with OTC and activated sludge surface functional groups.

As seen in Fig 2, it is noteworthy that pH influences the adsorption process in different ways depending on which metal ion is being tested. Maximum adsorption was obtained at pH 5.5 which is consistent with the above experiment.

3.3 FTIR analysis

A qualitative and preliminary analysis of the vibrational frequency changes of the main functional groups on the activated sludge that might be involved in adsorption process can be assessed by FT-IR. The peaks at 3345,2924, 2361, 1742, 1644, 1604,1542, 1440, 1193, and 883 cm^{-1} showed that native activated sludge had characteristic bands of chitin, acidic polysaccharides, lipids, amino acids proteins and cellulose [21], which may be able to react with OTC. Some similar functional groups were found in biosorption of nickel (II) onto activated sludge.

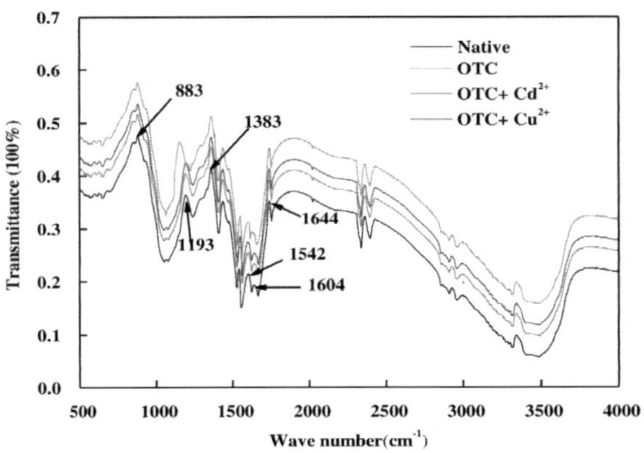

Fig.3. FTIR analysis of native activated sludge, after OTC biosorption, after OTC + Cd^{2+} biosorption, and after OTC+Cu^{2+} biosorption, indicating the involvement of various functional groups OTC and metal ions biosorption.

A significant change after the contact of OTC with activated sludge was a shift in the peak of C-O and phenolic -OH from 1193 $cm^{-1}$to 1190 cm^{-1}. This shift indicated that direct complexation between OTC and functional groups of activated sludge might have occured. What's more, a slight increase of wave numbers from 1542 cm^{-1} to 1550 cm^{-1} after reaction with OTC was probably caused by electronic induced effects, indicating a possible adsorption mechanism involving cation-π interactions between the OTC functional groups and activated sludge[18].

The presence of Cu^{2+}, during the OTC treatments, lead to a shift in the peak of-COOH from 1644 cm^{-1} to 1629 cm^{-1},a shift in the peak of C-O and phenolic-OH from 1190 $cm^{-1}$to 1150cm^{-1}, and enhancement of the peak that was assigned to aromatic C=O and carboxylate groups to1383cm^{-1}.What's more, a combination band from the -N-H bending mode and the C-N stretching mode of the amide (II) moiety at1604 cm^{-1}shifted to lower energy. These changes suggested that the carboxylic, amide (II) moiety of activated sludge and phenolic groups reacted with Cu^{2+} to form inner-sphere surface complexes, thus forming activated sludge-O≡Cu-OTC ternary complexes [15]. Therefore, Cu^{2+} could have played a bridging role, indicating a possible OTC adsorption mechanism of metal bridging and proving the above hypothesis. The presence ofCd^{2+}caused the peak of -N-H and C-N to shift from 1604 cm^{-1} to 1598cm^{-1}. The shift indicatesCd^{2+} interacts to some extent with amide (II). There was no evidence that Cd^{2+} interacts with aromatic compounds in FT-IR spectrum. Therefore, Cd^{2+}could not played a bridging role like Cu^{2+} but filled more adsorption points thereby suppressing the adsorption of OTC on activated sludge.

4. Conclusions

Activated sludge was an effective sorbents to OTC. Pseudo-second-order kinetic model was suitable for descriping adsorption kinetics of OTC. The initial rate attain to 100 mg $g^{-1}h^{-1}$ at MLSS concentration was 2000 mg/L. The presence of heavy metals had varying effects on OTC adsorption to activited sludge. Cu^{2+} enhanced the adsorption apparently while Na^+, K^+, Ca^{2+}, Mg^{2+} and Cd^{2+} suppressed adsorption. Multivalentions suppress adsorption onto activated sludge more than monovalent ions. According to FTIR analysis, the main adsorption mechanisms of OTC to activated sludge include cation exchange, surface complexation , metal bridging and cation-π interactions.

Acknowledgments

This work was financially supported by the Tianjin Science and Technology Research Projects (14TXSYJC00453) and (16YDLJSF00030).

References

[1]Sarmah, A.K., Meyer, M.T., Boxall, A.B. 2006. A global perspective on the use, sales, exposure pathways, occurrence, fate and effects of veterinary antibiotics (VAs) in the environment. *Chemosphere*, **65**(5), 725-59.

[2]Kulshrestha, P., Giese, R.F., Jr., Aga, D.S. 2004. Investigating the molecular interactions of oxytetracycline in clay and organic matter: insights on factors affecting its mobility in soil. *Environ Sci Technol*, **38**(15), 4097-105.

[3]Xu, X.R., Li, X.Y. 2010b. Sorption and desorption of antibiotic tetracycline on marine sediments. *Chemosphere*, **78**(4), 430-6.

[4]Lindsey, M.E., Meyer, T.M., Thurman, E.M. 2001. Analysis of trace levels of sulfonamide and tetracycline antimicrobials in groundwater and surface water using solid-phase extraction and liquid chromatography/mass spectrometry. *Anal Chem*, **73**(19), 4640-6.

[5]Avisar, D., Levin, G., Gozlan, I. 2009. The processes affecting oxytetracycline contamination of groundwater in a phreatic aquifer underlying industrial fish ponds in Israel. *Environmental Earth Sciences*, **59**(4), 939-945.

[6]Miao, X.S., Bishay, F., Chen, M., Metcalfe, C.D. 2004. Occurrence of antimicrobials in the final effluents of wastewater treatment plants in Canada. *Environ Sci Technol*, **38**(13), 3533-41.

[7]Boxall, A.B., Kolpin, D.W., Halling-Sørensen, B., Tolls, J. 2003. Peer Reviewed: Are Veterinary Medicines Causing Environmental Risks? *Environmental science & technology*, **37**(15), 286A-294A.

[8]Giger, W., Alder, A.C., Golet, E.M., Kohler, H.-P.E., McArdell, C.S., Molnar, E., Siegrist, H., Suter, M.J. 2003. Occurrence and fate of antibiotics as trace contaminants in wastewaters, sewage sludges, and surface waters. *CHIMIA International Journal for Chemistry*, **57**(9), 485-491.

[9]Jones, A.D., Bruland, G.L., Agrawal, S.G., Vasudevan, D. 2005. Factors influencing the sorption of oxytetracycline to soils. *Environmental toxicology and chemistry*, **24**(4), 761-770.

[10]Kong, W., Li, C., Dolhi, J.M., Li, S., He, J., Qiao, M. 2012. Characteristics of oxytetracycline sorption and potential bioavailability in soils with various physical-chemical properties. *Chemosphere*, **87**(5), 542-8.

[11]Kulshrestha, P., Giese, R.F., Jr., Aga, D.S. 2004. Investigating the molecular interactions of oxytetracycline in clay and organic matter: insights on factors affecting its mobility in soil. *Environ Sci Technol*, **38**(15), 4097-105.

[12]Kim, S., Eichhorn, P., Jensen, J.N., Weber, A.S., Aga, D.S. 2005. Removal of antibiotics in wastewater: effect of hydraulic and solid retention times on the fate of tetracycline in the activated sludge process. *Environmental science & technology*, **39**(15), 5816-5823.

[13]Zhang, C., Tezel, U., Li, K., Liu, D., Ren, R., Du, J., Pavlostathis, S.G. 2011a. Evaluation and modeling of benzalkonium chloride inhibition and biodegradation in activated sludge. *Water research*, **45**(3), 1238-1246.

[14]Liu, D., Tao, Y., Li, K., Yu, J. 2012. Influence of the presence of three typical surfactants on the adsorption of nickel (II) to aerobic activated sludge. *Bioresource technology*.

[15]MacKay, A.A., Canterbury, B. 2005. Oxytetracycline sorption to organic matter by metal-bridging. *Journal of environmental quality*, **34**(6), 1964-1971.

[16]Pils, J.R., Laird, D.A. 2007. Sorption of tetracycline and chlortetracycline on K-and Ca-saturated soil clays, humic substances, and clay-humic complexes. *Environmental science & technology*, **41**(6), 1928-1933.

[17]Wang, X., Chen, L., Xia, S., Zhao, J., Chovelon, J.-M., Renault, N.J. 2006. Biosorption of Cu (II) and Pb (II) from aqueous solutions by dried activated sludge. *Minerals engineering*, **19**(9), 968-971.

[18]Wessels, J., Ford, W., Szymczak, W., Schneider, S. 1998. The complexation of tetracycline and anhydrotetracycline with Mg^{2+} and Ca^{2+}: A spectroscopic study. *The Journal of Physical Chemistry B*, **102**(46), 9323-9331.

[19]Novák-Pékli, M., El-Hadi Mesbah, M., Pethö, G. 1996. Equilibrium studies on tetracycline—metal ion systems. *Journal of pharmaceutical and biomedical analysis*, **14**(8),

1025-1029.

[20]Zhao, Y., Geng, J., Wang, X., Gu, X., Gao, S. 2011. Adsorption of tetracycline onto goethite in the presence of metal cations and humic substances. *Journal of colloid and interface science*, **361**(1), 247-251.

[21]Vinod, V., Sashidhar, R., Sreedhar, B. 2010. Biosorption of nickel and total chromium from aqueous solution by gum kondagogu (<i> Cochlospermum gossypium</i>): A carbohydrate biopolymer. *Journal of hazardous materials*, **178**(1), 851-860.

IOP Conf. Series: Earth and Environmental Science **59** (2017) 012069 doi:10.1088/1755-1315/59/1/012069

Combustion characteristics and design of hot water boiler

Xuemin Liu[1], Dinghua Yang[2], Junfu Lu[2], Jian Guan[1], Guoli Qi[1]

1. China Special Equipment Inspection and Research Institute, Beijing 100029, China;
2. Key Laboratory for Thermal Science and Power Engineering of Ministry of Education, Tsinghua University, Beijing 100084, China

liuxuemin16@163.com

Abstract. In order to understand the combustion characteristics of biomass, a detailed comparison with coal was made. There are many differences between biomass and coal in combustion characteristics. The burning rate of biomass is much higher than coal. The burning rate of biomass char is also higher than coal char. During biomass combustion, HCl, SO_2 and NO_x emissions mainly concentrate in volatile combustion stage, while CO_2 emission mainly concentrates in char combustion stage. The slagging tendency of biomass ash is severer than coal ash and the adhesive force of biomass ash is higher. However, the wearing tendency of biomass ash is minor. Aiming at the particularity of biomass fuels, this paper briefly introduces the design of biomass hot water boilers. On this basis, a 2.8 MW biomass hot water boiler was developed and tested under the loads of 1.7 MW and 2.8MW. The running results show that the carbon content of bottom slag significantly decreases as the load increases. The boiler efficiency is higher than 85% under both loads.

1. Introduction

As the fossil energy such as coal, oil and natural gas gradually exhausts and the environment problems become serious, development of clean and renewable energy is imminent. Biomass has caused widespread attention because of its near zero emissions of greenhouse gas and acid gases [1]. At present, biomass accounts for about 9%~14% of the total energy supply in the industrialized countries [2, 3], while in the developed countries the proportion is about 1/5~1/3 [4], even as high as 90% in some countries. China is rich in biomass resources. The straw and other agricultural waste are up to about 300 million tons of standard coal each year [5]. And now the area with heating needs in winter accounts for more than 60% of the national area. Accordingly, the population is more than 700 million. Therefore, technology development of biomass hot water boilers is significant in solving winter heating problems and promoting energy conservation and emission reduction. However, the volatile matter content of biomass fuels is usually more than 70%~80% [6, 7]. The combustion characteristics are different from coal. In addition, biomass also has properties of serious fouling, high moisture and heat value fluctuations [8-10]. Therefore the design of biomass boilers has its particularity.

Biomass utilization started late in China. The main technology is Danish grate furnace technology. Because of short development time and the lack of theoretical and experimental research accumulation, biomass boilers currently used in China have many problems in fuel preparation, moisture adaptability and boiler efficiency.

In this paper, biomass combustion characteristics and pollutant emissions were studied through experiments. Ash composition and erosion mechanism were analyzed. Aiming at the particularity of

IOP Conf. Series: Earth and Environmental Science **59** (2017) 012069 doi:10.1088/1755-1315/59/1/012069

biomass fuels, the design of biomass hot water boilers was briefly introduced. Finally, a successful operation case of a 2.8 MW biomass hot water boiler was provided.

2. Biomass combustion characteristics

2.1. Combustion characteristics

The combustion process of fuels can be divided into volatile precipitation and combustion and residual char combustion. The volatile and char contents of biomass are different from coal, so the combustion characteristics are different. A STA409C thermogravimetric analyzer was used to study the combustion characteristics of wheat straw in this paper. The proximate analysis and ultimate analysis of wheat straw used in tests are given in Table 1. The heating rate was 50°C·min^{-1} and the gas flow rate was 100ml·min^{-1}. Volatile precipitation and fuel combustion characteristics were studied in nitrogen and air respectively. The DTG curves are shown in Figure 1 and Figure 2. The initial precipitation temperature of volatile is about 200°C and the temperature corresponding to the maximum release rate is about 330°C. The index of volatile release characteristics *D* is 1.1, much higher than that of coal. So the pyrolysis characteristics of biomass are better than coal, with volatile release peak appearing earlier and relatively concentrated, which is beneficial for ignition.

Table 1. The proximate analysis and ultimate analysis of wheat straw.

Proximate analysis (%)				Ultimate analysis (%)				
A	M	VM	FC	C	H	O	N	S
4.76	3.24	75.78	16.22	42.54	3.43	45.13	0.83	>0.06

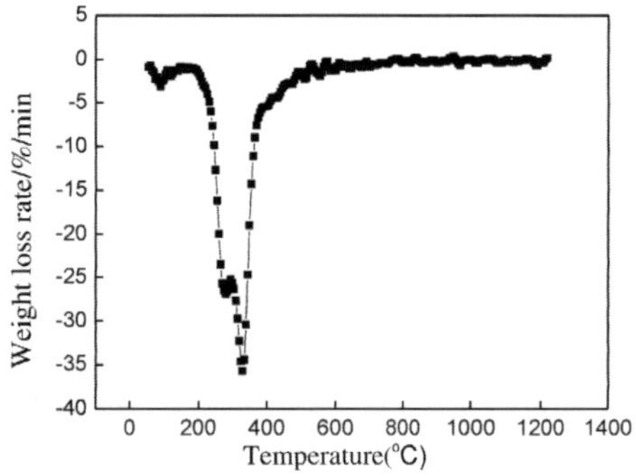

Figure 1. DTG curve in nitrogen

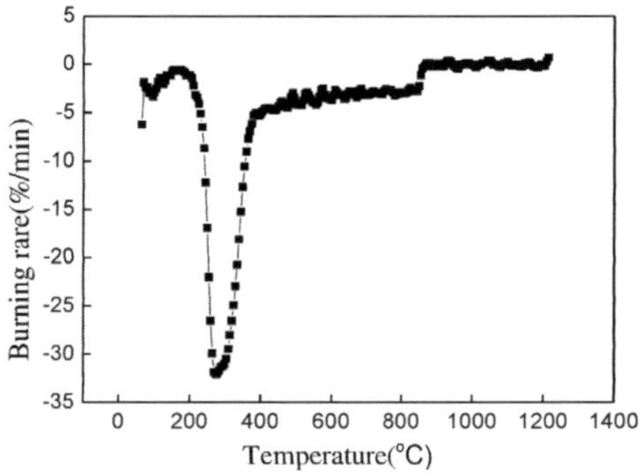

Figure 2. DTG curve in air

In order to further compare the combustion behavior, biomass char and coal char were heated to 900℃ in nitrogen and then to 1200 ℃ in air. The DTG curve is shown in Figure 3. It can be seen that the burning rate of biomass char is much higher than coal char.

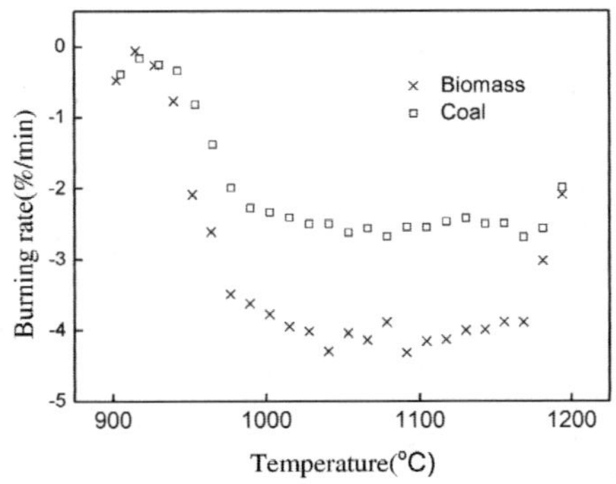

Figure 3. Burning rate of biomass char and coal char

Due to the high content of volatile matter in biomass, accounting for about 80%, as shown in Table 2, its burning rate is higher and burnout time is shorter. A comparison of different biomass and coal is shown in Table 3. The maximum burning rates of different biomass are similar, much higher than bituminous coal and anthracite coal.

Table 2. The proximate analysis and ultimate analysis of biomass

Item	Corm straw	Wheat straw	Korean pine	Sandbreak forest
M_{ad}	4.59	6.30	6.91	1.57
A_{ad}	3.67	3.87	5.57	1.34
V_{ad}	78.84	79.61	79.00	83.93
FC_{ad}	12.90	10.23	8.51	13.17
C_{db}	44.99	45.28	43.74	48.82
H_{db}	5.89	6.12	5.90	6.29
N_{db}	0.82	0.63	0.75	1.16
S_{db}	0.06	0.10	0.06	0.07
O_{db}	48.24	47.87	49.55	43.66
HHV_{ad}(kJ/kg)	17355	16967	17280	19750

Table 3. The maximum burning rate and burn out time of biomass and coal

	Maximum burning rate (s^{-1})	Burn out time (s)
Anthracite coal	0.0063	353
Bituminous coal	0.0396	136
Corm straw	0.0530	74
Korean pine	0.0620	73
Wheat straw	0.0600	78
Sandbreak forest	0.0506	69

2.2. Pollutant emissions

Compared with coal, biomass has great advantages in terms of pollutant emissions. In this paper, TGA-FTIR was used to study the pollutant emissions of wheat straw. The samples were heated from 30°C to 900°C in a mixture of N_2 and O_2 with mixing ratio of 4:1. The heating rate was 20°C/min. The flue gas components were recorded. As shown in Figure 4, HCl emission mainly concentrates in volatile combustion stage, with temperature range of 220 ~ 450 °C. The release curve has a single peak with temperature of 310°C. The release curves of SO_2 and NO_x present double peaks as shown in Figure 5 and Figure 6. The first peak is located in the volatile combustion stage with temperature of 320°C. CO_2 emission mainly concentrates in char combustion stage, depending on the carbon content of biomass.

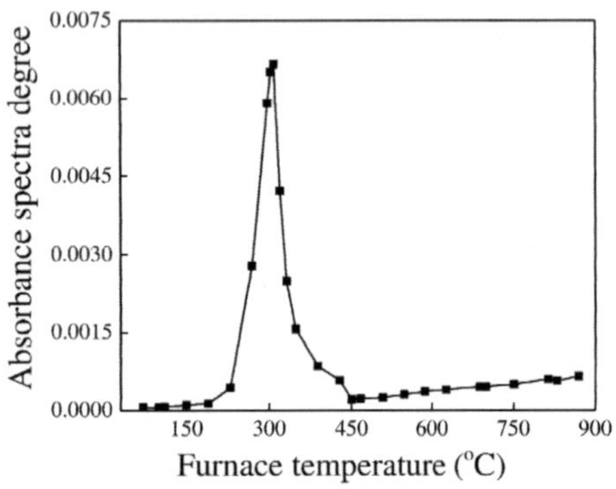

Figure 4. HCl release curve

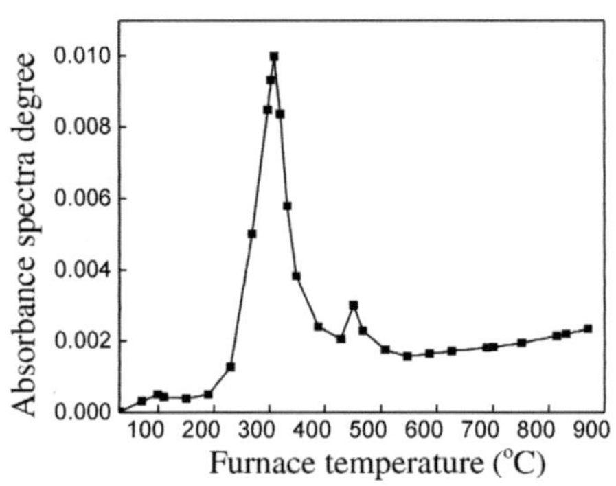

Figure 5. SO_2 release curve

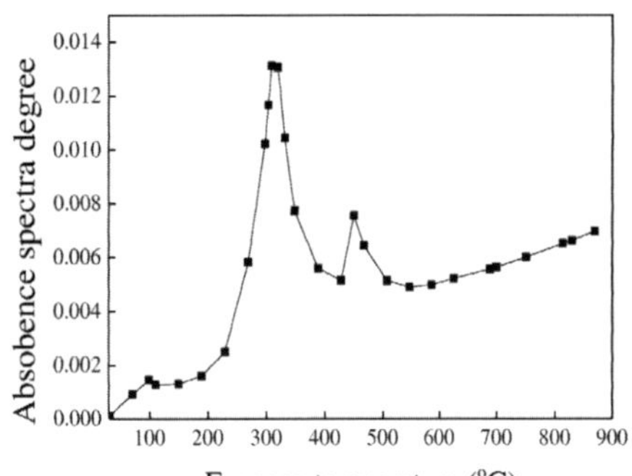

Figure 6. NO_x release curve

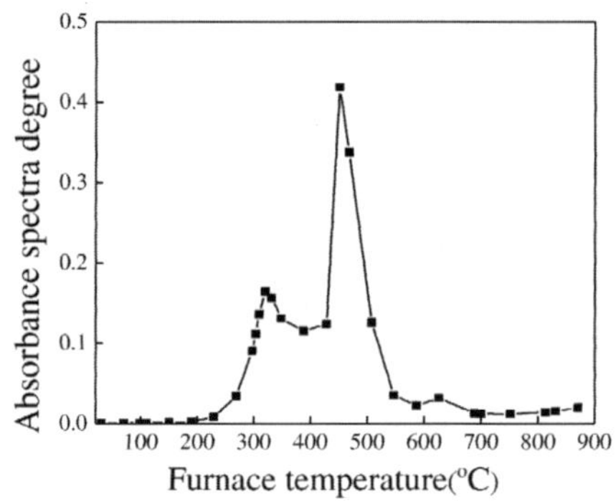

Figure 7. CO_2 release curve

2.3. Ash composition and erosion mechanism

In the process of biomass combustion, sintering may occur. The influencing factors of sintering include temperature and surrounding atmosphere, in which the temperature is the more important one. The size, number and hardness of sintering blocks increase as the temperature increases. The sintering temperatures of different biomass vary, corn straw about 740°C, rice straw and sorghum straw about 680°C. However, the melting points of sand and ash are higher than 1000°C. Generally, this is because that biomass is rich in K and Na. The compounds of these elements react with SiO_2 in sand, generating low melting point eutectic which can bond the sand [11, 12]. The reaction equations are as follows:

$$2SiO_2+Na_2O \rightarrow Na_2O \cdot 2SiO_2$$

$$4SiO_2+K_2O \rightarrow K_2O \cdot 4SiO_2$$

Ash slagging characteristics can be estimated by indexes such as Base/Acid ratio B/A, silicon ratio G, silicon aluminum ratio S/A, and iron calcium ratio. The fouling and wearing characteristics can be estimated by indexes H_W and H_M respectively.

$B/A = (Fe_2O_3+CaO+MgO+Na_2O+K_2O+P_2O_5)/(SiO_2+Al_2O_3+TiO_2)$.

$H_W = B/A \cdot Na_2O$. $H_M = A \cdot (SiO_2+0.8Fe_2O_3+1.35Al_2O_3)$ [13], in which A is ash content. The discriminant boundaries of these indexes are shown in Table 4 [14-16].

Table 4. Criterions of slagging, fouling and wearing characteristics

Index		Extent		
		Slightly	Moderately	Seriously
Slagging	B/A	<0.206	0.206~0.4	>0.4
	G	>78.8	78.8~66.1	<66.1
	SiO_2/Al_2O_3	<1.87	1.87~2.56	>2.56
	Fe_2O_3/CaO	<0.3or >3	0.3~3	Around1
Fouling	H_W	<0.2	0.2~0.5(Moderately)0.5~1.0(easily)	>1.0
Wearing	H_M	<10	10~20	>20

Components of different ash samples used in tests are shown in Table 5. Contents of alkaline elements in biomass ash such as K, Na, Ca, and Mg are higher than in coal ash, while contents of acid elements in biomass ash such as Al and Ti are lower than in coal ash. Although the content of Fe_2O_3 in biomass ash is lower, the Base/Acid ratio B/A of biomass ash is still much higher than that of coal ash. So the slagging tendency of biomass ash is severer, especially wheat straw, in which the content of K_2O is as high as 24.60%. The contents of Na in two kinds of biomass ash are similar. However, compared with coal ash, the wearing tendency of biomass ash is minor.

Table 5. Ash composition (%)

Samples	K_2O	Na_2O	Al_2O_3	Fe_2O_3	CaO	MgO	SiO_2	TiO_2
Wheat straw	24.60	1.71	2.48	1.53	7.21	2.54	54.82	0.08
Corn straw	4.40	1.29	6.20	2.80	7.59	8.45	48.99	0.22
Fuxin coal	1.07	0.58	22.26	13.03	3.17	0.53	56.41	1.16
Datong coal	0.91	0.35	27.04	8.09	2.66	0.40	57.70	1.25

Table 6. Slagging, fouling and wearing characteristics of ash samples

Samples	B/A	G	S/A	Fe/Ca	H_W	H_M
Wheat straw	0.66	82.83	22.10	0.21	1.13	2.58
Corn straw	0.44	72.22	7.90	0.37	0.57	1.79
Fuxin coal	0.23	77.13	2.53	4.11	0.13	21.31
Datong coal	0.14	83.81	2.13	3.04	0.05	36.36

The adhesive force of ash was measured. The value of corn straw ash is about $46mg/cm^2$, while coal ash is only $3mg/cm^2$. It can be seen that the adhesive force of biomass ash is significantly higher than that of coal ash. Adhesive force has dual influence on dust removal. If the adhesive force is too high, cleaning performance of bag filter will be affected. If the adhesive force is lower, the tendency of adhesion in the boiler will weaken, but flying up again of settled dust in rapping will reduce the ESP efficiency.

Microstructures of ash samples were observed through SEM. It can be seen that the microstructure of corn straw ash is small stick and floc with serious surface adhesion and coalescence. However, the microstructure of coal ash is irregular lumps mixed with many porous particles, as shown in Figure 8. This explains the reason of biomass's serious slagging and fouling tendency. It means that the fouling of the convective heating surface is more serious and it should be considered during the convective heating surface design.

Figure 8. SEM pictures of ash samples
a: Corn straw ash b: Coal ash

3. Design of biomass hot water boilers

As a fuel, biomass has features such as morphological heterogeneity, high moisture and volatile, low heating value and fixed carbon. It also has corrosive components and properties of heating value fluctuations. Therefore the design of biomass boilers has its particularity. Stable and complete burning should be ensured.

The main parameters of biomass grate-fired boilers are grate heat release rate q_R, furnace volume heat release rate q_V and grate ventilation ratio r. According to the laboratory and industrial research, the recommended value of q_R is 360~435 kW·m^{-2}, and q_V 160~195 kW·m^{-3}. The grate ventilation of biomass boiler is usually small and r can be chosen between 2%~5%. The recommended values of other parameters are shown in Table 7.

Table 7. Characteristic parameters of biomass grate-fired boilers

Item	Symbol	Unit	Value
Excess air coefficient of furnace outlet	a_f''	—	1.3~1.5
Heat loss due to unburned gas	q_3	%	0.2
Carbon loss	q_4	%	1~5
Fly ash content	a_{fa}	%	60~90
Supply air temperature	t_k	℃	20~200
Pressure under grate	P	Pa	200~1000

Since biomass burns in semi-suspended state, the secondary air is very important. It not only increases the disturbance in furnace, but also supplies air required for combustion. The secondary air ratio should be 30%~50% and the wind velocity should be 20~30m/s. Based on the experience of tangential pulverized coal boilers, the secondary air of biomass grate-fired boilers can also be arranged tangentially. The nozzle direction can be downdip. The angle is about 10°~25° when the nozzles are arranged on the front and rear wall, and 40°~45° when the nozzles are arranged in the corners. The position of nozzles is related to the downdip angle and airflow range. In order to ensure that the flue gas has sufficient burning time without disrupting normal burning of the bed surface, generally the distance of nozzles to the fuel layer surface should be 0.6~ 2m.

Due to the differences of fuel characteristics and boiler structures, the heat transfer calculation of biomass grate-fired boilers should not simply follow the method of coal grate-fired boilers. In this paper, the heat transfer model suggested by Tsinghua University was adopted to design the furnace [17]. As the moisture contents of biomass and coal are different, the heat transfer of flue gas also differs. Therefore, the convective heating surface was carefully arranged. The contribution of moisture is considered by introducing a correction factor C_r, which is more than 1 [18]:

$Re = 150 \sim 400$, for in-line tube bundles subjected to cross flow,

$$Nu = 0.2C_r C_z C_s Re^{0.65} Pr^{0.33}. \tag{1}$$

$$C_r = 1.18011 + 0.82808(r_{H_2O} - 0.25) - 2.06252(r_{H_2O} - 0.25)^2 \tag{2}$$

$Re = 150 \sim 400$, for staggered tube bundles subjected to cross flow,

$$Nu = 0.358C_r C_z C_s Re^{0.6} Pr^{0.33}. \tag{3}$$

$$C_r = 1.01025 + 0.56505(r_{H_2O} - 0.25) - 1.92401(r_{H_2O} - 0.25)^2. \tag{4}$$

$Re = 5000 \sim 13000$, for tube bundles subjected to longitudinal flow,

$$Nu = 0.023Re^{0.8} Pr^{0.4} C_t C_l C_r. \tag{5}$$

$$C_r = 1.13598 + 0.42449(r_{H_2O} - 0.25) - 1.02091(r_{H_2O} - 0.25)^2 \tag{6}$$

4. Operation performance

Based on the above results, a 2.8 MW biomass hot water boiler was developed and tested under the loads of 1.7 MW and 2.8MW. The boiler structure diagram is shown in Figure 9. The fuel is local corn straw bales. On the premise of ensuring the fuel fluidity and boiler stable operation, the fuel cost is reduced [19]. The proximate analysis and ultimate analysis of corn straw are shown in Table 8.

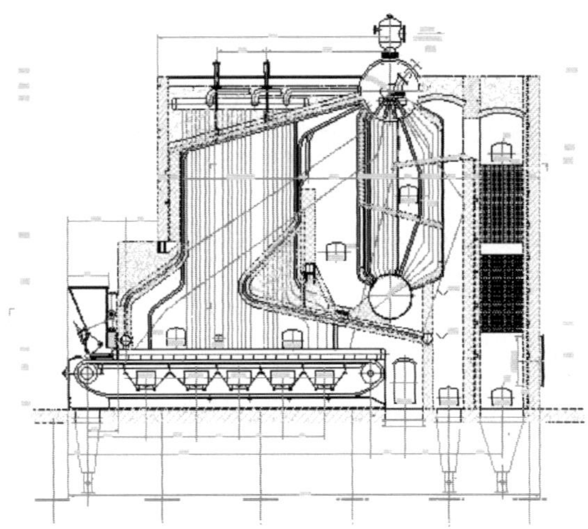

Figure 9. Boiler structure diagram

Table 8. The proximate analysis and ultimate analysis of corn straw (%)

War	Wad	Aad	Vad	Cf	Cad	Had	Nad	Sad	Oad	Qnet(MJ/kg)
39.6	8.8	5.3	65.9	20.0	31.9	3.8	1.7	0.2	62.5	16.3

As the load increases from 1.7MW to 2.8MW, the carbon content of bottom slag significantly decreases, as shown in Table 9. It is because that the total air volume and secondary air flow increase as the load increases. The secondary air jet flow has an important influence on the flow field of the lower region. With the primary air, it forms a larger vortex area, not only supplying oxygen better, but also increasing the fuel residence time in high temperature zone. Therefore, the overall furnace temperature increases, favourable to fuel burning out and efficiency improvement. The boiler efficiency is higher than 85% under both loads.

Table 9. The carbon content of fly ash and bottom slag

Load (MW)	Bottom slag (%)	Fly ash (%)
1.7	4.33	2.09
2.8	2.60	1.94

Table 10. Operation performance of 2.8MW biomass hot water boiler

Item	Results	
	1.7MW	2.8MW
Circulating water flow (kg/s)	19.89	19.89
Supply water pressure (MPa)	0.5	0.5
Supply water temperature (°C)	42	59
Backwater temperature (°C)	32	35
Feeding water pressure (MPa)	0.550	0.550
Air temperature of fan outlet (°C)	20	20
Blowdown rate (%)	0	0
Exhaust gas temperature (°C)	98	99
Oxygen content in flue gas (%)	9~11	9~11
Boiler efficiency (%)	86.19	88.21

5. Conclusions

Biomass has particularities on combustion characteristics and pollutant emissions. Compared with coal, the volatile matter content of biomass is much higher. The burning rates of biomass and biomass char are also similar. During biomass combustion, HCl emission mainly concentrates in volatile combustion stage, presenting a single peak as temperature increasing. The release curves of SO_2 and NO_x present double peaks. The first peak is located in the volatile combustion stage. CO_2 emission mainly concentrates in char combustion stage, depending on the carbon content of biomass. The slagging tendency of biomass ash is severer than coal ash and the adhesive force of biomass ash is higher. However, compared with coal ash, the wearing tendency of biomass ash is minor. Aiming at the particularity of biomass fuels, this paper briefly introduces the design of biomass hot water boilers. On this basis, a 2.8 MW biomass hot water boiler was developed and tested under the loads of 1.7 MW and 2.8MW. The running results show that the carbon content of bottom slag significantly decreases as the load increases. The boiler efficiency is higher than 85% under both loads.

Acknowledgements

This study was supported by the National Key Technology R&D Program (No.2014BAA02B02, No.2014BAA02B01).

References

[1] Liu S, Chen K and Zhang B 2002 *Renewable Energy* **4** 14-5

[2] Saidura R, Abdelaziza EA , Demirbasb A, Hossaina MS and Mekhilefc S 2011 *Renewable and Sustainable Energy Reviews* **15** 2262-89
[3] Shen D, Gu S, Luo K, Bridgewater AV and Fang M 2009 *Fuel* **88** 1024-30
[4] Khan AA, Jonga WD, Jansens PJ and Spliethoff H 2009 *Fuel Process Technol.* **90** 21-50
[5] Zhang Y 2009 *The applied basic research of biomass boiler with gasification and combustion* (Changchun: Jilin University)
[6] Su J, Zhao X and Wang W. 2009 *Energy Conservation Technology* **27** 117-20
[7] Fernández RG, García CP, Lavín AG and de las Heras JBL 2012 *Fuel Processing Technology* **103** 16-26
[8] Romeo LM and Gareta R 2006 *Engineering Applications of Artificial Intelligence* **19** 915-25
[9] Romeo LM and Gareta R 2009 *Biomass and Bioenergy* **33** 854-61
[10] Michelsen HP, Frandsen F, Dam-Johansen K and Larsen OH 1998 *Fuel Processing Technology* **54** 95-108
[11] Bashir MS, Jensen PA, Frandsen F, Wedel S, Dam-Johansen K, Wadenbäck J and Pedersen ST 2012 *Fuel Processing Technology* **97** 93-106
[12] Li Q, Ren J, Pan W, Wu J and Li Z 2007 *Journal of Shanghai University of Electric Power* **23** 333-6
[13] Pronobis M 2006 *Fuel* **85** 474-80
[14] Ai J, Qin Y, Zhu Q and He P 1994 *Power System Engineering* **10** 1-18
[15] Yan W and Chen Y 2007 *Journal of North China Electric Power University* **34** 49-54
[16] Xiao J, Duan J, Wang H and Zhuang X 2003 *Energy Engineering* **3** 9-13
[17] Zhang Y, Li H, Li Q, Wu Z and Wang B 2001 *Tsinghua Univ. (Sci & Tech)* **41** 95-8
[18] Zhang Y 2000 *Study on heat transfer of MSW incinerator* (Beijing: Tsinghua University)
[19] Mladenović R, Dakić D, Erić A, Mladenović M, Paprika M and Repić B 2009 *Energy* **34** 715-23

IOP Conf. Series: Earth and Environmental Science **59** (2017) 012070 doi:10.1088/1755-1315/59/1/012070

LWGeoWfMS: A Lightweight Geo-Workflow Management System

Ying Du[1,a], Pu Cheng[2,b]

[1] School of Computer and Information Engineering of Henan University, Kaifeng, 475004, China
[2]School of Software of Henan University, Kaifeng, 475004, China

[a]hd_duying@163.com, [b]chengpul@163.com

Abstract. The disadvantage of the existing workflow systems for spatial information processing flow management is need to extend the special resource managements for the data, models and operations in geographic information systems. It increases the difficulty of development for the systems, the complexity of the architectures and the demand of hardware and software resources. To solve the problem, a light-weight Geo-workflow management system is proposed in this paper. This system has the ability to manage the resource of Geography-related fields. This solution provides a resource sharing platform for different users, designs a script-based process engine for resolving and running the workflows quickly and achieves the functions of process monitoring and traceability at the same time.

1 Introduction

With the rapid development of computer technology and the gradual application of GIS, spatial information processing, which needs to be managed and controlled, become more and more complicated. Geo-Workflow is the workflow technology which can be used to express, manage, control and execute spatial information processing in many works [1]. The theoretical frameworks and technical solutions of Geo-Workflow have been applied in the field of Geographic Model Integration [2, 4], Land Resource Management [3], Spatial Information Processing Services [5] and so on.

Unlike the classical business process and scientific process, the process object of spatial information processing is spatial data with various types and large quantities, and space operations and spatial analysis models for spatial information processing require complex algorithms and intensive computations with many parameters and changing types. Moreover, the existing scientific workflow systems depend on DCIs (Distributed Computing Infrastructures), which contain a lot of components such as data resources, metadata directories, authentication and authorization methods, repositories [6]. These components need to specifically extend to support geospatial information. That greatly increase the difficulties of process management in spatial information processing. Obviously, the existing scientific workflow systems are not easily applied to spatial information processing.

In this paper, we propose a lightweight Geo-workflow management system (LWGeoWfMS), which can manage and control the spatial information processing and is able to complete the tasks in geosciences by scheduling different types of resources. The remainder of this paper is organized as follows. Section 2 describes the reference model of LWGeoWfMS. The design and implementation of

LWGeoWfMS are described in Section 3 and Section 4 respectively. Finally this paper is concluded in Section 5 with plans for future research.

2 Reference Model of LWGeoWfMS

Based on the Workflow Reference Model published by Workflow Management Coalition (WfMC) in 1994[7], we design the reference model of LWGeoWfMS, as shown in Figure 1.

This reference model of LWGeoWfMS has six important interfaces as follows:

- Interfaces1—Import and export of process definitions
- Interfaces2—Interaction with the client application of Geo-workflow
- Interfaces3—Scheduling the execution resources
- Interfaces4—Collaboration between different workflow management systems
- Interfaces5—Management and monitoring functions
- Interfaces6—Scheduling the data resources

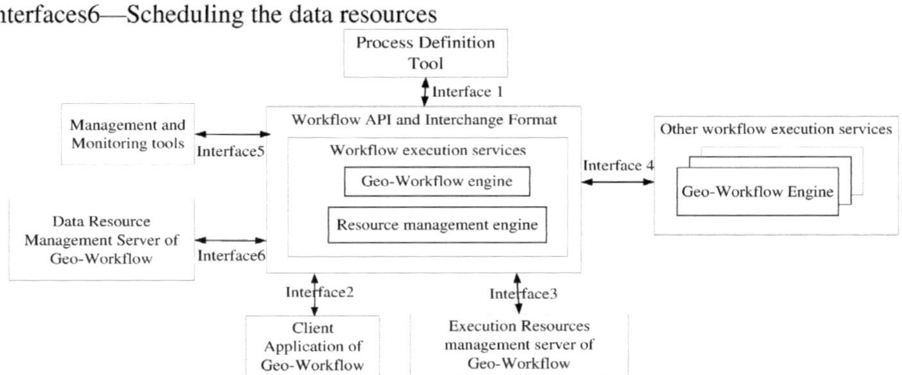

Fig1 Reference Model of LWGeoWfMS

The workflow execution service is the core of the whole system, including one or more Geo-workflow engines and resource management engines. The Geo-workflow engine provides the running environment for the spatial data processing such as interpreting process definitions, creating process instances, controlling process routes, changing or recording operational states and communicating with other engines. The process definition tool is used for process modeling which generate a formal description of a spatial data processing. This formal description can be passed to the Geo-workflow engine for analysis and processing through the interfaces 1. The resource management engine is used to schedule various data resources, execution resources and user resources through the interface 6, interface 3 and interface 2 respectively. While the system is running, the interface 2 provides an interaction channel between the system and users by calling the client application program, and the workflow execution service invoke external execution resources through interface 3 to accomplish a specific task. The workflow execution service also needs to interact with other workflow execution services via interface 4, which can make these systems interoperable. The management and monitoring tools perform monitoring management tasks of the process of the workflow execution through the interface 5.

3 Design of LWGeoWfMS

According to the guidance of the reference model, the architecture of LWGeoWfMS system is designed in the three parts: execution server, client application and monitoring server, as shown in Figure 2.

There are four types of users in the LWGeoWfMS: resource contributor, process model designer, task executor and system monitor. The resource contributors register their private resources in the system for the workflow instance to invoke. The process model designers define the workflow model in the visualization environment provided by the LWGeoWfMS system client application, and then they specify the default task executors and resources for the task when the system is executed. The task executors perform the assigned tasks in the workflow instance and the system monitors perform the monitoring functions while the workflow instance is running.

The execution server is the core of the whole system with the functions of resource management, process management and related scheduling. It consists of two components: the resource management engine and the Geo-workflow engine. The resource management engine is an important part of LWGeoWfMS. It includes the functions of resource registration, resource retrieval and resource scheduling. The resource contributors encapsulate their private resources as common resources and register them in the system through the registration function provided by the resource management engine. If the resources have been registered successfully, they can be retrieved and used in the process of scheduling. In order to organize, manage and schedule different types of resources, different resource templates based on the Resource Description Framework (RDF) are designed. The description of the resource is generated by the resource metadata at the time of registration.

The Geo-workflow engine implemented in Python language includes process analysis, execution and scheduling. According to the rule defined by the process model, the workflow engine resolves a process instance into a list of tasks and then inform the related task executor to carry out his tasks. Python scripts are small, flexible, and easy to scale, so the Geo-workflow engine can be easily deployed in a distributed environment combined with resource management engine, and can encapsulate itself as a resource model automatically.

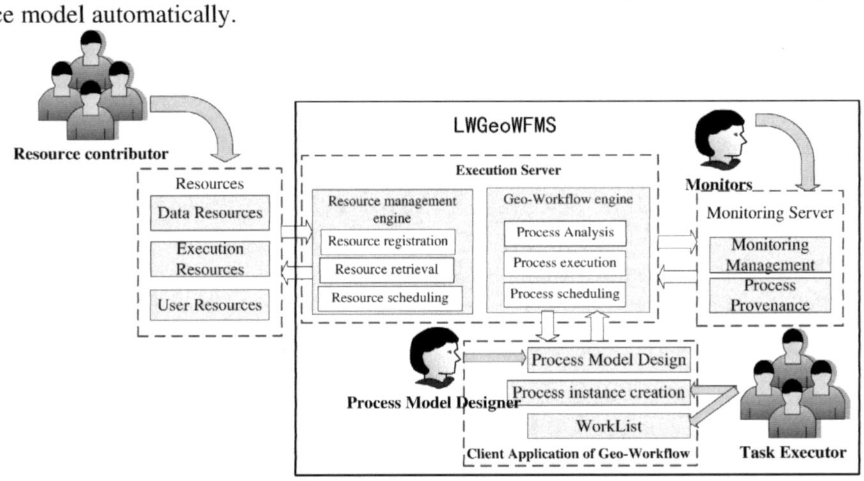

Fig2 the System Structure of LWGeoWfMS

The client application of LWGeoWfMS provides the users with a customizable visualization interface, which is not only a customized platform for process models, but also the display and operation interface of worklists.The Monitoring Server has the functions of management monitoring, provenance about the entire system.

4 Implementation of LWGeoWfMS

IOP Conf. Series: Earth and Environmental Science **59** (2017) 012070 doi:10.1088/1755-1315/59/1/012070

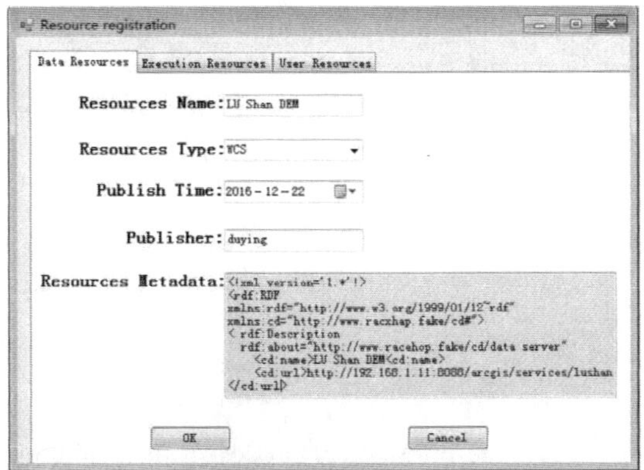

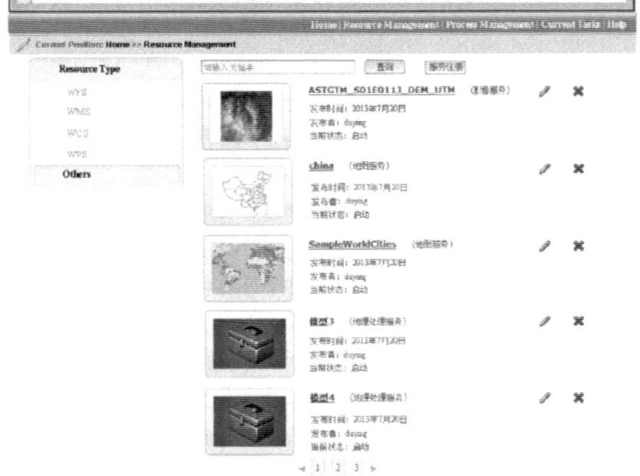

Fig3 the Interface of Resources Registration Fig4 the Interface of Resources Management

The resource management engine and the client application of LWGeoWfMS are developed on the Visual Studio 2012 platform, using the SQL Server 2008 as the database management system (DBMS) and Microsoft Silverlight which is a rich client Web rendering technology and can give users some novel experiences.

The monitoring server is implemented in C++ and the SQL Server 2008. Figure 3 shows the resource registration interface, through which the resource contributors register their private resources to the system for the workflow instance to invoke. Figure 4 shows the resource management interface, The interfaces of process design and process representation are showed in Figure 5 and Figure 6, respectively.

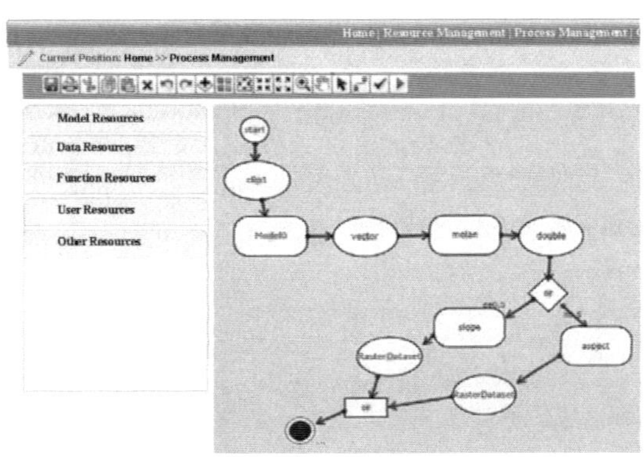

Fig5 the Interface of Process Design

Fig6 the Interface of Process Representation

5 Conclusions

In this paper, the LWGeoWfMS not only implements the basic components of the workflow reference model, but also provide the resource sharing platform for users. Moreover, it uses the Python language to implement the workflow engine and execute the process in an efficient interpreted way. But the research of this paper has many shortcomings: (1) workflow execution server is a centralized implementation, which does not take into account the process of distributed expression and only consider the implementation of distributed resources; (2) while using RDF for resource description, a resource description template has been developed for conforming to the standard only, and many types of resources have not to be further summarized; (3) the interoperation between the implementation and other heterogeneous workflow systems has not been realized. These are the next step to improve in the LWGeoWfMS system.

References

[1]Weske M, Vossen G, Medeiros C B, et al. Workflow Management in Geoprocessing Applications[C]// Proceedings of the 6th ACM international symposium on Advances in geographic information systems. ACM, 1998: 88-93.

[2]Hofer B, Mäs S, Brauner J, et al. Towards a knowledge base to support geoprocessing workflow development[J]. International Journal of Geographical Information Science, 2016:1-23.

[3]Wang Y. Research on the Typical GIS Service Chain Application in Land Resource Management [M]// Geo-Informatics in Resource Management and Sustainable Ecosystem. Springer Berlin Heidelberg, 2015.

[4]Yue P, Zhang M, Tan Z. A Geoprocessing Workflow System for Environmental Monitoring and Integrated Modelling[J]. Environmental Modelling & Software, 2015, 69: 128-140.

[5]Thakker S, Abdul J, B. Potdar M. GeoProcessing Workflow Models for Distributed Processing Frameworks[J]. International Journal of Computer Applications, 2015, 113(1): 33-38.

[6]Terstyanszky G, Kukla T, Kiss T, et al. Enabling scientific workflow sharing through coarse-grained interoperability[J]. Future Generation Computer Systems, 2014, 37(7):46-59.

[7]WfMC, The Workflow referencemodel 1994.

Comparative research on soil magnetic susceptibility and Chroma and grain of grassland soil and farmland soil in alpine region

YUAN Jie[1], CAO Guangchao[1], E Chongyi[1], JIANG Gang[1], YUAN Youjing[2], XIANG Cheng[1]

[1] Key Laboratory of Qinghai Province Physical Geography and Environmental Process, Qinghai Normal University, XiNing 810008,P.R. China
[2] Environmental Geological Prospecting Bureau of Qinghai Province, XiNing 810000,P.R. China

The corresponding author's e-mail: yuanjie8903@126.com

Abstract. This thesis aims at researching unchanged perennial farmland and grassland soil on the northern of Qinghai Lake basin and differences between soil magnetic susceptibility, chroma and soil particle size composition of grassland and farmland in lengthways profile(0-60cm). The result shows that the distinction of X_{lf} between grassland and farmland is smaller above 15cm, and farmland X_{lf} is finer than grassland below 15cm. The X_{fd} of grassland is finer than farmland above 30cm, which has a little difference below 30cm. Farmland chroma value is finer than the grassland generally. The lightness of grassland has increased trend and farmland has the decrease trend above 30cm, the lightness of grassland and farmland has no other changes below 30cm, the change of redness and yellowness in lengthways profile has a decrease trend from 0 to 60cm; the clay and silt content of grassland are finer than farmland except sand content; the sand maximum content of farmland in 0-10cm segment; 20 to 35 cm segment sand content decrease and began to increase from 35-60cm segment; the soil particle size composition has a big difference, in particular, when it is above 30cm, it has a little difference below 30cm.

1. Introduction

Qinghai Lake is regarded as an important part of the Qinghai-Tibet Plateau, and it's responsible for the ecological safety of the northeastern Qinghai-Tibet Plateau and even the inland, is to control the eastward spread of desertification in west natural barrier [1, 2]. The historical land-use patterns period of the lake mainly is animal husbandry. After the founding of our country, land-use pattern began to change significantly, and large areas from grassland to farmland is reclaimed [3]. Due to the artificial grassland reclamation into farmland, the area of desertification was increased rapidly. Although natural grassland vegetation was to some extent improved by the way of returning farmland back to grassland in some parts of lake region farmland later[3], the entire area around the lake still remains a serious deterioration of ecological environment in recent years. Then the soil quality and the evolution of traits in the lake surrounding area where it is filled with frequent human activities and natural ecosystem's fragile will become particularly important.

The magnetic susceptibility, chroma and particle size of soil were the most intuitive, obvious and important natural properties [4, 5]. The magnetic properties of the soil, chroma and particle size were closely related to the soil physical, chemical and biological properties [6, 7], which was the basis for the study of soil environment and entire ecosystem characteristics. At the same time, because of the simple, rapid and targeted measurement methods, they can be widely used in various sedimentary environment analysis [8, 9]. The current studies show that the soil magnetic susceptibility and chroma are shipped in two main aspects: one was that the soil magnetic susceptibility and chroma can be quantified as climate proxies because of sensitive to climate change, which also can be applied into the study of loess-paleosol to reflect the changes in the strength of warm, cold and dry soil of historical periods of climate and development [10]. On the other hand, the heavy metals of environmental pollutants produced by the industrial activity, chemical burns and automobile exhaust affect the magnetism and chroma characteristic of soil, so the combining of the soil magnetic susceptibility, chroma and elements of heavy metal could be regarded as an evaluation of soil pollution indicators to assess the degree of soil heavy metal contamination [11,12].Similar to the magnetic susceptibility and chroma of soil, the particle size indicators of soil, on the one hand, are currently used into aeolian deposits, lake sediments and river sedimentary environment [13,14,15], which can also indicate sediment provenance, condition wind power, water power environment, transport routes and etc.. On the other hand, the particle size indicators of soil can directly reflect the soil particles and the soil structure component to assess the degree of soil development. However, as for the progress in research stage, there are fewer studies of the magnetic susceptibility, chroma and particle size of soil under different land-use patterns in the same region. Furthermore, the magnetic characteristics, chroma, particle size of the soil profile distribution are closely related to the changes of the land use patterns made by human activities.

To sum up, this study was based on the five remote sensing interpretations from 1987 to 2010 in such area, and the grassland and farmland around Gangcha County in the northern part of Qinghai Lake, which have been unchanged for nearly 30 years, are selected as the research object, in order that a good research of natural grassland and farmland soil profile could be made to examine the influence of different land use patterns on soil magnetism, chroma and particle size. Furthermore, the soil succession laws under different human activities could be estimated correctly, and some references for sustainable use of area soil resources could be provided.

2. Study area

Gangcha County was located throughout the western Haibei state and belongs to Haibei Tibetan Autonomous Prefecture. The county is connected with the north shore of Qinghai Lake, and boasts a total area of nearly 8138km^2. It lies between 36°58'06 "~38°04'04" N, 90°20'44 "~100°37'24" E, at an altitude of 3200m~3800m between [16]. It is a typical plateau continental climate with long hours of sunlight, where large temperature differences between day and night, the seasons of this area are not clear and drought climate and some other characteristics are clear. In this area, the number of annual rainfalls is 370.5mm, which are mainly concentrated in the June-September. And the annual evaporation is 1500.6-1847.8mm, which forms a climate of cold winter and cool summer. The average temperature of January is -17.5 °C, while the average temperature of July is 11 °C. However, the annual average temperature is 0.2-5.7 °C. The dominant wind direction in winter and spring is from the west, while in summer and autumn the dominant wind direction is from the southeast. And the annual average wind speed is 3.7m/s; besides, the maximum wind speed is 30m/s [17]. Vegetation distribution shows the significant vertical zones, which can be divided into several typical types including alpine meadows, alpine grasslands, mountain pastures, meadows and plains [18]. Soil in this area mainly includes chestnut soil, chernozem, gray brown soil, alpine meadow soil, alpine cold desert soil, sand soil and etc.. The main type of soils around the lake is plain dark chestnut soil, which boasts a quality of higher soil organic matter content and soil fertile. The main types of grassland are rooted with alpine mountain temperate steppe class-based, and the native vegetation mainly conclude *splendens*, *stipa* and *poa crymophila*[18].

3. Experimental Methods

3.1 Soil plot choice

It is based on the five remote sensing image interpretation of the early Lake District from1987 to 2010, and five remote sensing images were superimposed. The grassland and farmland layers were intersected and interpreted in the GIS software, select the unchanged farmland and grassland nearly 30 years ministry of Qinghai lake north as study plots (Figure 1), and the unchanged grassland coordinate point is 37°16'43.29 "N, 100°17'37.12" E, the unchanged farmland coordinate point is 37°16'42.79 "N, 100°17'28.62 "E.

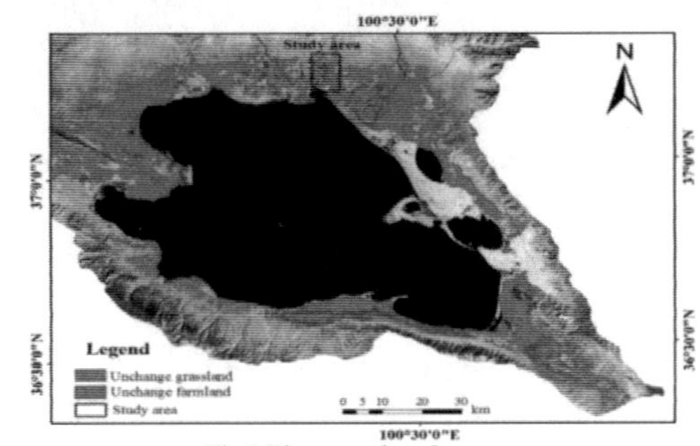

Fig.1 Diagram of study area

3.2 Soil Sampling Methods

Soil samples were collected in October, 2014. The sampling interval is 30m × 30m, five sampling points were respectively selected under two different land-use modes; each sample point includes 12 soil samples with a drill of 0-60cm depth (sampling interval of 5cm) with a diameter of 5cm of soil augers. 10 sampling points and 120 soil samples were selected under such two types of land use patterns. All soil samples were brought into laboratory and dried naturally and cleaned. After it was under grind through a 200 mesh sieve, it would be arranged, collected and set aside.

3.3 Determination of soil magnetic susceptibility

Magnetic susceptibility was measured with MS-2B type magnetic susceptibility instruments made by the British company. Measurement process mainly includes: the dried sample was poured into 10ml volume plastic cylindrical cartridge and weighed sample mass; the magnetic susceptibility meter background value was measured for a time, and the continuous measurements for three times was made after the samples were put. Meanwhile, the measurement sample is removed from its background value for one time. The change of both of the background measurement value should be under 0.3. The low frequency is 470hz, the high frequency is 4700hz, the unit gear of measuring for the SI is 0.1 tear [19]. Finally, the frequency and high-frequency magnetic susceptibility were calculated through magnetic susceptibility test results, samples low (high) frequency magnetic susceptibility value = (average of 3 measurements of the sample-around twice the average background value) × 10/sample mass, units $10^{-8}m^3kg^{-1}$. Frequency magnetic susceptibility (X_{fd}%) = (low frequency magnetic susceptibility X_{lf} - high-frequency magnetic susceptibility X_{hf}) / low-frequency magnetic susceptibility X_{lf} × 100. (Note: According to the measured low-frequency and high-frequency magnetic susceptibility magnetic susceptibility, the overall trend is consistent. However, the overall value of the low-frequency magnetic susceptibility values are higher than the high-frequency magnetic susceptibility. So, just low frequency magnetic susceptibility and the frequency magnetic susceptibility compared analysis were selected here .)

3.4 Determination of soil Chroma

Chroma was taken from Minolta spectrophotometric colorimeter measurement produced by Japan. When the sample was dried naturally in the laboratory, and then it was through 200 mesh sieve. Then, the sample was placed on the test board, flatten. Three areas of smooth surface were randomly selected to be measured for three times, and the instrument automatically determined three measurements of L *, a *, b * mean, and it was ensured that the error was less than 0.1. (L * value: psychometric lightness; a *value: psychometric red-green chromaticness; b * value： psychometric yellow-blue chromaticness)

3.5 Determination of Soil particle size
After all samples were dried naturally and through 2mm sieve, 0.4g soil samples were accurately weighed and placed in a beaker of 50mL. And then, the samples was processed as follows: 1) 10mL10% hydrogen peroxide was added and heated when it was to boil so that it could react calmly in order to remove soil organic matter ease of oxidation salts; 2) after the beaker was cool, 10mL10% hydrochloric acid (Hcl) was added and shaken sufficiently when it was to boil so that it could be reacted freely and carbonates could be removed; 3) the distilled water was put into the samples, and it would be kept for 12 hours; 4) 10mL of 10% sodium hexametaphosphate was added and it was placed in an ultrasonic shaker test. Among this, the particle size analysis instruments was Mastersizer2000 Malvern laser particle size analyzer produced by a British production company. In process of measurement, the degree of shading existed between 17% and 20%, and then, the same measurement would be done for three times, and the final result was accorded to the averaged value.

3.6 Soil data processing
In order to contrast and analyze the change features and otherness of magnetic susceptibility, chroma and particle size of soil between grassland and farmland intuitively, the measure numerical value of the soil magnetic susceptibility, chroma and particle size on five sampling sites of grassland and farmland were on the average from the surface layer 0-60cm to the deeper. And the standard deviation of the average value was accessed, which reflected the dispersion degree of the correspondent numerical value to the average value.

4. Results and analysis

4.1 Analysis of soil magnetic susceptibility difference sectional of grasslands and farmlands
The low frequency magnetic susceptibility (X_{lf}) and frequency dependent susceptibility (X_{fd}) of grassland and farmland were shown in Table 1: the X_{lf} of grassland between (38.83-54.73)$\times 10^{-8}m^3kg^{-1}$, the average value was 46.11 $\times$ $10^{-8}m^3kg^{-1}$, farmland X_{lf} between (45.6-64.2)$\times 10^{-8}m^3kg^{-1}$, the average value was 54.33$\times 10^{-8}m^3kg^{-1}$, we can see that the low-frequency magnetic susceptibility of grassland is less than farmland; X_{fd} of grassland between 5%-8.73%, the average value was 7.22%, farmland of X_{fd} between 4.64%-8.61%, the average value was 6.96%, it was obvious that the frequency of the magnetic susceptibility of grassland was less than farmland.

Fig.2 showed the trends of change on low-frequency and frequency magnetic susceptibility of the grassland and farmland in the profile. The low-frequency susceptibility of grassland was slighter than that of the farmland in the whole profile and had a small amplitude, while the low frequency magnetic susceptibility of farmland was larger in the profile. The difference between grassland and farmland's low frequency magnetic susceptibility in 0-15cm segment were not that significant, and 15-60cm segment of the low-frequency magnetic susceptibility of farmland was significantly greater than that of the grassland. The X_{fd} of grassland was greater than that of the farmland above 30cm, which also had a little differences below 30cm.

Table 1 The X_{lf} and X_{fd} content of grassland and farmland

Sampling point		X_{lf}($\times$10-8m3kg-1)	X_{fd}(%)
Grassland	Max value	54.73	8.73
	Min value	38.83	5.00
	Average value	46.11	7.22
Farmland	Max value	64.20	8.61
	Min value	45.60	4.64
	Average value	54.33	6.96

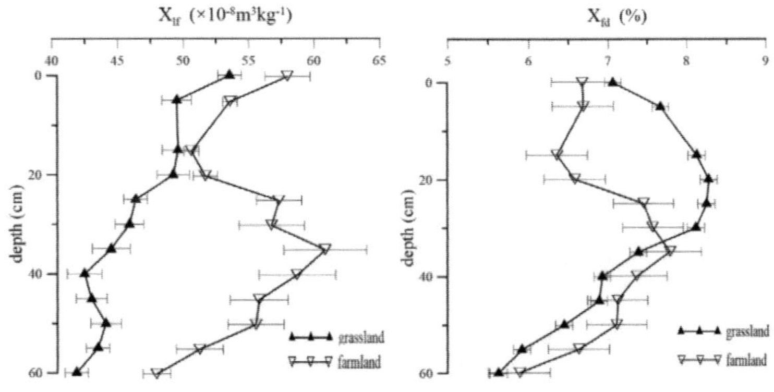

Fig.2 The X_{lf} and X_{fd} variation of grassland and farmland profile

4.2 Aanalysis of soil Chroma difference sectional of grasslands and farmlands

As Table 2 showed, the value of the lightness of the grassland was between 43.79 and 50.01, with an average value of 47.35. The value of farmland lightness was between 50.02 and 55.34, with an average of 52.37. The value of grassland redness was between 4.79and 6.03, with an average of 5.25. The value of redness of farmland was between 5.32and 6.10, with an average of 5.63. The value of yellowness of grassland was between 13.26 and 15.53, with an average of 14.02. According to these above, it was obvious that the lightness, redness, yellowness of farmland were greater than these of the grassland.

The chroma change trend of grassland and farmland in lengthways profile consistently showed that (Figure 3): the fluctuations extent of lightness and redness between grassland and farmland were greater than these of the yellowness. In addition, it was noticed that the lightness had the opposite trend with redness and yellowness. The lightness of grassland had increased trend and the lightness of farmland has decreased trend above 30cm. The lightness of grassland and farmland had no other changes below 30cm. The change of redness and yellowness in lengthways profile had decreased trend from 0 to 60cm. The redness of the grassland and farmland at 15cm kept unchanged basically, which below 15cm was of the less slowly decreasing trend. However, decreasing trend of grassland was significantly obvious compared with farmland. The yellowness of grassland and farmland over the entire lengthways profile was of the slowly decreasing range trend.

Table 2 The chroma content of grassland and farmland

Sampling point		L*	a*	b*
Grassland	Max value	50.01	6.03	15.53
	Min value	43.79	4.79	13.26
	Average value	47.35	5.25	14.02
Farmland	Max value	55.34	6.10	16.42
	Min value	50.02	5.32	14.46
	Average value	52.37	5.63	15.42

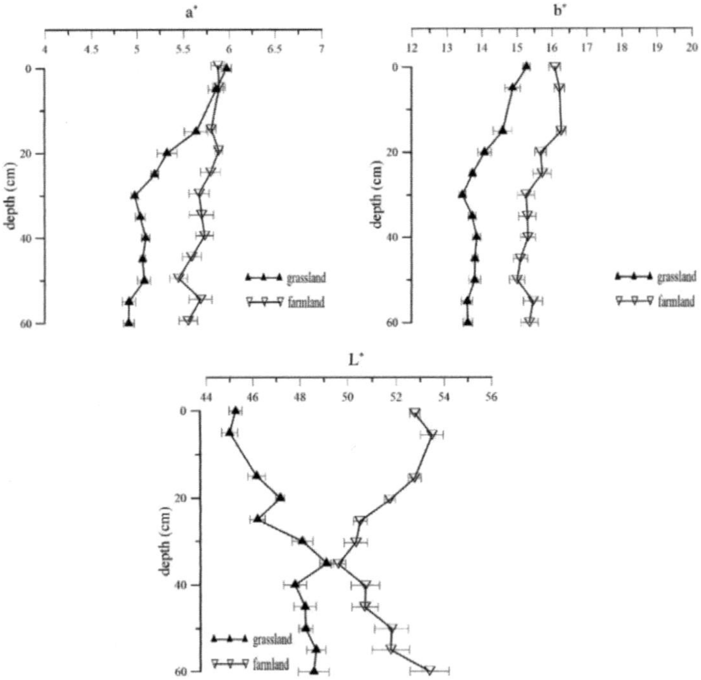

Fig.3 The chroma variation of grassland and farmland profile

4.3 Analysis of soil particle size sectional of grasslands and farmlands

According to analyzing particle size distribution of grassland and farmland(Table 3), it was obvious that grassland sand content was 5.62%~12.64%, with an average value of 8.98%, while the farmland of sand content was 9.81%~21.64%, with an average value of 15.25%. It was obvious that the sand content of farmland was 6.27% more than that of grassland, namely, it was nearly as twice as that of grassland; grassland silt content was 66.42%~73.18%, with an average value of 70.8%, while the farmland silt content was 60.45%~71.48%, with an average value of 66.65%, the different value between the silt contents was 4.15%; grassland clay content ranged from 17.92% to 21.99%, with an average value of 20.22%, while the farmland between clay content was 16.2%~19.82%, with an average value of 18.1%, the different value between the clay contents was 2.12%.

According to the soil particle size lengthways profile of grassland and farmland, it was obvious that the sand, silt and clay content of grassland and farmland changed with its depth(Fig. 4). The overall sand content of the farmland in the entire lengthways profile of the depth's degree was greater than that of the grassland. At 0~20cm segment, sand content of the farmland reached the maximum. Sand content at 20~35cm segment decreased, while when it was at 35~60cm or less, it started to increase again. The silt and clay contents of grassland on the entire profile were greater than these of the farmland. The soil particle size composition had a biggest difference when it was above 30cm, and it had an unobvious difference when it was below 30cm.

Table 3 The particle size content of grassland and farmland

Sampling point		Sand（%）	Silt(%)	Clay(%)
Grassland	Max value	12.64	73.81	21.99
	Min value	5.62	66.42	17.92
	Average value	8.98	70.8	20.22
Farmland	Max value	21.64	71.48	19.82
	Min value	9.81	60.45	16.2
	Average value	15.25	66.65	18.1

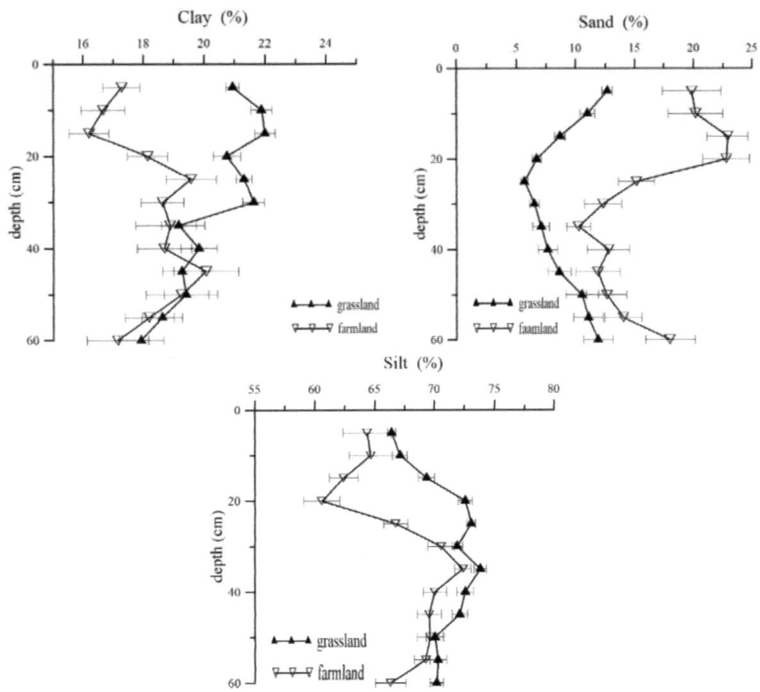

Fig.4 The particle size variation of grassland and farmland profile

5. Discussion

5.1 Effect on human activities of soil magnetic properties and chroma

Soil magnetic susceptibility was a proxy index for the degree of soil development, climate and environmental pollution, because its fast determination and visual had been widely used [20], which also was used to reflect the comprehensive information on global environmental change and human activities [5, 12]. This also could characterized the type and content of soil magnetic substance, so the process of soil formation and human factors were the main factors to affect soil magnetic susceptibility fluctuations in the vertical directions. Generally, the soil low-frequency magnetic susceptibility had been negatively correlated with clay and organic matter content and positively correlated with silt content [21].The low-frequency magnetic susceptibility of farmland in the whole was greater than the grassland, especially at 20cm. The value of low frequency magnetic susceptibility of farmland was obviously higher than that of the grassland, which was closely related to human activities such as the irrigation and long plowing of the farmland. According to the survey, planting rapeseed for many years in this area. After harvesting artificial, its roots still buried in the ground, which required human's plowing. In this process, on the one hand, because the root carried a portion of soil and leaded to loss of soil fine particles, on the other hand, from plowing to the sowing of next year spring, as abandonment leaded to the fine particle components in the way of wind erosion and it was exhausted. However, the low-frequency magnetic susceptibility were negatively related to the soil clay content, so the low-frequency magnetic susceptibility was greater than that of the grassland. In addition, the farmland would be irrigated when the sowing was finished. This process made the iron and other magnetic substances soil contain from the surface to the deep sides with a continuous migration, accumulation and transformation, which also made low-frequency magnetic susceptibility of farmland heighten. Relative to those farmland affected largely by human activities, grassland was of no frequent tillage and irrigation effect. In soil development process, the influenced frequency and intensity of human activities was much lower than farmland, which was essentially based on soil self-development. So the low frequency magnetic susceptibility trend of grassland along the whole profile was less than that of the farmland. Frequency magnetic susceptibility reflected the relative content of magnetic minerals, which could not only help people understand the transition state particles content in superparamagnetic (SP) to stable single domain (SSD) process[22], but also had a positive correlation with soil clay content and organic matter content in a way. It was not difficult to explain how 30cm above the grassland frequency magnetic susceptibility was greater than that of the farmland. However, 30cm below had no difference (grassland above 30cm was affected slightly by human activity, the mature degree of soil development was higher than farmland, below 30cm the farmland had a fewer and smaller difference with grassland because of less human activities affected).

Soil color was one of the most important characteristics of soil, which was widely used in semi-quantitative or quantitative description of regional climate change and the degree of soil development [23]. Generally speaking, the lightness of the soil color was mainly controlled by organic matter and carbonate content; the higher organic matter content was, the lower the lightness of the soil was, the higher carbonate content was, the higher the lightness was; redness was mainly controlled by the degree of hematite mineral content, the higher the hematite content was, the higher the soil redness was; yellowness was mainly controlled by goethite content, the higher the goethite content was, the higher the degree of yellow soil was[24]. Due to 35cm above the grassland, soil development degree was strong, soil fine particle content and the corresponding organic matter were high. Therefore, the brightness value of 0-35cm grassland was smaller than that of farmland and showed an increasing trend. Similarly, due to the effect of tillage in the 0-35 cm section, farmland soil development action was trended to be weak and its soil fine grained loss, so the corresponding organic matter was less, and the corresponding lightness values within this range was high; the human disturbance of farmland below 35cm was less, but the content of organic matter was still lower than that of grassland. Therefore, the lightness value was greater than the grassland and there was an increasing trend. The redness and yellowness indexes of farmland were bigger than these of grassland in the whole section, while the redness and yellowness were mainly related to hematite and goethite. Before the experimental analysis of hematite and goethite were performed, the conclusions could not be made optionally. So, the further research and analysis of this study would be made in the future.

5.2 Impact of human activities on soil texture

Soil particle size was the basic study of soil environment and the entire ecosystem characteristics, the particle size was closely related to the physical, chemical and biological properties of the soil [25]. The intensity of human activities would cause the changes in soil particle component. The study showed that there were differences in grain size between grassland and farmland. The sand particles, silt particles and clay particles had an obvious change as the bound of 30 cm. In order to explain this difference more intuitively, a definition of 30cm was made as the boundary, above which it was disturbance layer, below which it was non-disturbance layer. Moreover, the specific analysis of the particle size frequency curves was made: above the bound (Fig. 5-a), the grassland frequency curves were bimodal and the kurtosis was basically similar, and the main peak was between 10μm and 100μm which belonged to silty-sand loam; the minor peak was between 1μm and 10μm, which belonged to clay loam. Those indicated that the grassland was due to the less human interference, the vegetation was covered perennially and soil development was stronger, the soil particles were finer; the frequency curve of the disturbed layer of farmland was multimodal, the main peak was between 10μm and 100μm, the minor peak was between 100μm and 200μm, these indicated there was a considerable part of coarse particle joining in besides soil own development in the process of soil formation, because only less than 75μm dust usually existed in a suspended state [26], dust which was larger than 75μm was difficult to suspend in the air for a long time, and it could only saltation move. The third peak was between 2μm~10μm, which belonged to clay loam, but the volume percentage was small. Maybe, this was the results of the arable land under human short-term perennial crops and perennial tilling, and most fine particle was in erosion by the wind, so the soil particles became generally coarser; below bound (Fig. 5-b), the frequency curve shape of the grassland and farmland was substantially similar, it was almost close to normal distribution and had only one peak and lay between 4μm and 10μm. The peak of grassland was slightly higher than that of the farmland, which showed a relatively homogeneous and stable depositional environment and the sediment was very pure. The frequency curve shape of the grassland was the main components of the clay, which was not in much fluctuate. This indicated that human activities mainly occurred in the soil layer above 30cm which was under the impact of human activities. Grassland was regarded as a "natural land", the soil evolution would tend to be homogeneous without human disturbance or low intensity of human activity; the farmland was regarded as a "disturbance of land" in case of human's high intensive activity, and the particle size would be further coarsening, and the desertification process would be aggravated.

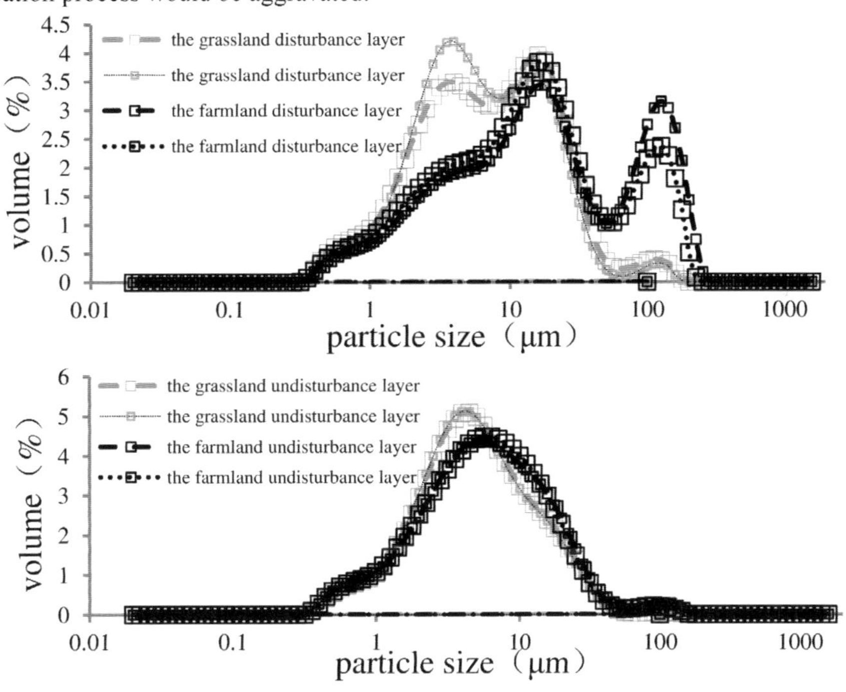

Fig.5 The particle size frequency curve of grassland and farmland
a. Disturbed layer, b. Undisturbed layer

6. Conclusion

(1) The magnetic and chromatic characteristics of soil would be obviously different under the disturbance of human activities, the low-frequency magnetic susceptibility of farmland along the entire profile was larger than grassland, especially in the 20cm-45cm segment; the X_{fd} of grassland was greater than that of the farmland above 30cm, and it had a little differences below 30cm. The lightness, redness, yellowness values of the farmland were greater than these of the grassland; the

lightness of grassland had a increasing trend and the lightness of farmland had a decreasing trend above 30cm; the lightness of grassland and farmland had no other changes below 30cm, and the redness and yellowness of soil had no fierce fluctuation in the entire profile.

(2) For the perspective of the composition, the particle size of grassland and farmland soil belonged to all sand-clay loam, grassland silt and clay contents were larger than farmland, and the sand content was less than that of the farmland; The components of the particle size of grassland and farmland along the entire profile had an obvious difference, and the soil particle size composition had a big differences especially above 30cm, but it had slight differences below 30cm, indicating that human activities mainly occurred in the soil layer above 30cm the impact of human activities, grassland as a "natural land" , without human disturbance or low intensity of human activity, the soil evolution will tend to be homogeneous, farmland as a "disturbance of land", in the case of human high intensity activity , particle size will be further coarsening, aggravate desertification process. This indicated that grassland was regarded as a "natural land", the soil evolution would tend to be homogeneous without human disturbance or low intensity of human activity; the farmland was regarded as a "disturbance of land" in case of human's high intensive activity, and the particle size would be further coarsening, and the desertification process would be aggravated.

References

[1] Xu J, Grumbine R E. Building ecosystem resilience for climate change adaptation in the Asian highlands[J]. Wiley Interdisciplinary Reviews: Climate Change, 2014, 5(6) : 709-718.

[2] CHEN K L, Li S C, ZHOU Q F, et al. Analyzing dynamics of ecosystem service based on variations of landscape patterns in Qinghai Lake area in recent 25 years [J].Resources Science, 2008, 30:274-280.

[3] QIAO Y M, WANG Z Q, DUAN Z H. Effects of differernt land-use types on soil carbon and nitrogen contents in the northern region of Qinghai Lake[J]. Acta Pratacul Tutae Sinica, 2009,18（6）： 105-112.

[4] KUKLA G, HELLER F, LIU X M, et al. Pleistocene climates in China dated by magnetic susceptibility [J]. Geology, 1988,16 :811-814.

[5] XIA D H,CHEN G H, MA J Y, et al.Magnetic characteristics of loess in the ILI area and their environmental implication[J]. Quaternary Sciences,2010,30（5）:902-910.

[6] SONG Y J, SHI Z T,FANG X M, et al. Loess magnetic properties in the ILI Basin and their correlation with the Chinese Loess Plateau[J]. Sci China Earth Sci,2010,40（1）:61-72.

[7] HELLER F, LIU T S. Magnetism of Chinese Loess deposits [J].Gephys J R astr soc, 1984, 77:125-141.

[8] YUAN S Y, LI C A, ZHANG Y F, et al. Grain sizes and magnetic susceptibility of the XiaoSi section in the Jianghan plain and their environmental significance[J].Transactions of Oceanology and Limnology,2011,（4）:169-176.

[9] DONG L, LI Y H. Comparative study on magnetic variations of eolian and alluvial-proluvial sediments from Lop Nur area[J].Journal of Arid Land Resources and Environment,2013,27（11）:111-116.

[10] E C Y, CAO G C, HOU G L, et al. The environmental change recorded in Jiang Xi Gou loess sections in Qinghai Lake region[J].Marine Geology & Quaternary Geology,2013,4(33):193-200.

[11] WANG L P, ZHANG M K. Release behaviors heavry metals from polluted soils with heavy metals of different sources[J].Research of Envionmental Sciences,2007,20（4）:134-138.

[12] XIA D S, WANG B, XU Q, et al.Magnetic records of heavy metals pollution in urban topsoil of Urumqi city[J].Journal of Lanzhou University (Natural Sciences),2013,49（2）:173-181.

[13] ZHANG Z E, DONG Z B. The effect of wind ersion on the surface particle size[J] . Journal of Arid Land Resources and Environment,2012,26（12）:86-89.

[14] LIU J L, LI L Q, LIN H N, et al. Characters of gain size of sediments from mangrove wetlands

of China[J]. Journal of Xiamen university(Natural Sciences).2008.47（6）:891-893.

[15] LIU H, HE Q, WANG Y Y, et al. Temporal and Spatial characteristics of surface sediment grain-size distribution in chang jiang estuary[J]. Acta Sedmentologica Snica,2007,25（3）:45-448.

[16] WEI Y L, XU C P, ZHANG S K, et al. Effect of climatic on biomass and eco-environments of natural grassland in Haibei region of Qinghai province. Pratacul Tural Science, 2008, 25(3): 12-17.

[17] ZHANG W, CHEN K L, DUO H R. Evaluatzon of eco-system stability and changing of land use in Qinghai Lake basin nearly 25 years[J].Qinghai Prataculture , 2009, 18(4): 8-12.

[18] LU H, YAN Y M. Reason of degradation grassland counter measures in GangCha county [J]. Qinghai Prataculture, 2007, 16(2): 50-51.

[19] ZENG F M, LAI Z P, LIU X J. The influence of mass on magnetic susceptibility during the measurement[J]. Journal of salt lake research, 2014, 3: 005.

[20] Thompson R, J Bioemedal, J A Dearing, et al. Environ-mental applications of magnetic measurements[J]. Science, 1980, 207 :481-486.

[21] LU S G. Relationship between frequency magnetic susceptibility and ferrimagnetic mineral grain size in soils and its environmental implication[J].Journal of basic science and engineerng, 2005, 8(1): 9-15.

[22] LIU Q S, DENG C L. Magnetic susseptibility and its environmental significances [J]. Chinese Journal of Geophysics, 2009 (4): 1041-1048.

[23] SHI P H, YANG T B, TIAN Q C, et al. Chroma chracteristics in the loss-paleosol at Jingyuan section and its signification to paleocliamate[J]. Journal of Lanzhou University (Natural Sciences), 2012, 48(2): 15-23

[24] DING M, PANG J L, HUANG C C, et al. Chroma chracteristics and its climatic significance in Holocene Liangcun profile in the western Guanzhong basin[J]. Journal of Shaanxi normal University (Natural Sciences), 2010 (5): 92-97.

[25] DANG X H, YU Y, GAO Y, et al. Analysis on impact of PLA barrier on soil particle sizes of sand dunes[J].Research of soil and water conservation,2014,21（3）:16-19.

[26] QING X G, WU J S, CAI B G, et al. The difference between dust-transporting wind systems of Beijing-Zhangjia kou area and the loess plateau since the LGM[J].Quaternary Sciences, 2004, 24(4): 430-436.

IOP Conf. Series: Earth and Environmental Science **59** (2017) 012072 doi:10.1088/1755-1315/59/1/012072

Research on Development of Low-carbon Tourism of Jiangxi

XIONG Wen-ping

Nanchang Normal University, Nanchang Jiangxi, China, 330029

lcshxwp@126.com

Abstract. Jiangxi is rich in tourism resources. A lot of tourists are attracted to Jiangxi for sightseeing and traveling, which has not only brought economic benefits, but also caused damage to the beautiful ecological environment of Jiangxi to a certain extent. This paper analyzes the rich resources and policies of Jiangxi and the problem of environmental disruption faced by the tourism development of Jiangxi. On this basis, the paper suggests the concept and measures for promoting the low-carbon tourism of Jiangxi so that the tourism industry of Jiangxi can develop sustainably.

1. Introduction

The burning of fossil fuel has been emitted an increasing volume of CO2 at an accelerating pace since the Industrial Revolution. The global emission of CO2 from fossil fuel had increased eight times from 34 tons in 1920 to 295 tons in 2007[1]. As estimated by the climate model under different situations, the global emission of green gas will continue to increase in the following decades. Tourism has always been known as a "smokeless industry"[2]. However, with the development of tourism economy, the tourism resources of some areas have been excessively developed. The misconducts of tourists and tourism enterprises have led to excessive carbon emission, which has caused a serious disruption to the environment. The contribution rate of artificial factors to the global warming in the entire tourism sector has accounted for about 5-14%[3]. In December 2009, the State Council issued *Suggestions on Accelerating Development of Tourism Industry*, which suggested cultivating the tourism industry into a strategic pillar industry in the national economy and a modern service industry more satisfactory to the people and required that tourism industry should save energy and water and reduce emission, manage energy based on contracts, reduce the emission of green house gases, actively develop cyclic economy and advocate low-carbon tourism[4]. Low-carbon tourism is a new mode for sustainable development of tourism which applies low-carbon technologies, adopts the mechanism of carbon sink and advocates low-carbon tourism consumption so as to achieve a higher quality of tourism experience and greater benefits of tourism economy and social environment. The core concept of low-carbon tourism development is to achieve greater benefits of tourism economy, society and environment with less carbon emission[5].

The tourism industry of Jiangxi is developing rapidly and has made wonderful achievements. Its development has faced many challenges such as the ones in ecological environment. For Jiangxi, an ecology-based province, the ecological environment is very important. However, with the development of tourism, the ecological environment has been disrupted. Each year Jiangxi attracts many tourists. For example, in 2015 the number of tourists to Jiangxi exceeded 380 million. Accordingly, a lot of carbon was emitted from the traffic and accommodation for the tourists. Besides, some accommodations, restaurants and scenic areas in Jiangxi are not deployed with sewage treatment system and oil-water separation system. Tourists to Jiangxi have generated garbage, sewage and wastes, which has caused pollution and disruption to the ecological environment of Jiangxi to a certain extent. Therefore, in the face of these conditions, to build a beautiful ecological environment, Jiangxi province has strived to develop low-carbon tourism, which is a necessary choice for the sustainable development of its tourism industry.

2. Advantages for Development of Low-carbon Tourism of Jiangxi

2.1. It has favorable ecological environment resources for development of low-carbon tourism.

(1) Jiangxi has a forest coverage rate of 63.1%, taking the second spot in China. Now Jiangxi has 220 forest natural reserves (including 11 national natural reserves), 177 forest parks (including 46 national ones, ranked top in China), 77 wetland parks (including 23 national ones and 54 provincial ones), 5 world heritage sites, 4 world geological parks, 1 key international wetland and 14 national tourist attractions[6].

(2) Boyang Lake is the largest fresh-water lake in China. As an important natural carbon sink, the vegetation of Boyang Lake Wetland has formed a huge biomass, which can fix and reduce CO2 in the atmosphere, increase O2 in the air, which plays an irreplaceable role in maintaining the dynamic balance of CO2 and O2 in the air, lowering the greenhouse effect and providing a living foundation for the mankind.

2.2. Favorable development opportunity

Jiangxi boasts distinctive advantages in regional strategies for it is covered by national strategies such as the Rise of Central China Plan, the Urban Agglomeration in the Middle Stream of Yangtze River, the Boyang Lake Ecological Economy Development Zone, the Revitalization Plan of Central Soviet Areas such as Gannan and the National Ecology Demonstrative Zone (Jiangxi). Jiangxi enjoys prominent advantages in ecological environment, so the development of low-carbon tourism has an important significance for the development of ecological economy of Jiangxi. The province has entered a new situation of integrated resource development and regional joint development, and ushered in a strategic opportunity period for development of low-carbon tourism industry.

3. Existing Problems in Development of Low-carbon Tourism of Jiangxi

3.1. Low-carbon tourism lacks funds

The development of low-carbon tourism of Jiangxi should depend upon high and new technologies and new energy, which require financial support. Funds are an important factor for the development of low-carbon tourism. In 2015, the economic development of Jiangxi was ranked 19th in China. Due to a relatively poor economic performance, the investment into tourism is insufficient, and the investment into low-carbon tourism is much less. Jiangxi does not provide enough support to the low-carbon industry, so it should strengthen the tax and finance for the low-carbon tourism industry.

3.2. Lacking low-carbon technology

The development of low-carbon tourism needs to depend upon energy-saving and environmentally friendly technologies and green energy such as solar energy and organic energy, replace cars with low energy consumption and carbon emission, use building materials free of pollution and saving energy and reasonably recycle and reuse garbage and wastes. However, the province is relatively weak in low-carbon environment protection technologies and still has a shortage in low-carbon tourism technologies.

3.3. Lacking talents of low-carbon tourism

Low-carbon tourism involves know-how with regard to tourism and energy. Highly professional, it is an emerging program rich in connotation. Its development requires high-quality composite talents who have mastered low-carbon and tourism know-how. Since low-carbon tourism is still a burgeoning program, Jiangxi province is relatively short of low-carbon tourism talents, especially in a serious shortage of high-quality managers in this area, which has affected the development of its low-carbon tourism industry.

3.4. Lacking a scientific planning

The low-carbon tourism resources of Jiangxi have been developed randomly, as a result of which the characteristics of its tourism resources cannot be well reflected, thus it has not formed a complete system of tourism products. The low-carbon tourism of Jiangxi lacks a reasonable planning. It has not guided the low-carbon tourism to develop scientifically and rationally following a concept of planning and piloting, nor well unified and planned its tourism routines, tourism products, tourism industry and tourism resources. It has no idea how to solve problems such as how to make up the shortage, how to develop advantages, how to set a clear positioning and where to find the breakthrough point.

3.5. Backward in low-carbon infrastructure

The low-carbon tourism of Jiangxi is still under a starting stage. The tourism facilities are still relatively backward in the tourist attractions and the traffic is also inconvenient. The infrastructure and service level involving tourism services such as food, accommodation and shopping are not complete enough, which has affected the development of low-carbon tourism of Jiangxi. The focus of low-carbon tourism of Jiangxi is low-carbon technical innovation, energy saving and environment protection. The low-carbon infrastructure in this area will affect the development of low-carbon tourism in Jiangxi to a very great extent.

4. Countermeasures for Development of Low-carbon Tourism of Jiangxi

4.1. Strengthening the publicity and education of low-carbon tourism

Only when the concept of low carbon enjoys popular support and becomes part of the sub-consciousness of the tourists can it gradually become a lifestyle of the tourists. Therefore, we need to publicize and spread the concept of low-carbon tourism among the public. Low-carbon education can enhance and change the outlook on environment and pattern of consumption of the tourism operators and tourists. It can enhance the awareness and quality of all people in low-carbon environment protection and energy saving and cultivate low-carbon production modes and lifestyles so that the whole society can consciously engage in the efforts to save energy, reduce emission and protect the environment and make concerted efforts to overcome the criss of resources and environment faced by the mankind[8] . In order to make low-carbon tourism of Jiangxi enjoy popular support, Jiangxi can fully exhibit its low-carbon tourism behaviors and concepts and the characteristics of its low-carbon tourism to the public by means of publicity videos, low-carbon products exhibition and display boards; it can hold seminars and forums on ecology of Jiangxi themed on low-carbon tourism and environment protection and organize lectures and training classes on low-carbon tourism lifestyles; it can organize lectures and courses on low-carbon know-how in the campuses to cultivate high-quality talents of low-carbon tourism; it can publicize low-carbon tourism via media such as radio, TV and network to enhance the influence of the construction of National Ecology Demonstrative Zone (Jiangxi) and establish a development benchmark for low-carbon tourism of Jiangxi. By adopting vivid and lively forms of publicity, it will make the concept of low-carbon tourism infiltrate into the brain and heart of the public, enhance their awareness and concept of low-carbon energy saving and emission reduction, actively guide the tourists to choose a low-carbon tourism mode, thereby encouraging the public and tourists to save energy and protect the environment and enhancing the popularity of low-carbon tourism of Jiangxi.

4.2. Scientifically planning the development of low-carbon tourism

A scientific tourism planning can avoid random development, the waste of tourism resources and disruption to the ecological environment due to lack of scientific and reasonable development. To scientifically plan the development of low-carbon tourism, it's required to investigate the aspects of the tourist destinations such as social environment, ecological environment, resource environment and traffic environment and the carbon sink resources of the region. It's required to confirm the development goal for low-carbon tourism on the basis of investigation, insist on integrating ecology, economy and social benefits in the low-carbon tourism, coordinate the development of the government, enterprises, tourists and communities, well design the boutique routines, projects and series products of low-carbon tourism of Jiangxi. It's required to construct a guarantee system for the development of low-carbon tourism of Jiangxi, including the organizational guarantee, technical guarantee, talent guarantee, policy guarantee, financial guarantee and marketing guarantee.

4.3. Striving to promote low-carbon tourism of Jiangxi

The traffic should be environment friendly, so we should encourage tourists to take an electronic car deployed by the tourist attraction to enter the scenic area rather than drive their own car. Electronic cars can also be arranged between scenic spots to make it convenient for the transfer of tourists. In the scenic areas, tourists are encouraged to ride a bike or walk. To make it convenient for tourists, public bikes can be deployed at the scenic areas. Information about energy saving and environment protection should be provided. A "low-carbon tourism service center of Jiangxi" should be established to provide the tourists with low-carbon tourism information, and information regarding the rental and borrowing location of bikes, low-carbon hotels and restaurants. Low-carbon behaviors should be encouraged among the tourists to ask the tourists to try to reduce garbage and carry a garbage bag to take their own garbage out of the scenic area. Tourists are suggested to carry environment-friendly tableware for having meals. Low-carbon guestrooms are advocated, which won't provide disposable toothbrush, slippers and comb, etc. and are installed with facilities that save energy and water[9]. Meanwhile, we should enrich the travel life, increase low-carbon tourism items, actively introduce low-carbon tourism routines such as strengthening the development of low-carbon tourism products such as forest tourism, wetland tourism, countryside tourism, sports tourism and technology tourism. In the scenic areas, we should actively advocate travelling by hiking or bike and design special low-carbon tourism routines for the tourists, set convenient bike rental stations and convenient stores for hikers, and issue *Guide to Low-carbon Tourism*, etc[10].

4.4. Creating low-carbon scenic areas and demonstrative zones

Ecology is a brand of Jiangxi and the characteristics of tourism of Jiangxi. To protect ecology, we should carry out ecological civilization construction around ecological and green Jiangxi, which is helpful for protection of ecology and bio-diversity. For the tourism industry of Jiangxi, to save energy and reduce emission and reduce carbon consumption, it's very necessary to construct a model of low-carbon tourism demonstrative zones, which involves "foods, accommodations, traveling and shopping" services for the tourists. Namely, we should build low-carbon restaurants, hotels, traffic, scenic areas and stores.

It's an important measure for implementing the low-carbon tourism to create low-carbon scenic areas. The scenic areas are the main areas of a tourist destination. The concept of low-carbon tourism should be carried out from the aspects such as tourism planning, development of tourism products and arrangement of tourism facilities. The primary direction is to build cyclic scenic areas, namely, to strictly follow the "3R" principles of cyclic economy "reduction, reuse and recycle", adopt the ecological law to guide the tourism development and activities, and achieve a close-loop feedback and cyclic process of "resources--products--renewable resources" in the development of tourism resources and activities, and integrate the clean production of scenic areas, comprehensive utilization of tourism resources, ecological design of tourism products and sustainable consumption of the tourists into one[11].

4.5. Increasing the funding, enhancing the technological content of low-carbon tourism

Funding is a key guarantee for the implementation of low-carbon tourism. Jiangxi should increase investment into the infrastructure for low-carbon tourism, particularly, perfect the construction of low-carbon tourism in the scenic areas, enhance the energy saving and consumption reduction of tourist destinations and scenic areas and protect its ecological environment. Since the foundation of low-carbon facilities is weak at the tourist destinations, it's very important to increase investment into low-carbon infrastructure. The government and competent department shall arrange funds for energy saving and emission reduction alone in their public fiscal budgeting, provide demonstrative cities of low-carbon tourism and pilot enterprises for environment protection with supports such as direct investment, fund subsidies and loans with discounted interest, and guide the social investment and private capital investment[12].

Technically, we should enhance the technological content of low-carbon tourism. We can introduce advanced low-carbon technologies from abroad and strengthen the research and development of low-carbon technologies of Jiangxi. Jiangxi should take energy saving and emission reduction technologies as its key strategic project, and concentrate the wisdom of technicians and increase the funding. It should promote the utilization of energy-saving and emission-reducing technologies and new energy such as wind energy, solar energy and biological energy in the tourism industry of Jiangxi. It should promote the implementation of new energy-saving products in the tourism of Jiangxi and the application of energy-saving electrical appliances in scenic areas and hotels involved in the tourism of Jiangxi.

References

[1] WRI. CAIT Database "GHG emission" [Z]. http: / /cait.Wri.org/cait. Php, 2011: 4-5.
[2] Pan Jiahua, Wang Mou, et al. Low-carbon Transformation--Fundamental Channel for Practicing Sustainable Development [M]. Beijing: Academy Press, 2010:51.
[3] Liu Xiaoming: Research on Low-carbon Tourism and Its Development [J]. Business Studies, 2011 (2)
[4] State Council. Suggestions of State Council on Accelerating Development of Tourism Industry [Z]. GF [2009] 41, 2009 (12).
[5] Cai Meng, Wang Yuming. Low-carbon Tourism: A New Development Mode of Tourism [J]. Tourism Tribune, 2010 (1).
[6] Ma Zhihui. Research on Development of Low-carbon Tourism in Boyang Lake Ecological Economy Area [J]. Boyang Lake Tribune, 2014 (3).
[7] Jiangxi, [EB/OL]. http://www.gov.cn/guoqing/2016-11/18/content_5134329.htm，2014 (12).
[8] Tang Chengcai, Zhong Linsheng, Cheng Shengkui. Research on Connotation of China's Low-carbon Tourism and Strategies for Its Sustainable Development [J]. Economic Geography, 2011 (5).
[9] Liu Xiaoming: Research on Low-carbon Tourism and Its Development [J]. Business Studies, 2011 (2)
[10] Huang Wensheng. On Construction of Low-carbon Tourism and Scenic Areas [J]. Ecological Economy, 2009 (11): 100-102.
[11] Li Tianyuan. Studies of Tourism [M]. Beijing: Higher Education Press, 2008: 62-64.
[12] Shi Peihua, Wu Pu, Feng Ling, Zheng Bin. Research on Framework Design and Strategic Measures for Emission Reduction Policies of Chinese Tourism Industry [J] . Tourism Tribune, 2010 (6).

Institute of Physics
Dirac House, Temple Back
Bristol BS1 6BE UK

ISSN: 1755-1307
ISBN 978-1-5108-3956-4

9 781510 839564

CURRAN ASSOCIATES INC.
proceedings
.com